MOLECULAR CELL BIOLOGY

MOLECULAR CELL BIOLOGY

By
H.S. Bhamrah
Kavita Juneja

ANMOL PUBLICATIONS PVT. LTD.
NEW DELHI - 110 002 (INDIA)

ANMOL PUBLICATIONS PVT. LTD.

H.O.: 4374/4B, Ansari Road, Darya Ganj,
New Delhi-110 002 (India)
Ph.: 23278000, 23261597

B.O.: No. 1015, Ist Main Road, BSK IIIrd Stage
IIIrd Phase, IIIrd Block
Bangalore - 560 085 (India)
Visit us at: www.anmolpublications.com

Molecular Cell Biology

First Edition, 2002

ISBN 81-261-1142-9

Reprint, 2007

PRINTED IN INDIA

Printed at Mehra Offset Press, Delhi.

CONTENTS

PREFACE

Life has an immense variety of forms that arose by the process of biological evolution, but all living organisms share a master plan of structural and functional organization. This book is about the building blocks—*cell* and *molecules*—that constitute the unity of the living world.

Molecular Biology is the excisting and fast moving field of science, and is condensed into but balanced review. In the last two decades the field of Molecular Biology has undergone a variable revolution leading to major advances in our understanding of cell structure and function at molecular level. The convergence of cytological, genetic and biochemical approaches has generated a rich panorama of detail, the significance of which we are still attempting to unrevel. The present title is being written as an introduction to this rapidly growing field. Our goal is to acquaint the undergraduate student who is encountering the subject for the first time with the fundamental principles that characterize the molecular organisation of cell.

Molecular Biology of the cell is chiefly concerned with eukaryotic cells, as opposed to Bacteria, and its title reflects the prime importance of the insights that have come from the molecular approach. The eukaryotic cells that form multicellular animals and plants are social organisms to an extreme degree, they live by cooperation and specialization.

We are pleased to thank those colleagues who graciously consented to contribute micrographs to this endeavour. The assistance of Dr. M.P. Arora was invaluable at virtually all stages of preparation of the book and is greatefully acknowledged.

We heartly thankful to Mr. J.L. Kumar, Publisher and other staff members of M/s Anmol Publications Pvt. Ltd. for their whole hearted cooperation in the publication of this book.

There can be no claim to originality except in the manner of treatment and much of the information has been obtained from the books and scientific journals available in different libraries.

The authors are painfully aware of the shortcomings, errors and misprints that have crept us, and shall be grateful to receive suggestions for improvements of the next edition from all the readers.

H.S. Bhamrah

&

Kavita Juneja

1

INTRODUCTION

The history of cytology is of considerable interest and the knowledge of it, is almost essential in order to understand its present status, problems and directions. *E.B. Wilson* (1925) has given a brief but complete sketch of its history in the introduction of his book— "The Cell". A more detailed description is provided by *Shleiden* (1953) and *Hughes* (1959). The discovery of the cellular structure of organisms is intimately related with the invention of the microscope (Gr. *Mikro*; small and ṣkopes; to see). The makers of spectacles improved the techniques for the polishing of short focus lenses. *Leeuwenhoek* (1674) was able to make some lenses of a magnification which were satisfactory for simple scientific observations. *Galileo* in 1610 was probably the first man who constructed a microscope by combining the lenses in a tube of lead, *Kihcher* in 17th century, was the first to employ the microscope systematically in the study of diseases. With the development of microscope, new scientific observations became possible. *Robert Hooke* presented the texture of cork before the "Royal Society of London" with the help of his microscope. This marked a revolution for all the knowledge concerning the microscopic organization of living substances. Thus *Robert Hooke* (1665) left to us the name of cells (Gr. *cella;* a small chamber).

In the same century some other workers live *Grew* and *Malpighi* repeated the observations of *Hooke* in different plants and recognized in them minute cavities which they called "utricles" or "vesicles". This knowledge remained stationary until the beginning of 19th century.

Cell Theory

It was proposed by *M.J. Schleiden* (German botanist) and *T. Schwann* (a German zoologist). According to this theory all animals and plants are made up of cells and their products. They were the initiators of the cell theory. Perhaps they were also the first to attach a meaning to the term "cell" which is also acceptable today. Long before *Schleiden* and *Schwann,* numerous investigators has previously sponsored more or less the same idea in the incomplete form. *Mirbel* (1808-09) came to the conclusion that, plants are formed by a membranous cellular tissue. Likewise *Lamark* (1809) stated, "Nobody can have life it its constituent parts are not cellular tissue or are not formed by cellular tissue". From 1824 to 1830 works of *Dutrochet* (1824), *Turpin* (1826), and many others, not only confirmed the universality of cell, but also hinted at the concept of cellular autonomy, in both, morphological and physiological sense. In 1831 *Robert Brown* described the nucleus, in 1832. *Dumortier* reported the cell division in algae, and form 1835 to 1839 *Van Mohl* sketched many features of mitosis. In spite of these facts "One of the most surprising print in the history of science, is that in many texts of biology, the writers consider *Schleiden* (1838), (Professor of Botany, founder of cell theory), as though he had discovered for the first time that all plant-tissues are composed of cells," *Conkein* (1940). The results of *Schleiden* were extended to animals by *Schwann* (1839). He carried out a minute investigation of the tissues of the animal body and of the development of cells and used the term "Cell theory" by stating, "The cells are organisms, and animals as well as plants are aggregation of these organisms, arranged in accordance with definite laws." His conclusion established the cell theory in a definite form.

Schwann conceived very definite conclusions regarding the morphology and physiology of cell. He divided the cellular phenomenon into two groups—the "Plastic phenomenon" and "Metabolic phenomenon." His plastic phenomenon corresponds with the modern term cellular morphology while metabolic phenomenon deals with results from chemical changes, whether in the particles that compose the cell itself or in the surrounding cytoblastema.

Schleiden, however, was wrong in many details, especially in regard to his conceptions concerning cellular replication. According to his theory the new cells were formed inside of old ones from the nucleus which he consequently referred to as the "cytoblast." This point of view seems to have been rather widely adopted despite of some criticism until the late 1840's. Between 1840 and 1860 the work of *Von Mohl, Nageli, Remak* and *Virchow* provided strong evidence for the idea that every cell is derived from a preexisting cell by some type of division. In 1882 *Flemming* described somatic division in considerable detail and applied the term "mitosis" to the process. *Louis Pasteur* also gave experimental evidences to support Virchow's view. It is obvious that there are two main components of cell theory, (i) living organisms are composed of cells and (ii) all the cells arise from pre-existing cells.

In recent years, large number of sub-cellular structures have been discovered and studied in detail so that the modern version of Cell-Theory states:

1. Cells are structural or morphological units of living organisms (i.e., body of all living organisms in an assemblage of cells).
2. Cells are physiological units of living organisms (i.e. the metabolic activities of living organisms are performed within the cells).
3. Cells are hereditary units (these maintain continuity through the hereditary material).
4. New cells originate from the pre-existing cells only.
5. Smallest unit of life is the cell.
6. All activities of living organisms are the outcome of the activities of constituent cells.

Exceptions of Cell Theory

The cell theory is one of the fundamental generalizations of biology and has universal application except for acellular protozoa; and the microbes (like bacteria, and blue green algae) in which the nuclear material does not form a distinct nucleus; and especially the viruses which are not complete cells. A virus has a central core of nucleic acid (DNA or RNA) enclosed in a protein covering. These do not have their own enzyme system and depend on the enzyme system of host cell for their metabolic activities. For this reason, viruses can multiply only inside the living cells and are called obligatory parasites. During infection only the nucleic acid or virus enters the host cell and takes the command of the synthetic machinery of host cells. The host cell now starts synthesizing viral nucleic acid and viral protein which form new viral molecules.

Viruses can be crystallized just like other inorganic and organic compounds. Therefore, the living nature of viruses is debatable.

Protoplasmic Theory

Max Schultze, in 1861, established the essential similarity which exists between sarcode and protoplasm of animal and plant cells, thus offering a theory which later was called by *O. Hertwig* (1892), the "Protoplasm Theory." This theory, more general in its conception than the cell theory, states that, the cell is an accumulation of living substances or protoplasm definitely limited in space and possessing a nucleus and a cell membrane." The primitive concept of one cell was thus changed. The cell became a delimited mass of protoplasm (Cytoplasm) surrounding the nucleus. For this unit the name "Protoplasm" is more appropriate than that of 'cell' although the later has persisted because of long use. The skeletal definition of cell which is mentioned above leads us into a certain amount of trouble. We must, for instance, assume either that coenocytic organisms and structures are single cells regardless of the number of nuclei or that a membrane is essential only where a cell is in contact with non-protoplasmic material. As noted later, even more coenocytic cells appear to be compartmentalized by membranes visible only with the electron microscope. This difficulty may be avoided by determine a cell as a nucleus and a portion of surrounding protoplasm upon which nucleus exerts a major effect. This description indicates that the nucleus is the main seat of organization or control center of a cell. The protoplasm without nucleus fails to carry life of a long time while decytoplasmated nuclei, however, are even more inadequate. *Huxley* and *O. Hertwig* referred protoplasm as the physical basis of life. Protoplasm is a complex colloidal system having many inorganic and organic compounds in such a combination that attributes the ability to perform life processes.

A number of workers in the late 1880's including *Heitzmann, Klein, Flemming* and *Butschli* described the vacuolar reticular texture of cytoplasm. Inclusions liked plastid, mitochondria, Golgi bodies, were recognized and studied in some detail by a number of workers before 1900. *Mayer, de Vries* and *Meves* discussed questions of origin and function of plastids. *Benda* used methods for differential staining of mitochondira. *Camillo Golgi* (1898) studied the net work of Golgi bodies in nerve cells. Golgi complex became a point of controversy for a very long time in the study of cell. The various workers primarily restricted their studies with nucleus or within the nucleus upto middle forties of this center. But in recent years, however, approach to cytological problems have been not only for more dynamic and quantitative but also more in terms of cell as a unit, continues with the study of its parts. A considerable progress in study of cytoplasm seems to have resulted probably from the development of new techniques of fixation and section cutting, electron micrographic and the modern methods of tissue homogenization and differential centrifugation of cell contents.

Organismal Theory

The early dissatisfaction with the conception of the cell as the primary and universal agent of organization led to the formation of the organismal theory; in which the emphasis was placed on the living mass as a whole rather than on the constituent of cells. According to this general interpretation, ontogeny is a function primarily of the organism as a whole and consists in the growth and progressive internal differentiation of a single protoplasmic individual, this differentiation often, but not always, involving the separation of the living mass into subordinate semi-independent parts, the cells. Since the septation is rarely complete, all parts remain in connection and the whole continues to act as a unit. Hence development is not primarily the establishment of an association of multiplying elementary units to form a new whole but rather the resolution of one persistent whole into newly formed metabolic units. It should be thought of not as primarily a multiplication and co-operation of cells but rather as the differentiation of growing protoplasm (*Wilson,* 1893). The real unity is that of the entire, organisms, and, as long as its cells remain in continuity, they are to be regarded not as morphological individuals but as specialized centers of action into which the living body resolves itself and by means of which the physiological division of labour is effected."

The phylogenetic implication of the organismal theory is that if multicellular organisms have arisen from unicellular forms, the process may have been the same as that seen in ontogeny: the differentiation and sub-division of a continuous growing mass of nucleated protoplasm into the system of uninucleate cells. If the septation coincided with nuclear division from the first, there was a direct transition of form with several cells. If the two processes did not so coincide, coenocytic or "plamodial" types arose. Some of which persisted as coenocytic organisms while others developed internal walls and became out multicelluair forms. The ability of isolated tissue cells to live independently, does not prove that independent cells have combined to form the multicellular body but only that such part can still carry on essential protoplasmic function. The phenomenon could as well be taken to prove the derived nature of Protista. The body is not an aggregation of elementary organisms, but a single organism which has evolved a multicellular structure. The cell colonies is certain green algae (*Bock* in 1926) described plasmodesms in *Pandorina, Eudorina* and *Gonium*) and the remarkable polyp colonies in the Siphonophora (*T.S. Huxley,* 1912) indicated the dependence of the part upon the whole even in a group formed by aggregation and they show the possibility of evolution by the combination of individuals; but it is not at all clear that they afford the key to the evolution of organisms in general. The protozoan is more properly compared with the whole man, both are organisms which have differentiated series of specialized internal regions or organs, the one without cellular sub-division or increase in size and the other into them.

Cell Biology and other Biological Sciences

Due to the wide application of cell biology in various branches of Science, many hybrid biological sciences, have spring up, some of them are given under the following sub heads:

Cytology and Taxanomy (Cytotaxonomy)

It has been found that characters useful in classification can often be recognized in the number and form of the chromosomes. The chromosomal data not only aids in grouping of species or variety, but also furnishes strong suggestions as to the manner in which certain taxonomic units have arisen during the course of evolution.

Cytology and Genetic (Cytogenetics)

Cytogenetics is that branch of cell biology which deals with the cytological and molecular bases of heredity, variation, mutation, phylogeny, morphogenesis and evolution of organisms. The Weismann's germ plasm theory, Mendel's laws of inheritance and the concept of gene could be well-understood only after the application of cytological concept to the genetics.

Cytology and Pathology (Cytopathology)

Since the diseases, particularly tissue diseases, are primarily because of the abnormal activity in cells and tissues, the close relations between cytology on the one hand and pathology and medicine on the other hand should be expected. One needs only to mention the growth of cancer cells—a dreadful disease or the effects of viruses on cell's structure are function of indicate the importance of co-operative studies on diseased tissues.

Cytology and Physiology (Cell-physiology)

The cell physiology is the study of life activities, viz., nutrition, metabolism, excitability, growth, reproduction or cell division and differentiation of the cell. The cell physiology has helped in under standing various complicated physiological activities at cellular level.

Cytology and Ecology (Cytoecology)

It has been found that the study of chromosomes in related plants led us to speculate that the differences in chromosome number and form show significant correlation with differences in geographical range of ecological habitat.

Cytology and Biochemistry (Cytochemistry)

The cytochemistry is that branch of cytology which deals with the chemical and physio-chemical analysis of living matter. For example, the cytochemical analysis has revealed the presence of carbohydrates, lipids, proteins, nucleic acids and other organic and inorganic chemical compounds in the cells.

Cytology and Evolution

The chromosomes can be made to alter by treatment with some radiations (such as X-rays) and also some chemicals. These alterations manifest themselves in the morphology and physiology of the organisms; it must indicate to some extent how the evolution has occurred.

Thus, the cell biology has given impetus to early and modern biologists to explore new vistas in biological science. The modern cytological studies are helpful in understanding various chronic human diseases and in improving the breeds of plants and animals.

Table 1.1. List of Important Contributions by Scientists to the Study of Cell Biology.

	Name	*Year*	*Contribution*
1.	Robert Hooke	1665	Discovered cell and coined the term cell.
2.	Anton Van Leeuwenhock	1722	Studied blood cells, spermatozoa, protozoa and micro-organisms.
3.	Charles Bonnet	1745	Discovered natural parthenogenesis.
4.	Rene Dutrochet	1824	Suggested that all plants and animals are aggregates of globular cell.
5.	P. Provost and J.B.A. Dumas	1824	Described cell division for the first time by studying cleavage of frog's egg.
6.	Turpin	1826	Reported occurrence of cell division.
7.	R. Brown	1823	Described the characteristic Brownian movement of cytoplasmic particles.
8.	G.B. Amici	1830	Described fertilization in plants.
9.	Robert Brown	1831	Described presence of nucleus in plant cells.
10.	Dumortier	1832	Studied cell division in algae.
11.	Felix Dujardin	1835	Described protoplasm under the heading sarcode.
12.	H. Von Mohl	1838	Described cell division in animals.
13.	M.J. Schleiden	1838	Described nucleolus, which was first observed by *F. Fontana* in 1781.
14.	M.J. Schleiden	1838	Formulated 'Cell Theory'.
15.	T. Schwann	1839	Applied Cell Theory to animals.
16.	J.E. Purkinje	1840	Proposed the term protoplasm for the cell contents.
17.	Kolliker	1846	Reported the ova and spermatozoa are single products.
18.	A. Donne	1845	Studies spermatozoa and used photomicroscopy for the first time.
19.	Hugo Van Mohl	1846	Described importance of protoplasm in the cell and cellular activities.
20.	W. Hofmeister	1848	Drew figures of chromosomes of the pollen mother cells of *Tradescantia* and described nuclear division.
21.	R. Virchow	1855	Stated that all cells arise from the pre-existing cells.
22.	Pringsheim	1855	Described fertilization process in the alga.
23.	R. Virchow	1858	Recognized cellular basis of disease and gave his famous expression 'cellula a cellula' i.e. all cells arise from pre-existing cells.
24.	Schultz	1861	Proposed 'Protoplasm Theory' and emphasized that cell is a living substance limited by a cell membrane and containing a nucleus.
25.	W.E. Waldeyer	1863	Used haematoxylin for staining tissue cells.
26.	E. Haeckel	1866	Named plastids.
27.	L. St. George	1867	Described Golgi apparatus.

28. W. His	1870	Developed microtome for cutting sections of tissue for cell study (Preservation of tissue dates back to 1663 when *Boyle* used alcohol as a preservative).
29. H. Fol	1870	Described spindle and astral rays.
30. F. Miescher	1871	Separated nucleoproteins from the pus cells and named them nucleic acids.
31. Anton Schneider	1871	Described chromosomes as nuclear filaments.
32. Strasburger	1875	Described chromosomes.
33. O. Hertwig	1876	Demonstrated that fertilization was the fusion of two cells.
34. L. Pasteur	1876	Discovered anaerobic release of energy from cells in yeast and moulds.
35. E. Abbe	1877	Discovered the oil immersion objective.
36. H. Fol	1879	Observed the penetration of sperm in an ovum.
37. W. Flemming	1879	Introduced the term chromatin and described the splitting of chromosomes.
	1882	Described cell division (mitosis) in animals cells and suggested correlation between nucleic acid and chromatin.
38. E.G. Balbiani	1881	Described giant salivary gland chromosomes.
39. W. Pfitzner	1882	Discovered chromomeres on the chromosomes.
40. F. Strasburger	1882	Described cell division in plant cells and introduced the terms cytoplasm and nucleoplasm.
41. Schimper	1883	Named choloroplasts to those green bodies to the cells which were described by *Sach* (1865) and were called green granules by *Comparatii* (1971).
42. W. Roux	1883	Proposed that chromosomes contain hereditary units.
43. E. van Benden	1883	Demonstated that the number of chromosomes is reduced to half during gamete formation in *Ascaris*.
44. Metschnikoff	1883	Observed and named phgocytosis in cells.
45. O. Hertwig and E. Straburger	1885	Proposed the role of nucleus in heredity.
46. R. Altmann	1886	Studied mitochondria and suggested their role in respiration.
47. MacMunn	1886	Discovered cytochromes.
48. Van Benden	1887	Discovered contriole.
49. T. Boveri	1888	Described the centriole.
50. Waldeyer	1888	Introduced the term chromosome.
51. A Weismann	1892	Proposed 'Germplasm Theory.'
52. T. Boveri	1892	Described spermatogenesis and oogenesis in *Ascaris*.
53. Granier	1897	Named and described ergastroplasm.
54. R. Altmann and C. Benda	1897	Observed the 'mitochondrion.'

55.	B. Benda	1898	Introduced the term 'mitochondrion.'
56.	C. Golgi	1898	Described Golgi complex in the nerve cells of owl.
57.	J. Loeb	1900	Discovered artificial parthenogenesis.
58.	T.H. Montgomery	1901	Showed that homologous chromosomes undergo pairing during reduction division.
59.	Strasburger	1901	Introduced the term plasmodesmata.
60.	C.E. McClung	1902	Discovered sex chromosomes in grasshopper.
61.	W.S. Sutton	1903	Showed that chromosomes occur in homologous pairs in diploid cells. Propounded "Chromosome Theory" of heredity.
62.	E. Buchner	1903	Received Nobel Prize for the discovery of enzyme.
63.	F. Meves	1904	Demonstrated the presence of mitochondria in plant cells.
64.	J.B. Farmer and J.E. Moore	1905	Coined the term meiosis.
65.	C. Golgi and Ramon Y. Cajal	1906	Received Nobel Prize for discovery of Golgi complex.
66.	M. Tswett	1906	Discovered chromatography.
67.	R.G. Harrison	1907	Developed tissue culture technique.
68.	F.A. Janssens	1909	Proposed chiasma-type theory for exchange of chromosomal segments.
69.	A. Kossel	1910	Received Nobel Prize for describing the chemistry of nucleus.
70.	T.H. Morgan	1910	Discovered sex-linkage in *Drosophila*.
71.	C. Carrel	1912	Described technique of tissue culture.
72.	L. Michaelis and M. Menton	1913	Showed that during enzyme action enzyme and substrate form enzyme substrate complex.
73.	R.M. Willstater	1915	Received Nobel Prize for the investigation of cholorophyll.
74.	G. Hevesy	1923	Discovered the technique of tracing the path of labelled isotopes through metabolic path way.
75.	F. Pregl	1923	Received Nobel Prize for micro-analysis of organic substance.
76.	A. Feulgen and H. Rossenbeck	1924	Described Feulgen test for location DNA in the cell.
77.	T. Svedberg	1926	Received Nobel Prize for ultra-centrifugation technique.
78.	F. Griffith	1928	Discovered transduction in bacteria.
79.	O.P. Warburg	1931	Discovered cytochrome oxidase, the respiratory enzyme which catalyses the oxidation of cytochromes.
80.	W.H. Lewis	1931	Discovered pinocytosis.
81.	M. Knoll and R. Rusaka	1932	Produced first electron microscope.
82.	T.H. Morgan	1933	Received Nobel Prize for the discovery of role of chromosomes in the transmission of heredity.
83.	R.R. Bansley and H.L. Hoerr	1934	Isolated mitochondria from the cell.

84.	W.M. Stanley	1935	Isolated Tobacco Mosaic Virus in crystalline form.
85.	Hans. A. Krebs	1937	Discovered citric acid cycle or T.C.A. Cycle which after his name is called Krebs' cycle.
86.	W.M. Hawonth and P. Karrer	1937	Discovered molecular structure of carbohydrates, Vitamin C, carotenoids, flavins, Vitamin A and B received Nobel Prize.
87.	A.E. Blakeslee	1937	Induced polyploidy by blocking cell division with the drug colchicine.
88.	T. Caspersson	1938	Developed ultraviolet photomicrography for the study of nucleic acids.
89.	M. Kunitz	1940	Crystalized ribonuclease.
90.	G.W. Beadle and E.L. Tatum	1941	Induced biochemical mutations in bread mould *Neurospora* by X-ray *irradiation* and propounded 'one gene one enzyme' theory.
91.	A. Claude	1941	Isolated mitochondria by ultracentrifugation.
92.	A. Claude	1944	Isolated cell components-ribosomes, nuclei and mitochondria by differential centrifugation.
93.	C.F. Robinow	1944	Demonstrated nucleus in bacteria.
94.	O.T. Avery, C.M. Macleod and MacCarty	1944	Discovered transformation in bacteria showed the significance of DNA in heredity.
95.	K.R. Porter	1945	Discovered endoplasmic reticulum.
96.	F. Lipmann	1945	Discovered coenzyme A. Demonstrated its importance in intermediary metabolism and received Nobel Prize.
97.	H.J. Muller	1946	Production of mutations by X-ray radiation.
98.	J.B. Sumner	1946	Crystallized the first enzyme.
99.	J.H. Northop and W.M. Stanly	1946	Prepared enzymes and virus proteins in pure form.
100.	A. Boivin, R. Vendrely and C. Vendrely	1948	Demonstrated quantitative constancy of DNA material in various cells of the same organism.
101.	L. Pauling	1949	By studying sickle cell anemia demonstrated that protein structure is under genic control.
102.	J. Casperson	1950	Demonstrated the role of RNA in protein synthesis.
103.	E. Chargaff	1950	Discovered that in DNA amount of purine is equivalent to the amount of pyrimidine.
104.	M. Chase and A.D. Hershey	1952	Showed that gene is DNA.
105.	C. de Duve	1952	Identified lysosome.
106.	G.E. Palade	1952	Analysed the fine structure of mitochondria and showed the presence of mitochondrial cristae.
107.	F. Zernike	1953	Received Nobel Prize for the discovery of phase contrast microscope.
108.	H.A. Krebs	1953	Received Nobel Prize for the discovery of citiric acid cycle.
109.	H.D. Watson and F.H.C. Crick	1953	Demonstrated double helical model for DNA and received Nobel Prize in 1962.
110.	F. Sanger	1954	Gave first complete structure of protein molecule.

111.	H. Fraenkel, Conrat and R.C. Williams	1955	Analysed the chemical nature of TMV virus.
112.	Grumberg, S. Ochoa and Manago	1955	Discovered polynucleotide phosphorylase enzyme.
113.	S. Ochoa and A Kornberg	1955	Biologically synthesized nucleic acid of bacterium *Escherichia coli,* and received Nobel Prize.
114.	A.R. Todd	1957	Got Nobel Prize for the discovery of nucleotides and nucleotidic coenzymes.
115.	F. Sanger	1957	Presented first time the complex structure of protein molecule-insulin.
116.	Vogei, Magasanik and others	1957	Described genetic repression of enzyme and other synthetase.
117.	Hoagland	1957	Isolated transfer RNA and prostulated its function.
118.	Skou	1957	Discovered Na^+, K^+ stimulated ATPase and postulated its rule in transport of Na^+ and K^+ across the cell membrane.
119.	Seymour Benzer	1957	Initiated the concept of cistron (gene) the unit of function.
120.	M. Meselson and F. Stahl	1958	Provided the experimental confirmation of the Watson and Crick model of semi-conservative replication of DNA.
121.	G.W. beadle and E.L. Tatum	1958	Received Nobel Prize for one gene one enzyme theory.
122.	F.H.C. Crick	1958	Proposed the central dogma of a molecular biology that DNA determines the sequence of amino acids in a polypeptide chain.
123.	V. Ingram	1958	Showed that normal haemoglobin differs from sickle cell haemoglobin in single amino acid residue in one of the β chain.
124.	A. Kornberg and S. Ochoa	1959	Received Nobel Prize for in vitro synthesis of polydeoxyribonucleotides.
125.	C.E. Ford P.A. Jacob and J.H. Tijo	1959	Discovered chromosomal basis of certain genetic abnormalities.
126.	J. Hartwitz, A. Stevens and S. Weiss	1960	Demonstrated the role of enzyme RNA polymerase in the synthesis of RNA from DNA template.
127.	Park and Pon	1960	Discovered quantasomes in chloroplasts.
128.	Hirs, Moore and Stelu	1960	Determined the amino acid sequence of ribonuclease.
129.	Raker and his collegues	1961-68	Isolated F_1 ATPase from mitochondria and reconstituted oxidative phosphorylation in submitochondrial particles.
130.	F. Jacob and J. Monod	1961	Proposed operon hypothesis and postulated functions of mRNA.
131.	M.W. Nirenberg and Mathaei	1961	Reported that polyuridylic acid code for polyphenylalanine.

132.	M. Calvin	1961	Received Nobel Prize for work on photosynthesis.
133.	F.H.C. Crick *et. al.*	1961	Showed that the genetic code is a triplet one.
134.	Beevers	1961	Discovered glyoxysomes.
135.	J.C. Kndrew and M.F. Perutz	1962	Structure of protein.
136.	Menke	1962	Reported thylakoid in plants.
137.	H.M. Temin	1963	Discovered RNA-Directed RNA synthesis in certain tumour viruses.
138.	Chace and Parson; Smith and H. Fernandez-Moran	1963	Discovered elementary particles in mitochondria.
139.	F. Jacob, J. Monod and A. Lwoof	1965	Discovered activities of the regulatory genes.
140.	P. Rous and C. Huggins	1966	Received Nobel Prize for their work on cancer.
141.	M.W. Nirenberg, H.G. Khorana and R.H. Holley	1968	Discovered base sequence of tRNA Received Nobel Prize.
142.	M. Delbrueck, A.D. Hershey and S.E. Luria	1969	Reproduction in viruses. Received Nobel Prize.
143.	Tolbert	1969	Discovered peroxisomes.
144.	B. Ephrusli	1970	Cell hybridization.
145.	D. Ephrusli and H. Harris	1970	DNA-RNA polymerization. RNA-RNA hybridization.
146.	H.G. Khorana	1970	Synthesized an artificial gene from DNA nucleotides.
147.	Knippers, Kornberg and Gefter, Moses and Richardson	1970	Isolated DNA-polymerase-II enzyme.
148.	E.W. Sutherland	1971	Role of cycling AMP.
149.	W.H. Stein, S. Morre and C.B. Anfinsen	1972	Structure of ribonuclease enzyme. Received Nobel Prize.
150.	C. de Duve	1974	Structure and function of lysosomes. Received Nobel Prize.
151.	A. Claude and G. Palade	1974	Ultrastructure of cell and received Nobel Prize.
152.	H.M. Temin and D. Baltimore	1975	Discovered reverse transcription.
153.	R. Dulbecco, T.M. Temin and D. Baltimore	1975	Viruses are causative agents of cancer, Received Nobel Prize.
154.	C. Gajdusek and B.S. Blumberg	1976	Cancer research, Got Nobel Prize for discovering the cause of 'Kuru disease'.
155.	R. Talow, R. Guillemin and A. Schally	1977	These three American endocrinologists got Nobel Prize in Physiology and Medicine for developing the techniques of radioimmunoassay for the estimation of hormones in blood and for isolating, characterising and synthesising three polypeptides involved in the control of the anterior pituitary gland from brain.
156.	D. Nathans, H.O. Smith and W. Arber	1978	These three microbiologists were awarded Nobel Prize in Physiology and Medicine for their "discovery of restriction enzymes and their application to problem of molecular biology."

			The restriction enzyme provide the "chemical knives" that split chromosomes in their component genes.
157.	Peter Dennis	1978	Received Nobel Prize in chemistry for his work in bioenergetics which concerns the chemical processes responsible for the energy supply in living cells.
158.	P. Berg, F. Sanger and W. Gilbert	1980	Received Nobel Prize in chemistry for their investigation in genetic engineering. Half of this year's prize was shared by *Sanger* and *Gilbert* for developing techniques for decoding about 1000 alphabets of heredity from a DNA strand in one day. Another half of prize was awarded to *Berg* for his invention of technique of gene—splicing which literally means joining together of the various pieces of gene or DNA.
159.	G. Sneil, J. Dausset, B. Benacerraf	1980	Received Nobel Prize in Physiology and Medicine for their research work in establishing and links between genetics and immunology and to show ways to cure many human genetic diseases.

2

HOW A CELL IS STUDIED

The human eye is capable of discerning an object on smaller than 0.1 mm. One of the smallest cells in the human body is the red blood cell, and it is about 8 m in diameter (1m=1/1000 mm). One of the largest human cell is the egg, and it is about 120m in diameter. It is no wonder that the science of cell biology was virtually non-existent until the development of the microscope. With the microscope, structure smaller than 0.1 mm could be studied for the first time. First came the simple magnifying glasses, then compound microscopes were devised. More recently developed electron microscope gradually extended the resolving power 100,000 times greater than those of the light microscope.

Resolving Power of the Microscope

In the light microscope, as in any other type of microscope, the resolving power (the capacity of the instrument for showing distinct images of points very close together) depends upon the wavelength (λ) and the numerical aperture (NA) of the objective lens. The limit of resolution, defined as the minimum distance between two points that allows for their discrimination as two separate points is the limit of resolution. The objective lens determines the resolving power of a microscope. The resolving power R, in terms of size of object that may be seen is determined from the formula.

$$R = \frac{\lambda}{Na}$$

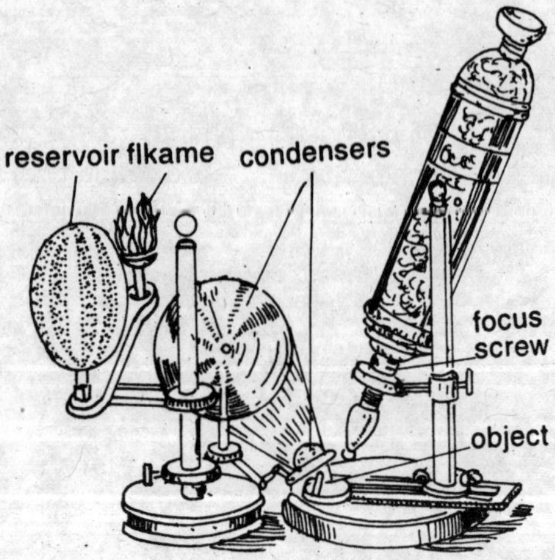

Fig. 2.1. Hooke's compound microscope.

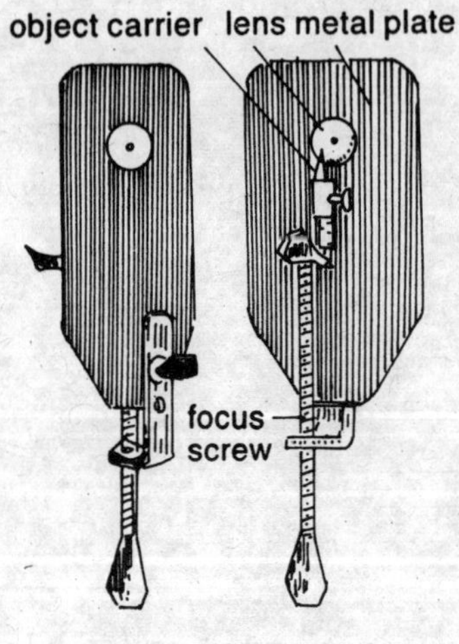

Fig. 2.2. Leeuwenhoek's microscope.

In this equation λ is the wave length of light use, Na is the numerical aperture of objective lens.

If a condenser is used then the formula changes to

$$R = \frac{\lambda}{Na + na}$$

Here na stands for the numerical aperture of the condenser.

The value 0.61 is usually multipled with the value of # because it is a constant, representing the minimum difference in contrast to that is detectable.

The above formula denotes that with objectives of the equivalent numerical aperture, the resolving power of a microscope is inversely proportional to the wavelength of the illumination and the factor which limits the resolution power of any microscope is the wavelength of light or electron.

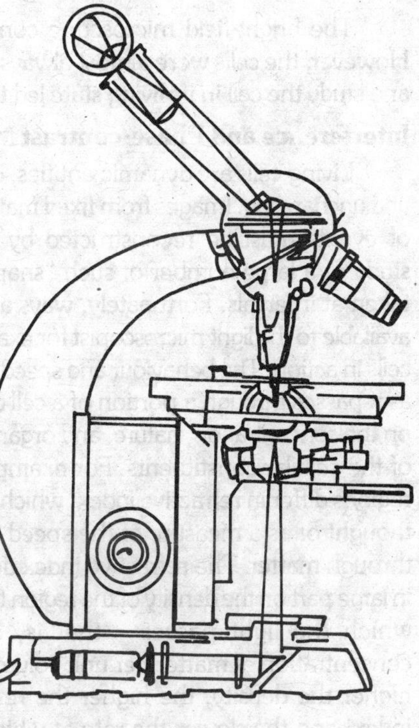

Fig. 2.3. Sectional diagram through a compound microscope.

Light Microscopy

The light microscope was first developed in the sixteen century, but because of problems with the design and construction of good lenses, the usefulness of the early instruments was limited. Around the middle of the eighteenth century, lenses were developed that permitted construction of high-resolution compound microscope. The basic design of the compound microscope has remained remarkably constant since that time.

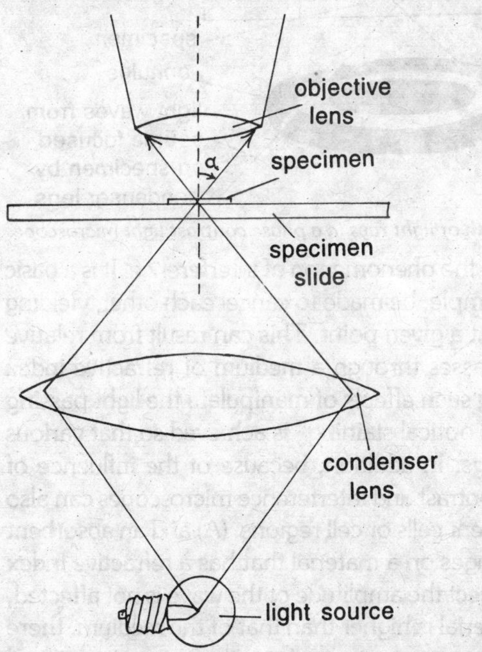

Fig. 2.4. Light path from the condenser to the objective lens for maximum resolution of a light microscope.

There are three major lenses in the compound light microscope: the condenser, the objective, and the eyepiece. The condenser gathers the light and concentrates it onto the specimen. The objective makes a magnified images. The eyepiece further magnifies the image of the objective.

There are numerous forms of light microscopy. The first developed was brightfield microscopy, which relies entirely on the absorptive properties of the cells. Cells without pigment have no light-absorbing tendencies and, therefore, are totally transparent to the condenser light. As a result, they are not clearly visible to the observer. It would be like passing a beam of light through a pane of glass. To facilitate viewing the cell, many specific stains were developed that bind selectively to parts or regions of the cell. When subjected to light from a condenser, these cells and the structures within them become visible by virtue of the transmittance of the unabsorbed light. In other words, the stains colour parts of the cells, rendering them like the image of a stained glass window. Although the cells are still transparent, the selective absorption of coloured light makes them visible. The contrast between the cell and its background and the various structures and their backgrounds becomes great, thus making them detectable.

The bright-field microscope contributed much to our knowledge and understanding of the cell. However, the cells were almost always killed before or during the staining process. The desire to observe and study the cell in its living state led to the development of phase contrast microscopy.

Interference and Phase-contrast Microscope

Living cells are dynamic entities, continually moving and changing; fixed and stained cells are stable and unchanging. Images from fixed material are at best only "snapshots" of dynamic cells. Often a sequence of events must be reconstructed by careful study of a large number of such "snapshots" taken at intervals. Fortunately, ways are now available to the light microscopist for examining cells in action. The behaviour and speed of light as it passes through a portion of a cell depend on the concentration, nature, and organization of the cellular constituents. For example, cell regions differ in refractive index, which can be thought of as a measure of the speed of light through matter. The refractive index depends in large part on the density of the region through which the light passes— that is, on the concentration of matter per unit volume. The higher the density, the higher the refractive index, and the slower the rate at which light passes. Thus light travelling through the nucleus will usually be affected differently from light travelling through the cytoplasm because of differences in the concentration of material. Differences in light paths due to factors of this type are not easily detectable in the ordinary microscope. The phase-contrast and interference microscope are light microscope modified to translate refractive index differences and related optical effects arising at boundaries between structures into visible contrast, either in brightness or in colour. This is done by taking advantage of the phenomenon of interference. It is a basic physical principle that two identical light waves can, for example, be made to cancel each other, yielding darkness rather than brightness when they arrive together at a given point. This can result from relative shifts in the "phase" of vibration occur when one wave passes through a medium of refractive index different from that encountered by the other wave. By using such effects of manipulate the light passing through the cell rather than manipulating the cell itself, an "optical staining" is achieved so that various organelles stand out in sharp contrast to their surroundings. In addition, because of the influence of concentration on the optical properties of cells, the phase-contrast and interference microscopes can also be used to estimate the amounts of material present in different cells or cell regions. (A) and an absorbent transparent material (C) on a light ray. In A, the wave impinges on a material that has a refractive index different from that of the medium. In passing through the object the amplitude of the wave is not affected, but the velocity is changed. If the refractive index of the material is higher than that of the medium, there is a delay or retradation, also called a phase change. After the wave emerges from the object, its original velocity is reestablished, but retardation is maintained. This retardation implies a phase change, which can be measured in fractions of a wavelength. The phase change increased in direct proportion to the difference

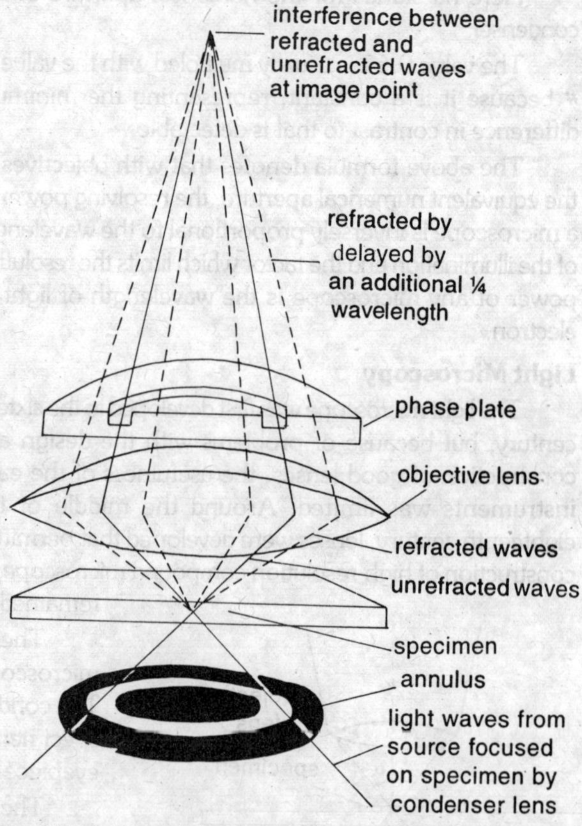

Fig. 2.5. Path of light rays in a phase contrast light microscope.

between the refractive indices of the object and the surrounding medium and to the thickness of the object.

Phase microscopy is used routinely to observe living cells and tissues, and is particularly valuable for observing cells cultured in vitro during mitosis.

Darkfield Microscopy

Darkfield microscopy, also called ultramicroscopy, is based on the fact that light is scattered at boundaries between phases having different refractive indexes. The instrument is a microscope in which the ordinary condenser is replaced by one that illuminates the object obliquely. With this darkfield condenser, no direct light enters the objective; therefore, the object appears bright because of the scattered light, and the background remain dark. Consequently, details such as small structures surrounded by a less organized solution show up as bright against a dark background (the effect is similar to the "shining" of dust particles in an intense light beam in a darkened room, such as a movie theatre). With such a microscope objects with dimensions smaller than the limit of resolution can be seen; the limit, however, severely restricts the accuracy with which the actual shape and size of such structures can be determined. In a living cell in the tissue culture, for example, the nucleolus, nuclear membrane, mitochondria, and lipid droplets appear bright, and the background of cytoplasm appears dark.

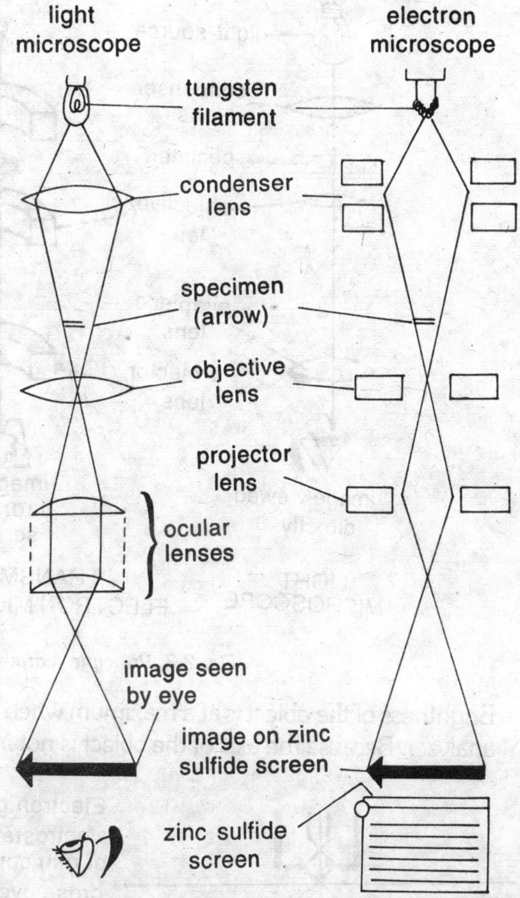

Fig. 2.6. A comparison of the basic features of the light microscope and transmission electron microscope.

In recent years darkfield microscopes have proved especially useful in observing patterns of movement of structures too small to be viewed effectively with other microscopes, such as bacteria or bundles of the fine filamentous structures, bacterial flagella by which bacteria propel themselves.

Polarization Microscopy

The polarizing microscope is similar in principle and constitution to a compound microscope. The polarizing microscope has a polarizer in the condenser and an analyzer in the occular. The polarizer is a prism or polaroid plate which transmits plane polarized light instead of ordinary light. The polarizer is mounted below the substage condenser and admits linear polarized light vibrating in a single plane into the optical system of the microscope. The analyzer is rotabel and is mounted above the objective lens in the body tube of the microscope. When the analyzer is rotated through 360^0, the field of view appears alternatively bright and dark for every 180^0 turn, that is, for every complete rotation of the analyzer there are two positions at which maximum transmission of light occurs, and two positions at which no transmission occurs. Maximum light transmission by the analyzer results when its axis of transmission is parallel to that of the polarizer. If polarizer and analyzer are crossed at an angle of 90^0, no light is transmitted. This condition is said to be "extinction" stage of crossed position. If an object is placed between crossed polarizer and analyzer, the effect of object on the natural path of polarized light will become immediately apparent. When an object is placed on the stage of the polarizing microscope and rotated between crossed polarizer and analyzer, a birefringent object shows varying degrees of brightness and darkness depending on the position of its axis with respect to those of the analyzer. This is usual test for birefringence of an object.

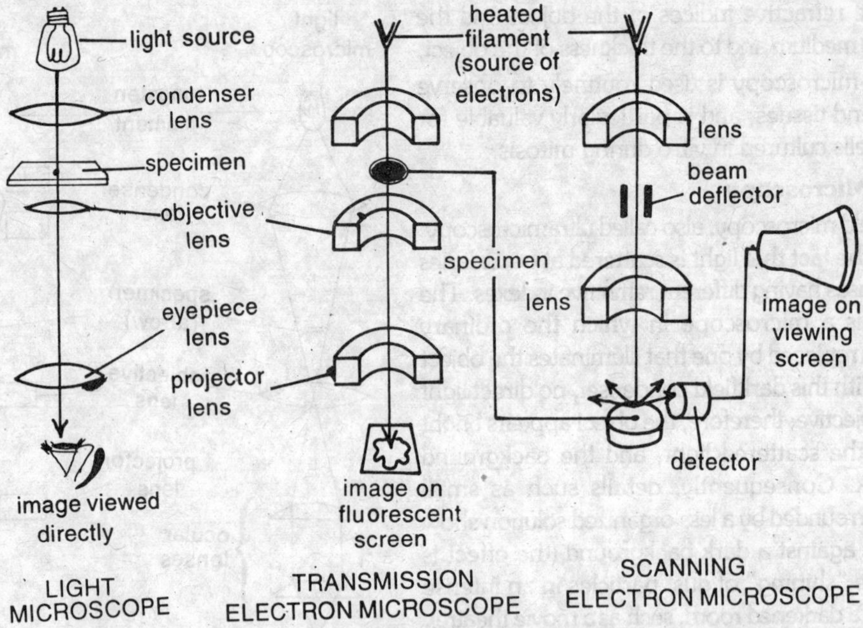

Fig. 2.7. Principle features of light microscope, TEM and SEM.

Brightness of the object is at a maximum when its axis makes a $\pm 45^0$ angle with those of the polarizer and analyzer. Because the axis of the object is not in line with that of the analyzer, neither of the two polarized components resolved by the object and oscillating in mutually perpendicular planes are exactly crossed or at right angle with the axis of the analyzer. Both polarized components of the recombined ray coming from the object are partially transmitted by the analyzer, with the results that the object appears maximally bright.

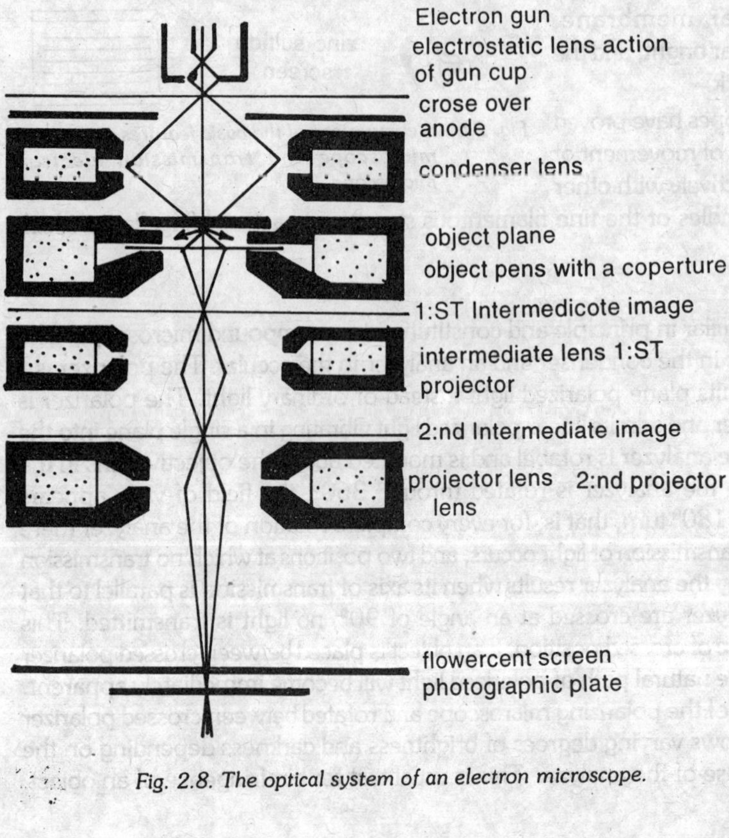

Fig. 2.8. The optical system of an electron microscope.

Electron Microscopy

The electron microscope was developed on the eve of World War Second. A considerable increase in resolving power has been possible by its development. It was designed by *Knoll* and *Ruska* of Germany in 1932, *Morton* of Belgium in 1934 and *Prebus* and *Miller* of Canada in 1934. Since then improvements were made continually by several workers in different countries and very high degree of perfection has been reached in modern design.

The size of objects resolved with the light microscope is a direct function of the wavelength of light used to illuminate the specimen. The smallest object resolved with the light microscope is half the wavelength of the light. Therefore, for blue light (wavelength of 4000Å), the smallest resolvable unit is 2000 Å or 0.20 µ.m. (1 mm = 10^7 angstroms). The electron microscope is an instrument that used electrons to illuminate the specimen. These electrons are analogous to light with an extremely short wavelength. As a result, the size of the smallest resolvable structure is 1 to 5Å, or 100 times smaller than that observed with the light microscope. With the electron microscope, it became possible to examine previously described structures in far greater details. In addition, structures not previously detected with the light microscope were discovered.

The filament or cathode of the electron microscope emits the stream of electrons. By means of a magnetic coil, which acts as a condenser, electrons are focused on the plane of the object and then are deflected by another magnetic coil, which acts as an objective lens and gives a magnified image of the object. This is received by a third magnetic "lens", which acts as an ocular or projection lens and magnifies the image from the objective. The final image can be visualized on a fluorescent screen or recorded on a photographic plate.

In summary, the electron microscope, in conjunction with other biological techniques provides the investigator with powerful methods of visualizing cells structures and biological molecules.

Use of Tracers

Several biological process may be studied by the use of appropriate tracers which are revealed by their electron opacity. For example, the uptake of macromolecules into the cells by pinocytosis or phagocytosis or the transport across cellular barriers can be studied with various opaque particles.

The ideal tracer should be: (a) non-toxic and physiological inert; (b) composed of uniform particles of known size; (c) preserved *in situ* during the processing of the tissue: and (d) composed of particles of a small size. Several colloidal substances, ranging in size from 1.0 to 10 nm, such as gold, mercuric sulfide, iron oxide, thorium dioxide, and colloidal lanthanum have been used. Ferritin, as well as dextrans of various sizes and other polymers, are now being widely used. A variety of tracers are represented by enzymes that are able to produce a reaction that greatly enhances the contrast of biological specimens. Various peroxidases are demonstrated by their reaction with peroxide and 3, '3'-diaminobenzidine, for example. One of the smallest tracers of this kind is the so-called microperoxidase, which as a molecular weight of only 1900 daltons.

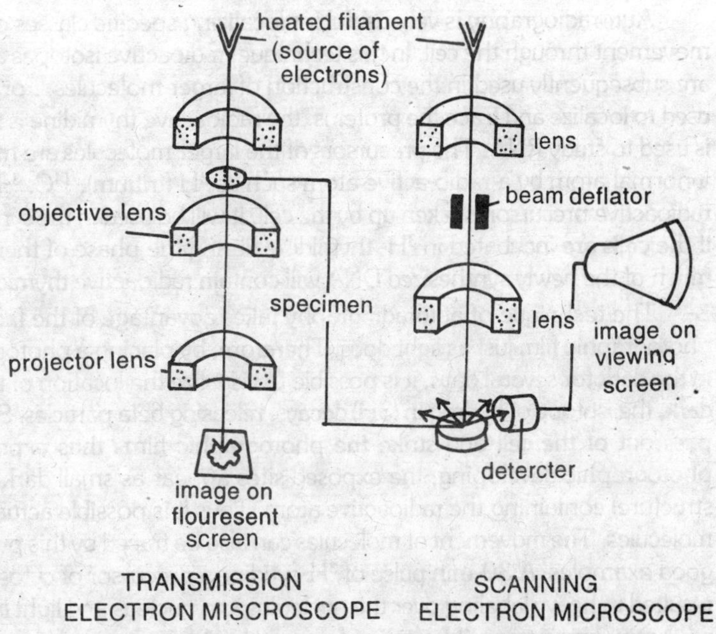

Fig. 2.9. Transmission and scanning electron microscope.

High Voltage Electron Microscopy

While most electron microscope use accelerating voltages between 50 and 100 kilovolts, there are instruments now in opeation that greatly surpass this voltage and reach 500 to 3000 kV. The design of these newer instruments is essentially similar to the older models, but the construction is much more massive to permit higher acceleration, greater magnetic excitation for the lenses, and the shielding needed to protect against X-radiation. While its main application is in metallurgy, the high-voltage electron microscope is being used increasingly in studies of biological material to examine very thick sections (upto 5 μm), with less radiation damage resulting from ionization and temperature effects. With this instrument there is the possibility of examining living cells. The practice of obtaining stereomicrographs by tilting the specimen also gives a greater amount of three-dimensional information.

Video Microscopy

In the last few years the capabilities of the optical microscope have been vastly expanded by combining it with sensitive television cameras and manipulating the image by means of the electronic features of television camera tubes. Two methods have been described by *S. Inoue* and *R.D. Allen*. Both methods make use of the video camera's dynamic range of response to light and the ability to transform electronically a small difference detected by the camera into a greatly enhanced image on a television screen. These methods appear to result in images with better resolution and contrast than are normally detectable through the microscope.

The technology of video systems has advanced so rapidly that it is now possible to improve the images further by computer. In the laboratory, microscope video images are digitized and fed into an image-processing computer. The computer can be used to accentuate edges (edge enhancement), bring out suitable differences in contrast (contrast enhancement), or replace various gray tones with different colours (colour enhancement), in addition, the speed and storage capacity of the computers permit such complicated tasks as tracking cells or organelle movements and measuring morphometry (cell size, area and volume) and a rate and with an accuracy not previously possible.

Autoradiography

Autoradiography is very useful for localizing specific classes of molecules as well as for tracing their movement through the cell. In this technique, radioactive isotopes are put into "precursor" molecules that are subsequently used in the construction of larger molecules. For example, radioactive amino acids are used to localize and trace the proteins, the radioactive thymidine is used to study DNA, radioactive uridine is used to study RNA. The precursors of the larger molecules are made radioactive by the replacement of a normal atom by a radio-active atom such as ^{3}H (tritium), ^{14}C, ^{32}P, or ^{35}S, in their structure. When the radioactive precursor is taken up by the cell, it will be used in the synthesis of a larger molecule. Therefore, if the cells are incubated in ^{3}H- thymidine during the phase of their life cycle when DNA is being made, much of the newly synthesized DNA will contain radioactive thymidine.

The technique of autoradiography takes advantage of the fact that radio-active decays will expose photographic film just as light does. Therefore, by placing a photographic film over the cell and leaving it in the dark for several days, it is possible to visualize the location of the radioactive thymidine. While in the dark, the isotope of hydrogen (3H) decays, releasing beta particles. Some of these high-energy particles will pass out of the cell and strike the photographic films thus exposing it. When subjected to standard photographic developing, the exposed sites appear as small dark spots directly over the cell region (or structure) containing the radioactive atom. Thus it is possible actually to visualize the location of specific molecules. The movement of molecules can also be traced by this procedure. Studies of RNA synthesis are good examples. A 30-min pulse of ^{3}H-cytidine a precursor of cytosine, followed immediately by fixation, resulted in heavy labeling over the nucleolus and nucleus and light labeling over the cytoplasm. However, a 30 min exposure to ^{3}H-cytidine followed by 4 hr in normal non-radioactive medium resulted in a light nucleolar and nuclear label and a heavy cytoplasmic label. These experiments were used to demonstrate that RNA is made in the nucleus and subsequently moves to the cytoplasm. Autoradiography can also be applied at the ultrastructural level using the transmission electron microscope. The living cells are treated essentially as in light microscope autoradiography except that the radioactive molecules have a higher specific activity of isotope. In addition, the sectioned material must be kept in the dark to expose the photographic emulsion for several months instead of for a couple of weeks as in the light microscope procedures. This permits localization at the subcellular level. Further modification of this method permits recovery of single cells viewed with the light microscope.

Cytochemistry

The procedures of cytochemistry aim at producing a colour or special contrast, such as enhanced brightness or darkness, at this sites of specific constituents within cells. In cells photographed in ultraviolet light, for example, the regions where nucleic acids are concentrated appear darker than their surroundings, since nucleic acids selectively absorb light with wavelengths around 260 nm.

In the Feulgen procedure, sections are treated with dilute hydrochloric acid. The DNA, because of the unique sugar it contains (deoxyribose), is the only constituent changed in the right way (aldehyde group form) to react with the Schiff's reagent, a colourless form of the dye basic fuchsin. The modified DNA reacts with this reagent to resore the bright red colour of the dye. This Feulgen reaction generally strains only the nuclei, where almost all cellular DNA is found. The DNA is seen to be localized in the chromatin, of which chromosomes are composed. The small amount of DNA present in some cytoplasmic organelles usually does not show because its concentration is below the limit needed for visibility.

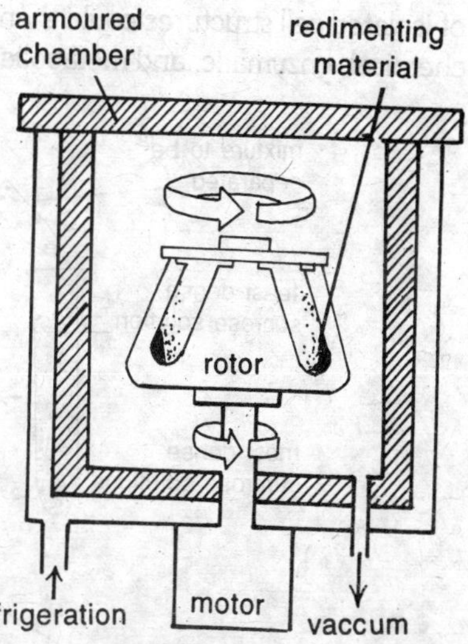

Fig. 2.10. The preparative ultracentrifuge. The sample (i.e., homogenate) is contained in tubes that are inserted into a ring of cylindrical holes in a metal rotor. Rapid rotation of the rotor generates immense centrifugal forces, which cause particles in sample to sediment. The vaccuum tends to reduce friction, preventing heating of the rotor and allowing the refrigeration system to maintain the sample at 4⁰C.

If tissue sections are exposed to dyes that are soluble only in specific types of lipids, the cell will absorb the stain only in the areas that contain these lipids. Staining procedures that can be used to demonstrate RNA are also available. The show, for example, the localization of much RNA within a discrete intranuclear body, the *nucleolus*. Stain also appears in areas of the cytoplasm, which when viewed in the electron microscope are seen to contain many small RNA—protein granules, the *ribosomes*.

Such staining techniques can given quantitative information. For instance, the amount of dye taken up may be directly proportional to the amount of stained component, and the amount of dye present at a given site within a cell can be measured through the microscope by determining how much light is absorbed by that site (microspectrophotometry). This quantitative approach using the Feulgen stain was important in establishing that the amount of DNA in cell nuclei is characteristic of given species and constant from cell to cell in a given organism; these were historical steps in evaluating the role of DNA.

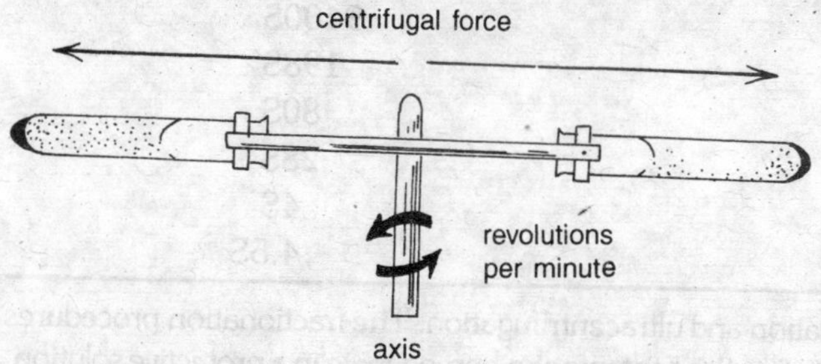

Fig. 2.11. Centrifugation. Extremely rapid rotation of liquids in centrifuge tubes produces a force in the tubes analogous to that of gravity, but much more powerful. Thus, suspended particles more dense than the liquid in a centrifuge tube separate (sink) rapidly. Following centrifugation, the supernatant fluid in the tube can be decanted (poured off), leaving the pellet of dense particles in the bottom of the tube.

Cell Fraction and Centrifugation

Microscopy whether light, electron, or video, reveals the cell as a unit composed of a myriad of "Subcellular" structures (organelles). Although much can be learned about the morphological and functional properties of cell structures by microscopic procedures, the fact that only a few cells can be studied at a time imposes severe limitations on biochemical studies. The birth of molecular biology in the early 1950s shered in the widespread application of cell fraction and ultracentrifugation. These procedures permitted the large-scale isolation and collection of subcellular structures from literally millions of cell at a time. The fractions of isolated cell structures could then be subjected to the full range of biochemical analysis is uncover their chemical, enzymatic, and metabolic characteristics.

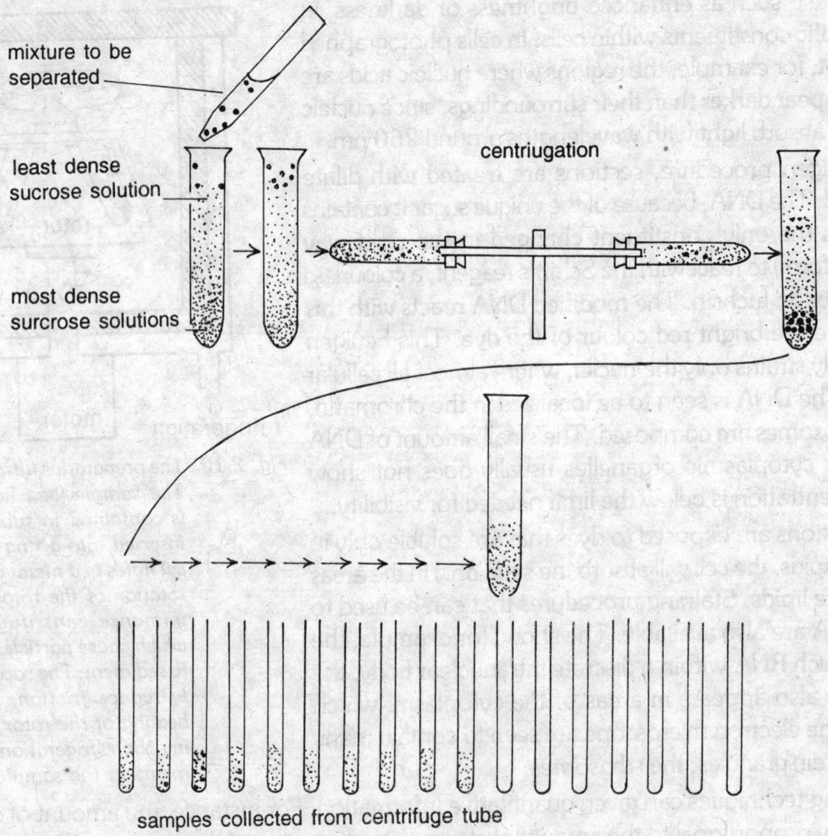

Fig. 2.12. Equilibrium Centrifugation.

Table 2.1. Some Typical Sedimentation Coefficients

Particle or Molecule	*Sedimentation Coefficient*
Lysosome	9400S
Tobacco mosaic virus	198S
Ribosome	80S
Ribosomal RNA molecule	28S
tRNA molecule	4S
Hemoglobin molecule	4.5S

There are several methods of cell fractionation and ultracentrifugation. The fractionation procedures involve mechanically rupturing the cells and collecting the contents as a homogenate in a protective solution. Sucrose or other sugar solution are used because their concentration can be readily adjusted to maintain the structural integrity of the released cell components. The homogenate of subcellular structures may next be subjected to low-speed centrifugation of about 1000 times the force of gravity (1000 x g) for 10 min. This treatment separates out a pellet containing nuclei, unbroken cells, and large species of cell membranes. This fraction may be subjected to further subfractionation, or it may be discarded. The supernatant resulting from this low-speed centrifugation is next subjected to a higher "differential" centrifugation of 10,000 x g for 30 min. This procedure produces a pellet containing a group of medium sized organelles (mitochondria, choloroplasts, lysosomes, and peroxisomes) and a supernatant containing smaller structures (endoplasmic reticulum, cell membrane, ribosomes) and soluble molecules. Both the pellet and supernatant of the 10,000 x g centrifugation can be further centrifuged at 100,000 x g in a solution of sucrose with varying densities. This procedure is called density-gradient centrifugation, and it permits the separation of cell components on the basis of their densities. With this procedure, it is possible to obtain relatively pure subfraction of the structures that were originally separated into two bulk groups.

Cell Culture

Many of the procedures discussed thus far involve the use of either dead cells or cells that are killed within a few hours following treatment. Because in the natural state the cell is living, methods for studing living cells are necessary. Indeed, an entire set of procedures specifically designed of maintaining and growing cells outside the organism in a fortified nutrient culture medium has evolved. The procedures of cell culture are numerous and differ widely among laboratories. However, several features are generally common. If the cells are taken directly from the organism and are grown in culture for several generations, they are called *primary cultures*. The cells in these cultures contain the same chromosome complement as the organism form which they were derived. However, it the cells are grown for numerous generations, they may become "established" in such a way that they grow indefinitely, going through a replication phase with periodicity. Frequently these "established" cells undergo a genetic change in the form of an increase in chromosome number, and they often lose the "differentiated" characteristics of the tissue from which they came. This is not always the case, however, as "cell lines" have been established that do maintain some characteristics of the parental tissue.

The ability to grow large numbers of viable cells over long periods of time has permitted extensive studies of such basic cellular problems as metabolism, genetic expression and mutation, cell communication macro molecular synthesis, and cell division. In addition, many biologists are using cell culture to study the induction and treatment of cancer. In fact, it is possible to take normal non-cancer cells and treat them with various agents and/or viruses and bring about a "transformation" from the normal to the cancerous state. It is the hope of many scientists that studying basic cellular mechanisms of proliferation *in vitro* (in cell culture) will unlock the mysteries of cancer.

Table 2.2. Some landmarks in the Development of Tissue Culture

Year	*Name*	*Contribution*
1885	Roux	showed that embryonic chick cells could be maintained alive in a saline solution outside the animal body.
1907	Harrison	cultivated amphibian spinal cord in a lymph clot, thereby demonstrating that axons are extensions of single nerve cells.
1910	Rous	induced a tumor by using a filtered extract of chicken tumor cells, later shown to contain an RNA virus (Rous sarcoma virus).
1913	Carrel	showed that cells could grow for long periods in culture. provided they were fed regularly under aseptic conditions.
1948	Earle	and colleeagues isolated single cells of the L cell line and showed that they formed clones of cells in tissue culture.
1952	Gey	and colleagues established a continuous line of cells derived from human cervical carcinoma, which later became the well-known HeLa cell line.
1954	Levi-Montalcini	and associates showed that nerve growth factor (NGF) stimulated the growth of axons in tissue culture.

1955	Eagle	made the first systematic investigation of the essential nutritional requirements of cells in tissue culture and found that animal cells could propagate in a defined mixture of small molecules supplemented with a small molecules supplemented with a small proportion of serumproteins.
1956	Puck	and assiciates selected mutants with altered growth requirements from cultures of HeLA cells.
1958	Temin and Rubn	developed a quantitative assay for the infection of chick cells in culture by purified Rous sarcoma virus. In the following decade, the characteristics of this and other types of viral transformation were established by *Stoker, Green* and other virologists.
1961	Hayflick and Moorhead	showed that human fibroblasts die after a definite number of divisions in cultures.
1964	Littlefield	introduced HAT medium for the selective growth of somatic cell hybrids. Together with the technique of cell fusion, this made somatic-cell genetics accessible.
1964	Kato and Takeuchi	obtained a complete carrot plant from a single carrot root cell in tissue culture.
1965	Ham	introduced a defined, serum-free medium able to support the clonal growth of certain mammalian cells.
1965	Harris and Watkinns	produced the first heterocaryons of mammalian cells by the virus-induced fusion of human and mouse cells.
1968	Augusti-Tacco and Sato	adapted a mouse nerve cell tumor (neuroblastoma) to tissue cutlure and isolated clones that were electrically excitable and that extended nerve processes. A number of other differentiated cell lines were isolated at about this time, including skeletal-muscle and liver cell lines.
1975	Kohler and Milstein	produced the first monoclonal-antibody-secreting hybridoma cell lines.
1976	Sato and associates	published the first of a series of papers showing that different cells lines require different mixtures of hormones and growth factors to grow in serum-free medium.

3

THE CELL

The cell is a unit of biological activity delimited by a semipermeable membrane and capable of self-reproduction in a suitable non-living medium. The body of all living organisms (bacteria, blue green algae, plants and animals) except viruses has cellular organization and may contain one or many cells. The organisms with only one cell in their body are called acellular organisms (e.g., bacteria, blue green algae, some algae, protozoan, etc.). The organisms having many cells in their body are called multicellular organisms (e.g., most plants and animals). Any cellular organism may contain only one type of cell Prokaryotic cells, Eukaryotic cells.

It was *Robert Hooke* who first of all in 1665 observed under a microscope certain honey comb like structures in cork, and he applied the term 'cell' (L., *cella;* small room) for the same. Previously it was believed that there is only one component of the cell which is cell wall. But in 1831, one more peculiar structure was observed by *Robert Brown*. He gave the name 'nucleus'. Later on the 1835 the word 'Sarcode' was proposed for the jelly like material present inside the cell by *Dujardin*. In 1840 *Purkinje* replaced the word sarcode by Protoplasm which is now used universally.

PROKARYOTIC CELLS

The prokaryotic (Gr., *pro*-primitive, *Karyon*-nucleus) cells are the most primitive cells from morphological point of view. They occur in bacteria and blue green algae. Prokaryotes are small, single cells organism, usually less than a micrometer (abbreviated μm; 1000 μm = 1 millimeter, abbreviated mm) are generally not longer than 3 μm. Their structural organization is simple than that of the eukaryotes. They do not have a distinct nucleus, and they are lacking in many membranous structures found in the more complex cell types. The hereditary information is contained in a single circular structure that lacks many of the molecular components of the eukaryotic hereditary structure called the chromosome. In addition, the prokaryotes seem to lack any kind of special apparatus for cell division that is why they divide amitotically or by fission.

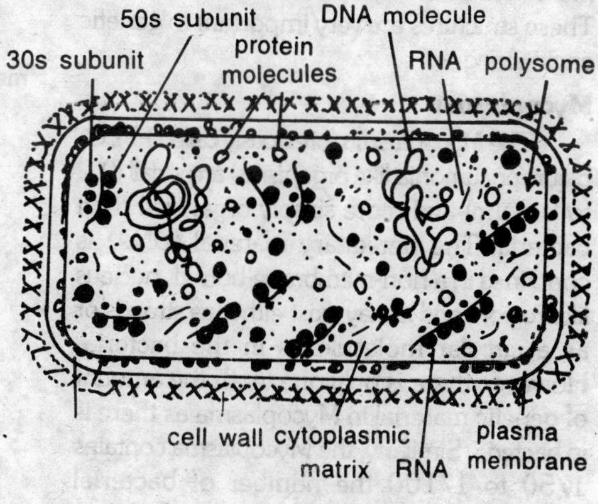

Fig. 3.1. A prokaryotic cell.

The prokaryotes are characterized by:

(i) the absence of a membrane around the nuclear material.

(ii) the absence of clearly defined membrane-limited organelles like mitochondria, chloroplast, Golgi complex and lysosomes.

(iii) the genetic material is located on a single chromosome which consists of a circular double strand of DNA.

(iv) the basic protein—histones, which are one of the most important constituents of chromosomes of eukaryotic cells, are absent in prokaryotic chromosomes.

(v) the absence of nucleolus and mitotic apparatus.

(vi) the cell wall is non-cellulosic, being formed of carbohydrates and amino acids.

(vii) the plasma membrane which lies below the cells wall is produced into the cytoplasm and acts as the mitochondrial membrane carrying respiratory enzymes, and

(viii) the cytoplasm neither exhibits streaming nor the amoeboid movement.

The prokaryotes include bacteria, blue green algae, spirochaete, *Mycoplasma* or pleuropneumonia like organism (PPLO).

Bacterial Cell

The bacteria are microscopic, cellular, achlorophillous, asexually reproducing prokaryotes which occur in fresh and salt water, in soil and in plants and animal. They lead either a saprophytic or parasitic mode of existence. The saprophytic species of bacteria are of great economic significance for man. Some parasitic species of bacteria are pathogenic (disease producing) to plants, animals and man.

Bacteria are very small having an average diameter of 1.25μ.

The smallest bacterium is *Dialister pneumosintes* (0.15 to 0.3μ) and largest bacterium is *Spirillum volutans* (13-15μ) in length.

One of the best studied bacteria is *Escherichia coli.* It is about 2 μm long and 1μm wide and contains several thousand different kinds of molecules. Structurally, *E. coli* is rod like and has a typical cell wall and underlying cell membrane. The membrane contains the special regions called *mesosomes,* which apparently have a role in cellular respiration and cell division. There may be one or several "nucleoid" areas, which contain a single, circular molecule of hereditary information (DNA).

The rest of the bacterial cytoplasm is full of structures called ribosomes, which differ slightly from the ribosomes of eukaryotic cells. These structures are the sites of protein synthesis. In addition, recent studies have revealed the presence of small circular DNA molecules in the cytoplasm called *plasmids.* These structures are very important in genetic engineering.

Mycoplasma

The *Mycoplasma* (previously called PPLO, pleuropneumonia like organisms) are small (0.2 to 0.3 μm) and more simply organized than bacteria. The hereditary material (DNA) is contained in a non-membrane-bound, nucleus like region, and it may exist either as strands or as a circular molecule as in the bacteria. However, there is only one tenth the amount of genetic material in Mycoplasma as there is in bacteria. Similarly, the Mycoplasma contains 1/50 to 1/100 the number of bacterial ribosomes per cell. A delimiting cell membrane exists, but no cell walls have been detected. A variety of other cytoplasmic inclusions, such as

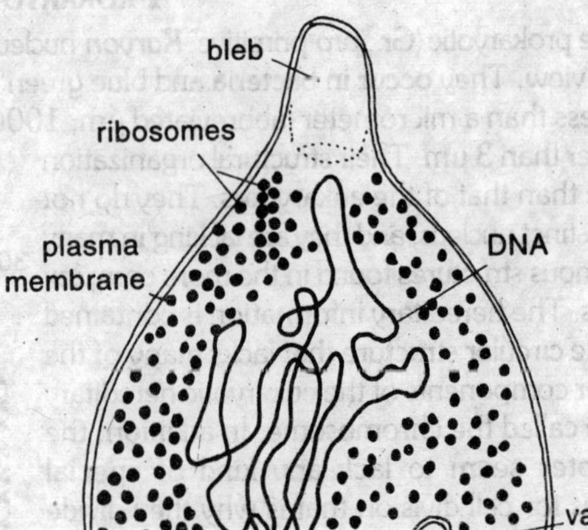

Fig. 3.2. Diagram of a Macoplasma.

vacuoles and granules, have been detected, but their function is not known. As in other prokaryotes, there are no intracellular membranous structures.

Mycoplasma has been known to cause certain respiratory diseases in animals and man. They also occur as routine contaminants in eukaryote tissue culture and currently are of much concern to researchers who use these culture procedures in their studies.

Blue-Green Algae

The blue-green algae are the third major group of prokaryotic organisms. The cellular organization is more comples than that of the other prokaryotes, primarily because of the existence of photosynthetic machinery. The blue-green algae contain chlorophyll as well as other more unusual pigments called *phycobilins.* The chlorophylls are bound to membranes in flattened rows called lamellae, which give the algal cytoplasm a rather organized and structured appearance. The phycobilins appear to be located in granular structures called *cyanosomes,* which apparently are attached to the outer lamellar membrane surface. Like the other prokaryotes, the blue-green algae do not have a membrane-bound nucleus or other membrane-delimited organelles. The genetic material exists as free strands of DNA, and the ribosomes are free in the cytoplasm as in the bacteria and *Mycoplasma.* There appears to be other cytoplasmic granules of a protein and phosphate nature, and the entire cell is surrounded by a cell wall with an underlying plasma membrane.

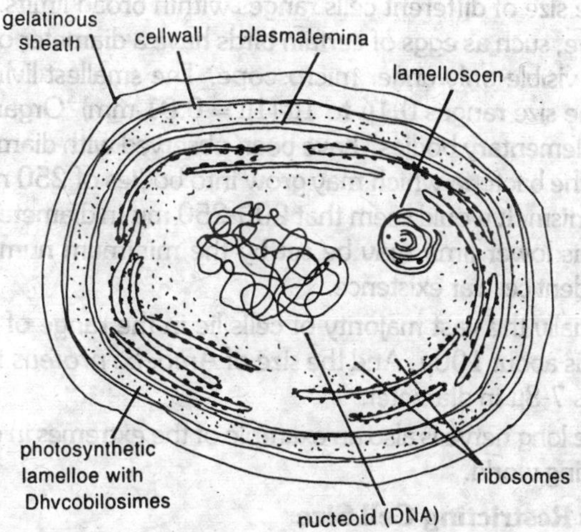

Fig. 3.3. Diagram of a blue green alga.

Eukaryotic Cells

Eukaryotic cells may be acellular organisms, such as protozoans and acellular algae, or they may be cells that make up the tissues and organs of multicellular organisms. Though the eukaryotic cells have different shape, size, and physiology but all the cells are typically composed of plasma membrane, cytoplasm and its organelles, viz., mitochondria, endoplasmic reticulum, ribosomes, Golgi complex, etc., and a true nucleus.

Morphology of Eukaryotic Cell

Shape

Cell shapes are almost as numerous as cell types: there is not typical shape. The cells of certain unicellular forms, such as *Amoeba, Diatoms, Acetabularia* and bacteria exhibit a number of shapes. But generally the cells are rounded or spherical. Besides this the cells like oval, cuboidal, cylindrical, flat, discoidal, polygonal have also been observed.

The shape of cells depends mainly on functional adaptations and partly on the surface tension and viscosity of the protoplasm, the mechanical action exerted by the adjoining cells and the rigidity of the cell membrane. Some cells, such as *Amoeba* and !eucocytes, can change their body shape very frequently.

Number

Some organisms like protozoans are single celled and others are multicellular. The body of human being is composed of about 26 trillions of cells. In human blood the number of erythrocytes is about five million per cubic ml of blood. About 10 billion neurons constitute the nervous system in human being.

Size

The size of different cells ranges within broad limits. Some plant and animal cells are visible to the naked eye, such as eggs of certain birds have a diameter of several centimeters. But the great majority of cells are visible only under microscope. The smallest living cells are found among bacteria and viruses where the size ranges 0.1μ to 1μ (1μ = 0.01 mm). Organisms like Pleuropneumonia-like organisms so called "elementary bodies" have been observed with diameters of 100 mμ. These appear to be a resting from of the bacteria which may grow into bodies of 250 mμ in diameter during the active metabolism of the organism. It would seem that 200-250 mμ in diameter is the lower limit for the size of inactive, living cells. This lower limit may be set by the minimum number and size of the component necessary for independent cellular existence.

Usually the vast majority of cells lie in the range of .5 to 20μ in diameter. The length of acellular diatoms is about 100μ. And the size of *Amoeba proteus* is 1 m.m. (1000μ) in length. The size of human R.B.C. is 7-8μ in diameter.

The long nerve is also an example of the extremes in cell sizes, the range of which is quite remarkable in the living world.

Factors Restricting Cell Size

In general, the size of the cell is correlated with its function. In no way it is associated with the size of the organism. For instance, a microscopic cell of bread mould can be several metres long, whereas cells in the body of whale or elephant may be as small as 10μ. Two major factors seem to restrict cell size.

1. The nucelocytoplasmic ratio

The nucleus controls activities of the cell by protein synthesis through RNA synthesis. In a large cell cytoplasm requires more proteins and consequently more RNA. The DNA content of a cell is constant, which can control the activities of the cell upto a specific size, beyond which proper control is not possible. For this reason a cell after attaining a specific size stops growing.

2. The surface area

As shown in fig, the small cells have more surface area per unit volume and with an increase in size, this ratio between surface area and volume gradually decreases. However, a large cell naturally requires more nutrients and oxygen for its metabolism, which diffuse into the cell through its surface. It means, the surface area of a cell affects its ability to exchange materials with its environment and puts a check on cell volume. For this reason, the cells that are metabolically more active are small in size.

Similarly, large cells face problem in the diffusion of excretory products from cell cytoplasm to the exterior.

Structure of Fixed Cells

Examination of the living cell is limited to light microscopy and is based mainly on the differences in refractive index of the different cell components. Sometimes the use of stains that act on the living organism (vital staining) facilitates observation of the living cell. However, more important in the morphologic study of the cell are methods of fixation, by which cell death results in such a way that physiologic structure and chemical composition are preserved as much as possible.

The complexity of the structural organization in a cell of the higher plants and animals is most impressive. Although there are great differences between the primitive forms of life, and the higher plant and animal cells, the similarities between primitive and advanced cells are also notable. Eukaryotic cells are characterized by a true nucleus with a nuclear membrane or envelope which divides the cell into two main compartments: nucleus and cytoplasm. The cytoplasm is turn is limited by the plasma membrane. In a plant cell the plasma membrane is covered and protected on the outside by a thicker cell wall through which there are tunnels, the *plasmodesmata,* by which the cell intercommunicates with neighbouring cells via fine cell process. In animal cells, parts of the plasma membrane are covered by a thin layer of material, which is generally described as the extraneous coat of the plasma membrane. The co-called *basement membranes* correspond to this extraneous coat.

In every stained cell usually one nucleus is present. It is central organelle of cell and is the main site of hereditary material. The nucleus is surrounded by a nuclear membrane but in bacteria and blue green algae it is without nuclear membrane. Within the nucleus a single nucleolus and network of chromatin is present. These two materials remain suspended in the nucleoplasm.

A small cytoplasmic organoid 'centriole' is present in the majority of animal cells and in some cells of the lower plants. The position is generally fixed for each type of cell. Generally the cell center or centriole occupies an axial position behind the nucleus. In some cells, it has the tendency to occupy the geometrical center i.e., in the leucocytes it occupies this position within the nucleus which is horseshoe shaped. The morphology of centriole changes with the functional state of the cell.

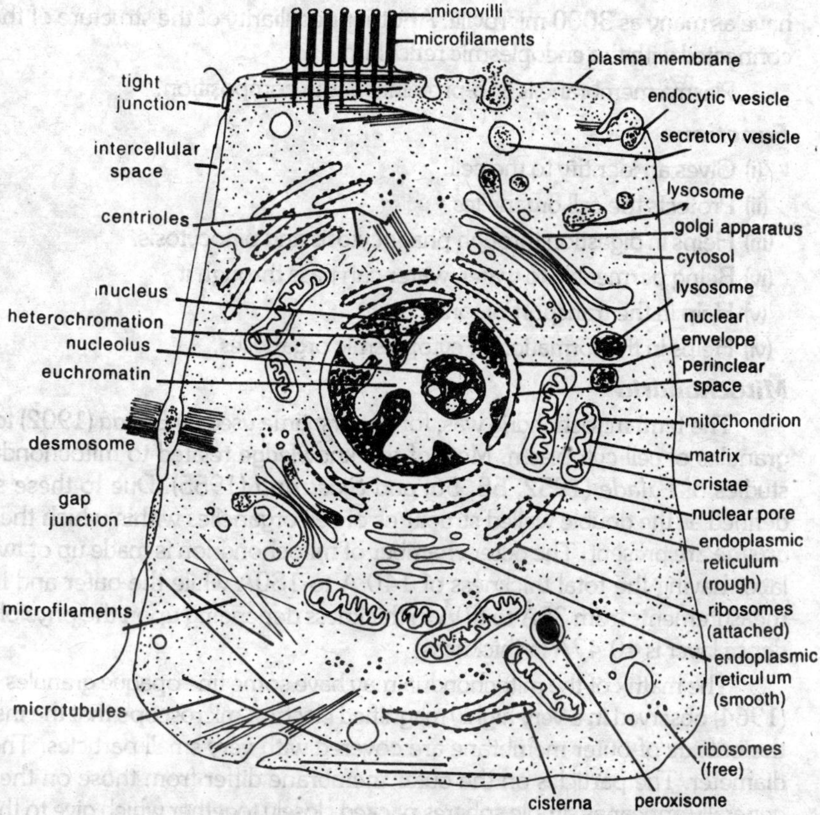

Fig. 3.4. Electron microscopic structure of an animal cell.

It is cytoplasm of the cell, there are several organelles out of them, only mitochondria and Golgi bodies are visible under light microscope. Mitochondria or chondriosomes are scattered in the cytoplasm of all kinds of cells. Their shape in the fixed cells depends upon the fixation and handling of the tissues before preparation they may appear as long rods, short rods or as small granules. Golgi material appears in its very prominent form in nervous and secretory cells in animals. Plastids are cytoplasmic organelles of plant cells may or may not contain pigments. In green plants chlorophyll is present in these organelles and thus termed as chloroplasts.

Electron Microscopic Structure of a Typical Cell

Under ordinary light microscope only few cell organelles like mitochondria, Golgi complex, chloroplasts and nucleus are visible. However, under electron microscope, several other cytoplasmic organelles such as endoplasmic reticulum, ribosomes, lysosomes, nuclear membrane, plasma membrane, etc. are seen.

Some important cellular components are being described here under the following heads:

Plasma Membrane

The outermost boundary of the animal cell is called the 'plasma membrane'. It is invisible under a light microscope but under electron microscope, it appears to be composed of two dense layers which are called the outer dense layer and inner dense layer. Both these layers are about 20Å in thickness, and are separated from each other by a less dense area of about 35Å in thickness. By this way the total thickness of plasma membrane is about 75Å. Before the advance of electron microscope people used to think of the plasma membrane has been stretched tightly over the cell. But now we know that a single plasma membrane can have as many as 3000 microvilli. Another peculiarity of the structure of the membrane is that, it remains connected with the endoplasmic reticulum.

Plasma membrane is lipo-proteinaceus in composition.

Functions

(i) Gives an identity to the cell.
(ii) Protects the cell organelles.
(iii) Helps in digestion through phagocytosis and pinocytosis.
(iv) Being permeable to water, water can pass through it.
(v) Help in the transport of ions.
(vi) Helps in the formation of cytoplasmic organelles.

Mitochondria

The term mitochondria was, for the first time used by *Benda* (1902) to designate the filamentous and granules of cell cytoplasm. Most of the knowledge related to mitochondrial fine structure is due to the studies of *Palade* (1952, 55, 56) and *Sjostrand* (1955). Due to these studies, now mitochondria are defined as the double walled structures of cell organelles within which the membranous structures called *cristae* are present. The outer chamber of mitochondrion is made up of two dense layers and a less dense layer having the total thickness of 140 Å to 180Å while the outer and inner dense layers vary in their measurements from 35Å to 60Å in thickness depending upon the physiological state of the cell. The less dense layer is 40-47Å in thickness.

The matrix of the mitochondria may have some fine opaque granules of various size. *David E. Green* (1964) observed in a very highly magnified electron microscope that the inside of the inner membrane and the outside of outer membrane are covered with very small particles. They measure about 90-100Å in diameter. The particles on the outer membrane differ from those on the inner membrane. The former generally appear as simple spheres packed closely together which give to the surface of the mitochondrion, a rough or pimpled appearance. The particle on the inside of the inner membrane has the more complex configuration. In some cells, but not all, this particle is seen to have a base, a stalk and a spherical head. Mitochondria represent the power generating centres of the cell. They can bring about oxidation of the intermediates of carbohydrates, fat and amino acid metabolism in the presence of O_2 releasing energy. The released energy is trapped in the form of adenosine triphosphate (ATP).

Golgi Complex

Golgi comples have probably been first observed earlier in 1867 by *La. Vallete St. George,* but any how credit goes to *Camillo Golgi* (1898) who described it as internal reticulate apparatus' and after it, was named *Golgi apparatus.* It is variously called Golgi element, Golgi complex, Dalton's complex etc. Each Golgi complex is composed of many lamella, tubules, vesicles and vacuoles. The membranes of Golgi complex are of lipoproteins and these are supposed to be originated from the membranes of endoplasmic reticulum. The function of the Golgi complex is storage of proteins and enzymes which are synthesized by ribosomes and transported by endoplasmic reticulum to them. Further the Golgi complex has most important secretory function. It secretes many secretory granules and lysosomes. In plant cells the Golgi complex is known as dictyosome and secretes necessary materials for cell wall formation during cell division.

Lysosomes

The lysosomes (Gr. *lyso;* digestive+*some;* body) are also found scattered in the cytoplasm. By gently centrifuging lysosomes can be separated from other cell organelles. They are usually spherical but may be irregular in shape, covered with a membrane of lipoprotein. Inside structure is always variable. Lysosomes contain hydrolysing enzymes, such as acid phosphatase, acid ribonuclease, acid deoxyribonuclease and cathespins. But they do not have oxidative enzymes. This property distinguishes them from the mitochondria. Lysosomes are polymorphic in nature and are commonly known as suicide bags and are responsible for postmortem changes in cells.

Endoplasmic Reticulum

The term "Microsome" was introduced to the modern cytology by *Claude* (1943) to represent one of the submicroscopic cellular components isolated by centrifugation. These are now known to be the broken parts of the endoplasmic reticulum after the studied of *Kollman* (1953). *Porter* (1945) was the 1st man to describe their electron microscopic structure in cultured cells. He described them as lace-like reticulum. These are membrane bounded sacs in the form of double membrane or cisternae. The term cisternae was introduced by *Sjostrand* (1953) to describe long or elongated rod like parts of the endoplasmic reticulum measuring 50-200Å. *Wiess* (1953), described the vesicles of the endoplasmic reticulum which usually has a diameter of 25-500Å while *Bradfield* (1913) came across of another type of endoplasmic reticulum termed them as tubules, they measures from 50 to 100Å in diameter.

On the basis of presence or absence of ribosomes or RNP particles the endoplasmic reticulum (E.R) is distinguished in two varieties. The *rough walled endoplasmic reticulum*, and *smooth walled endoplasmic reticulum.*

Rough Walled Endoplasmic Reticulum

Rough walled endoplasmic reticulum (R.E.R.) is that variety which bears the ribosomes at the external surface of the cisternae. The distribution of the ribosomes can be circular, spiral, or rosette type. These particles may be induced to leave the surface of the cisternae without damaging the structure of the cisternae or the capacity to associate again with the particles. Usually the rough walled endoplasmic reticulum is richly distributed in those cells which are engaged in the synthesis of the protein.

Smooth Walled Endoplasmic Reticulum

The other large division of the system owes its identity in parts of the absence of the particles and therefore commonly referred to as the smooth surfaced a granular form. Like the R.E.R., the S.E.R. shows a characteristic morphology which is tubular rather than cisternae. The tubules have the diameter nearly of 50-100 mμ. As studied by *Fawcett* (1960) S.E.R. is richly distributed in those cells which are engaged in the synthesis of steroids.

Continuity between the S.E.R. of a particular cell type possess a characteristic pattern of structure. Thus it is typical for rat liver cells to show groups of eight or ten slender profiles in parallel array.

Plastids

Plastids are present only in plant cells (not in animal cells). These are double-membrane

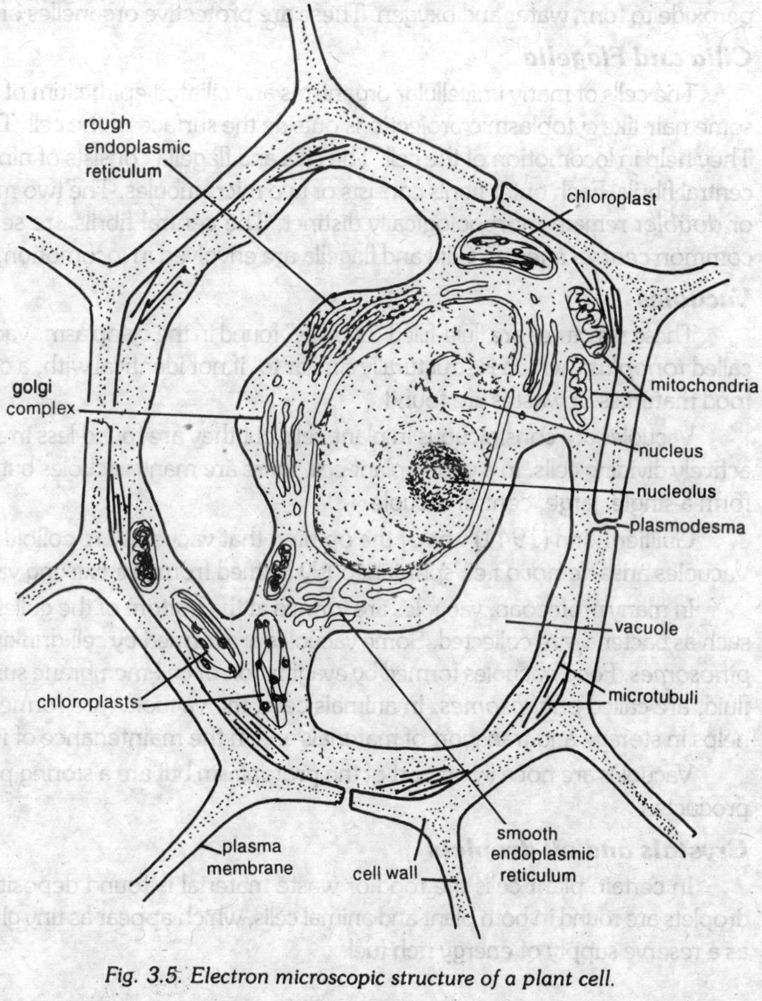

Fig. 3.5. Electron microscopic structure of a plant cell.

structures relatively larger than the mitochondria. Some of the plastids are colourless and store starch (Leucoplasts). Other contain various pigments and impart different colours (red, brown, yellow, etc.) to different parts of the plant. These are collectively called chromoplasts. Those containing green pigment—cholorophyll, are called choloroplasts. These contain several enzymes in their matrix necessary for photosynthesis. The chloroplasts capture light energy of the sun and convert it into the chemical energy—ATP, which is used in the biosynthesis of glucose from CO_2 and H_2O.

Ribosomes

Many minute, spherical structures known as *ribosomes.* They remain attached with the membrane of endoplasmic reticulum and forms granular or rough type of E.R. The ribosomes are originated in the nucleolus and consist of mainly the ribonucleic acids (RNA) and proteins. Each ribosome is composed of two structural units, small subunit known as 40S subunit and a larger subunit known as 60S subunit. The ribosomes remain attached with the membranes of endoplasmic reticulum by the 60S subunits. The 40S subunits occur on the larger sub unit and form a cap-like structure. The ribosomes consists of 3 types of RNA's known as ribosomal RNAs or rRNA, viz., 5S, 18S and 28S rRNAs. The 28S and 5S rRNAs occur in large (60S) subunit, while 18S rRNA occurs in the smaller ribosomal subunit. The ribosomes also occur freely in the cytoplasm. They are the sites of protein synthesis.

Peroxisomes

These are also bounded by a single membrane. These contain enzymes for the breakdown of hydrogen peroxide to form water and oxygen. These are protective organelles of the cell.

Cilia and Flagella

The cells of many unicellular organisms and ciliated epithelium of multicellular organisms consists of some hair-like cytoplasmic projections outside the surface of the cell. These are known as cilia or flagella. They help in locomotion of the cell. The cilia and flagella consists of nine outer fibrils around the two large central fibrils. Each outer fibrils consists of two microtubules. The two microtubules of each peripheral pair or *doublet* remain morphologically distinct. The central fibrils are separate but remain enveloped in a common *central sheath.* Cilia and flagella are envolved in locomotion, feeding, cleansing etc.

Vacuoles

These structures are fluid-filled "bubbles" found in the cytoplasm. Vacuoles are bounded by a membrane, called *tonoplast,* which is structurally similar to, if not identical with, a cell-membrane. Inside the vacuoles, food materials or wastes are found.

Vacuoles are consipicuous in plant-cells but they are found less in animal cells. Vacuoles are absent in actively dividing cells. In a young plant-cell, there are many vacuoles but, when a cell matures, they unite to form a single, large, central vacuole.

Guilliermond (1941) was of the opinion that vacuoles are colloidal with a strong imbibing capacity. Vacuoles arise *de novo* i.e., spontaneously formed from pre-existing vacuoles.

In many protozoan, vacuoles are formed at the bottom of the gullet. In these vacuoles, food-materials, such as bacteria, are collected. Some vacuoles are formed by cell-drinking. This type of vacuoles are called *pinosomes.* Food-vacuoles formed by evagination of cell-membrane surrounding food-particles with some fluid, are called *phagosomes.* In animals cells, the vacuoles are formed by a lipoprotein membrane that helps in storage and transport of materials and in the maintenance of internal pressure of the cell.

Vacuoles are not a living part of the protoplasm but are a storing place of reserve materials and waste products.

Crystals and oil droplets

In certain plant cells the food or waste material is found deposited in the form of crystals. The oil droplets are round in both plant and animal cells, which appear as tiny glistening white spheres. These serve as a reserve supply of energy rich fuel.

Nucleus

It is the most vital part of cell which is invariably present in all the cells of higher animals and plants. Normally one nucleus is present in a cell but more than one may be present in certain cells.

Structurally it is enveloped by a porous *nuclear membrane* that separates it from the surrounding cytoplasm. It has the characteristic permeability. During cell division, it breaks up and soon after the mitosis it is reconstituted. The inside of nucleus is filled with *nuclear sap* or *karyolymph*. In nuclear sap remain embedded other nuclear constituents. The *chromatin* is the Feulgen positive material in the interphase nucleus which during division becomes the definite chromosomes. In short it represents the chromosomes substance. The sex-chromatin are the specific heterochromatic bodies, found generally near the periphery of the nucleus. In nucleus usually one sex chromatin body is found but more than one can also be observed. *Nucleolus* is relatively large, generally spherical ball-like body. It has a high concentration of RNA and thus is an important structure involved in protein synthesis. Its number in a nucleus vary in different cells and depend on either the species or the sets of chromosomes.

The nucleus co-ordinates many cell activities. DNA molecules are the source of coded instructions, which pass to RNA molecule which then act as messenger, moving out of the nucleus to one of the various structures of the cytoplasm. It also controls the cell reproduction. Here too, DNA is the source of coded instructions.

The Cytosol, Hyaloplasm or Ground Substance

The material that remains suspended under conditions sufficient to sediment even the ribosomes is referred to by biochemists as the *soluble fraction* or *cytosol*. This fraction usually contains some material lost from the organelles in the course of homogenization and centrifugation. However, it also contains other molecules, including some enzymes, that are not known to be part of the intracellular structures indentifiable in the microscope. Such molecules are usually thought to derive from the *hyaloplasm* (also called ground substance, cell matrix, or cell sap), which was originally defined by microscopists as the apparently structureless medium that occupies the space between the visible organelles. As will be evident in subsequent chapters, with the improved techniques less and less of the cell appears structureless. Most investigators continue to believe that there does exist in the cytoplasm a "soluble phase"—a solution containing inorganic ions and small molecules, materials in transit from one cell region to another and dissolved or suspended enzymes that function as individual molecules or perhaps or parts of small groups of molecules rather than as components of large, microscopically recognizable organelles. But there is a lively debate going on over proposals that many components once though to be suspended in this solution are actually bound, perhaps loosely, in structured arrangements that do not survive the disruption involved in cell fractionation.

Table 3.1. The Relative Volumes Occupied by the Major Intracellular Compartments in a Typical liver cell (Hepatocyte)

Intracellular Compartment	*Percent of Total Cell volume*	*Approximate Number per Cell*
Cytosol	54	1
Mitochondria	22	1700
Rough endoplasmic reticulum cisternae	9	1
Smooth endoplasmic reticulum cisternae plus Golgi cisternae	6	
Nucleus	6	1
Peroxisomes	1	400
Lysosomes	1	300

Table 3.2. Comparison of Prokaryotic and Eukaryotic Organisms

	Prokaryotes	*Eukaryotes*
Organisms	bacteria and cyanobacteria	protists, fungi, plants, and animals
Cell Size	generally 1 to 10 μm in linear dimension	generally 10 to 100 μm in linear dimension
Metabolism	anaerobic or aerobic	aerobic
Organelles	few or none	nucleus, mitochondria chloroplasts, endoplasmic reticulum, etc.
DNA	circular DNA in cytoplasm	very long DNA containing many non-coding regions; organized into chromosomes and bounded by nuclear envelope
RNA and protein	RNA and protein in same compartment	RNA synthesized and processed in nucleus; proteins synthesized in cytoplasm.
Cytoplasm	no cytoskeleton, cytoplasmic streaming, endocytosis, or exocytosis	cytoskeleton composed of protein filaments; cytoplasmic streaming; endocytosis and exocytosis
Cell division	by binary fission	by mitosis (or meiosis)
Cellular organization	mainly acellular	mainly multicellular, with differentiation of cells.

Table 3.3. Comparison of plant and animal cells

Plant Cell	*Animal Cell*
1. Generally larger than the animal cell.	1. Usually smaller than the plant cell.
2. Plant cells are surrounded by a rigid cellulose wall.	2. Cellulose wall is not found in animal cell.
3. Majority of the mature plant cells have a large central sap vacuole.	3. Vacuoles in animal cells are small.
4. Plastids, especially chloroplasts are found usually in cells of green plants.	4. Chloroplast is not found in animal cell except in some Protozoa.
5. In the cell division, the division of cytoplasm of plant cells take by formation of a partition called place cell plate.	5. Animal cells divide by constriction during cell division.

Table 3.4. Functions of Cellular Organelles

Cell Organelles	*Functions*
Cell or Plasma membrane	Serves as a differentially permeable membrane through which extra-cellular substances may be selectively sampled and cell products may be liberated. Gives external form. Provides mechanical support and thus protects the cell.
Cell Wall (in plants)	Thick cellulose wall around the cell membrane gives strength and rigidity of the cell.
Cytoplasm	Provides greatly expanded surface. Endoplasmic recticulum-area for various biochemical reactions which normally take place at or across membrane surfaces.

Ribosomes	Sites of protein synthesis and known as "Protein factories on the cell"
Mitochondria	Oxidation of food substances, changes potential energy into a form of energy to be used by cell. Energy production (Krebs' cycle, electron transport chain, beta oxidation of fatty acids etc.). Power house of cell.
Centrioles (usually not seen in plant cells)	Form poles for the cell division. Capable of replication usually.
Golgi body or apparatus	Produce cellular secretions known as dictyosomes in plants.
Lysosome (in animals)	Produce intracellular digestive enzymes which help in disposal of bacteria and other foreign bodies. If ruptured, may cause cell destruction.
Plastids (in plants)	Serve for storage of starch, pigments and other cellular products. Photosynthesis takes place in chloroplasts.
Vacuoles	Store excess water, waste products, soluble pigments etc.
Hyaloplasm	Contains enzymes for glycolysis and structural materials, viz., sugars, amino acids, water, vitamins, nucleotides etc.
Cilia and flagella	Serve to move the animals or to move the material.
Nucleus	Provides selective continuity.
Nuclear membrane	between nuclear and cytoplasmic materials.
Nucleoplasm	Contains materials, for building DNA and messenger molecules which serve as intermediates between nucleus and cytoplasm.
Nucleolus	May synthesize ribosomes. Disappears during cellular replication. Function uncertain.
Chromosomes	Bearers of hereditary instructions. Regulation of cellular processes (evident only during nuclear division). Control the nucleus and metabolism.

4

PLASMA MEMBRANE

The outer most limit of a cell is composed of living or non-living material or both. Whatever their chemical constitution and structural organization, their function includes regulation of the flow of materials into and out of the cell and they provide some degree of support and maintenance of cell shape and cell size. Plasma membrane is variously called as *cell membrane, plasmalemma, plasma membrane* etc. The term plasma membrane was given by *Nageli* in 1855. On account of the technical difficulties it is not possible to investigate the structure of plasma membrane in every kind of cell. Most of the informations on the subject has accumulated through the study of mammalian R.B.C. and the myelin sheath of the nerve fibres.

This membrane is so thin that it cannot be resolved with the light microscope, but in some cells it is covered by thicker protective layers that are within the limits of microscopic resolution. For example, most plant cells have a thick cellulose wall that covers and protects the true plasma membrane. Some animals cells are surrounded by cement-like substances that constitute visible cell walls. Such layers, also called *extraneous coats*, generally play no role in permeability, but do have other important functions.

For the study of their structure, chemical composition and for knowing the physiological and biochemical nature of the plasma membrane or any other cytoplasmic organelle the first thing which we face is their isolation in pure state.

Methods of Isolation

The ideal isolated membrane component is that which does not have any non-membrane contamination. The isolation of the membrane can be regarded as the removal of other cell components from the cell. This operation is somewhat less difficult in the non-nucleate mammalian red blood corpuscles or erythrocytes. During hemolysis the membrane structure loosen sufficiently to allow the cell contents to escape without rupturing completely. After this hemolysis the isolation of erythrocyte plasma membrane is easy. However, the product obtained depends upon the method of lysis and the *pH* of the liquid.

The other procedure of getting the plasma membrane is introduced by *Wolport* and *O' Neill* in 1962 by treating the *Amoeba* with buffered 45% glycerol or 2.4 M sucrose solution, caused the cell contents of shrink away from the cell membrane, which retained its characteristic shape.

The membranes were fused from the cell contents by gently homogenization and were subsequently collected in bulk by centrifuging density.

Chemical Composition

The plasma membrane is composed mainly of protein, lipid, and a small percentage (1 to 5 percent) of oligosaccharides that may be attached to either the lipids (glycolipids) or the proteins (glycoproteins). There is a wide variation in the lipid protein ration between different cell membranes. Myelin is an exception, in the sense that the lipd predominates; in the other cell membranes there is higher protein/lipid ratio. In myelin the area occupied by the protein is insufficient to cover that of the lipids, whereas in a red blood cell ghost the opposite situation is found.

Lipids

The main lipid components of the plasma membrane are the phospholipids, cholesterol and glactolipids; their proportion varies in different cell membranes. Myelin differs, with regard to lipid composition, from

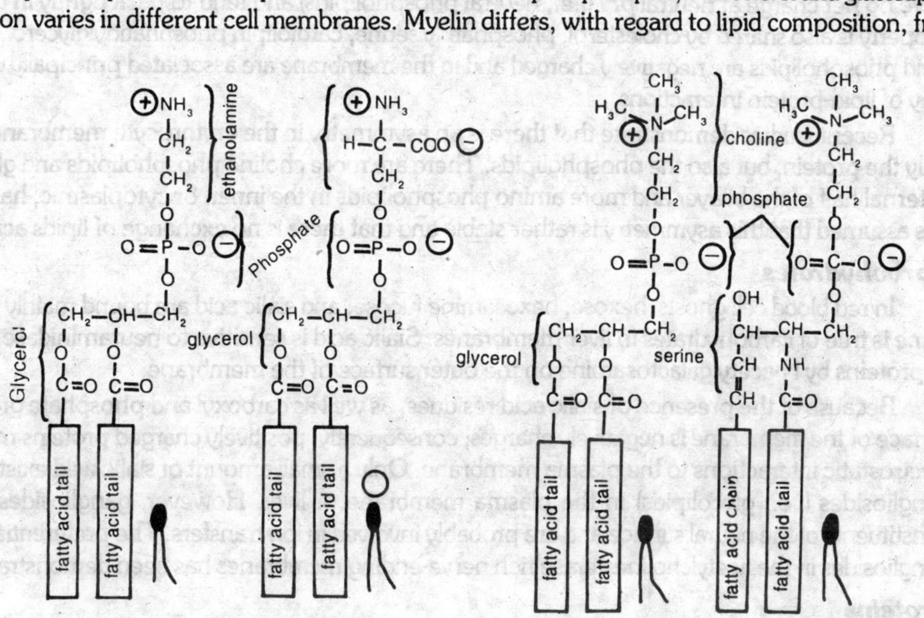

(a) Phosphatidyl-ethanolamine (derived from glycerol) (b) Phosphatidylserine (derived from glycerol) (c) Phosphatidylcholine (derived from glycerol) (d) Sphingomyelin (derived from serine)

Fig. 4.1. Four representatives with generalized structure of a phospholipid molecule.

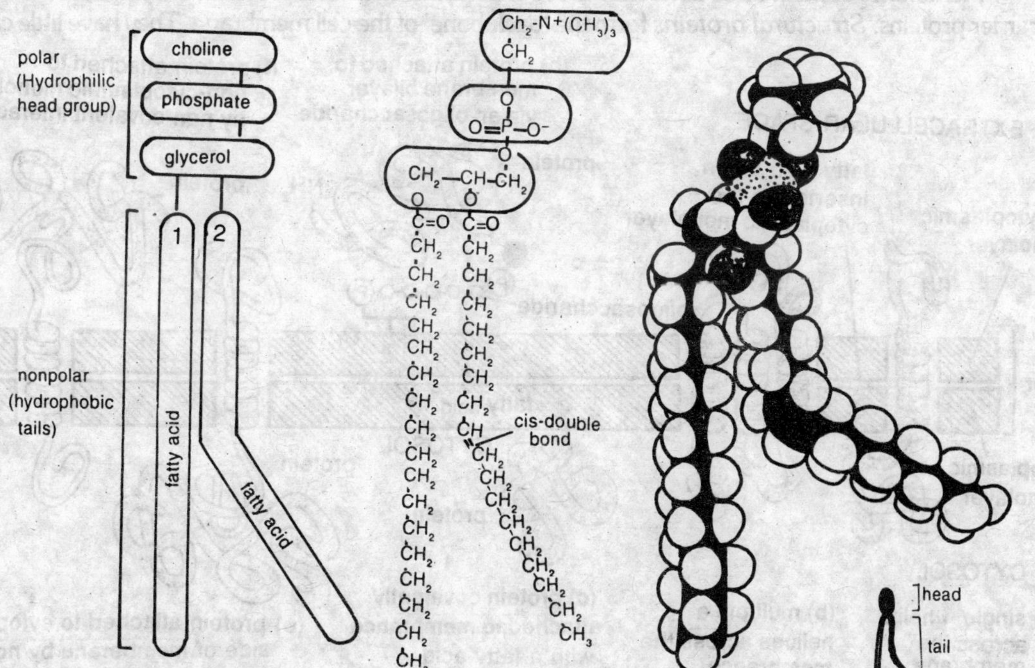

(a) schematic representation (b) structural formula (c) space filling model (d) symbolic representation

Fig. 4.2. A lipid molecule showing a head and two tails.

other cell membranes found in the brain. The major proportion of membrane phospholipids is represented by phosphatidylserine phosphatidylcholine, phosphatidylethanolamine, and sphingomyelin, all of whcih have no net charge at neutral *pH* (i.e., neutral phospholipids) and tend to pack tightly in the bilayer. This property is also shared by cholesterol, phosphatidylserine, cardiolipin phosphatidylglycerol, and sulfolipids. Acid phospholipids are negatively charged and in the membrane are associated principally with proteins by way of lipid-protein interactions.

Recent studies demonstrate that there is an asymmetry in the erythrocyte membrane regarding not only the protein, but also the phospholipids. There are more choline phospholipids and glycolipids in the external half of the bilayer and more amino phospholipids in the inner, or cytoplasmic, half of the bilayer. It is assumed that this asymmetry is rather stable and that there is no exchange of lipids across the bilayer.

Carbohydrates

In red blood cell ghosts hexose, hexosamine fucose, and sialic acid are bound mainly to proteins; the same is true of carbohydrates in liver membranes. Sialic acid is sensitive to neuraminidase and is attached to proteins by N-acetylgalactosamine on the outer surface of the membrane.

Because of the presence of sialic acid residues, as well as carboxyl and phosphate groups, the outer surface of the membrane is negatively charges; consequently, positively charged proteins may be bound by electrostatic interactions to the plasma membrane. Only a small amount of sialic acid exists in the form of gangliosides (i.e., glycolipids) in the plasma membrane of liver. However, gangliosides are important constituents of the neural surface and are probably involved in ion transfers. The preferential localization of gangliosides in the acetylcholinesterase rich nerve-ending membranes has been demonstrates.

Proteins

Proteins represent the main component of most biological membranes. They play an important role, not only in the mechanical structure of the membrane, but also as carries or channels, serving for transport; they may also be involved in regulatory or ligand-recognition properties. In addition, numerous enzymes, antigens and various kinds of receptor molecules are present in plasma membranes.

Membranes contain three different classes of proteins, structural proteins, enzymes (functional) and carrier proteins. *Structural proteins* form the 'back-bone' of the cell membrane. They have little catalytic

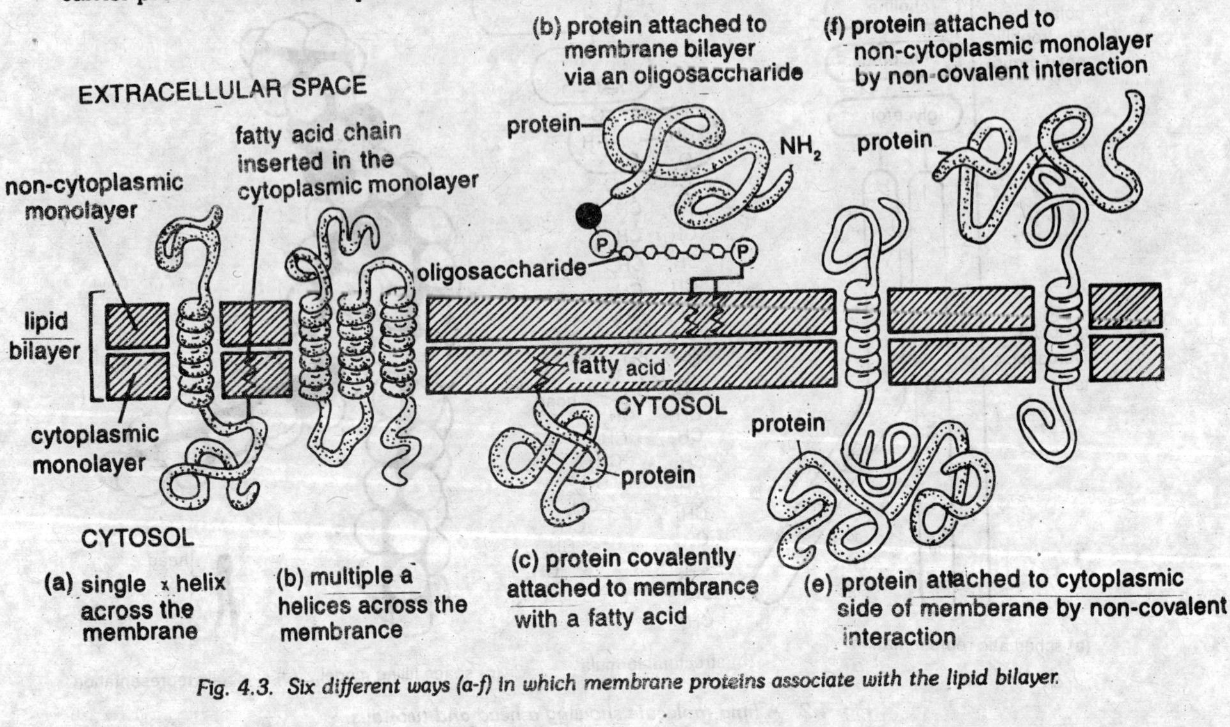

Fig. 4.3. Six different ways (a-f) in which membrane proteins associate with the lipid bilayer.

activity and are extremely lipophalic. Their amino acid composition is nearly the same. The plasma membrane consists largely of structural proteins. The average molecular weight of structural proteins is 3×10^4. Enzymes form the major component of many membranes and are catalytic proteins. The endoplasmic reticulum, mitochondria, and plasma membrane containmany enzymes. The structure of the enzymes varies from membrane to membrane. Some 30 enzymes have been detected in isolated plasma membranes. These most constantly found are 5' nucleotidase, Mg^+ activated ATPase, Na^+—K^+ activated $-Mg^{2+}$ ATPase, alkaline, phosphatase, adenyl cyclase, acid phosphomonesterase, and RNAase.

Some enzymes have a preferential localization; for example, alkaline phosphatase and ATPase are more abundant at the bile capillaries, while disaccharidases are present in microvilli of the intestine. A specific localization with a mosaic arrangement has been postualted for some of these enzymes. Disaccharidase form 5 to 6 nm globular units coating the membrane of the microvilli. The plasma membrane lacks the respiratory chain and glycolytic activity.

Of all the enzymes mentioned, Na^+—K^+ activated $-Mg^{2+}$ ATPase is one of the most important concentration gradient. The molecular weight of structural and carrier proteins are rather similar.

The amount of protein varies in different cell membranes. Myelin contains only 20% protein, plasma membranes about 50% and the inner mitochondrial membrane 75%.

The plasma membrane proteins fall in two main categories, *intrinsic* or *integral proteins* and *extrinsic* or *peripheral proteins*. The former are firmly associated with the membrane, while the latter have a weaker association and are bound by electrostatic interaction. Human erythrocyte glycoprotein consist of three chemically distinct regions: (1) N-terminal part external ot the membrane and containign all the carbohydrates, (2) a middle hydrophphobic regin located withinthe membrane and (3) a hydrophilic C-terminal portion, rich in proline but lacking carbohydrate, on the internal side of the membrane. The middle intramembranous part appear to be intimately associated with the membrane phospholipids.

The glycoprotein of human erythrocyte membrane has a polypeptide backbone to which are attached two types of oligosaccharide chains, straight chain and a longer brached one at about 10 sites.

Molecular Structure of Plasma Membrane

All biological membranes, including the plasma membrane and the internal membranes of eukaryotic cells, have a common overall structure: they are assemblis of lipid and protein molecules held together by non-covalent interactions. The lipid molecules are arranged as a continuous double layer 4 to 5 nm thick. This lipid bilayer provides the basic structure of the membrane and serves as a relatively impermeable barrier to the flow of most water-soluble molecules. The protein molecules are "dissolved" in the lipid bilayer and mediate the various functions of the membrane; some serve to transport specific molecules into or out of the cell; others are enzymes that catalyze membrane-associated reactions; and still others serve as structural links between the cell's cytoskeleton and the extracellular matrix, or as receptors for receiving and transducing chemical signals from the cell's environment.

All cell membranes are dynamic, fluid structures. Most of their lipid and protein molecules are able to move about rapidly in the plane of the membrane. Membranes are also asymmetrical structures. The lipid and protein compositions of the two faces differ from one another in ways that reflect the different functions performed at the two surfaces.

Although the specific lipid and protein components vary greatly from one type of membrane to another, most of the basic structural and functional concepts discussed in this chapter are applicable to intracellular membranes as well as to plasma membranes. After considering the structure and organization of the main constituents of biological membranes—the lipids, proteins and carbohydrates—we will discuss the mechanisms cell employ to transport small molecules across their plasma membranes and the very different mechanisms they use to transfer macromolecules and larger particles across this membrane.

The Lipid Bilayer

The first indication that the lipid molecules in biological membranes are organized in a bilayer came from an experiment performed in 1925. Lipids from red blood cell membranes were extracted with acetone and floated on the surface of water. The area they occupied was then decreased by means of a movable barrier until a monomolecular film (a monolayer) was formed. This monolayer occupied a final ara about twice the surface area of the original red blood cells. Because the only membrane in a red blood cell is the plasma membrane, the experimenters concluded that the lipid molecules in this membrane must be arranged as a continuous bilayer. The conclusion was right but it turned out to be based on two wrong assumptions that fortuitously compensated for each other. On the one hand, the acetone did not extract all of the lipid. On the other, the surface area calculated for the red blood cells was based on dried preparations and was substantially less than the true value seen in wet preparations. Nonetheless, the conclusions drawn from this experiment had a profound influence on cell biology; as a result, the lipid bilayer became an accepted part of most models of membrane structure, long before its existence was actually established.

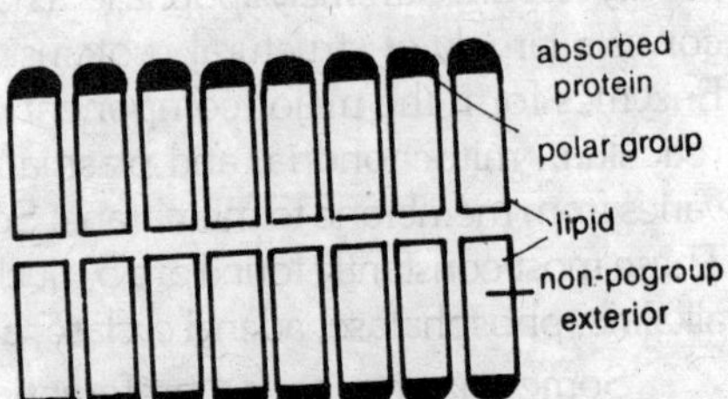

Fig. 4.4. The lipid membrane, of Gorter and Grendel (1925).

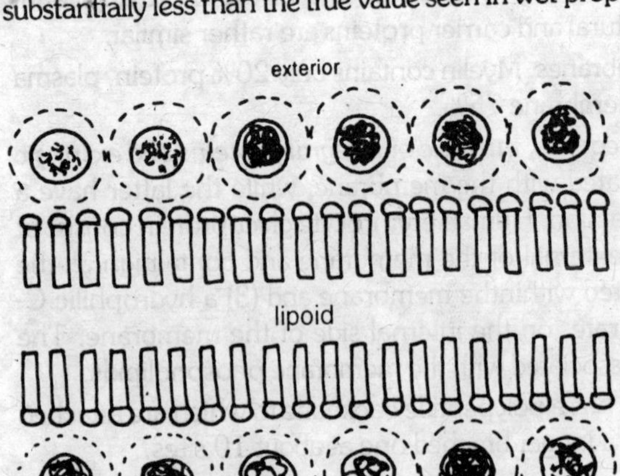

Fig. 4.5. The original Danielli-Davson model.

Danielli-Davson model or Protein-lipid-protein or Sandwich Model

Harvey and *Cole* (1931) indicated the existence of protein by studying the surface tension of cells. This led them to propose a lipo-protein model of the cell membrane. According to this modle the plasma membrane consists of two layes of lipid molecules as shown in the lipid bilayer model. The lipid molecules have their polar regions on the outer side. Globulin proteins are thought to be associated with the polar groups of the lipids. The non-polar hydrophobic ends of the two layers of lipids face each other, whereas their polar hydrophilic ends are associated with protein molecules by electrostatic interaction, Protein-linked polar pores are present in the membrane. These pores are formed by periodic continuity of outer and inner layers of proteins of plasma membrane.

Modifications of Danielli-Davson Membrane Model. Several modifications of the above arrangement have been described:

(i) Some plasma membranes have folded b-chains of proteins on both the surfaces of lipid bilayer.

(ii) Coiled a-chains of helical protein on the surfaces of lipid bilayer.

(iii) With globular proteins on both the surfaces.

(iv) With folded proteins on both the surfaces and helical proteins extending into the pores.

(v) With folded b-chain protein on one side and globular protein on the other side.

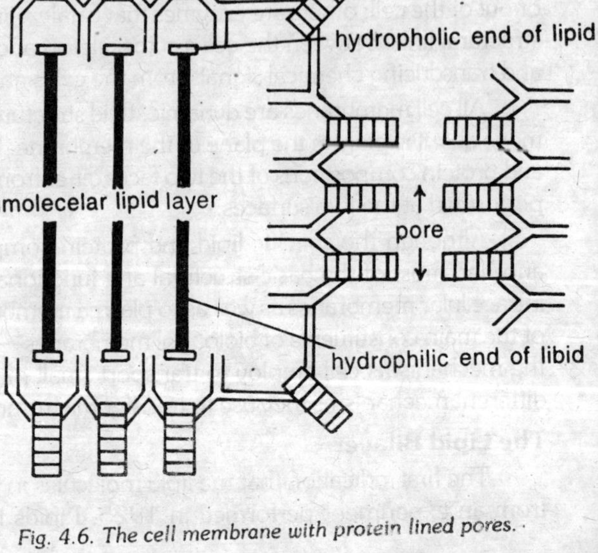

Fig. 4.6. The cell membrane with protein lined pores.

Robertson's Unit Membrane Model

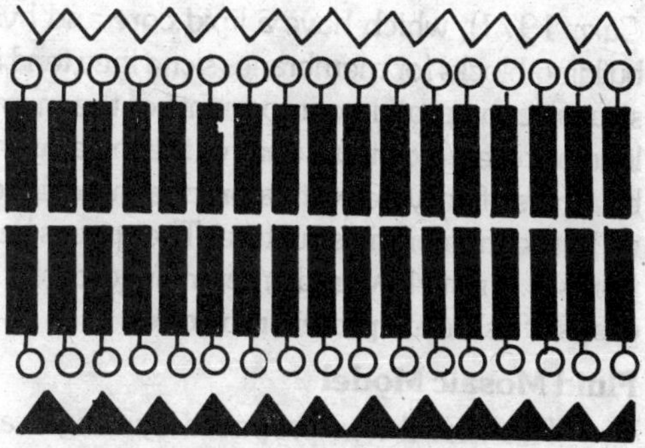

Fig. 4.7. Robertson's unit membrane model.

Unit membrane model was put forward in the year 1953 while studying cell under electron microscope. The basic unit membrane structure was considered to be general for a wide variety of plant and animal cells. All cell organelles such as Golgi body, mitochondria, endoplasmic reticulum, nuclear membrane etc., have the unit membrane structure. The unit membrane is considered to be trilaminar with a biomolecular lipid layer between two protein layers. Two parallel outer dense osmiophilic layers each of 20Å which correspond to the two protein layers. The middle light coloured osmiophobic layer is about 35Å in thickness corresponding to the hydrocarbon chains of the lipids. Thus the unit membrane is about 75Å in thickness. In this respect it reassembles the Danielli-Davson model. It, however, differs fromthe Danielli-Davson model in that the protein is asymmetrical. On the outer surface is mucoprotein, while on the inner surface is non-mucoid protein.

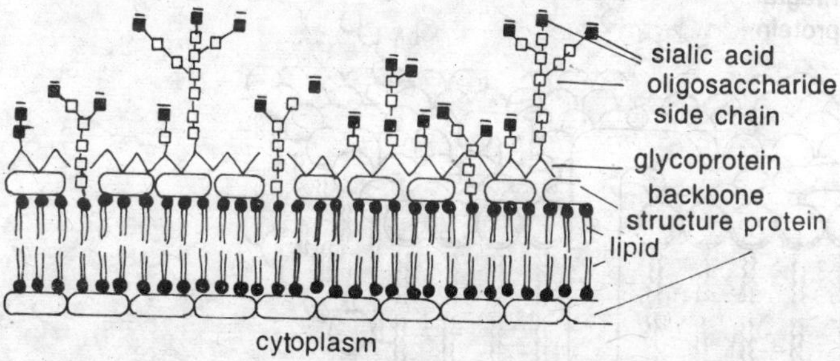

Fig. 4.8. The greater membrane model of plasma membrane.

Objections of unit membrane theory

Objections to the unit membrane model increased during the 1960s and this led to re-examinations not lipid-ptotein interactions and to new models. Studies of *F.S. Sjostrand* (1963), of smooth endoplasmic reticular, mitochondrial and chloroplast membranes underscored the differences between observed features of membranes and the uniformity that was required by unit membrane concept. Mitochondrial and chloroplast membranes contain displays of particulate units in or on the membrane. The plasma membrane did not present the same appearance as mitochondrial or chloroplast membranes. It seemed that different models might be needed to described different functional types. This unsuitable approach became unnecessary when a mosaic membrane model was proposed.

Greater Membrane Model

Like the trilaminar model here too the lipid layer is sandwiched between two layers of structural proteins. *Robertson* described the different nature of outer and inner surfaces of the membrane. The inner surface was thought to be covered with unconjugated protein, and the outer surface with glycoprotein, which is super imposed on the structural protein oligosaccharide chains with negatively charged sialic acid terminals are attached to the glycoprotein.

Micellar Model

An alternative interpretation of molecualr structure of plasma membrane has been postulated by *Hilleir* and *Hoffman* (1953). They have suggested that biological membranes may have a non-lamellar pattern, consisting of a mosaic of globular subunits known as *micelles* (*Toner* and

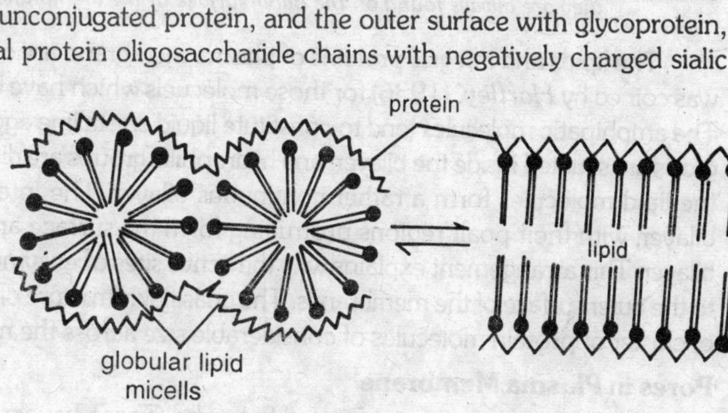

Fig. 4.9. Transformation between micellar and lamellar states of plasma membrane.

Carr, 1971), which have a lipid core and hydrophilic shell of polar groups. Lipid micelles are possible building-blocks for membranes sicne they tend towards spontaneous association. In this model of membrane structure the protein components of the membrane may form a monolayer on either side of the plane of lipid micelles. Individual units of the micellar mosaic might be replaced by individual enzymes molecules or by arrays of enzymes with a precise-three dimensional organization, enabling specific functions to be 'built in' to the membrane structure. The spaces between the globular micelles are thought to form water-filled pores 0.4 nm (4 Å) in diameter, lined partly by the polar groups of the micelles and partly by the polar grops of associate protein molecules.

Fluid Mosaic Model

This model was proposed by *Singer* and *Nicolson* (1972). According to this concept, the lipid molecules are arranged to form a rather continuous bilayer that forms the structural frame work of plasma membrane. The protein molecules are arranged in two different manners. Some proteins are located exclusively adjacent to the outer and inner surfaces of lipid bilayer and are called *extrinsic proteins.* Other proteins penetrate lipid bilayer partially or wholly and form *integral* or *intrinsic proteins.*

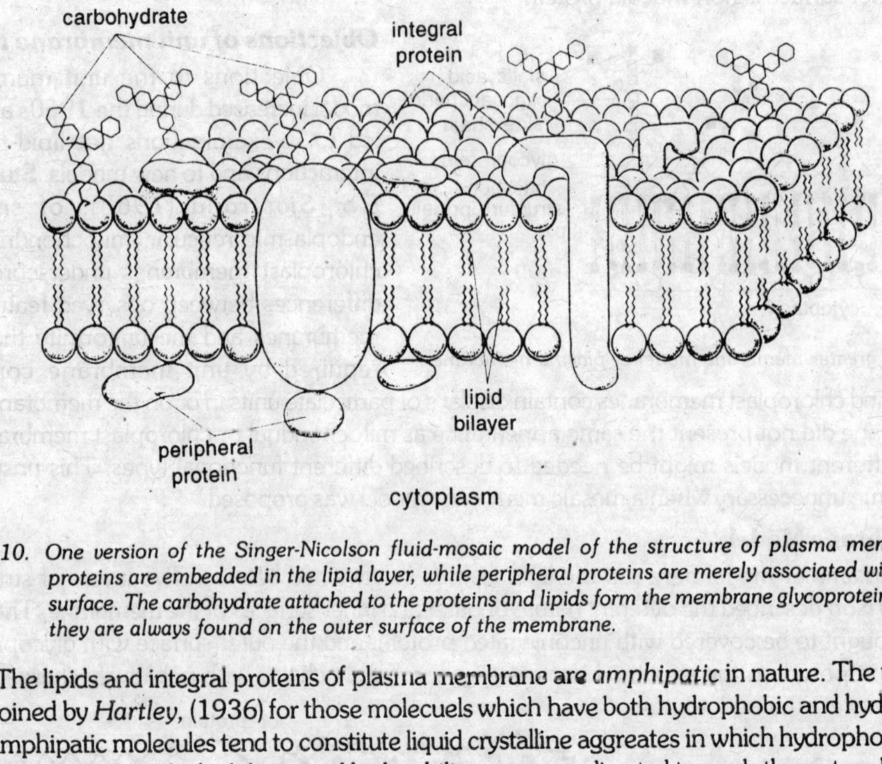

Fig. 4.10. One version of the Singer-Nicolson fluid-mosaic model of the structure of plasma membranes. Integral proteins are embedded in the lipid layer, while peripheral proteins are merely associated with the membrane surface. The carbohydrate attached to the proteins and lipids form the membrane glycoproteins and glycolipids; they are always found on the outer surface of the membrane.

The lipids and integral proteins of plasma membrane are *amphipatic* in nature. The term amphipaty was coined by *Hartley,* (1936) for those molecuels which have both hydrophobic and hydrophilic groups. The amphipatic molecules tend to constitute liquid crystalline aggreates in which hydrophobic or non-polar grops are situated inside the bilayer, and hydrophilic groups are directed towards the water phase. Therefore, the lipid molecules form a rather continuous bilayer. The integral protein are interecalated in the lipid bilayer, with their poalr regions protruding form the surface and non-polar regins embedded in the lipid bilayer. This arrangement explains why the active sites of enzymes and antigenic glycoproteins are exposed to the outer surface of the membranes. The quasifluid structure of plasma membrane explains the movement of cluster of protein molecules of considerable size across the membrane.

Pores in Plasma Membrane

Plasma membrane is perforated by pores. These have a diameter of about 0.35 nm (nanometer), slightly larger than the sodium ions. Less than 0.1 per cent of the plasma membrane is perforated by pores

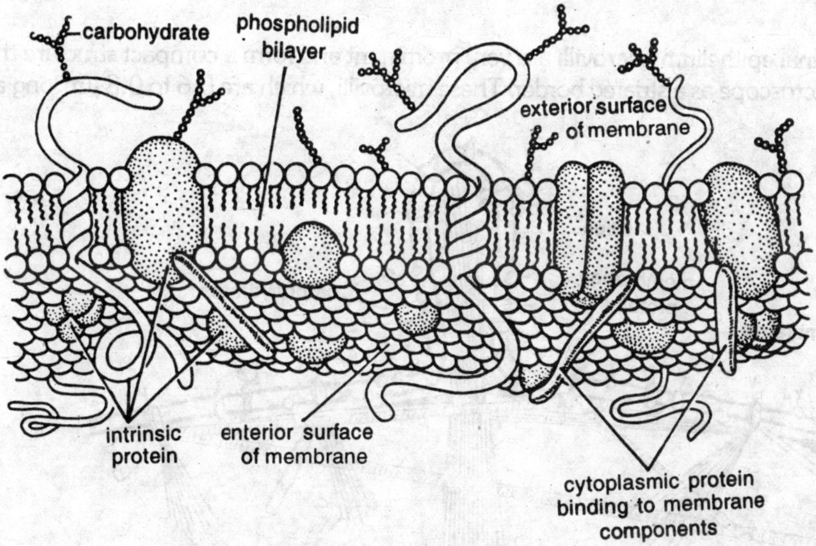

Fig. 4.11. Fluid mosaic model of cell membrane.

while 99.9 percent of the cell surface is impenetrable for ions. Several models of structures of pores have been proposed. Some of them are:

1. *Structural pores.* These are permanent cylindrical holes that interrupt the otherwise continuous bilayer sheet.
2. *Dynamic pores.* These pores are transient cylindrical hoels rather than being permanent. They appear only at the time of intake.
3. *Paving block pores.* According to this concept the pores are regarded to be the corners of the closely filled nearly hexagonal paving blocks of lipid and protein subunits.
4. *Protein channel pore.* These pores are considered to be parts of the lipid-globular protein mosaic model. These form small channels of specific proteins embedded in the membrane through which ions and small molecules can diffuse.
5. *Ionophore.* The ionophores are small polypeptides whose oen end is hydrophobic and other hydrophilic. The hydrophobic (outer) end dissolves in the membrane while the hydrophilic end (inner side) picks up ions or water-soluble materials and dumps them on the other side. The ionophores help in exchange of substances from or into the cell.

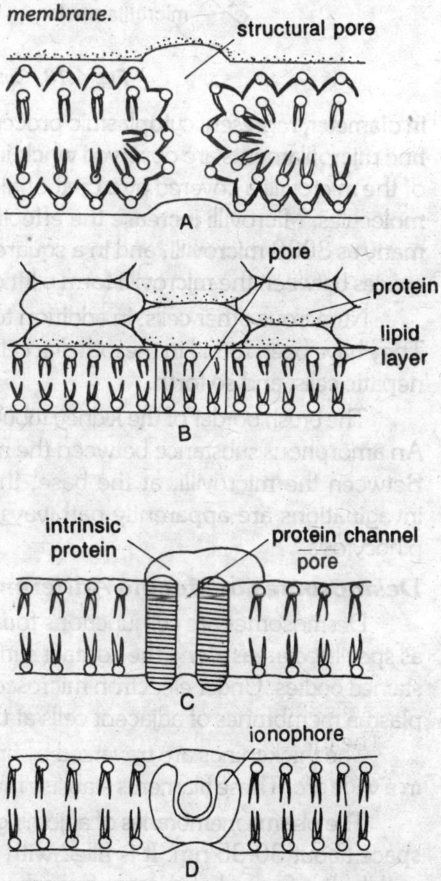

Fig. 4.12. Different models of structure of pores. A—Structural pore; B—Paving block pore; C—Protein channel pore; D—Ionophore.

Specialization or Modifications

With the increased resolution of the electorn microscope numerous specialization of cell surface have been recognized (*Sjostrand* 1956; *Fawcett* 1958). Following description of *Fawcett* (1958) deals with the various specializations of plasma membrane studied topographically.

Microvilli

In the intestinal epithelium microvilli are very prominent and form a compact structure that appears under the light microscope as a striated border. These microvilli, which are 0.6 to 0.8 μm long and 0.1 μm in diameter, represent cytoplasmic processes covered by the plasma membrane. Within the cytoplasm core fine microfilaments are observed whch in the subjacent cytoplasm form a terminal web. The outer surface of the microvilli a covered by a coat of filamentous material (fuzzy coat) composed of glycoprotien macromolecules. Microvilli increase the effective surface of absorption. For example a single cell may have as many as 3000 microvilli, and in a square millimeter of intestine there may be 200,000,000. The narrow spaces between the microvilli form a kind of sieve through which substances must pass during absorption.

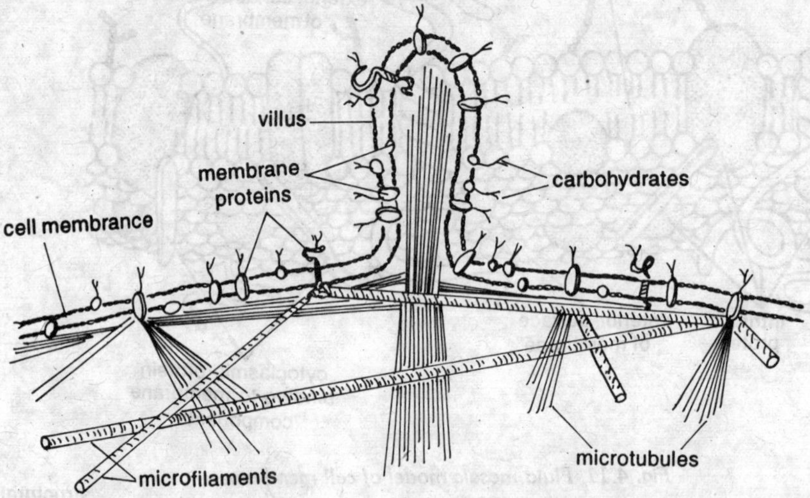

Fig. 4.13. Cytoskeleton underlying the cell membrane.

Numerous other cells, in addition to intestinal epithelium, have microvilli, although fewer in number. They have been found in mesothelial cells, in the epithelial cells of the gall-bladder, uterus, and yolk sac, in hepatic cells, and so forth.

The brush border of the kidney tubule is similar to the striated border, although it is of larger dimensions. An amorphous substance between the microvilli gives a periodic acid-Schiff reaction for polysaccharides. Between the microvilli, at the base, the cell membrane invaginates into the apical cytoplasm. These invaginations are apparently pathways by which large quantities of flow enter by a process similar to pinocytosis.

Desmosomes or Macula Adherens

Desmosomes are cell junctions found mainly in the cells of simple columanr epithelium. These occur as specilized areas along the contact surfaces. Under light microscope the desmosomes are seen as darkly stained bodies. Under eleectron microscope these appear as button-like thickenings on the inner surface of plasma membranes of adjacent cells at the point of contact.

The thickenings are traversed by fine cytoplasmic fibrils called *tonofibrils*, which form a kind of loop in a wide arc. These filaments stabilise the junction and act as anchoring sites for the cytoplasmic structures.

The plasma membranes of adjoining cells in the regions of desmosomes are separated by an intercellular space about 30-35 nm. It is filled with intervening dense coating material that forms a dark line in the middle. It is formed of mucopolysaccharides and proteins.

The desmosomes are primarily concerned with cell adhesion, but also help in maintaining cell-shape, providing it rigidity and cellular support. The former is brought about by the intercellular coating substance and latter by the tonofibrils.

Plasmodesmata

Sometimes, the cells are joined by bridges of cytoplasm passing between pores of cell wall or plasma membrane between the adjacent cells, such connections are called *plasmodesmata*. They are usually simple but anastomosing plasmodesmata may also found. Their distribution and number may also very considerably. They were discovered by *Tangl* (1879) and were named as such by *Strasburger* (1882). Endoplasmic reticulum often is closely associated with the cell surface, at the points where plasmodesmata ae present. Through them cytoplasmic continuity is often maintained among the adjacent cells. They provide a mean for interaction between adjacent cells which are separated in other regions. Through them the material can pass form cell to cell.

It is not known whether all plasmodesmata are similar to each other or not. There exist some difference because they are not only produced at the time when cell divides but also formed spontaneously between cells that have grown into contacts with one another, e.q. tyloses in xylem vessel elements. They may occur singly or they can be aggregated into groups. In many primary walls, the plasmodesmata are usually asssociated with a reduced deposition of wall material, and the area then is known as primary pit or field.

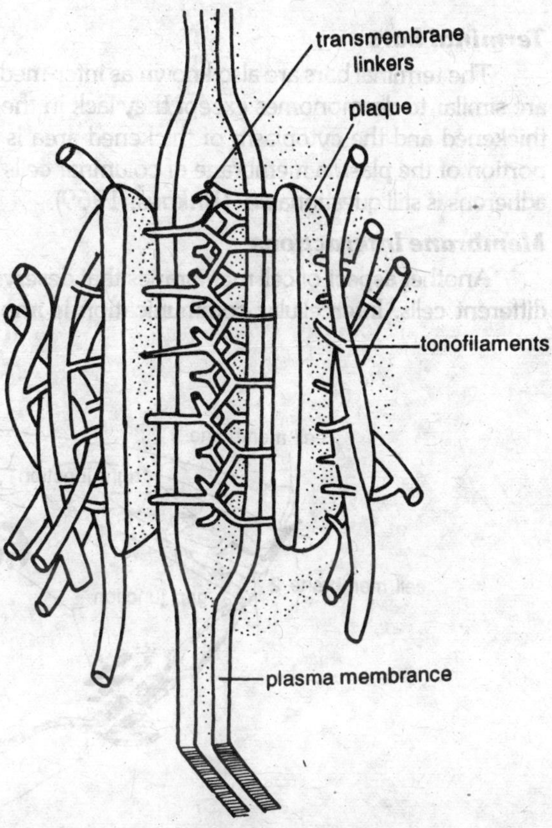

Fig. 4.14. Spot desmosome.

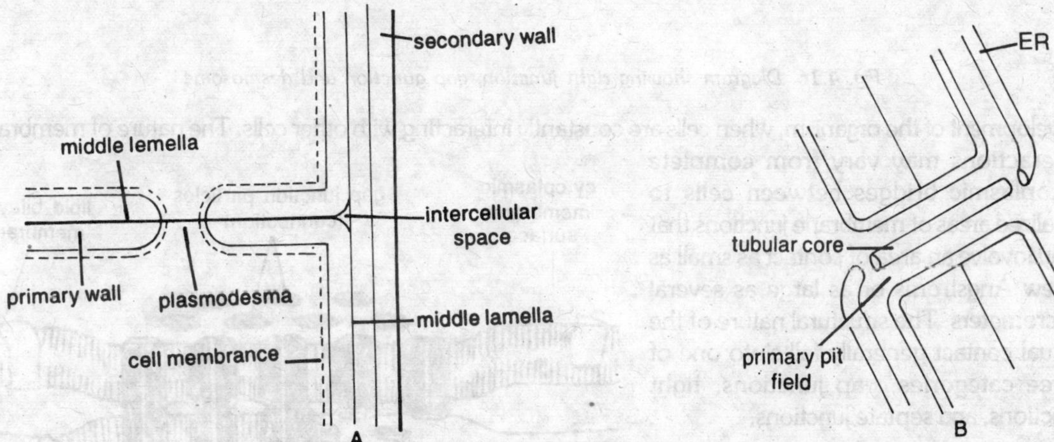

Fig. 4.15. A-B. A part of cell wall showing primary pit field and details of plasmodesma.

Hemidesmosomes

These are found in the basal surface of some eptihelial cells. Their structure is similar to desmosomes but these are represented by one half; their counterpart is usually represented by collagen fibrils.

Terminal bars

The terminal bars are also known as inter-mediary junctions or *zonula adherens*. The terminal bars are similar to desmonomes except they lack in the tonofibrils. In terminal bar the plasma membrane is thickened and the cytoplasm of thickened area is dense. The terminal bars occur in the intermediary portion of the plasma membrane of columnar cells just below the surface. The correct identity of zonula adherens is still questionable (*Trinkaus*, 1969).

Membrane Interactions

Another aspect of cell membranes that deserve discussion is the interaction between membranes of different cells. Intercellular communication is important in many cell functions and especially during

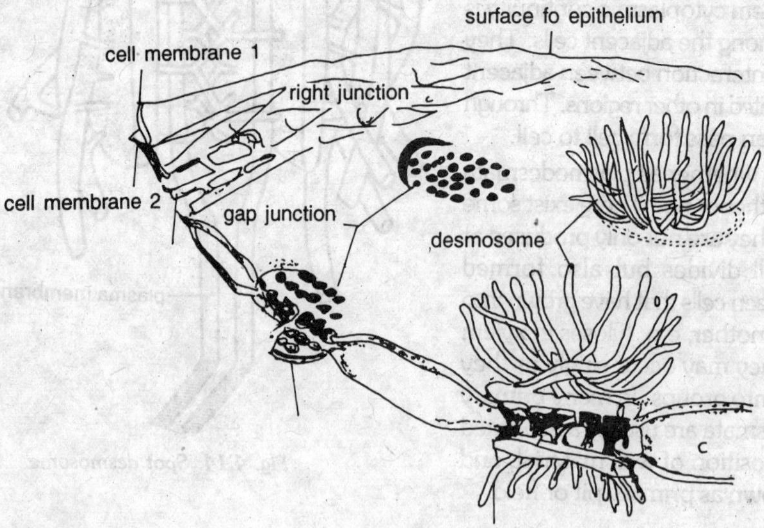

Fig. 4.16. Diagram showing tight junction, gap gunction and desmosome.

development of the organism, when cells are constantly interacting with other cells. The nature of membrane interactions may vary from complete cytoplasmic bridges between cells to localized areas of membrane junctions that may involve an area of contact as small as a few Angstroms or as large as several micrometers. The structural nature of the actual contact generally fall sinto one of three categories; gap junctions, tight junctions, and septate junctions.

(a) *Gap junctions* appear as multilayered structures when observed with the electron microscope. They appear to be two unit membranes closely apposed to each other with a 20 to 40Å gap between. The total thickeness of the entire gap junction

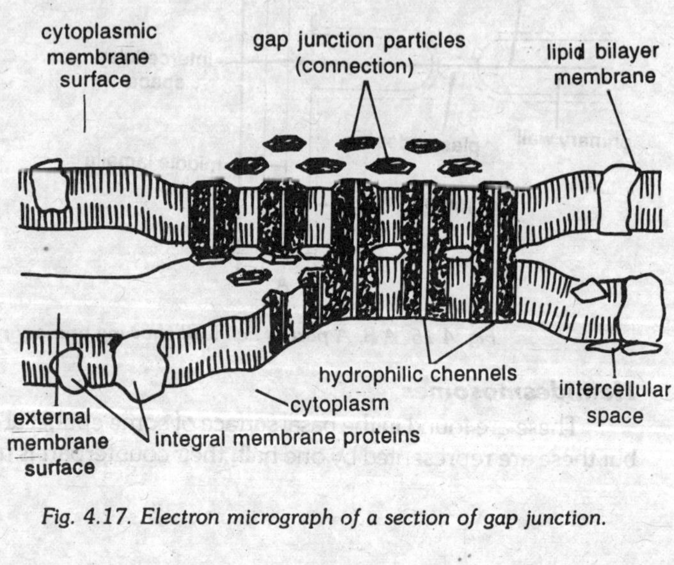

Fig. 4.17. Electron micrograph of a section of gap junction.

is 170 to 190Å, and they are found in both vertebrates and invertebrates. They are not found in skeletal muscle fibers or red blood cells.

(b) *Tight junctions* are found only in vertebrates and occur in cells such as epithelial cells. These junctions appear to be true fusions between the two membrane, and they are 100 to 140Å thick.

(c) *Septate junctions* have been found only in invertebrates. They are much larger than the other types of junctions and are characterized by electron-dense cross bridges that extend between the two cells membranes.

PREMEABILITY OF MEMBRANES AND TRANSPORT ACROSS CELL MEMBRANES

The membranes of a cell pass small ions and molecules through them. The passage of ions or molecules may occur as passive diffusion, or active transport evolving the expenditure of energy. In passive diffusion, membranes may be classed according to their degree of permeability.

1. *Impermeable.* A membrane of this kind allows nothing to pass through it. Certain unfertilized fish eggs, such as trout, are permeable only to gases and water, labelled with deuterium does not penetrate the egg.
2. *Semipermeable.* No cell membranes are in this category. A model membrane may be constructed to allow passage of water molecules, but no solute particles.
3. *Selectively permeable.* Most membranes of the cell belong to this category. Such membranes allow water and certain selected ions and small molecules to pass through, but prohibit other ions as well as small and large molecules.
4. *Dialyzing membranes.* The endothelial cells and their basement membranes of the capillaries and nephron can act as a dialyzer across the membrane down their concentration gradients while restricting the passage of colloids.

Transport of substances across the plasma membrane into the cytoplasm of cell can be achieved by following methods:

Osmosis

Osmosis is special type of diffusion which involves the movement of water or other solvent molecules through a semipermeable or differentially permeable membrane from an area of high potential (pure solvent) to an area of low potential (more concentrated solution).

The entry of water into the cell from its medium is called *endosmosis,* the reverse process in which water leaves the cell is called *exosmosis*. An *osmotic pressure* is maintained by the salts present in cytoplasm.

The cell always remains in a liquid or fluid medium for physiological exchange of gases, nutrients, etc. this fluid is usually designated by Extra Cellular Fluid (ECF). In protozoans and other lower organisms it is water. Depending upon the concentration, ECF may be

(i) *Isotonic Solution.* If the concentration of ECF in which the cell is present is similar to that of intracellular fluid of the cell, it is known as isotonic solution. The shape of cell remains normal.

(ii) *Hypotonic Solution.* If the concentration of ECF is less concentrated than the intracellular fluid, it is known as hypotonic solution. In such a solution the cell swells up due to water reaching the cell by *endosmosis.*

(iii) *Hypertonic Solution.* If the concentration of ECF is higher than the intracellular fluid of cell, the soltuion is called hypertonic solution. In such a case water diffuses out of the cell by *exosmosis*. As a result the cell undergoes *plasmolysis*.

Passive Transport

Passive transport is straight diffusion of water, ions or molecules of various substnaces move through the plasma membrane from a region of higher concentration to the low concentration. Transport of molecules takes place along concentration gradient so that no energy is required for diffusion.

(a) Simple diffusion

According to a large numer of evidence many substances move through plasma membrane at rate of free diffusion that are directly proportional to their solubility in lipid. Water molecules are a notable exception to this rule, since they freely diffuse through membranes regularly and rapidly. Plasma membrane supposed to contain two types of pores.

(i) *Fine aqueous channels.* These are present through a protein or between clustered integral proteins. These pores are 10 nm in diameter and permanent in nature. These extend through the entire lipid bilayer. These pores act as valvular openings. Some pore are charged positively other are negatively charged.

(ii) *Statistical pores.* These pores are unstable. They keep on appearing and disappearing. These are formed as gaps in the highly fluid lipid bilayer. These are formed due to random thermal movement of membrane phospholipids. Substances readily pass through these pores if they are solube in lipids (*Overton*). That is why the hydrophobic substances of both low and high molecular weight can pass through plasma membrane.

The relative rate of diffusion of molecules across the membrane depends upon size of molecules; concentration gradient across the membrane; and the solubility in lipids or hydrophobic nature of molecule.

Collander and *Barlund* in their classical experiments with cells of plant, *Chara,* demonstrated that the rate at which substrate penetrate depends on their solubility in lipids and their size of molecules.

The permeability (P) of molecules across the membrane is represented by a formula

$$P = \frac{KD}{t}$$

where K is partition coefficient; D is the diffusion coefficient (depends on molecular weight), and t is the thickness of membranes. The partition coefficient in cell membranes is similar to that of olive oil and water.

The partition coefficient can be measured by mixing the solute with an oil-water mixture and their waiting until the phases are separated. The partition coefficient (K) is the concentration of solute in oil, divided by the concentration of solute in aqueous phase. The diffusion coefficient (D) can be determined by using radioactive solutes and measuring their rate of entry into the cytoplasm at various external concentrations.

(b) Facillitated Diffusion

The diffusion of a substance across a membrane always occurs from a region of higher concentration on one side to a region of lower concentration on the other side. But it is not always the case sicne

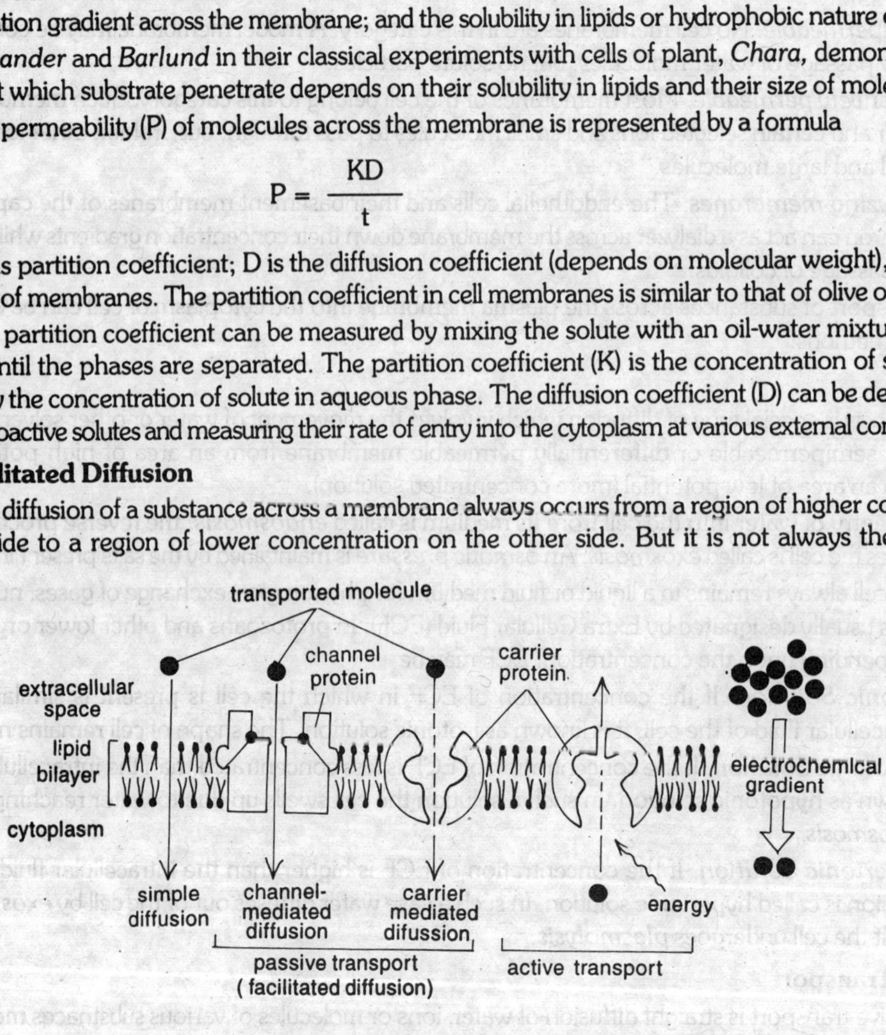

Fig. 4.18. Schematic diagram showing various types of transports across the membrane: simple diffusion, passive transport (down an electrochemical gradient) and active transport (against an electrochemical gradient).

numerous examples have been uncovered where a protein, *permease* exists within the plasma membrane that facilitates the diffusion process. This mechanism is called *facilitated diffusion.* This process is most common in the movement of sugars and amino acids. The presence of permease in the membrane offers a pathway through the membrane which is an alternative to the lipid layer. The binding of solute on the outer surface of the membrane would trigger a conformational change in the permease, exposing the solute to the inner surface of membrane from which it can diffuse into the cytoplasm down its concentration gradient.

The characteristics of faciliated diffusion are:

(i) The rate of transport of molecules across the membrane is far greater than expected from simple diffusion.

(ii) The permeases are very specific and each transport only a single species of ions or molecules or group of closely related molecules.

(iii) With an increase in the concentration gradient there is a corresponding increase in the rate of transport.

As in the case of enzymes, permeases that facilitate diffusion show saturation type kinetics. If a substnace (S) is present initially outside the plasma membrane, its transport inside can be represented by the following equation:

S(out) + Permease Km = S-permease complex Vmax→S have S is substrate, Km is binding constant for substrate to permease and Vmax is maximum rate of transport.

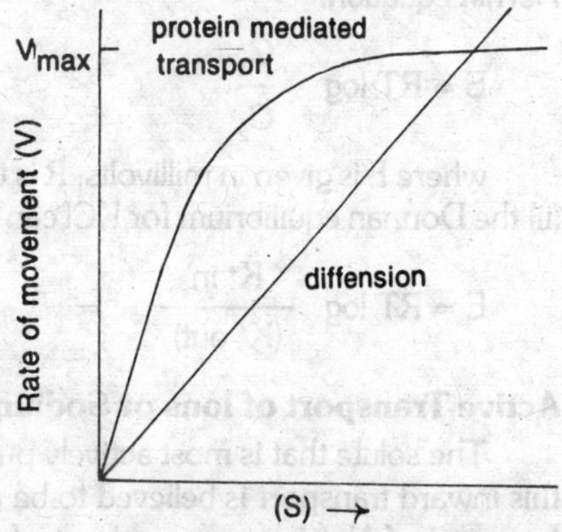

Fig. 4.19. A comparison of the kinetics of transport when mediated by a carrier as opposed to simple physical diffusion.

If the concentration of S outside is C, the rate of transport can be calculated as

$$V = \frac{Vmax}{1 + C/Km}$$

Active Transport

The diffusion of ions across the membranes is even more difficult because it depends not only on the concentration gradient but also the electrical gradient present in the system. Because active transport is a proces that it requires the expenditure of energy. The process involves the use of carrier molecules within the cell membrane proper. These carrier molecules apparently shuttle back and forth between the inner and oter cell membrane surfaces and either pick up or release a particualr ion being regualted. The energy requried for this process is obtained from adenosine triphosphate (ATP), which produced mainly by oxidative phosphorylation in mitochondria.

From the Table 4.1. shown below, it is evident that within a cell these is a large concentration of non-diffusible anions and that an electrical gradient across the membrane is established.

Table 4.1. Showing Ionic Concentration and Steady Potential in Muscle

Ions	*Interstitial fluid →*		*Intra cellular fluid*
Cations	Sodium (Na^+)	143	12
	Potassium (K^+)	4	155
Anions	Chloride (Cl^-)	120	3.8
	Bicarbonates (HCO_3^-)	27	8.0
	Anions and others	7	155
Potential		0	– 90mV

Donnan (1911) predicted, if a cell having a non-diffuisble negative charge inside, is put in a solution of KCl^-, K^+ will be driven into the cell by both the concentration and the electrical gradient. Cl^- ions on the other hand, will be driven in by the concentration gradient, but will be repelled by the electrical gradient. According to *Donnan*, the equilibrium concentrations will be exactly reciprocal

$$\frac{K^+ \text{ in}}{K^+ \text{ out}} = \frac{Cl^- \text{ out}}{Cl^- \text{ in}} \quad \text{(i)}$$

The relationship between the concentration gradient and the resting membrane potential is given by *Nernst* equation.

$$E = RT \log \frac{C_1}{C_2} \quad \text{(ii)}$$

where E is given in millivolts, R is universal gas constant and T is absolute temperature. From (i) and (ii) the Donnan equilibrium for KCl can be expressed as :

$$E = RT \log \frac{K^+ \text{ in}}{(K^+ \text{ out})} = RT \log \frac{(Cl^- \text{ out})}{(Cl^- \text{ in})} \quad \text{(iii)}$$

Active Transport of Ions or Sodium Pump

The solute that is most actively pumped into the cells are potassium (K^+) ions. The driving force for this inward transport is believed to be a sodium (Na^+) gradient across the membrane, created by active transport of Na^+ ions pumped out of the cell. The concentration of Na^+ ions outside the membrane becomes high where as the internal concentration becomes low. Energy, required to pump out Na^+ ions, is provided by ATP. In the presence of Mg^{++} activated ATPase, the ATP molecule is hydrolyzed and it is believed that ATPase is situated within the membrane.

The Na^+ pump was discovered by *Hodgkin* and *Keynes* (1955) and associated with ATP hydrolysis *in vitro* by *Skou* (1957). Two distinct mechanisms of Na^+ pump have been described for animal cells. These are:

(i) Sodium-Potassium exchange pump

In such a Na^+ pump, the outward pumping of Na^+ ions is linked to the inward transport of K^+ ions. Since Na^+ and K^+ are exchanged in a compulsory way, outward movemnt of Na^+ is always accompanied by inward movement of K^+. Such pump occurs in nerve cells and muscle cells.

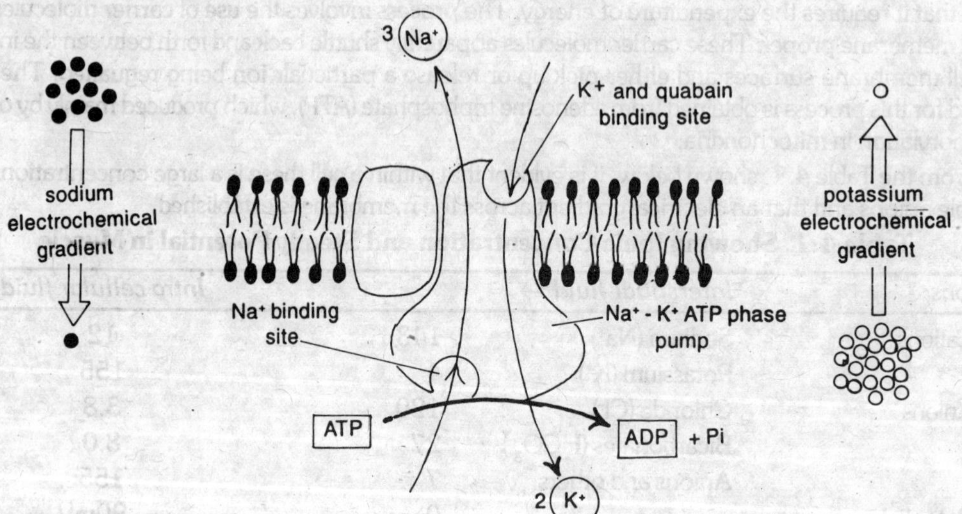

Fig. 4.20. Diagram of the Na^+ - K^+ ATPase in the cell membrane. Observe that for each molecule of will introlyzed at the inner part of the membrane 3 that limits are transported outside and 2 k^+ ions are imported inside.

(ii) Electrogenic Na^+ pump

In this pump ther is no compulsory exchange of inwardly moving K^+ ions and out wardly moving Na^+ ions. In this pump a gradient of electrochemical potential may be generated when exit of Na^+ ions is not compensated by one to one entry of K^+.

A high intercellular concentration of K^+ ions, regardless of the external concentration of Na^+ and K^+, is needed by aerobic cells. The high K^+ concentration in side the cell is necessary for protein synthesis and glycolysis. The high K^+ concentration of cell must be balanced by loss of some cations such as Na^+, otherwise excessive swelling would cause the cell to burst by creating a condition of higher internal osmotic pressure.

Active transport of glucose into cell is another consequence of the action of an electrogenic Na^+ pump. Extrusion of Na^+ from the cell generates a gradient of low internal and higher external concentration of Na^+. Sugar active transport accomplished under conditions in whcih outside Na^+ concentration is kept high enough to create an appropriate gradient whose energy drives metabolities into the cell from very dilute outside solution of sugar. The accumulation of sugar is compled to Na^+ extrusion and is also assisted by specific carrier proteins.

Translocation Across the Membrane

The carrier proteins assist hydrophilic molecules across a membrane thickness of 6 to 10 nm. The metabolites are considerably smaller than 6 nm, so it is imporant to know how these molecules are translocated acorss this relatively large distance by the carriers. Several alternatives have been proposed, but two have been studied more intensively than the other possibilities.

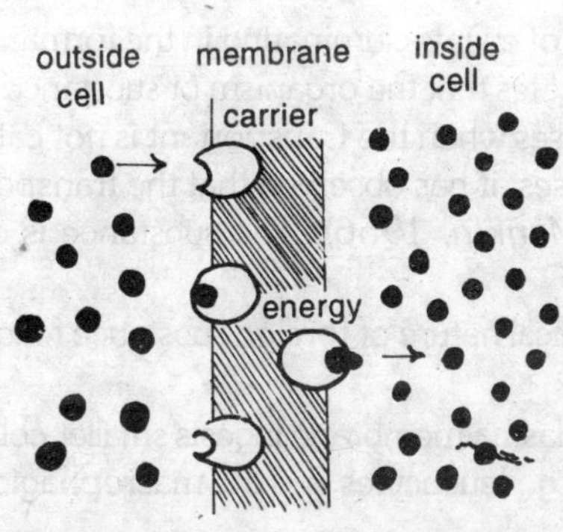

Fig. 4.21. Showing translocation across the Membrane by the carrier mechanism.

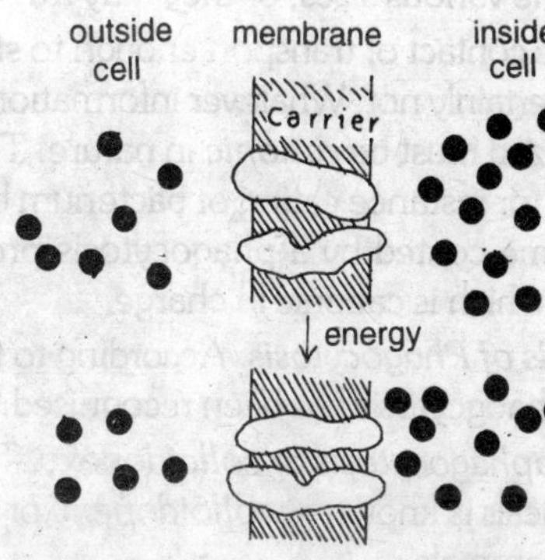

Fig. 4.22. Showing the fixed pore mechanism for selective transport.

One alternative hypothesis postulate that the carrier binds with the hydrophilic molecule and that then the entire transport protein rotates across the membrane and delivers its bound metabolite to the other side. The second alternative proposes that the carrier is fixed in place within the membrane, and that the carrier molecule undergoes a conformational change that translocates the binding site across the membrane and the bound metabolite along with it at the same time. Once the metabolite has ben translocated the bindign site is freed and restored to its orginal conformation, ready to bind another hydrophilic molecule in another transport event. This second alternative has been referred to as the *fixed pore mechanism*. The first alternative is known as the *carrier mechanism*.

Transport by Vesicle Formation

Most cell membranes can enclose material in vesicles and bring the substances into the cell in this way, or package materials for discharge from the cell in a reverse process. The process is called *endocytosis* when materials are brought in, and *exocytosis* when vesicle discharge takes place. Endocytosis has features that are analogous to active transport.

For example, substances enter along an "uphill" concentration gradient, and energy is required to support the process. Endocytosis will stop if poisons that stops energy production in the cell are added, and the process can be stimulated by the addition of ATP to cells in suspension.

Various cell secretions are discharged from the cell by exocytosis. The mucus droplets of intestinal goblet cells, digestive enzyme precursors in granule packages from pancreatic cells, and other cell secretions leave the cells in which they are produced by this general pathway. Although more difficult to see by light microscopy, numerous endocytotic infolded regions can be seen along the plasma membrane in suitable electron micrographs.

According to the natur of intake of food or foreign substances, the endocytosis may be classified into following types:

(i) Phagocytosis

In some cases, cells may ingest larger, solid particles through plasma membrane. This activity is called *phagocytosis* (Gr. *Phagein*=to eat; *kyots*=cell). It is found in large number of protozoans and among certain cells of the metazoans. Among protozoans, phagocytosis is intimately associated with amoeboid movement and is the only means for obtaining nutrition. Among metazoans it is generally a means of defense than serving the cell nutrition. This permits the ingestion of bodies which are foreign to the organisms, like bacteria, dust particles and various colloids. Among mammals this property is found very highly developed in the granular leucocytes. The particles become absorbed at the surface of membrane and later on they are taken into the cytoplasm by infolding of the plasma membrane which soon piched off forming vesicles consisting of the particles envelope by a membrane. The vesicle so formed may then fuse and assume various sizes, or they may fragment.

Is the contact of transport enough to stimulate the sequence of events cluminating in the formation of vesicle? Certainly not. Whatever information available clearly indicates that the organism or substance to be phagocytized must be *cationic* in nature. There are, however, cases when the transportant is not cationic in nature (for instance eating of bacterium by R.B.C.). In such cases, it has observed that the transportant first become coated by a phagocytosis-promoting substance (*Menkin,* 1956). The substance is called *opsonin,* which is cationic in charge.

Kinds of Phagocytosis. According to the physical and chemical nature of foreign substance following types of phaogcytosis has been recognised:

(a) *Ultraphagocytosis or colloidopexy.* The process in which plasma membrane ingests smaller colloidal particals is known as *colloidopexy* or *ultraphagocytosis,* e.g., leucocytes and the macrophagic cells of mammals.

(b) *Chromopexy.* When the cell ingests colloidal chromogen particles phagocytotically the process is known as *chromopexy,* e.g., some mesoblastic cells.

(ii) Piocytosis

When the ingestion of fluid material is bulk takes place by the cell through the plasma membrane, the process is known as *pinocytosis* (Gr. *pinein*=to drink). The process of pinocytosis was first of all observed by *Edward* in *Amoeba* and by *Lewis* (1931) in the cultured cells.

In the process of pinocytosis, the particle free globules of fluid are surrounded and ultimately engulfed by the clasping folds of the cytoplasm. The plasma membrane forms membranous vacuoles around the fluid globules. Such membrane bound vacuoles are called *pinosomes* (*Lewis*, 1931). The pinosomes, later on, are transported to the interior of the cell where they are fused with the secretory granules or lysosomes. In the food vacuoles the digestion of food substance takes place and the digested food is diffused to the surrounding cytoplasm. The pinocytosis can be observed ordinarily in the cells by light microscopy.

(iii) Micropinocytosis

Electron microscopic observations have been made on the pinocytotic process at sub-cellular or sub-microsocopic level in the cells. The pinocytosis which occurs at sub-microscopic level is known as

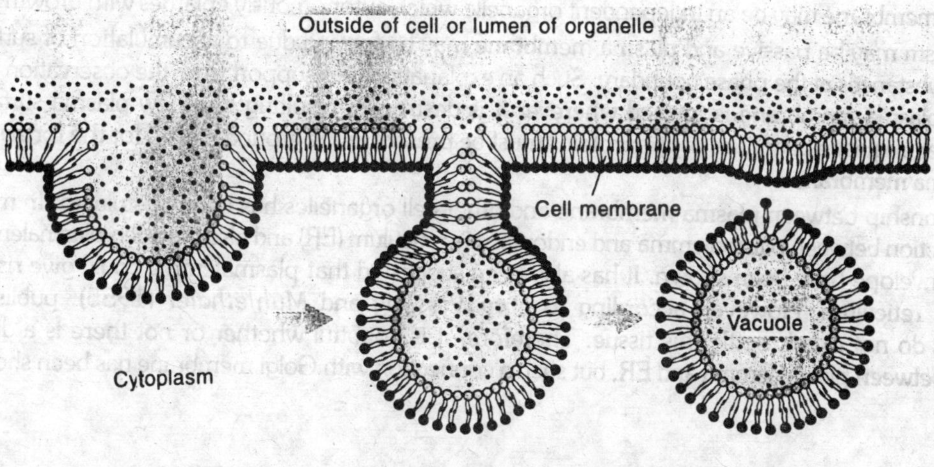

Fig. 4.23. A model of Endocytosis.

micropinocytosis. In the process of micropinocytosis, the plasma membrane invaginates to form small vesicles of 650Å diameter. These vesicles have openings on both outer and inner surface which suggest the possible transportation of fluids through these vesicles to the cell. The micropinocytosis have been observed in endothetial cells, Schwan and satellite cells of nerve ganglion, macrophages, muscles cells and reticular cells, etc.

(iv) Emeiocytosis or Exocytosis or Cell Vomitting

The process of exudating the secretory products to outside the cell cytoplasm is known as *emeiocytosis*. In cells of pancreas, the vacuoles containing enzymes move from the interior of the cytoplasm towards the surface. Here they fuse with plasma membrane and discharge their contents to the exterior.

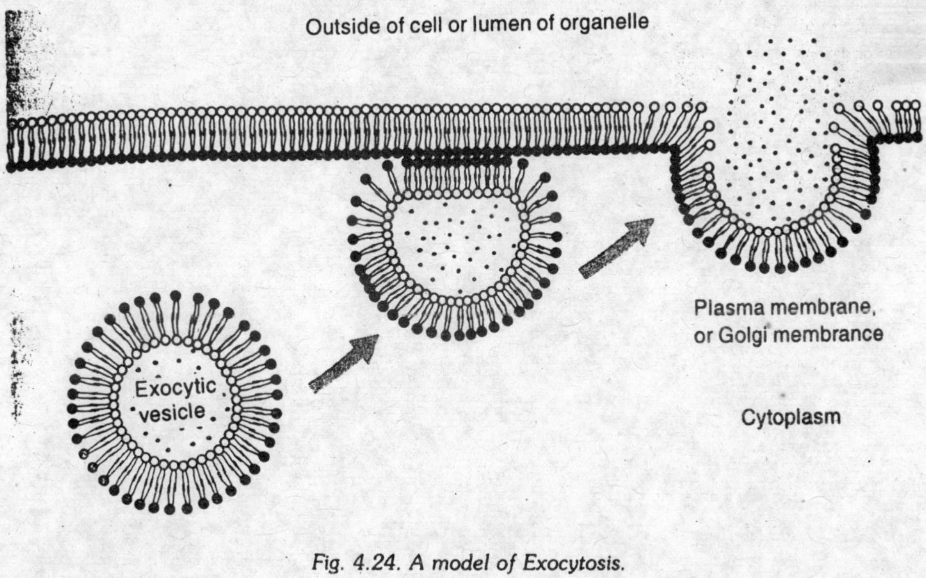

Fig. 4.24. A model of Exocytosis.

(v) Cytopermphis

In this process, a material enters the cell by endocytosis and then it comes out of the cell by exocytosis without any change. Water molecules move through the epithelial cell by this method.

Origin of Plasma Membrane

There are two interpretations to explain the nature and origin of plasma membrane:

1. Plasma membrane may be an independent organelle which autonomously enlarges with growth.
2. Cytoplasm may be passive and plasma membrane may be formed due to accumulation of surface active substances at the phase boundary. Such an explanation gets support from the observation that a new plasma membrane is formed wherever it is destroyed, resulting into new plasma surface. However, requisite quantities of lipoproteins must be present in the ground plasm, for the formation of plasma membrane.

A relationship between plasma membrane and other cell organelles has been postulated. In many cases, connection between plasmalemma and endoplasmic reticulum (ER) and also between plasmalemma and nuclear envelope have been shown. It has also been suggested that plasmalemma could give rise to endoplasmic reticulum. However, according to *Frey-Wyssling* and *Muhlethaler* (1955), published micrographs do not clearly settle this tissue. Therefore, it is doubtful whether or not there is a direct connection between plasmalemma and ER, but such a connection with Golgi membrane has been shown.

5

CELL WALL

In plants as well as in bacteria, a cell is always surronded by a cell wall lined throughout with plasmalemma. The plant cell wall is a non-living structure made up of cellulose, hemicellulose and compounds of pectic acid. The cell wall of bacteria is made up of peptidoglycans, proteins, polysaccharides and teicholic acids. Presence or absence of cell wall is a distinctive character to distinguish between a plant and animal cells.

STRUCTURE

Cell wall is differentiated into following layers:

1. Middle Lamella

It is the intercellular matrix present between two adjacent cells and connects two cells. It is primarily composed of gel-like magnesium and calcium pectates, along with a protein components. It is optically inactive and colloidal in nature. It can be dissolved in strong acids. This is really the first formed cell wall, which may persist in many cells as the only wall.

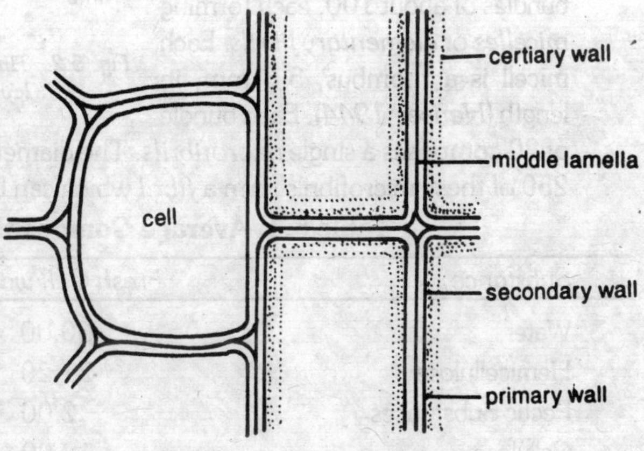

Fig. 5.1. Structure of cell wall.

2. Primary Wall

The first product of cell wall synthesis by the protoplast is a thin, and elastic primary wall, bordering the middle lamella. It is about 1-3/4 in thickness. During the development of the primary wall and middle lamella, openings are left at places in the walls of adjacent cells. These pores are called *plasmodesmata.* Through plasmodesmata cytoplasmic continuity is maintained in the neighbouring cells. It is made up of cellulose, hemicellulose, proteins and polysaccharides.

3. Secondary Wall

It is deposited after the primary wall. It is much thicker and consists of several layers. It is the secondary wall that gives the plant cell its definite shape and structure. In tracheids of conifers, secondary wall can be divided into three layers termed as S1, S2 and S3. Because of its thickness and rigidity, it provides great tensile strength to the cell. The secondary wall is made up of cellulose, hemicellulose and polysaccharides. In addition, it may be deposited with inorganic salts, tannins, waxes, calcium salts etc.

4. Tertiary Cell Wall

Usually it is rarely present, whereas present, it occurs on inner side of secondary wall. It seen in xylem tracheids of gymnosperms. It is composed of hemicellulose xylan.

Utrastrucrture

When cell wall is studied under electron microscope, it reveals two parts:

(i) *Matrix.* It is a ground substance in which the fibrils are embedded. It is amorphous deposited with lignin or hemicellulose and pectin.

(ii) *Fibrils.* The fibrils are made up of cellulose. Cellulose molecules are long chains formed by the condensation of about 3000 *b. D-glucose units.* These units are arranged in a linear order froming cellulose strand. These cellulose strands occur in parallel bundles of about 100, each forming *micelles* or *elementary fibrils.* Each micell is a rhombus, 5-6 mm, in length (*Henser, 1944*). Each bundle of 20 comprises a single *microfibrils*. The diameter of each microfibril is about 100-250Å. About 250 of these microfibrils form a *fibril* which can be seen under light microscope.

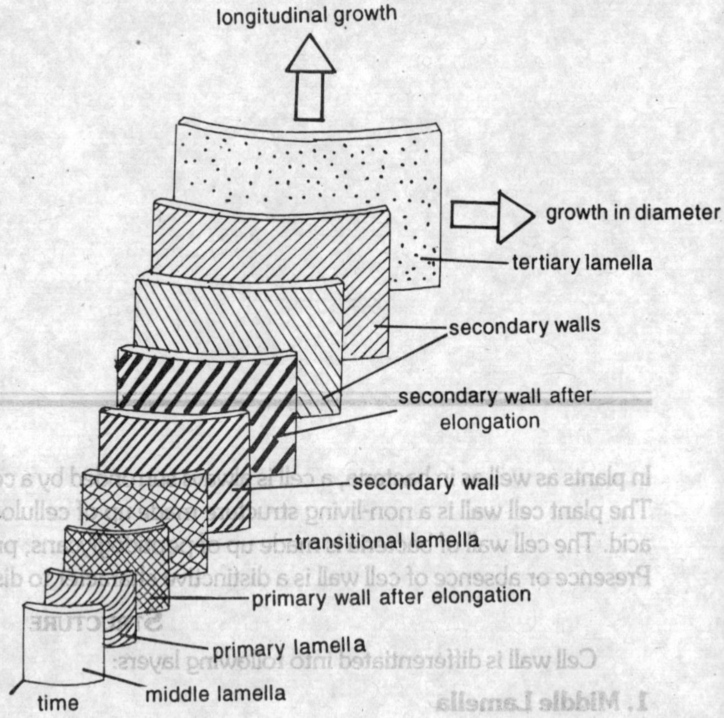

Fig. 5.2. Time sequence of the formation of various types of cell wall layers in a tracheid.

Table 5.1. Average Composition of a primary cell wall

Substance	*Fresh Cell wall wt. %*	*Dry wt. %*
Water	60.00	—
Hemicellulose	21.20	53.00
Pectic Substances	2.00	5.00
Cellulose	12.00	30.00
Protein	2.00	5.00
Lipids	2.80	7.00

Growth of Cell Wall

The cell wall is very thin and delicate in the beginning, as the cells grow, it is stratched and new substances are deposited on the primary cell wall.

Growth of cell wall takes place in two methods:

(i) *By intussusception.* In this method as the cell wall stratches in one or more directions, new cell wall material (secreted by protoplasm) gets embedded within original wall.

(ii) *Growth by apposition.* In this method new cell wall material secreted by protoplasm is deposited as definite thin plates one after the other, on the inner surface of original wall.

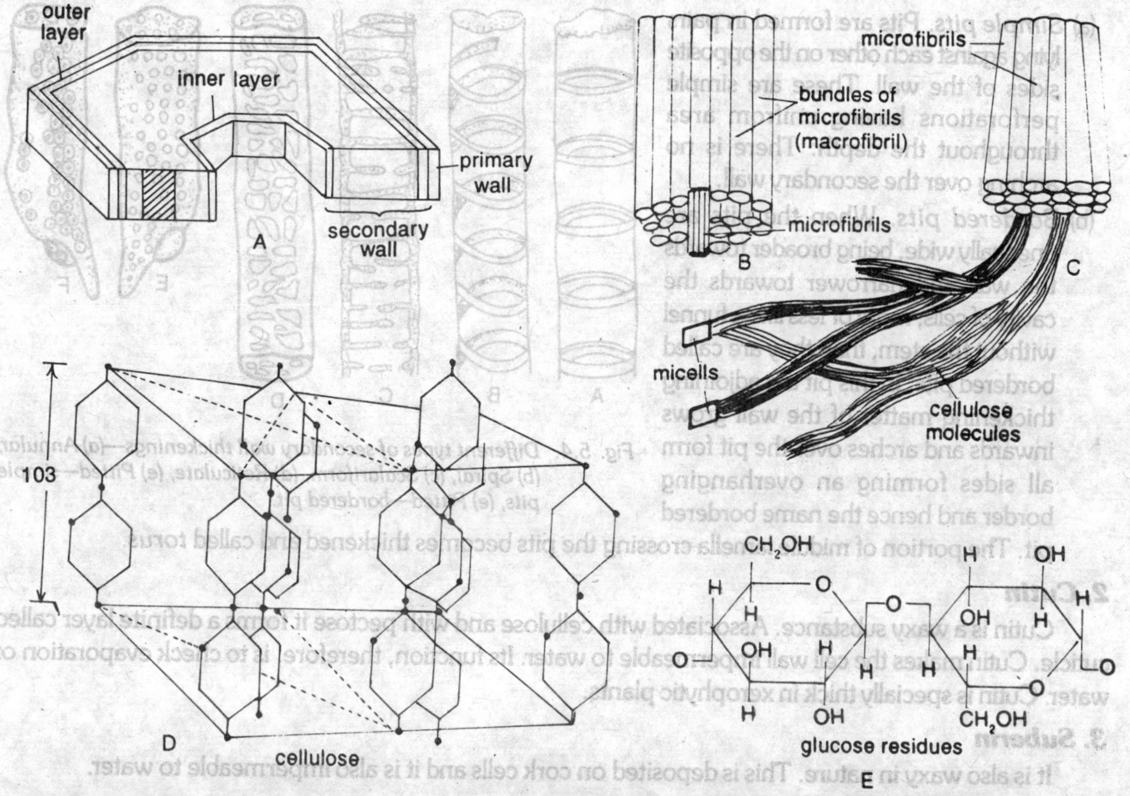

Fig. 5.3. Diagrammatic representation of the submicroscopic structure of the cell wall: A—portion of a cell with secondary wall layer; B—bundles of microfibrils; C—portion of a microfibril showing micelles; D—fragment of a micelle showing glucose residues; E—two glucose residues connected by glycosidic linkage.

Thickening of Cell Wall

During maturation phase, the cells are modified according to functions they have to perform. The thickening material is secreted by the protoplasm and is laid down on the primary wall, more or less in uniform manner giving a stratified appearance. The thickening materials are of following types:

1. Lignin

Due to development of lignin, the cell wall of xylem becomes hard and thick. The lignin is deposited in secondary wall and with the lignin deposition the protoplasm is lost. Lignin is deposited at specific places of the cell walls due to which xylem tracheids and trachea take up a number of forms. These are:

(i) *Annular Thickenings.* In this case deposition of lignin takes place in the form of rings. These rings are placed, one above the other on the inner surface of protoxylum cells. Some space is left between two neighbouring rings where the cell wall remains unthickened.

(ii) *Spiral Thickenings.* In such a case the thickening of lignin takes place the form of a spiral band. It is found in tracheids and trachea of protoxylem.

(iii) *Scalariform Thickenings.* In this type of cell thickening, the lignin is deposited in the form of transverse rods of the ladder. The unthickened areas between the successive thickenings appear as elongated transverse pits.

(iv) *Reticular Thickenings.* Here the lignin is deposited in the form of network or reticulum leaving irregular meshes of unthickened area is the wall. These are found in metaxylem.

(v) *Pitted Thickenings.* Here entire inner surface of the cell wall is more or less uniformly thickened, leaving here and there some small unthickened areas or cavities called *pits*. These pits are of two types:

(a) *Simple pits.* Pits are formed in pairs lying against each other on the opposite sides of the wall. These are simple perforations having unifrom area throughout the depth. There is no arching over the secondary wall.

(b) *Bordered pits.* When the pits are unequally wide, being broader towards the wall and narrower towards the cavity of cells, more or less like a funnel without the stem, than they are called bordered pits. In this pit the adjoining thickening matter of the wall grows inwards and arches over the pit form all sides forming an overhanging border and hence the name bordered pit. The portion of middle lamella crossing the pits becomes thickened and called *torus.*

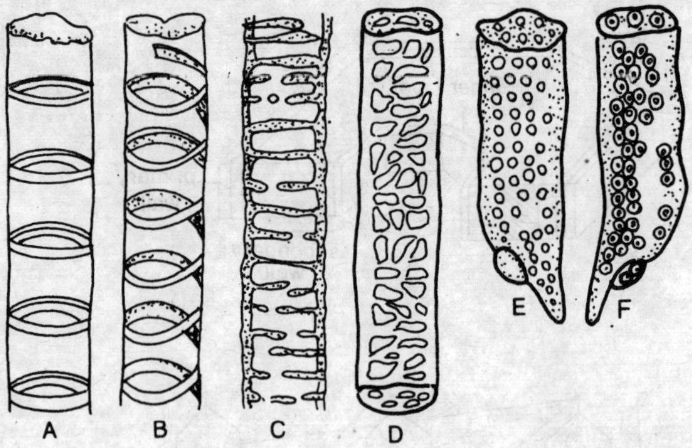

Fig. 5.4. Different types of secondary wall thickenings—(a) Annular, (b) Spiral, (c) Scalariform, (d) Reticulate, (e) Pitted—simple pits, (e) Pitted—bordered pit.

2. Cutin

Cutin is a waxy substance. Associated with cellulose and with pectose it forms a definite layer called cuticle. Cutin makes the cell wall impermeable to water. Its function, therefore, is to check evaporation of water. Cutin is specially thick in xerophytic plants.

3. Suberin

It is also waxy in nature. This is deposited on cork cells and it is also impermeable to water.

4. Mucilage

Mucilage is a slimy substance. Chemically it is a complex carbohydrate. Its property is that it absorbs water and retain for a long time. Mucilage is found in the leaves of *Aloe,* flowers of *Hibiscus,* seeds of *Linum* etc.

5. Mineral Crystals

These may be introduced into cell wall as crystals of silica, calcium oxalate and calcium carbonate. In the majority of fungi, the cell walls are made up of a substance called chitin.

Function

The main function of cell wall in plants is provide mechanical strength. Being hydrophilic in nature it is capable of imbibing water and thus helps in the movement of water and solutes towards protoplasm.

Tensile strength of cell wall and turgor pressure in the cell

The cell wall also contains a fluid having more solutes than in the plant's external milieu (e.g. soil), but still hypotonic relative to cell's interior. This difference in osmotic pressure inside the cell (higher) and in the cell wall (lower) leads to the development of turgor pressure against the cell wall. This turgor pressure increases to the limit, where the cell is in osmotic equilibrium and on net influx of water takes place. This turgor pressure is actually responsible for cell expansion during growth. This also provides the desired rigidity to plant organs. If we compare a wilted leaf with a turgid leaf, we can recognize the role of cell wall in providing tensile strength, which is partly based on the cellulose microfibrils, hemicellulose, pectin and structural proteins found in the cell wall.

Function of cell wall

The function of cell wall in plants is believed mainly to provide mechanical strength. Particularly in higher plants having vascular system, cell walls provide main supporting framework. Since in animals, some sort of exoskeleton or endoskeleton is present, perhaps there is hardly any necessity of definite cell walls. Therefore besides others, main function of the cell wall is the same as that of a skeleton in animals.

ORIGIN

The *cell plate* is the precursor of cell wall. It is the structure formed between two daughter cells after karyokinesis during cell division. Cell wall formation takes place at *plasmodesmata* (*Scott et. al.*, 1956; *Wardrop*, 1958). Formation of new cell wall takes place by *phragmoplast*. Phragmoplast appears in the region of equatorial plane of dividing cell. These are barrel-shaped bodies. The spindle and phragmplast gradually flatten and assume the form of biconvex lens. The phragmoplast is formed from Golgi complex and it gives rise to cell plate. Cell plate than grows and fuse with mother cell wall. In the initial stage cell plate consists of three layers—central middle lamella, and primary walls on both the sides of middle lamella.

6

THE CELL SURFACE

Multicellular organisms are composed of two compartments: the cellular and the extracellular. Although the cells of an organism are obviously the basis for its being alive, the "nonliving" parts generally have very important functions. The cell surface is the boundary between the cellular and non-cellular compartments, and we can distinguish three categories of cell surface materials: (1) extracellular surface-adherent materials, and (3) extraneous structures.

THE CELL COAT AND EXTRACELLULAR STRUCTURES

The cell coat is generally considered to be equivalent to the carbohydrate chains attached to the proteins of the outer leaflet to the plasma membrane. This layer is visualized in the electron microscope by treating the cells or sections of the cells with heavy metal strains that form complexes with the carbohydrates of the coat. Even though these materials are actually part of the structure of the membrane, they are on the outside of the cell and they can be removed without dealing the cell a mortal wound. The analogy has been made by *John Luft* between the plasma membrane—cell coat complex and the pelt of an animal. The skin of the pelt is analogous to the membrane—it forms a cover for the living contents that cannot be removed, although it can heal if damaged. The fur is like the cell coat: both are useful and necessary under many environmental conditions, but both can be removed without fatal consequences and both can be resynthesized by the underlying layer.

In many cells there is a separate layer of extracellular surface-associated material present outside of the cell coat as a tightly adherent matrix composed primarily of carbohydrate-containing macromolecules secreted by the cell. This extracellular material is very prominent in a few types of cells, such as those of the intestinal epithelium of mammals and single-celled amoebas, where it forms a filamentous, or "fuzzy," layer. In most cells, however, the extracellular adherent material is not nearly as evident. Since extracellular adherent materials are generally visualized by the same types of stains that reveal the cell coat, it often becomes impossible to see where one ends and the other begins. For example, when the stain ruthenium red (which binds to negatively charged carboxyl and sulfate groups) is used, a thick dense layer at the cell surface is usually seen, but it is generally impossible to determine how much of this material is actually covalently bonded to membrane components (as either integral glycoproteins or glycolipids) and how much is an independent extracellular layer. The term *glycocalyx* is often used to refer to the entire complex of material at the outer edge of the cell membrane. The function of the extracellular material is not always evident. In addition to its role in mediating cell-cell and cell-substrate interactions, it has been suggested that it provides mechanical protection to cells and/or serves as a barrier to particles reaching the membrane, i.e., as a sieve.

The third category of cell surface materials includes a great variety of structures ranging from jelly coats and thick capsules around eggs to the basement membranes upon which epithelia rest and the materials in which cartilage cells and bone cells are embedded. These structures, and many other are outside the limits of a text in cell biology, but two features of these extracellular materials bear a special note. One is the remarkable self-assembly properties that these materials seem to possess, and the other is the major influences they can have on the activities occurring within the cells they surround.

THE NATURE OF EXTRACELLULAR MATERIALS (ECMs)

Extracellulary matrix is really made up of macromolecules produced locally by cells which are embedded in the ECM itself. The orientation of these ECM macromolecules is actually determined by the orientation of the components of cytoskeleton of these embedded cells giving birth to ECM. In most connective tissues, the ECM molecules are secreted by the fibroblasts or by other cells of fibroblast family including *chondroblasts* in cartilage and *osteoblasts* in the bone.

Most ECMs are made up of the following two main classes of extracellular macromolecules, (i) *polysaccharide chains* of the class *glycosaminoglycans* (GAGs), which are covalently linked to proteins in the form of *proteoglycans* (except in case of *'hyaluronan'*) and (ii) *fibrous proteins* of two main types—*structural* (e.g., *collagen* and *elastin*) and *adhesive* (e.g., *fibronectin* and *laminin*). Sometimes, these extracellular macromolecules are also classified as small glycoproteins (e.g., collagen and proteoglycans) and large glycoproteins (e.g., fibronectin, laminin and elastin/entactin). These molecules are found in a wide variety of shapes and sizes performing a variety of functions. In case of connective tissues, glycosaminoglycans (GAGs) and proteoglycans are high hydrated and form gel-like 'ground substance,' in which fibrous proteins facilitate rapid diffusion of nutrients, metabolites and hormones between the blood and the tissue cells; collagen provides strength and helps organize the ECM; elastin fibres give it resilience (elasticity); the adhesive proteins help cells attach to appropriate parts of ECM; e.g., fibronectin promotes attachment of fibroblasts and various other cells to ECM in connective tissue, while laminin promotes attachment of epithelial cells to the basal lamina.

GLYCOSAMINOGLYCANS (GAGs) AND PROTEOGLYCANS

Glycosaminoglycans (GAGs)

GAGs in ECM are equal to only 10% of the fibrous proteins by weight. They are unbranched polysaccharide chains consisting of disaccharide units. One of the two sugar residues of a disaccharide unit, is often sulfated, having an amino group (N-acetylglucosamine or N-acetylgalactosamine), hence the name glycosaminoglycans (*glycose* is the general name of a monosaccharide and *glycan* is the general name of polysaccharide. The second sugar is often a uronic acid (glucuronic or iduronic). Due to the presence of sulphate and carboxyl groups, GAGs are negatively charged.

GAGs can not fold and take up globular shape; instead they remain highly extended, occupying huge volume relative to their mass and thus filling most of the extracellular space. They form gel even at low temperature and, due to negative charge, attact cations like Na^+ which are osmotically active, thus helping in sucking large quantities of water into the matrix. The ECM thus swells causing turgor to withstand forces. Due to this property of GAGs, the cartilage at our knee joints can withstand pressure of hundreds of atmospheres. The deficiency for a GAG can cause severe diseases. For instance a deficiency in the synthesis of *'dermatan sulphate'* causes dwarfness, premature aging and generalized defects in skin, joints, muscles and bones.

Except hyaluronan, all GAGs are covalently linked to proteins giving rise to *proteoglycans* found in majority of animal cells. *Hyaluronan,* itself is the simplest of the GAGs, consisting of 25,000 non-sulphated disaccharide units. It is found in all tissues and fluids in adult animals and is especially abundant in early embryos. Hyaluronan expands with water to occupy large volumes and therefore, is a space filler during embryonic development. It can be quickly and cheaply produced and, therefore, is also used in wound repair. Hyaluronan also help in cell migration during formation of heart, cornea and other organs. After cell migration, excess hyaluronan is degraded by *hyaluronidase* enzyme. All the other GAGs occurring as proteoglycans have following common features (i) they contain sulphated sugars; (ii) different disaccharide units are arranged in a complex manner in the same GAG; (iii) they are much shorter than hyaluronan, having fewer than 300 residues; (iv) they are covalently linked to proteins forming proteoglycans; (v) they are synthesized inside the cell and released by exocytosis, while hyaluronan is spun out directly from the cell surface by an enzyme embedded in plasma membrane (as in case of cellulose for plant cell walls).

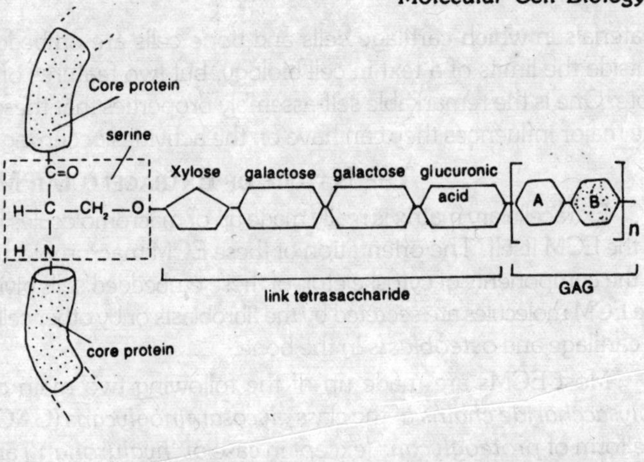

Fig. 6.1. A tetrasaccharide linkage between serine residue of the core protein and GAG in a proteoglycan molecule.

Proteoglycans

Proteoglycans (protein + glycans, i.e. carbohydrates) are specific types of glycoproteins, which are characterized by the presence of carbohydrate component, which far exceeds the protein content and consists of linear chains of repeating disaccharide units. Usually one of the sugars of a disaccharide monomer has an amino group, so that the polysaccharide with this repeating unit is called a *glycosaminoglycan*. The interconnections of protein and carbohydrate components lead to the formation of a web-like matrix. Following are some functions of proteoglycans: (i) In many motile cell types, the proteoglycan surrounds the cells and prohibit the cells from coming together. (ii) Proteoglycans also mediate connections between adjacent tissues of an organ. For instance, heparan sulphate (a glyconsaminoglycan) proteoglycans bring together chondrocytes, the cartilage producing cells. (iii) In still other cells, proteoglycans secreted by one cell type are essential for the growth of neighbouring cells, so that if proteoglycans (heparan sulphate) are removed from the surface of axons of the dorsal root ganglia, the neighbouring Shawann cells do not proliferate. (iv) Proteoglycans also help in binding of growth factors to their corresponding cell receptors.

Fibrous Structural Proteins

Collagens

Collagens are the abundant proteins (25% of total protein mass) in mammals, found as a major component of skin and bone. They are glycoproteins, which are characterized as long, stiff, triple stranded structures, each strand being a collagen polypeptide chain called α-chain. The three strands are wound around one another in a superhelix. These collagen molecules are assembled into collagen fibrils, 10-300μm in diameter and hundreds of micrometers (nm) long in mature tissues. These collagen fibrils may further aggregate into larger cable like bundles, described as *collagen fibres* (0.5-3 nm in diameter). The molecules (α helices) within a triple helix are cross-linked and the different triple helices are also cross-linked to each other due to intra- and inter-molecular cross-links.

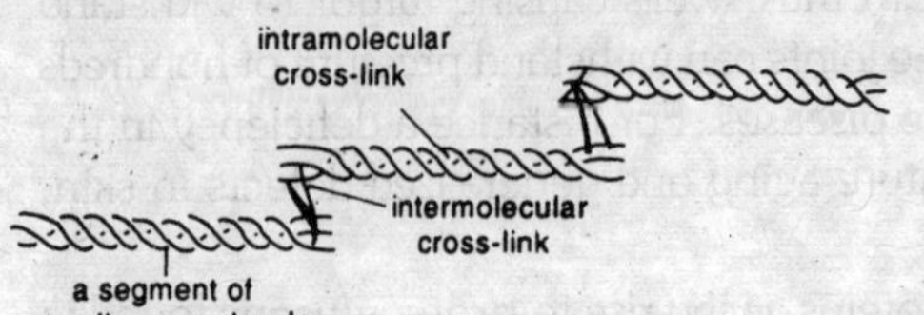

Fig. 6.2. Intramolecular and intermolecular cross-links within a collagen fibril.

Collagens are also characterized as glycoproteins which are extremely rich in proline and glycine, both of which help in the formation of triple-stranded helix. Glycine (the smallest amino acid) is found at every third position in the central region of chain and proline, due to its ring structure, stabilizes the helical structure. The collagen fibrils form structures that resist tensile forces, unlike GAGs which resist compressive forces. The collagen fibrils are, therefore, organized in different ways in different tissues: (i) In mammalian skin, they are interwoven in wickerwork pattern (pattern of twigs used in baskets, chairs, mats, etc) so that tensile stress in many directions is resisted. (ii) In tendons, they are arranged in parallel bundles aligned along a major axis of tension. (iii) In mature bone and cornea, they are arranged in orderly plywood like layers, with fibrils in each layer parallel to each other but at right angle to those in the layer on either side.

In animal tissue 15 different types of collagen molecules, involving 25 different types of α chains are known.

Collagen polypeptide α chains are synthesized on rough ER and are released as *pro-α chains* which not only have an amino-terminal signal peptide to direct the polypeptide to ER, but also have propeptides on both its ends. Three pro-α chains combine to form triple helical molecule called *pro-collagen*, which changes to collagen by removal of pro-peptides at the two ends. The collagen molecules first assemble in *collagen fibrils* (10-300 nm) and then into collagen fibres (0.5-3 mm).

Elastin

Elastin is a highly hydrophobic protein (about 750 amino acids), which is also proline and glycine rich (like collagen) but is not glycosylated. It contains little hydroxyproline and no hydroxylysine. The elastin protein is composed of (i) hydrophobic segments and (ii) alanine and lysine rich helical segments, the former providing elasticity and the latter forming cross-links between adjacent molecules. These two types of segments alternate along the length of polypeptide chain. Elastin protein makes the elastic fibres, to which integrity is provided by another protein fibrillin. Mutation in fibrillin leads to *Marfan's syndrome* disease, where aorta (whose wall is rich in elastin) is prone to rupture.

Elastic molecules are secreted into the extracellular space and assemble into elastic fibres close to the plasma membrane. After secretion, elastin molecules become cross-linked to generate an extensive network of elastic fibres and sheets which are very common in tissues like skin, blood vessels and lungs. The elastic fibres are atleast five times more extensible than rubber band of the same thickness. In elastic tissues, collagen fibres are interwoven with elastic fibres to limit the extent of stretching, so that the tissue may not tear apart.

Fibrous Adhesive Proteins

ECM contains a variety of other specialized large glycoproteins (e.g. fibronectin, laminin, and tenascin), which are involved in arranging cells into organized structures through their function as substrate adhesive molecules. In the ECM, collagen and proteoglycan function as substrates, which utilize other substrate adhesive molecules for linking cells. These substrate adhesive molecules are also utilized for migration of one cell type on another cell type both having these substrate adhesive molecules on their surface as components of ECM. The substrate adhesive molecules also provide sites for attachment of colonizing microorganisms.

Fibronectin

Fibronectin is a very large glycoprotein dimer synthesized by cells like fibroblasts, chondrocytes, endothelial cells, macrophages and certain epithelial cells (e.g., hepatocytes and aminocytes). Two large subunits of a dimer fibronectin are joined by two disulphide bonds near their carboxyl termini.

Fibronectin is a general substrate adhesion molecule and has binding sites, whose interaction with appropriate molecules results in the proper alignment of cells with the ECM.

Fibronectin also helps in cell migration by making roads on which certain migrating cells travel, as illustrated by the following examples: (i) During gastrulation, certain mesodermal cells migrate on fibronectin surfaces of other cells and cease to migrate when fibronectin is removed. (ii) In chick embryos, the precursors of heart i.e., the pre-cardiac cells migrate on fibronectin to travel from lateral side to the midline, so that if fibronectin is blocked (due to injection of antibodies, which will occupy the binding sites of fibronectin), migration will not take place and two hearts will develop.

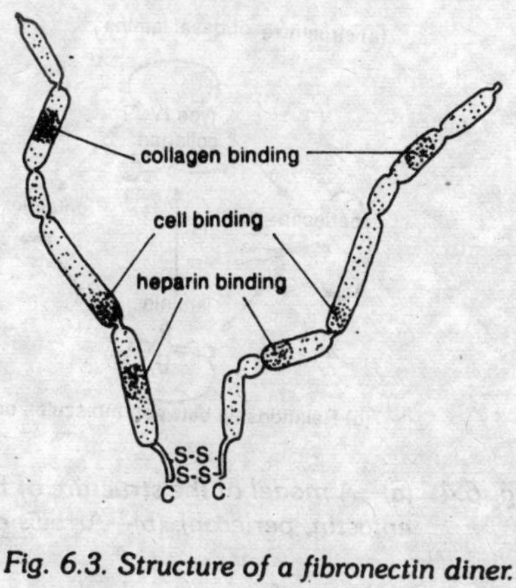

Fig. 6.3. Structure of a fibronectin diner.

Laminin

Laminin is a major component of basal lamina and provides a link between the cells and the structural components of *basement membrane* (found between epithelial and mesenchynatous cells). It is a large multi-chain glycoprotein and consists of three polypeptide chains. These three polypeptide chains are repeated, but exhibit diversity in different organisms. The three chains are also linked via disulphide bonds to form an asymmetric cross structure. Each of these three chains of laminin, have a number of functional domains, responsible for binding separately to type IV collagen, heparan sulfate, entactin and laminin receptors.

Laminin is the first ECM protein to appear during development. It is known to influence cell attachment, movement, differentiation and growth with the help of several laminin receptors, which have been identified as a class of cell surface receptors. Integrins and glycosyltransferases are two classes of such receptors.

Tenascin (also called cytotactin)

Tenascin resembles fibronectin for about half its length and is transiently for about half its length and is transiently found in several ECMs during embryonic development. However, different cells react differently to tenascin, so that it does not always promote cell adhesion.

Basal Lamina and ECM

At the interface between all epithelial cells sheets and connective tissues the matrix is organized as a thin but tough mat that is known as *basal lamina.* Basal lamina also surrounds individual muscle cells, fat cells and Schwann cells, which are separated from the underlying or surrounding connective tissue with the help of this basal lamina. In kidney glomerulus and lung alveolus also, a basal lamina lies between two cell sheets and functions there as a highly selective filter. Basal lamina influences cell behaviour in a variety of ways. It determines *cell IV collagen molecules* which have more flexible structure than the collagen found in collagen fibrils, polarity, influences cell metabolism and organizes proteins in adjacent plasma membrane. Basal lamina also induces cell differentiation and serves as specific highways for cell migration.

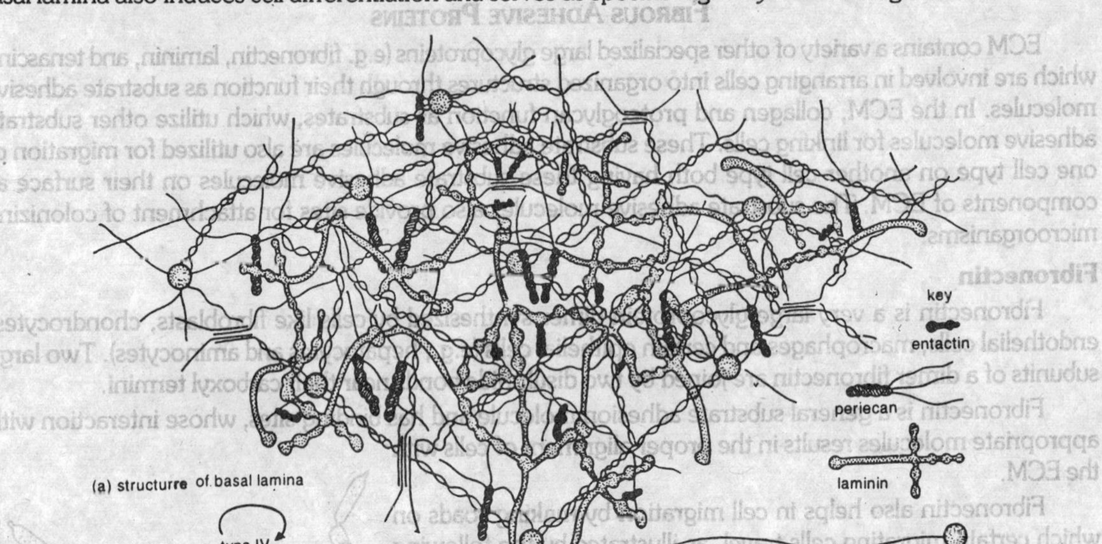

Fig. 6.4. (a)—A model of the structure of basal lamina showing organization of different proteins (IV collagen, laminin, entactin, perlecan); (b)—Arrows connect molecules that can bind directly to each other.

The basal lamina strikingly differs from other examples of ECMs which mainly fill up the space between cells (e.g., connective tissue). The connective tissue itself also consists mainly of ECM, but basal lamina depends on some specialized types of ECM molecules including *type.*

Most basal laminae consist of two distinct layers (i) *lamino lucida* or *rara* lying towards plasma membrane of say epithelial cells and (ii) *lamina densa* lying just below the lamina lucida. Sometimes a third layer (*lamina fibroreticularis*) is also found, which connects the basal lamina with underlying connective tissue. These three layers together, are sometimes also described as *basement membrane,* a term which has been used interchangeably for *basal lamina.*

The precise composition of basal lamina differs in different tissues or even in different regions of the same tissue, but the most important components of all basal laminae include type IV collagen, '*perlecan*' (a heparan), laminin and entactin (large glycoproteins).

Basal lamina performs following functions: (i) *Filtration of macromolecules* in kidney glomerulus prevents passage of macromolecules from blood into urine and is facilitated by 'heparan sulphate' in basal lamina, so that if GAG chains of heparan sulphate are removed, the filtering properties of lamina are destroyed. (ii) *A barrier* provided by basal lamina does not allow fibroblasts from coming in contact of the epithelial cells, although macrophages, lymphocytes and nerve processes can pass through basal lamina. (iii) *Wound repair,* through regeneration, is facilitated by basal lamina, which provides a scaffold, along which regenerating cells can migrate to reconstruct the damaged tissue including muscle, nerve and epithelia. Injuries in the nerve or the muscle at the neuromuscular junction (synapse) is also facilitated by the basal lamina. (iv) *Cell migration* during embryonic development is also regulated by basal lamina, so that mutation in a gene encoding components of basal lamina (e.g. laminin) disrupts the path of mesoderm cells or nerve axons that migrate over basal lamina (migration only along the dorsoventral axis is affected and that along the anteroposterior axis is not affected).

7

PROTOPLASM

Protoplasm is a highly complex mixture of some elements and compounds found in the bodies of living beings. The protoplasm is variously knonw as the *living matter, living substance* or *physical basis of life.* It is the basic fundamental substance exhibiting all the vital processes of the cell. The protoplasm was first observed by *Corti* in 1772. In 1835 *Dujardin* a Frenchman described it as a soft and gumy substance and named it as *scarcode* i.e. *flesh. Purkinje*, a Bohemian physiologist (1839), was the first biologist, who gave the name *protoplasm* to this living substance. *Hugo von Mohl,* a German botanist, in 1839, also suggested the name *protoplasm* for the granular and viscous substance found in plants similar to that found in the animals. The above-mentioned German botanist popularized the word protoplasm as a name given to the living matter found in plants and animals. Protoplasm is the most complex and interesting substance. It is not to be thought of a chemical compound but rather as very complex organized system. Protoplasm varies somewhat in its nature from cell to cell and from organism to organism, but basically it must be the same, as evidenced by its common manifestations of metabolism, growth, reproduction and by some other peculiarities.

PHYSICAL NATURE OF PROTOPLASM

Different workers have proposed different theories to explain the physical nature of protoplasm as given under the following heads:

1. Granular Theory

This theory was propounded by *Altmann* in 1893. According to this theory, protoplasm consists of numerous tiny granules as shown in *Amoeba. Henle, Maggi,* etc., considered these protoplasmic granules *plastidules. Altmann* reconginzed them as "elementary organisms," or *bioplasts* (or cytoplasts).

2. Alveolar Theory

The alveolar nature of protoplasm was suggested by *Butschilli* in 1892. According to him, protoplasm consists of many suspended droplets or alveoli or minute bubbles, resembling to foam of emulsion.

3. Fibrillar Theory

This theory was put forward by *Flemming.* According to him, protoplasm consists of fibres embedded in the inner mass of matrix. The fibrillae are called mitome or spongioplasm and ground substance is temed paramitome or hyaloplasm.

4. Reticulate Theory

This theory was proposed by *Klein, Cornoy* etc., suggesting that the protoplasm consists of a network or reticulum of fibres in the ground substance.

5. Colloidal Theory

Proposed by *Wilson* in 1925. According to this theory the protoplasm having any of the aforesaid appearances is always a fluid-colloidal system having various chemical inclusions in *gel phase*. A gel is a semi-solid condition which presents jelly-like appearance. Here the molecules are held together by vrious bonds depends upon the imature and strength. By absorbing water the gel changes to more liquid-like phase. This is known as *sol* and the process is termed as *solation*. The sol can stream easily and by losing water again changes into gel-state. Both these states of colloidal matrix are interchangeable according to the various physiological, mechanical and biochemical activities of the cell. This represents phase reversal in the colloidal system. These sol-gel conditions of colloidal system are the basis for the mechanical behaviour of protolplasm.

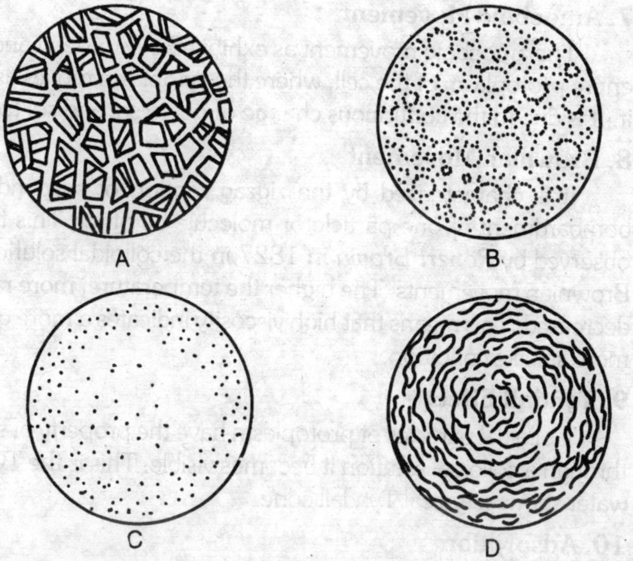

Fig. 7.1. Physical appearance of protoplasm—A. Reticular, B. Alveolar, C. Granular, D. Fibrillar.

Properties of Protoplasm

1. Cohesiveness

The various particles or molecules of protoplasm are adhered with each other by forces, such as Van der Waal's bonds, that hold long chains of molecules together. These Van der Waal's bonds are weak and non-specific forces between non-polar groups of atoms. This property varies with the strength of these forces.

2. Contractility

This property is significant in various stomatal operations in plants. The contractility of protoplasm is important for the absorption and removal of water as they generally occur in protoplasm.

3. Electrical Charge

Protein molecules repel each other because their overall charge is similar. All molecules are either positively charged or negatively charged. However, if the molecules approach one another close enough so that valency forces can act, then they may be attracted to each other.

4. Precipitation

Addition of certain amount of electrolyte in a colloidal suspension causes its dispersed particles to colloide, aggregate and finally to precipitate as suspension. For example, the addition of HCl to a colloidal system of arsenic sulphide causes precipitation.

5. Viscosity

The viscosity of the ground substance of the cell varies greatly. It may be as low as that of water, or may be very high in the gelating cytoplasm of pseudopodia of *Amoeba*.

6. Streaming Movement or Cyclosis

The protoplasm exhibits various sorts of streaming movements inside the cell. These have been studied in *Amoeba, Paramecium* etc. The movement involves only the localized portions with no visible changes in the protoplasm. No complete explanation of this has been given as yet. But it is seen that its rate depends upon the rate of cell metabolism. It is due to the fact that the energy is supplied by respiration.

7. Amoeboid Movement

The amoeboid movement as exhibited by *Amoeba* and other protozoans involves the movement of entire protoplasm of the cell, where the cytoplasm moves as one mass carrying the various inclusion with it. It is due to the continuous change of gel to sol and sol to gel.

8. Brownian Movement

It is characterized by the zigzag motion of suspended colloidal particles, occurring due to the bombardment of one particle or molecule by other. This type of movement of particles was first of all observed by *Robert Brown* in 1827 in the colloidal solution and hence such movements are known as Brownian movements. The higher the temperature, more rapid the movement and thus viscosity of cell is decreased. This means that high viscosity indicates a more gel-like state of protoplasm and low viscosity, a more sol-like condition.

9. Tyndall Effect

Colloidal particles of protoplasm have the property of scattering light. When a beam of light is passed through a colloidal solution it becomes visible. This is the Tyndall effect. A colloidal solution of proteins in water shows a typical Tyndall cone.

10. Adsorption

The tendency of particles, molecules or ions to adhere to the surface of certain solids or liquids is known as adsorption and is exhibited by the particles of colloidal system. The phenomenon helps the matrix to form protein boundaries.

11. Biological Properties

Protoplasm has all the biological properties of a living organism. It is capable of nutrition, respiration, excertion, metabolism, growth and reproduction. It has the property of irritability, e.g., it responds to stimuli like heat, light and chemicals. It also has the property of conductivity, i.e., of conducting impulses produced by stimuli.

Chemical Nature of Protoplasm

For the detailed study the chemical components of the cell can be classified as inorganic (mineral salts) and organic (proteins, carbohydrates, nucleic acids, lipids and so forth) substance. Although the most prominent constitutent of protoplasm is water—the substance whcih gives protoplasm its characteristic structure is protein. Lipids are important in all membranes and carbohydrates serve as nutrient stores. The protoplasm of a plant or animal cell contains 75 to 85% water, 10-20% protein, 2-3% lipids, 1% carbohydrates and 1% inorganic material. The following Table 7.1. gives approximate active protoplasm.

Table 7.1. Chemical Analysis of Protoplasm

Substance	*Percent*	*Average molecules weight*	*Number of molecues in relation to protein*
Water	85	18	180
Protein	10	36000	1
DNA	0.4	10	—
RNA	0.7	4.0×10^4	—
Lipid	2.	700	10
Other Organic matter	0.4	250	20
Inorganic substances	1.5	55	100

Water

Water is a substance that is justifiably called the *fluid of life*. It comprises over 90% of the chemical content of most organisms. In human body about 55% of the water (20-22 litres) is intracellular water and remains confined to the cells and the rest is found in extracellular fluids such as blood, tissue fluid and lymph. Water participates directly or indirectly in all metabolic reactions. The biologically active conformations of macromolecules and arrangement of phospholipids in the lipid bilayer of membranes are dependent on water. Water helps to keep minerals ionised in body fluids. It ionises itself to provide hydrogen ion (H^+) concentration to body fluids. It also helps in maintaining the constancy of the internal environment of an organism.

Water is a remarkable compound with unique properties that result from its molecular configuration and hydrogen bonding. Some important properties of water are as follows.

1. Polarity

The water molecule is composed of two hydrogen atoms covalently bonded to one side of an oxygen atom. Since the mean angle (105^0) between the hydrogen atoms is not rigid, water can absorb large quantities of heat and be subject to other physical stresses without breakage of the bonds. Water is a polar molecule and as with other polar molecules it has a surface charge. Water is a dipolar substance in that the hydorgen pole is positively charged and the other pole is negatively charged due to electrophilic (electron attracting) properties of oxygen. The polar nature of the water molecule causes salts to be held in solution through charge interaction (ionization; salt dissolved in water exist in the form of positive and negative charges).

2. Cohesion and adhesion

Because of the asymmetrical distribution water molecules associate with each other (cohesion) and 'wet' other substances (adhesion). The attraction of the positive hydrogen atom of one water molecule for the negative oxygen atom of another water molecule results in a *hydrogen bond*. Hydrogen bonding of water produces a dipolar molecule and favours the formation of a lattice-like structure that enables the packing of many atoms into a small area and stabilizes the molecular structure of water. Fluidity of water is maintained by very rapid formation and dissociation of hydrogen bonds between water molecules.

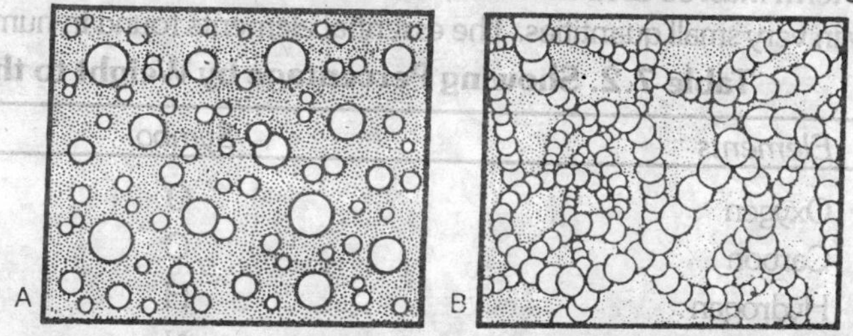

Fig. 7.2. Two water molecules showing polarity and formation of a hydrogen bond between them.

3. Latent heat and specific heat

The presence of hydrogen bonds is directly responsible for the high heat of fusion, high specific heat, and high heat of vaporization of water. The energy required to break hydrogen bonds for melting of ice, heating of water, and evaporation of water is considerably greater than the energy needed to overcome the Van der Waal's forces that are normally found in the weak association of molecules of ethane, ether and benzene. Due to high specific heat, water affords protection against sudden temperature changes in living organisms. High latent heat of evaporation of water causes elimination of excess heat through evaporation of sweat. This maintains a constant body temperature.

4. Density of freezing properties

The density of water decreases below 4^0C and ice therefore tends to float. It is the only substance whose solid form is less dense than its liquid form. The fact that water below 4^0C tends to rise helps to maintain circulation in large bodies of water. This may result in nutrient cycling and colonisation of water to greater depths.

5. Solvent action

A property of water that is very important to the living cell is its solvent action. Because it forms a solution with a vast array of compounds, water is sometimes referred to as the *universal solvent*. The solvent action of water is an effect of its ability to form hydrogen bonds due to the asymmetrical distribution of its charges. In solution with water, compounds like sugars, alcohols, and amino acids—which contain oxygen atoms, hydroxyl (–OH) groups, amino (–NH_2) groups—form hydrogen bonds with the molecules of water.

The solvent action of water is of tremendous importance for the living system. The essential elements necessary for normal growth, the compounds necessary for energy transfer and storage, and the components of structural compounds all require water as a translocation and reaction medium. Indeed, physiological processes in dilute solutions and suspensions; and the reactions are, therefore, under control of the physical and chemical laws that govern the activities of dilute solutions and suspensions.

INORGANIC COMPOUNDS

A number of inorganic salts occur both in free as well as in the ionised state. The elements that occur in quantity in protoplasm are oxygen, carbon, hydrogen and nitrogen. Always present but in much smaller amounts are potassium, phosphorus, calcium, sulphur, magnesium and iron. These ten are generally referred to as *essential elements.* Chlorine and sodium are necessary components of most animal protoplasm but are apparently not essential for plant protoplasm. Copper, boron, iodine, maganese, zinc and several other elements which are present in all protoplasm but only in minute quantities, are called *trace elements.* This term must be used with care, because these elements are still important even though they are present only in very small quantities. The essential elements found in human protoplasm are listed below:

Table 7.2. Showing Percentage by Weight to the Element in Protoplasm

Elements	*Symbol*	*Percent*
Oxygen	O	65.04
Carbon	C	18.24
Hydrogen	H	10.05
Nitrogen	N	3.15
Potassium	K	1.60
Phosphorus	P	0.84
Calcium	Ca	0.25
Sulphur	S	0.20
Magnesium	Mg	0.04
Iron	Fe	0.01
Chlorine	Cl	0.27
Sodium	Na	0.26

The elements are almost always present in protoplasm in the form of chemical compounds rather than elements. Many of these compounds inorganic salts, whch are usually in solution. These salts have numerous functions. They serve as source of elements to be built into other compounds, and some act as *buffer* in maintaining the proper acidity or alkalinity in the protoplasm. Because salts in solutions are electrical conductors, they also function in some way in connection with the electrical properties of protoplasm, however, this is not well understood at present.

ORGANIC COMPOUNDS

Proteins

Protein is an indispensable constituent of the diet because it is the only source of the amino acids. Amino acids are needed to built up new tissue during he period of growth; to maintain the structure of every tissue cell including its content of protein—containing enzyme systems; to provide raw material for the manufacturing of digestive enzymes and certain hormones; and to maintain the normal concentrations of plasma proteins and haemoglobin.

The proteins are macromolecules of very high molecular weight. The polymer molecules of protein are formed by linking together of a number of amino acids. About 20 odd different kinds of amino acids are known. Each amino acid has an amino (NH_2) and a carboxyl (COOH) group. A bond is formed between the amino group of one amino acid with the carboxyl group of another amino acid. The bond is called peptide bond. With the help of such peptide bonds a variety of amino acids can form a long chain molecule called polypeptide.

The nutritional value of a protein depends upon its amino acid composition. Synthesis of amino acids from organic keto-acids is of common occurrence in bacteria and plants. Animals can synthesize only about half of the naturally occurring amino acids (20). Thus there are two categories of amino acids:

1. Essential amino acids

The amino acids which cannot synthesized by an organisms i.e. they mut be obtained from the dietary proteins are called essential amino acids. These are *arginine, histidine, isoleucine, leucine, lysine, methconine, theronine, phenylalanine, tryptophan* and *valine.*

2. Non-essential amino acids

Although these amino acids are required by the animal as they are found in the protein of the tissues, but they can apparently be synthesized from alpha-keto acids, by the process of amination. These amino acids are *tyrosine, serine, alanine, asparagine, proline, hydroxyproline, asparatic acid, glycine, glautamine, glutamic acid* and *cystine.*

People who eat diets lacking only one of the essential amino acids are unable to synthesize normal body proteins and become protein deficient, despite having a normal nitrogen intake.

Most animal proteins, are *complete or first class proteins;* that is they contain all the necessary amino acids in appropriate proportions for ultilization by man as eggs, meat, kidney, fish, liver, poultry, milk etc.

Plant proteins are *incomplete* or *second class proteins,* since they lack certain essential amino acids. as. cereals, nuts, legumes etc. Thus people living on a strict vegetarian diet may become protein deficient.

Associated with protein deficiency is *Kwashiorker,* which affects millions of children in developing countries. It must commonly occurs in infants after weaning and particularly when given an inadequate and predominantly carbohydrate diet. Many of the children also have an inadequate energy intake and the disorder is known as *protein-energy malnutrition* (PEM). The affected child stops growing and loses weight. The skin and hair may become depigmented, and oedema associated with low plasma levels of albumin often develop. Untreated Kwashiorker is often fatal.

Structure of Protein

The critical determination of biological function of a protein is its conformation, which is defined as the three-dimensional arrangement of the atoms of a molecule. Four basic structural levels are assigned to proteins. These are:

1. Primary Structure

Arrangement of amino acid in a polypeptids chain is the primary structure of a protein. Amino acid sequence of the smallest protein known, that is of insulin, was studied by *F. Sanger* (1954). Insulin consists of two chains, A and B, joined together by two disulplide bonds (A chain conains 21 and B chain 30 amino acids). Normal human haemoglobin consists of four polypeptide chains (2α and 2β) to be held together by non-covalent forces. Each α-chain is composed of 141 amino acids, and each β chains 146 amino acids.

2. Secondary Structure

The polypeptides are held together to give rise a definite shape (helical pattern) of the protein molecules by hydrogen bonds. Hydrogen bonds can be broken by many physical and chemical treatments, for example, by excessive heat, pressure, *pH*, electricity, heavy metals and other agents that changes the environment of protein.

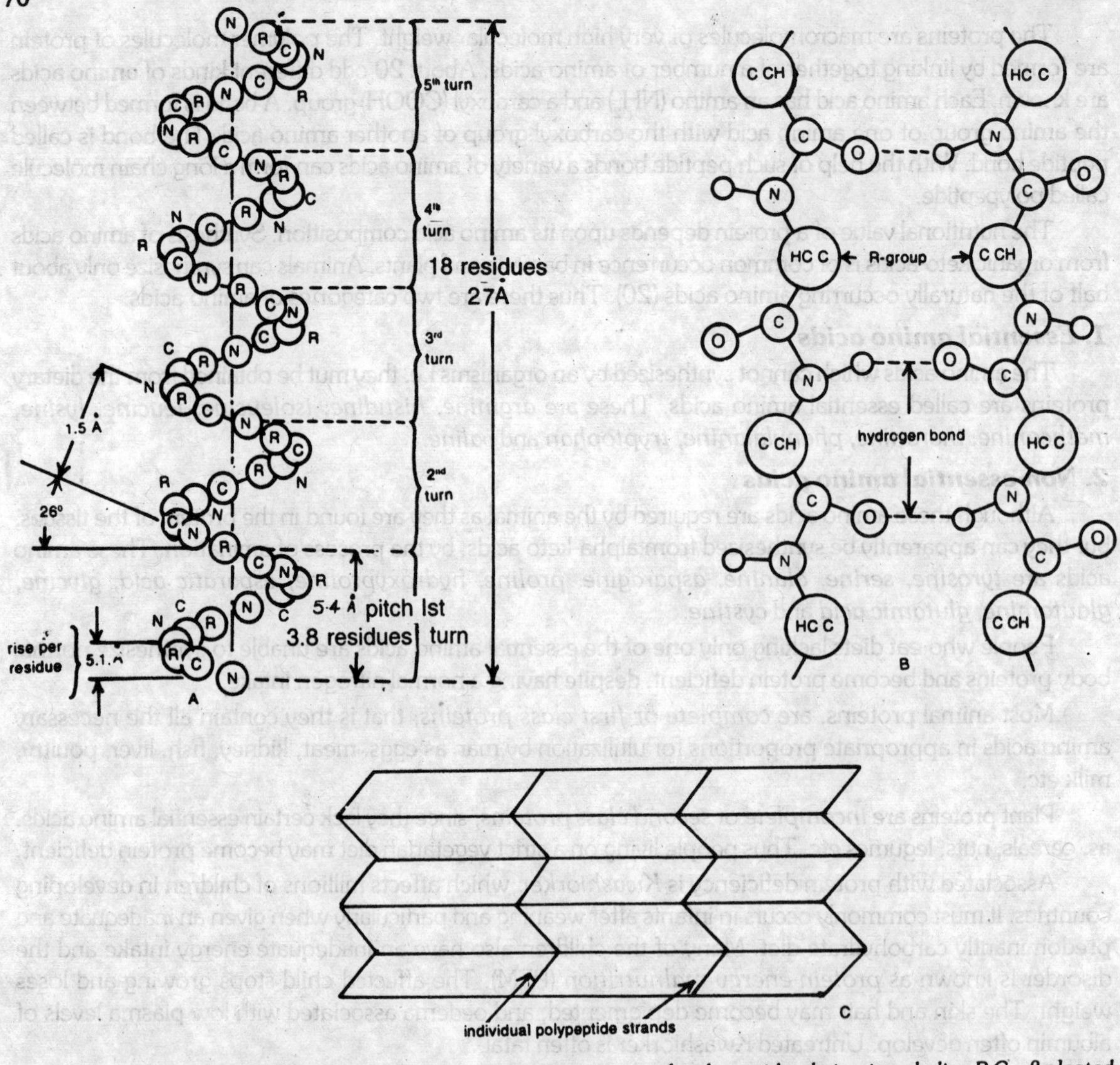

Fig. 7.3. A-C—Secondary structure of proteins : A—arrangement of polypeptide chains in α-helix; B-C—β-pleated sheets; B—extended β- (antiparallel) polypeptide strands; C—schematic of pleats (top view).

The most common type of secondary structure is α-helix. This is formed by the bending of polypeptide chain to form hydrogen bonds in the same chain. The bending of the chain is very regular and chain assumes the shape of a helix.

Pauling and *Corey* have considered an alternate type of secondary structure. In this case the polypeptide chains are lead linearly either parallel or antiparallel with respect to one another. This is called β-structure of β-helix.

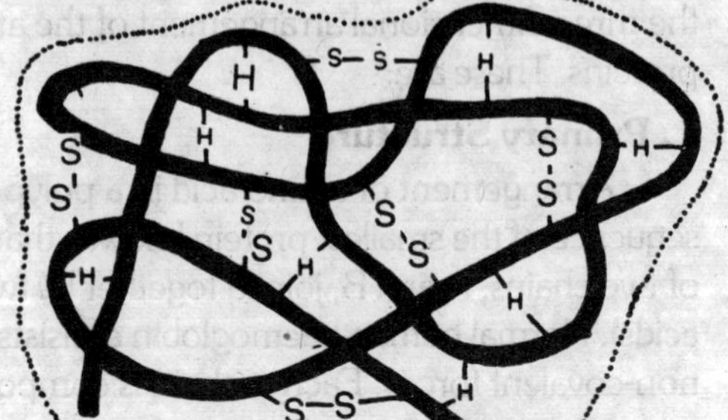

Fig. 7.4. Tertiary structure of a hypothetical protein molecule; note the extensive folding of the secondary structure which imparts a globular shape to the protein.

3. Tertiary Structure

The arrangement and inter-relationship of the twisted chains of protein into specific loops and bends in called the tertiary structure. Such a structure enables the proteins to form specific layers, crystals or fibres. The tertiary structures is maintained by hydrogen bonds disulphide bonds, Van der Waals interaction, hydrophobic interactions and ionic bonds. The tertiary structure is important and found in globular proteins. If this structure is disrupted, the biological activity of protein would be lost.

4. Quaternary Structure

This defines the degree of polymerisation of a protein unit. This structure is found in only oligomeric proteins. They are composed of subunit peptide chains linked together by any or all of the forces that can act between amino acid side chains. Haemoglobin is a fine example of quaternary structure and was studied by *Kendrew* and *Perutz* (1963).

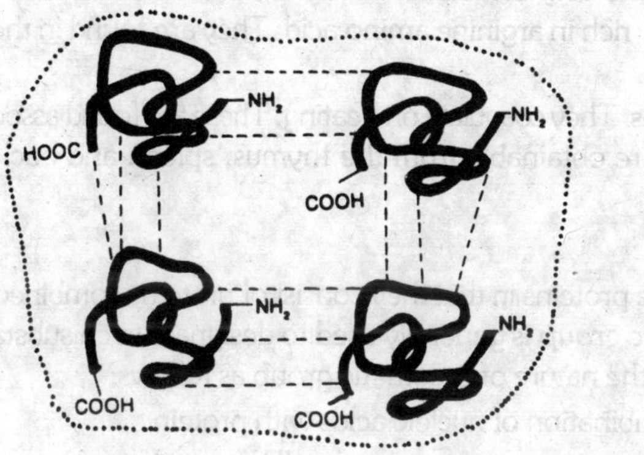

Fig. 7.5. Quaternary level of protein structure; note a diagrammatic representation showing four sub-units which are polymerised in the functional protein molecule.

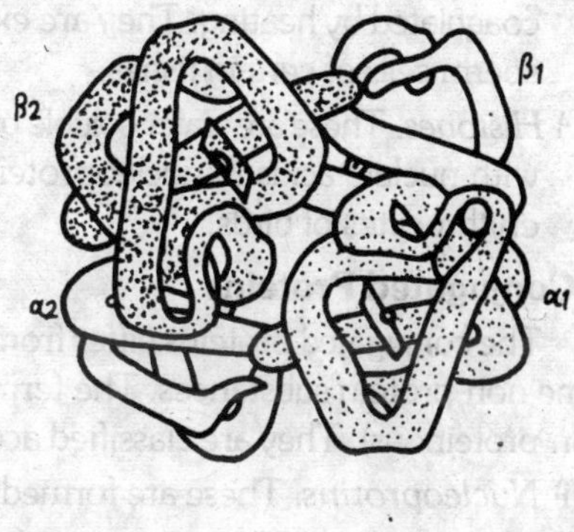

Fig. 7.6. Structure of hemoglobin; the molecule consists of four chains: two α-chains and two β-chains.

Classification of Protein

The proteins are mainly classified into two—simple and conjugated proteins.

1. Simple Proteins

These are consisting mainly of peptide chains. On hydrolyses they yield only amino acids. These are of two types:

(A) Fibrous or insoluble proteins.
(B) Globular or soluble proteins.

(A) Fibrous (insoluble) proteins

Fibrous proteins (scleroproteins) comprise all insoluble proteins which have a supporting or protective function in the animals. Secondary structure is the final structure of these proteins. The fibrous proteins can be subdivided into following types:

(i) *Collagens*. Approximately 30% of the total protein in mammalian body is collagen. They are the principle proteins of skin, tendon and bones and are resistant to peptic and tryptic digestion. They contain large amounts of hydroxyproline. They can be converted into soluble gelatin by boiling with water, dilute acids or alkalies.

(ii) *Elastins*. They are present in ligament, arteries and other elastic tissue. They cannot be converted into gelatin.

(iii) *Keratins*. They are present in animal skin, nails, hairs, horns, hooves, claws, feathers etc. they are derived from ectoderm.

(iv) *Myosin*. The major protein of muscles is also a fibrous protein.

(B) Globular or Soluble proteins

These are soluble simple proteins with more or less definite molecular weight. These are speherical or ovoid in shape. They are generally soluble in water. This group includes enzymes, oxygen carrying proteins (globulins), hormones like oxytocin, vassopressin, insulin, glucogon etc.

(i) *Albumins.* They are soluble in water, dilute acids and bases. They coagulate when heated. Common examples are albumins of egg white, blood serum and lactalbumin.

(ii) *Globulins.* They are insoluble in water but soluble in dilute salt solution. They are also coagulated when heated. Examples are serum globulin, fibrinogen, ovoglobulin in egg-yolk etc.

(iii) *Glutelins.* These are insoluble in water but soluble in dilute acids or base solution. They aré commonly found in various plant seeds. Examples are wheat, cereals etc.

(iv) *Protamins.* These are basic proteins. They are very soluble, small, stable proteins and cannot be coagulated by heating. They are exceptionally rich in arginine amino acid. They are found in the ripe sperm cells of certain fishes.

(v) *Histones.* These are water-soluble basic proteins. They coagulate on heating. These are found associated with nucleic acids in nucleoproteins. They are obtainable from the thymus, spleen and nucleated erythrocytes of birds.

2. Conjugated Proteins

The conjugated proteins differ from the simple proteins in that they consist of proteins combined wiht some non-protein substances. The term *prosthetic group* is generally used to designate such substances (non-proteinous). They are classified according to the nature of prosthetic group as follows:

(i) *Nucleoprotins.* These are formed by the combination of nucleic acids with protein.

(ii) *Mucoproteins.* They contain carbohydrate bound to protein molecules. The protein components of *mucoids* are combined with large amount (more than 4%) of carbohydrate. Examples are ovomucoid-b from egg white; orosomucoid from blood serum.

(iii) *Glycoproteins.* The carbohydrate percentage is less than 4% in glycoproteins. The carbonhdrate in these is usually the mucopolysaccharids. Mucin of saliva, chorionic gonadotropins and some other hormones of pituitary such as follicle stimulating hormone and luteinizing hormones are glycoproteins.

(iv) *Lipoproteins.* They are compounds of lipid and protein. The protein part is water soluble and lipid unsoluble. Examples are cholesterol, phosphatidylglycerides, lipovitelline of egg-yolk and serum lipoporteins.

(v) *Chromoproteins.* These are the protiens which have a matelloprophyrin or some similar substnace. Examples are haemoglobin which contains a basic protein globin with an iron porphyrin as Heme. Haemocyanin, haemoerythrin, erythrocuorins are other examples.

(vi) *Phosphoproteins.* In thse the protein molecule is linked to phosphoric acid when treated with dilute sodium hydroxide yield inorganic phosphate. Examples are casein of milk, vitellin of egg-yolk and pepsin.

(vii) *Metalloproteins.* These proteins contain metals as an inherent part of their molecules. They include tyrosinase (Cu), carbonic anhydrase (Zn), arginase (Mn or Mg) etc.

(viii) *Flavoproteins.* The prosthetic group of flavin component remains permanently attached to the protein. Flavoproteins act as coenzymes which catalyze oxidation-reduction reactions.

Derived Proteins

These include the products obtained from proteins by the action of heat and other physical agents or by hydrolysis.

1. Denatured and Coagulated proteins

Upon treatment with certain chemial agents or heat treatment the proteins become insoluble on account of some sort of intramolecular rearrangement and it is said to have denatured and when it gets separated in the form of precipitate, it is said to have undergone coagulation.

2. Peptides

These include the various fragments of the proteins which are broken off during hydrolysis of the big protein molecules either the presence of the acids or enzymes. The larger fragments had previously been named as the proteoses and the smaller fragments as peptones. This terminology is being abandoned now.

Properties of Proteins

1. ***Colour and Taste.*** Proteins are colourless and usually tasteless. These are homogenous and crystalline.
2. ***Molecular weight.*** The proteins generally have large molecular weights ranging between 5×10^3 and 1×10^6. It might be noted that the values of molecular weights of many proteins lie close to or multiples of 35,000 and 70,000.
3. ***Colloidal nature.*** Because of their giant size the proteins exhibit many colloial propertis such as (a) their diffusion rates are extremely slow, (b) they may produce considerable light— scattering in solution, thus resulting in visible turbidity (Tyndall effect).
4. ***Denaturation.*** In refers to the changes in the properties of a protein. The process of denaturation is followed by coagulation. Denaturation may be brought by a variety of agents. The physical agents, include mechanical action, heat treatment, cooling and freezing operations, ultraviolet rays etc. The chemical agents include X-rays, acetone, alcohol (solvents), salicyclates etc.
5. ***Amphoteric nature.*** The proteins are amphoteric i.e., they act as acids and alkalies both. These migrates in an electric field and the direction of migration depends upon the net charge possessed by the molecule. The net charge is influenced by the *pH* value. Each portion has a fixed value of isoelectric point (pl) at which it will move in an electric field.
6. ***Solubility.*** It is influenced by *pH* solubility is lowest at isoelectric point and increases with increasing acidity or alkalinity.
7. ***Hydrolysis.*** The proteins are hydrolyzed by a variety of hydrolytic agents.

(a) *By acidic agents.* Proteins, upon hydrolysis which come HCl (6-12 N) at 100-110°C for 6-20 hrs yields amino acids in the form of their hydrochlorides.

(b) *By alkaline agents.* Proteins may be hydrolyzed with 2N NaOH. This process is less used as it is highly disadvantageous.

(c) *By proteolytic enzymes.* Under relatively mild conditions of temperature and acidity, certain proteolytc enzymes such as pepsin and trypsin hydrolyze the proteins. Enzyme hydrolysis is used for the tryptophan.

8. ***Reactions involving—COOH group***

(a) *Reaction with alkalies* (salt formation). The carboxylic group of amino acids release a H^+ ion and forms carboxylae (Coo^-) ions. These may be neutralized by cations like Na^+ and Ca^{++} to form salts.

(b) *Reaction with alcohols (esterification).* With alcohols, corresponding esters are produced.

$$\mathrm{R{-}\overset{\overset{\displaystyle H}{|}}{\underset{\underset{\displaystyle NH2}{|}}{C}}{-}COOH + HOC_2H_5 \xrightarrow[\text{catalyst}]{\text{Acid}} R{-}\overset{\overset{\displaystyle H}{|}}{\underset{\underset{\displaystyle NH_2}{|}}{C}}{-}COOC_2H_5 + H_2O}$$

Ethyl ester of AA.

(c) *Reaction with amines.* Forms amides.

$$\mathrm{R{-}\underset{\underset{\displaystyle NH_2}{|}}{CH}{-}COOH + HHN\text{-}R \longrightarrow R{-}\underset{\underset{\displaystyle NH_2}{|}}{CH}{-}CO{-}NH{-}R + H_2O}$$

9. ***Reaction involving NH2 groups***

(a) *Reaction with mineral acids* (Salt formation). When AA or proteins are treated with HCl the acid salts are formed.

(b) *Reaction with formaldehyde.* With formaldehyde, he hydroxy methyl derivatives are formed which are insoluble in water and resistant to the attack of microorganisms.

(c) *Reaction with Nitrous acid* (*Van Slyke reaction*). The amino acids react with HNO_2 to liberate N_2 gas and to produce α-hydroxy acid.

(d) *Reaction with fluro-dinitrobenzene* (*FDNB*) or *Sanger's reagent*. In mildly alkaline solution, 1-fluro - 2, 4-dinitrobenzene reats with a-amino acids to produce yellow coloured derivative, DNB-amino acid.

10. *Reaction involving both COOH and NH_2 groups* (a) Reaction with triketohydrindene hydrate (*Ninhydrin reaction*). Ninhydrin in a powerful oxidizing agent causes oxidative decarboxylation of a-amino acids producing CO_2, NH_3 and an aldehyde. The reduced ninhydrin then reacts with the liberated NH_3 forming blue-coloured Ruheman's complex.

11. ***Reactions involving—R group or side chain***

(a) *Biuret test.* Compounds containing peptide bonds produce a purple colour when treats with Biuret reagent (.2% alkaline $CuSO_4$). A substance Biuret is formed. Dipeptides do not respond to this reaction.

(b) *Xenthoproteic test.* Yellow colour develops when proteins are boiled with conc. HNO_3 due to presence of benzene ring. This reaction is due to nitration of phenyl rings. This is the test for tyrosine, tryptophan, phenylalanine.

(c) *Million's test.* When proteins are heated with $HgNO_3$ in HNO_3, a red colour develops. This reaction is specific for tyrosine. Tryptophan also responds to this reaction.

(d) *Hoplin's—Cole test* (or *Glyoxylic acid test*). Violet ring develops on addition of Conc. H_2SO_4 at the junction of protein and glyoxylic acid solution. This test is specific for tryptophan.

(e) *Folin's test.* Blue colour develops with phosphomolybdotungstic acid in alkaline soln-due to phenol group. Specific test for tyrosine.

Biological Importance of Proteins

Proteins constitute a large part of he cell-structure and are present in all the tissue. Many proteins have special physiological functions.

1. *Membrane proteins.* The integral proteins include translocases, the peripheral proteins include cytochrome—C and monamine oxidase.
2. *Enzymes.* The enzymes are biocatalysts to influence the rats of a chemical reaction. All the enzymes are proteins.
3. *Hormones.* Some of the messangers of our body, the hromones, are proteins.
4. *Blood proteins.* 6 major plasma components are albumen, α-1 globulin, α-2 globulin, β-globulin, γ-globilin and fibrinogen. All are proteins.

CARBOHYDRATES

The class of substances, known as carbohydrates, is comprises of a large number of relatively heterogenous compounds. These are synthesized form carbon dioxide and water in chlorophyll bearing cells during the process of photosynthesis. They are especial constituents of plants (cellulose, starch etc.) but also occur and serve important functions in animals. They serve as the chief source of energy in the food of many animals. Their energy vlaue is 4 cal per gram and this energy is provided to the various synthetic needs of a cell. In carbohydrate carbon, hydrogen and oxygen are generally found in 1:2:1 ratio.

Structure

Structurally the carbohydrate are the polyhydric alcohols of carbon possessing active aldehyde or ketonic groups, which on hydrolysis yield to such products.

CLASSIFICATION

The carbohydrates may be classified according to their complexity. They are classified into following groups:

1. Monosaccharides
2. Oligosaccharides
3. Polysaccharides.

Monosaccharides

Monosaccharides are those sugars which cannot be hydrolysed into a simple form. Their genral formula is $(CH_2O)n$. The value of n ranges from 3-7. If the monosaccharide has an aldehyde group—CHO it is called an aldose e.g. glucose and if a keto group > C = O is present it is called a ketose e.g. fructose.

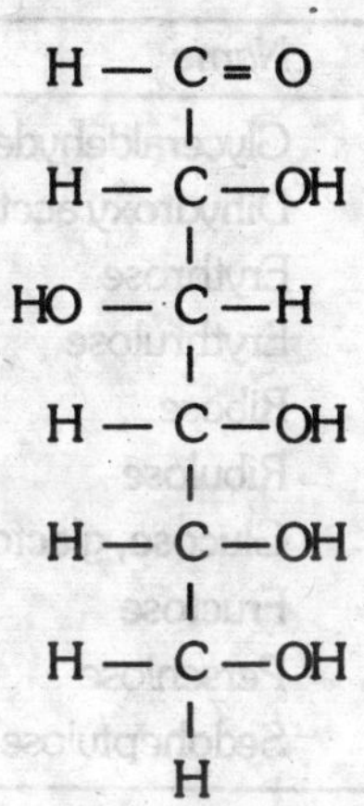

D-glucose

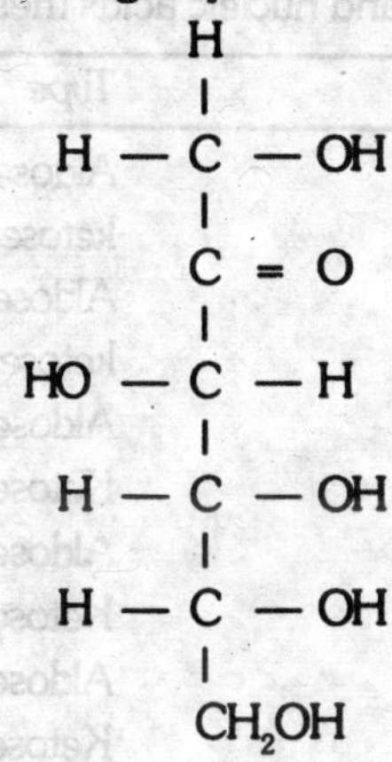

D-fructose

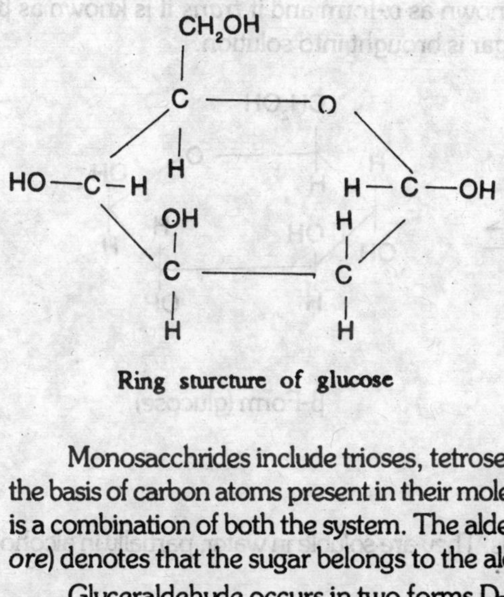

Ring sturcture of glucose

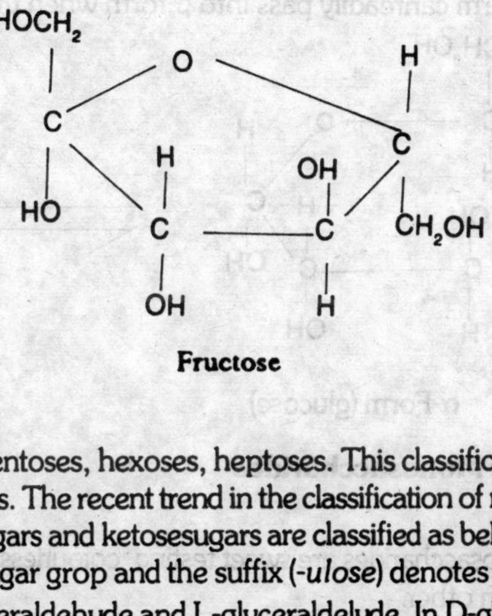

Fructose

Monosacchrides include trioses, tetroses, pentoses, hexoses, heptoses. This classification is based on the basis of carbon atoms present in their molecules. The recent trend in the classification of monosaccharides is a combination of both the system. The aldeosugars and ketosesugars are classified as below. The suffix ('-*ore*) denotes that the sugar belongs to the aldosugar grop and the suffix (-*ulose*) denotes the ketosugas.

Glyceraldehyde occurs in two forms D-glyceraldehyde and L-glyceraldelyde. In D-glyceraldelyde the hydroxyl group (OH) on right side. In L-glyceraldelyde the position of the hydroxyl group is on the left on the asymmetric carbon atom.

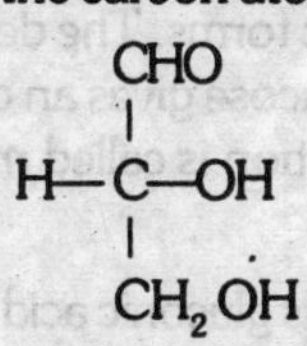

D-glyceraldehyde

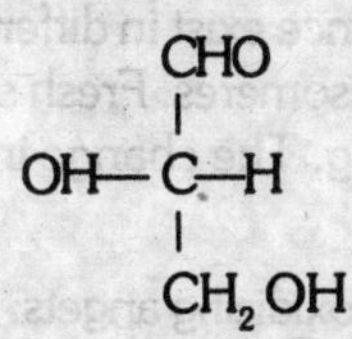

L-glyceraldehyde

It has now been well-establsihed that in the crystalline form, the monosaccharides containing 5 or more carbon atoms exist in the *tautomeric* ring forms. Even in solutions, these sugars mostly occur in the ring forms. The free sugars mostly contain six membered ring in which five members are carbon atoms and the remaining one member is an atom of oxygen. The six membered ring is known as *pyranose ring.* In adition, there is evidence of the existence of five membered rign in the sugar in which four members are carbon atoms and the remaining one member is the atom of oxygen. The five membered rign is known as *furanose rang.* Trioses and tetroses do not possess ring structure. The free pentose sugars are largely found in pyranose forms but in glycosides and nucleic acids these exist in the pyranose form.

Monosaccharides			*Type*	*Name*
1.	Triose	($C_3H_6O_3$)	Aldose	Glyceraldehyde
			ketose	Dihydroxy acetone
			Aldose	Erythrose
2.	Tetrose	($C_4H_8O_4$)	ketose	Erythrulose
3.	Pentose	($C_5H_{10}O_2$)	Aldose	Ribose
			Ketose	Ribulose
4.	Hexoses	($C_6H_{12}O_6$)	Aldose	Glucose, glactose
			Ketose	Fructose
			Aldose	Persenlose
5.	Heptose	($C_7H_{14}O_7$)	Ketose	Sedoheptulose

In the pyranose type of hexoses, carbon at 'I' position is asymmetrical. If the hydroxyl group at position 'I' is *cis* to the hydroxyl group at position 2, it is known as α-form and if *trans* it is known as β-form. The α-form canreadily pass into β-form when the sugar is brought into solution.

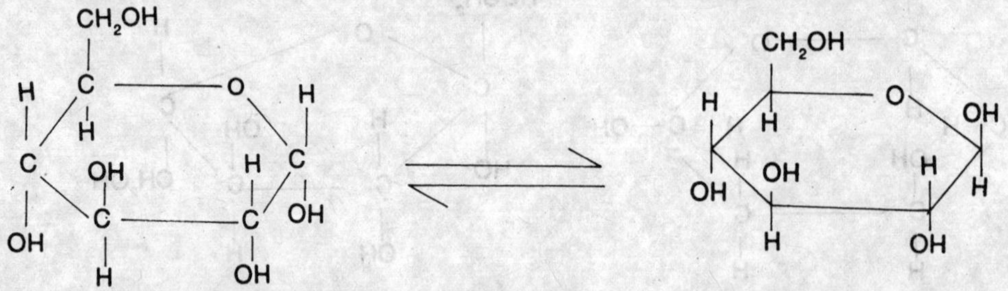

α-Form (glucose) β-Form (glucose)

Properties of Monosaccharides

(a) Physical

The monosaccharides are sweet testing, colourless solids. They are soluble in water, partially in alcohols and insoluble in other.

When a polarized light is passed through a solution of these carbohydrates, the plane of light is rotated to either right or left side.

They contain asymmetric carbon atom hence exist in different isometric forms. The degree of optical rotation may change due to interconversion of isomeres. Fresh solution of glucose gives an optical rotation of $+112^0$ which change to $+ 52.7^0$ on standing. The change in optical rotation is called *mutarotation.*

(b) Chemical

(i) *Oxidation.* They can easily be oxidized by oxidizing angets. Glucose yields gluconic acid after oxidation with Tollen's reagent (ammonical Ag_2 O) or Fehlings solution (alkaline $CuSO_4$). The reduction of

Tollen's reagent yields silver as polishing on the surface of tube whereas with the Fehling solution red ppt are obtained.

When strong oxidizing agent is used like conc HNO3, gluconic acid, is ultimately oxidizes to discarboxylic saccharic acid.

Glucuronic acid is obtained in animal body on slow oxidation of glucose. Glucuronic acid combines with hormones. It is a major component of hyaluronic acid, heparin, mucoitin sulphate and chondriotine which are found in blood, skin and cartilage.

(ii) *Reduction*. Free aldehyde and ketone groups of monosaccharides are reduced to alcoholic hydroxy groups by sodium—mercury amalgam and water. D-glucose after reduction yields a mixture of sorbitol and mannitol.

(iii) *Condensation*. Aldehyde groups of monosacharides condense with primary amines to form Schiff's base, when treated with hydroxylamine, glucose forms glucose oxime.

(iv) *Esterification*. The hydroxyl groups of alcohols in the carbohydrate may be converted to esters by treating with the appropriate acetylating agents. When D-glucose is treated with acetic anhydrine in the presence of pyridine, penta-acetyl glucose is formed.

(v) *Methylation*. Methylating agents such as Ag_2O, CH_3OH react with monosaccharide to yield glycoside.

(vi) *Fermentation*. Monosaccharides such as glucose and fructose undergo alcholic fermentation by micro-organisms such as yeasts and produce ethanol and carbon dioxide.

Oligosaccharides

The oligosaccharides are those carbohydrates which on hydrolysis give two to five simple monosaccharide molecules. The oligosaccharides are composed of two to five monosaccharide units. During union of monosaccharde units water molecule is eleminated and the units are linked through an oxygen bridge. It is a glycosidic linkage. Their general formula is $(CH_2O)_{n-1}$.

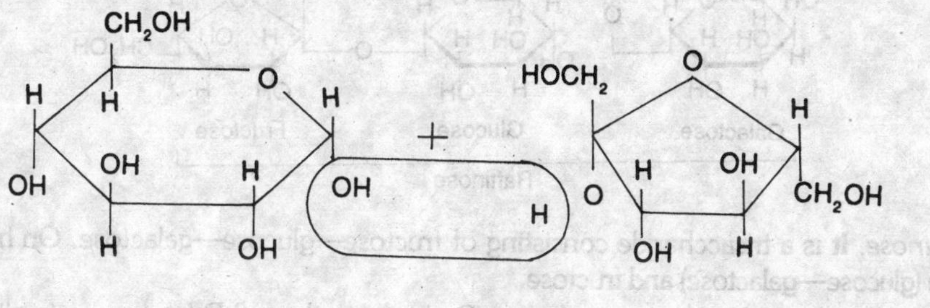

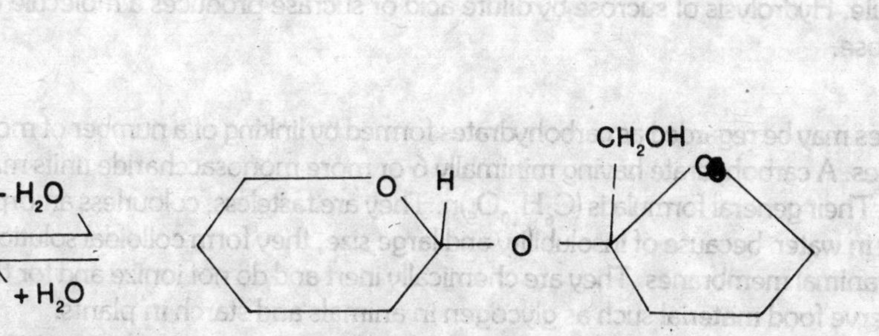

sucrose (diasaccharide)

The disaccharides have been classified into two groups namely, reducing and nonreducing. The reducing disaccharides are maltose, lactose, cellobiose, gentibiose and the non-reducing are sucrose and trehalose.

Maltose. It is composed of two units of D-glucose joined together through their 1 and 4 carbon atoms. It is obtained as a hydrolytic product by the action of amylase on strach

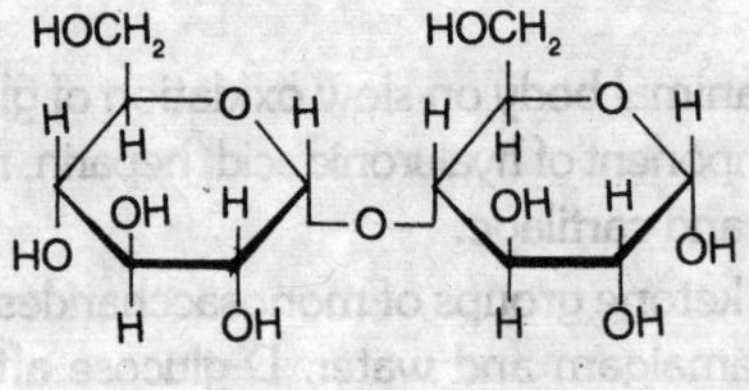

Lactose. Lactose is a disaccharide consisting of glucose and galactose which is synthesized in the mammary glands.

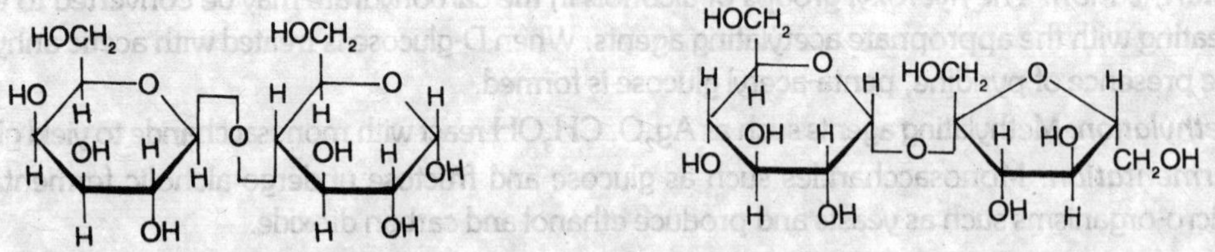

Cellobiose. Incomplete hydrolysis of cellulose given cellobiose. It is composed of two molecules of glucose linked by β-1, 4 glucosidic bond.

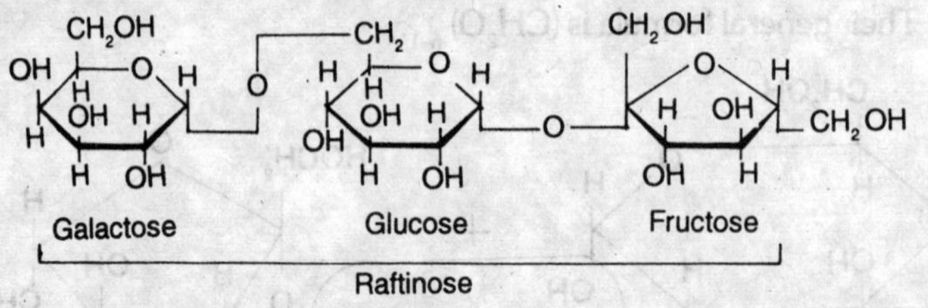

Raffinose. It is a trisaccharide consisting of fructose—glucose—galactose. On hydrolysis it yields *Mélibiose* (glucose—galactose) and fructose.

Sucrose. It is formed by the union of one α-D-glucose and one β-D-fructose units with the elemination of one H_2O molecule. Hydrolysis of sucrose by dilute acid or sucrase produces a molecule of glucose and a molecule of fructose.

Polysaccharides

Polysaccharides may be regarded as carbohydrates formed by linking of a number of monosaccharides by glucosidic linkages. A carbohydrate having minimally 6 or more monosaccharide units may be regarded as polysaccharides. Their general formula is $(C_6H_{10}O_5)n$. They are tasteless, colourless amorphous powders which are insoluble in water. because of insolubility and large size, they form colloidal solutions and will not pass across natural animal membranes. They are chemically inert and do not ionize and for this reason very much suited as reserve food material such as glycogen in animals and starch in plants.

Polysaccharides consisting of only one type of monosaccharide units are called *homopolysaccharides* while those with different types of monosaccharde derivatives are *heteropolysaccharides*.

I. Homopolysacchardes

(i) *Glycogen*. It forms the carbohydrate reserve of the animal tissues and it is mainly stored in the muscles and liver. Fungi and yeast do also contain glycogen. It is a branched chain polymer having 6000—30,000 glucose units.

(ii) *Starch.* It is the most important food source of carbohydrate and is found in potatoes, rice, cereals etc. It is hydrolysed in gut yielding dextrins and maltose and eventually glucose.

Treatment of starch with hot water dissolves amylose, while amylopectine remains as such:

Amylose. Contains about 200-500 glucose units which are arranged in the form of a straight chain. The molecular wt. is about 150,000. It gives intense blue colour with iodine.

Amylopectin. It has about 1000 glucose resideus and its molecular wt. is about 200,000-1,00,000. It has a branched structure.

(iii) *Agar.* It is a galactan consisting of both D and L glactose. It is used as a bacteriological culture medium.

(iv) *Pectins.* These are abundant in fruits, particularly in the rim of citrus fruits like orange and lemons. They contain arabinose, glactose and galacturonic acid.

(v) *Xylan.* In addition to cellulose all plants contain xylan. It is a hemicellulose and consists of D-xylose.

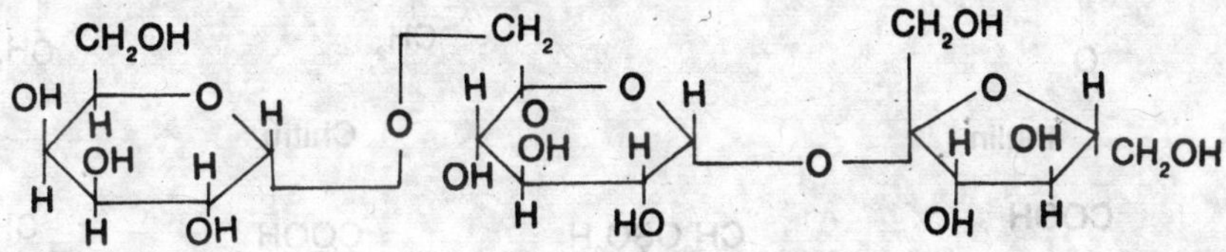

Glactose Glucose Fructose

Raffinose

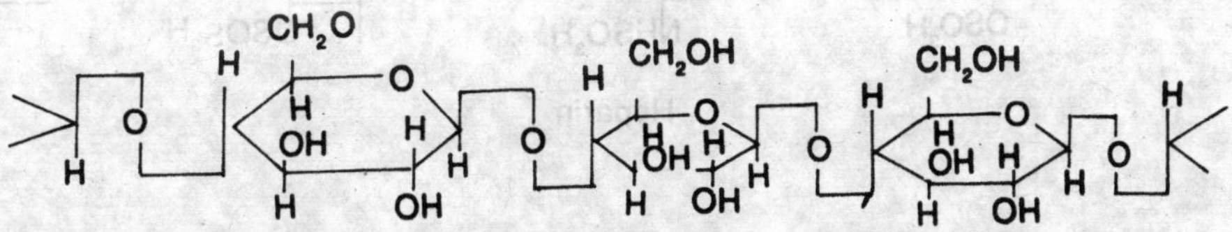

Cellulose

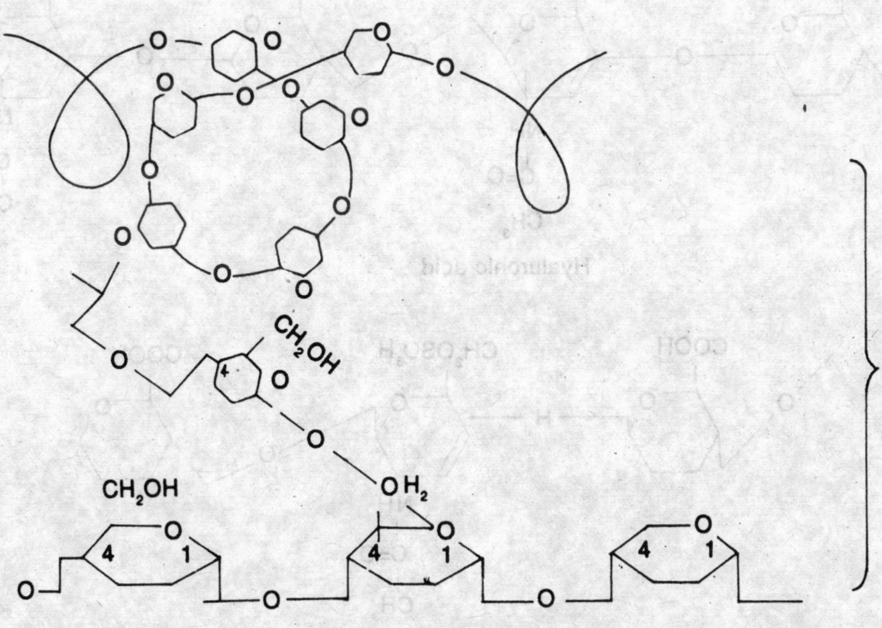

Two units of starch

Amylopectine

(vi) *Inulin.* It is a starch found in the tubers and roots of *Dahlia*, *Dandelions* etc. On hydrolysis it yields fructose. M. Wt. is about 5,000 and about 30-35 fructose units per mole are present.

(vii) *Dextrins.* They are the intermediate products formed during hydrolysis of starch to the glucose.

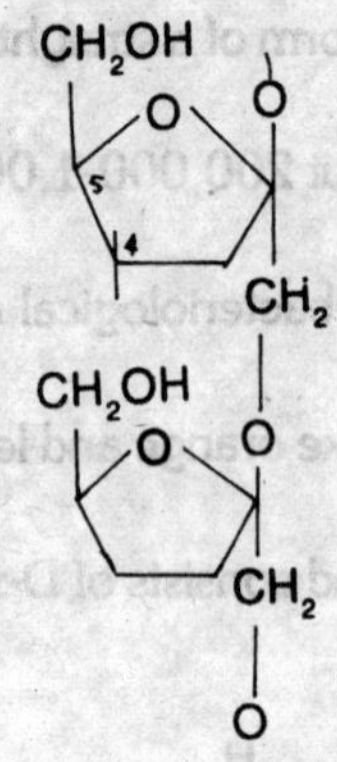

Inulin

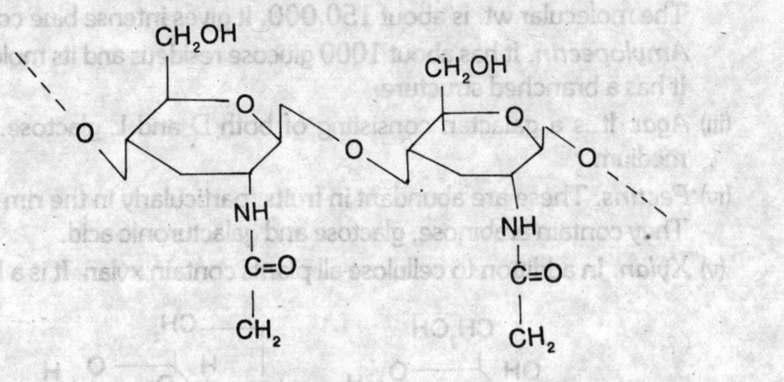

Chitiu

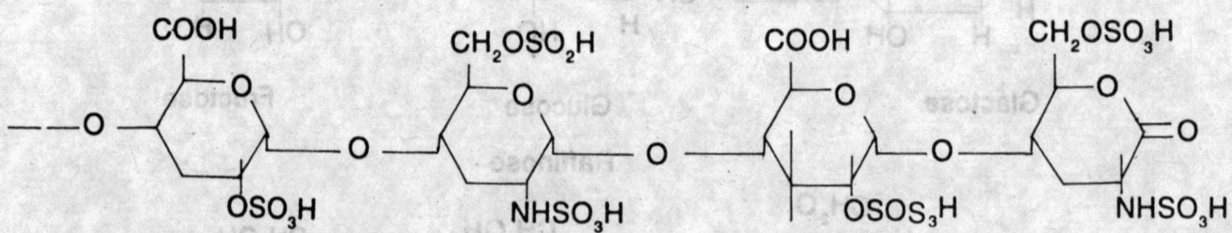

Heparin

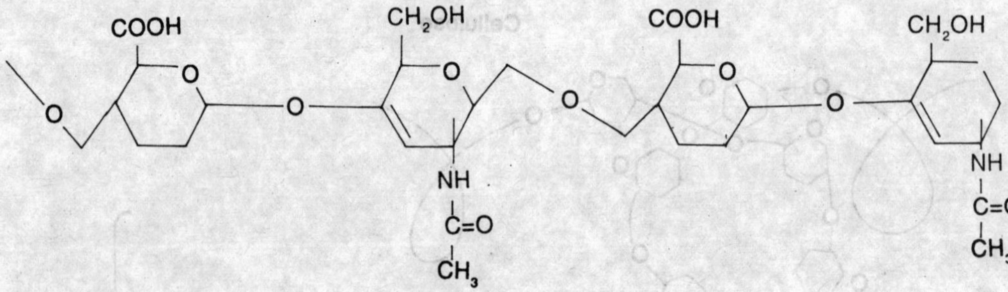

Hyaluronic acid

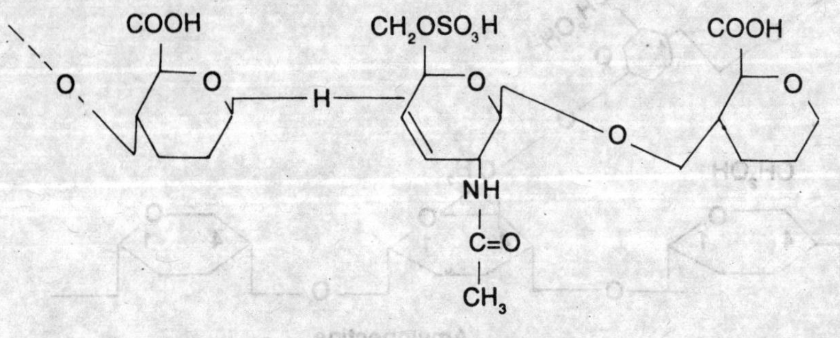

chondriotin sulfate

(viii) *Cellulose.* The cell wall of plants is formed by cellulose. The cellulose consists of a lenier chain of glucose units ranging 900-2000. It does not occur in the animal tissue and it cannot be utilized by man for energy production purposes. Herbivorous animals can, however, utilize cellulose with the help of microorganisms as they can digest cellulose.

(ix) *Chitin.* It is an important structural polysaccharide in invertebrates particularly in arthropods. Structurally, it consists of N-acetyl-D glucosamine units.

II. Heteropolysaccharides

A. *Glycoproteins.* These are protein-polysaccharide compounds occuring in tissue, particularly in mucus secretion. Examples are ovalbumin, fibrinogen, γ-globulin of serum, human chorionic gonadotropins, luteinizing hormones and the blood group substances of RBC (antigen).

B. *Mucopolysaccharides.* The most important mucopolysaccharides from biological point of view are:

(i) *Heparin.* It is present in the liver, lungs and spleen having a molecular wt. about 20,000. The *exect* structure of heparin in unknown. It is found to be constituted by glucuronic acid, glucosamine and sulfuric acid. Sulfuric acid is linked to hydroxyl group of sugar derivatives as well as to amino group of glucosamine. Heparin prevents clotting of blood in the vessels.

(ii) *Hyaluronic acid.* It is a mucopolysaccharide present in connective tissue and acts as an intercellular connecting material. It is abundently present in the umbilical cord, vitreous fluid of eyes and in the synovial fluid present at the joints. This polysaccharide consists of N-acetyl glucosamine and glucuronic acid.

(iii) *Chondrotin sulfates.* They are found in the cartilage, adult bones, skin, cornea, tendons and heart valves. It conists of N-acetylgalactosamine and glucuronic acid.

Functions

The carbohydrates are very important biologically as a source of energy for the cell. The main source of energy is glucose. It the main form of crbohydrates whch is transported form cell to cell by blood in animal and by sap in plants. Carbohydrates also help in the formation of cell wall in plants. Some carbohydrates may serve as the prosthetic group of certain proteins. Glycogen and starch are major storage materials in animals and plants respectively.

Lipids

Term lipid includes fats and fat-like substances. Strucrurally, the different substances of this group may not be similar to fats, but all are soluble in fat-solvents like alcohols, ether, chloroform, carbon tetrachloride, acetone etc. They are insoluble in water. Lipids, like carbohydrates also, contain carbon, oxygen and hydrogen. The carbon and hydrogen are present in larger quantity than oxygen. Some lipids have phosphorous and nitrogen. The lipids in the cell may serve as condensed reserve of energy as well as form the membranous structures.

Classification of Lipids

Lipids are classified into:

I. Simple Lipids
II. Compound Lipids
III. Derived Lipids

I. Simple Lipids

These are esters of fatty acids with various alcohols.

1. Natural Fats

These are the esters of fatty acids with glycerol. The chemical term for a natural fat is *triglycerides.* The triglycerides are formed by the combination of three molecules of fatty acids joined to one molecule of glycerol. As already stated triglycerides are neurtral fats, neutral because the acid ions are neutralized during their unification with glycerol; fats because the bulk of the molecule is devoid of electro-negative elements which can unite with hydrogen to form water molecule.

CH_2OH		HOOC—R_1		CH_2OCOR_1
|				|
CHOH	+	HOOC—R_2	→	$CHOCOR_2$
|				|
CH_2OH		HOOC—R_3		CH_2OCOR_3
Glycerol molecule		3 fatty acids molecules		Triglyceride + $3H_2O$

Triglycerides can exist in the solid or liquid form. It is based on the kind of fatty acid residues in its structure. The triglycerides which are blow 20°C called *oils* containing a large proportion of short-chained unsaturated fatty acids like *oleic, linoleic acid* etc. The triglycerides which are solid above 20°C are called *fats.* They contain long chained saturated fatty acids like *palmitic acid* and *stearic acid.*

2. Fatty acids

These are obtained by the hydrolysis of fats. These are monocarboxylic acids. The molecules of a fatty acid has a polar carboxyl group soluble in water and a non-polar hydrocarbon chain soluble only in fat solvents.

The fatty acids are classifed under three groups based on their degree of saturation and unsaturation.

(i) *Saturated fatty acids.* A saturated fatty acid contains as many hydrogen atoms as its carbon chain can hold. General formula for saturated fatty acid is R—COOH, where R is CH_3 (CH_2)n. n is varying from zero in acetic acid to 86 in mycolic acid. The most abunant saturated fatty acids in nature *palmitic* (C_{18}) and *Stearic* (C_{16}) *acids.*

(ii) *Unsaturated fatty acids.* They have one or more double bonds in their carbon chain. When there is a single double bond, results inthe loss of 2 hydrogen atoms, such fatty acids are called *monous saturated fatty acids.* Their genral formula is CnH_{2n}—1—COOH, examples are *palmitoletic acid* and *oleic acid.* When there ae 2,3,4 or more double bonds in the carbon chain with the consequent absence of 2,6,8 or more hydrogen atoms, such fatty acids are *polyunsaturated fatty acids.* The general formula is CnH_{2n}—2 COOH or CnH_{2n}—3 COOH. Examples are *linoleic acid* and *archidonic acids.*

Linoleic, linolenic and archidonic acid are often termed as *essential fatty acids.*

Table 7.3. Showing a few common fatty acids found in lipids.

Fatty acid		*Formula*
Saturated fatty acid	→	
Butyric acid	→	$CH_3(CH_2)_2$ COOH
Caproic acid	→	$CH_3(CH_2)_4$ COOH
Palmitic acid	→	CH_3 $(CH_2)_{14}$ COOH
Steraric acid	→	CH_3 $(CH_2)_{16}$ COOH
Unsaturated fatty acids		
Palmitoleic acid		CH_3 $(CH_2)_5$ (H=CH $(CH_2)_7$ COOH
Oleic acid		CH_3 $(CH_2)_7$ CH=CH $(CH_2)_7$ COOH
Linoleic acid		CH_3 $(CH_2)_4$ CH=CH CH_2 CH=CH $(CH_2)_7$ COOH
Linolenic acid		CH_3 CH_2 CH=CH CH_2 CH=CH CH_2 CH=CH $(CH_2)_7$ COOH

3. Waxes

Wexes are another class of simple lipis containing one molecule of fatty acid and one high molecular weight alcohol. The constituent acids and alcohols have usually 24-36 carbon atoms. The waxes has high melting points. They are chemically inert as they do not have double bonds in their hydrocarbon chains and are highly insoluble in water. They serve as protective converings on leaf surface (plants). *Bees wax* is an ester of palmitic acid with myricyl alcohol ($C_{30}H_{61}OH$) and *spermaceti* from the sperm whale is an ester of palmitic acid with cetyl alcohol ($C_{16}H_{33}OH$).

II. Compound lipids

These are esters of fatty acids and alcohol with additional compounds, such as phosphoric acid, sugars, proteins etc. These are classified as follows:

(i) Phospholipids

Lipids containing phosphorous are phospholipids. They also have nitrogen containing bases and other substituents. The phospholipids are abundant in brain and nervous tissues. The different types of phospholipids are:

(a) *Phosphatidic acids.* These are compounds consisting of glycerol, two fatty acids and a phosphate group.

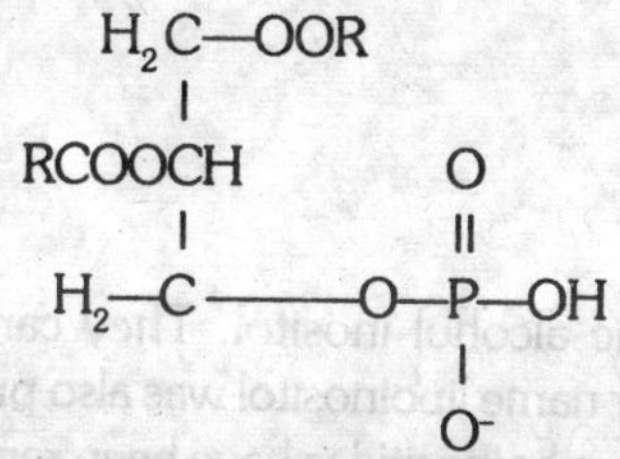

Here R_1 and R_2 represent the residues of the molecules of fatty acids. The phosphatidic acids do not found in any great quantity in tissues but these are important as in intermediate biosynthesis of triglycerides and other phospholipids.

(b) *Lecithins.* Lecithins are choline esters of phosphatidic acid. On hydrolysis they yield one molecule of glycerol, 2 molecules of fatty acids and one molecule of phosphoric acid to which a nitrogenous base choline is attached.

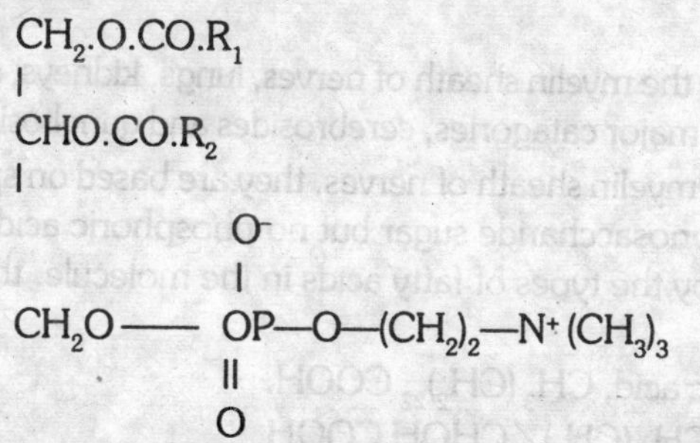

The fatty acids are palmitic, stearic, oleic, linolic and arachidonic acids. Lecithins are yellowish grey solids, soluble in ether and alcohols but insoluble in acetone. Lecithins get broken down by the enzyme lecithins to lysolecithin. This product (lysolecithin) has the ability to hemolyze the red blood corpuscles. Enzyme lecithinase occurs in the venom of snake cobra. and bee.

Lecithin is an important constituent of lipoproteins. Egg-yolk is a rich source of lecithin which also has an important role in fat metabolism in the liver.

(c) *Cephalins.* They resemble lecithins in most properties. The fundamental difference between the lecithins and cephalins is the nature of nitrogenous base. Cephalins contain ethanolamine, colamine and sometimes serine in place of choline.

$CH_2O.CO.R_1$
|
$CHO.CO.R_2$
|
| O
| |
$CH_2—O—P—O—(CH_2)_2\ N+H_3$
‖
O

(d) *Plasmalogens.* These phospholipids are abundant in brian and muscle. They resemble lecithins and cephalins in structure but possess an aldehyde group in place of one of the fatty acids in typical phospholipid molecule.

$CH_2OCH = CH_2R$
|
$CHOOCR_2$
|
| O
| ‖
$CH_2\ O—P—O—CH_2\ CH_2\ NH_3$
|
O

(e) *Phosphoinositides.* Phosphoinositides contain hexahydric alcohol inositol. They can be either monophosphoinositides or diphosphoinositides. An another name lipoinositol was also proposed for these substances. They occur in animal and plants but diphosphoinositides have been reported from brain tissue only.

(f) *Phosphosphingosides.* Phosphosphingosides have been found to occur in high concentration in brain, nerves, lungs and spleen tissue. These contain a nitrogenous base sphingosine or its derivative which remains attached to long-chain fatty acid by its amino group. On hydrolysis they yield fatty acids, phosphoric acid, choline and sphingosine.

(ii) Glycolipids

The glycolipids occurs in brain tissue and in the myelin sheath of nerves, lungs, kidneys, spleen, liver, retina, egg-yolk etc. The glycolipids include two major catagories, cerebrosides and gangliosides.

(a) *Cerebrosides.* They occur in the brain and myelin sheath of nerves. they are based on sphinge mine and have in addition a fatty acid and a monosaccharide sugar but no phosphoric acid or glycerol. Individual cerebrosides are differentiated by the types of fatty acids in the molecule. the important types are:

Kerasin containing the saturated lignoceric acid, $CH_3\ (CH_2)_{22}\ COOH$.

Cerebron containing the cerebronic acid, $CH_3\ (CH_2)_{21}\ CHOH\ COOH$.

Nervon containing the nervonic acid $CH_3\ (CH_2)_7\ CH = CH\ (CH_2)_{13}\ COOH$.

Oxynervone containing hydroxy derivatives of nervonic acid $CH_3\ (CH_2)_7\ CH{=}CH\ (CH_2)_{12}\ CHOH\ COOH$.

(b) *Gangliosides.* These contain N-acetylneuraminic acid (sialic acid), fatty acids, sphingosine and 3 molecules of hexose.

III. Derived Lipids

These lipids include hydrolytic products of lipids as well as other lipid like compounds like sterols, carotenoids, hydrocarbous etc.

The *sterols* are solid wax-like substances chemically, they are alcohols and occur either as such or as esters of fatty acids. They are highly soluble in fat solvents. They all contain a cyclopentanophenanthrene (sterane) nucleus (made up of three cyclohexane rings, in the phenanthrane type of arrangement and a terminal cyclopentane ring).

Cholesterol is animal origin occuring in bile, brain, spinal cord etc. It is obtained form human gall bladder stones which are deposited in the bile duct.

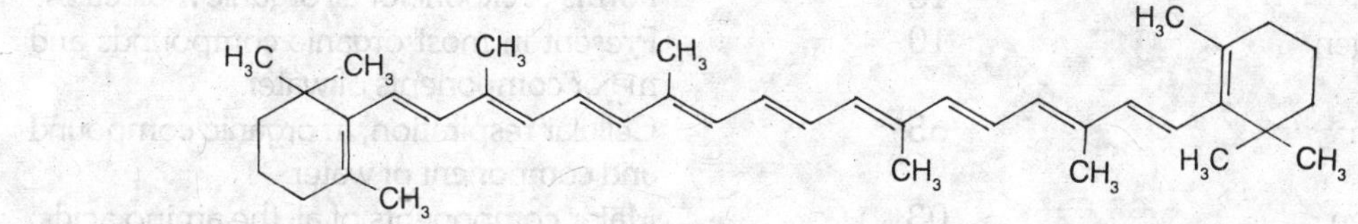

Ergosterol. It is similar to 7-dehydrocholesterol but differs in the side chain. It can be converted into vitamin D on exposure to the ultra violet light and, therefore, also called provitamin D.

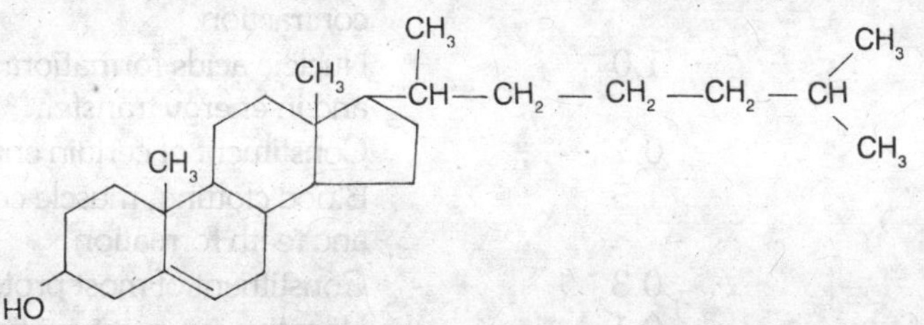

Carotenoids. These are also included in lipids because of their solubility in fat solvent. Carontenes consists of carbon and hydrogen only where as xanthophyll contain oxygen in addition. Vitamin A is derived from carotene. *Estrogens, progesterone, testosterone* and *anderosterone* are steroidal hormoens.

Prostaglandins

Prostaglandins are derivatives of fatty acids. They were first discovered in human seminal fluid secreted by prostate gland, hence the name. It has been shown now that prostaglandins are synthesized and released by many other tissues such as kidneys, testis, placenta, lungs, liver, uterus, gastrointestinal tract, brain and heart.

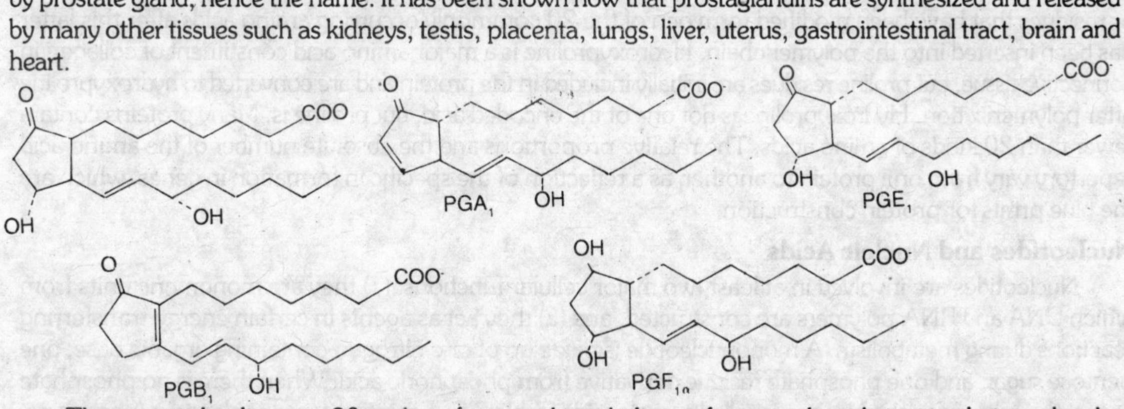

The prostaglandins are 20 carbon fatty acids including a five membered ring in their molecular structure. Different prostaglandins differ with one another in the number and position of double bonds and hydroxyl group substituents.

Prostaglandins PGE, PGE_1, PGE_3, $PGF_{1\alpha}$, $PGF_{2\alpha}$ and $PGF_{3\alpha}$ are considered as *primary prostaglandins.* Most of their action appear to be binding of hormones to membrane.

Recently, new type of prostaglandin has been isolated from human seminal fluid which has been designated as PGX^2.

Table 7.4. Showing Elements that Occur in Protoplasm

S.No.	*Name of elements*	*Approximate percentage*	*Function*
Major Elements			
1.	Carbon	18	Forms backbone of all organic molecules.
2.	Hydrogen	10	Present in most organic compounds and major components of water.
3.	Oxygen	65	Cellular respiration; in organic compound and component of water.
4.	Nitrogen	03	Major components of all the amino acids, proteins and nucleic acids.
Tracer Elements			
5.	Sodium	0.2	For water balance, conduction of nerve impulse.
6.	Potsium	0.5	Conduction of nerve impulse, muscular contraction.
7.	Phosphorus	1.0	Nucleic acids formation; bones formation and in energy transfer.
8.	Magnesium	0.1	Constituent of certain enzymes (ATPase).
9.	Calcium	1.5	Blood clotting, muscle contraction, bones and teeth formation.
10.	Sulfur	0.3	Constituent of most proteins.
11.	Chloride	0.1	Negative ion of interstitial fluid.
12.	Iron	0.01	Component of haemoglobin and certain enzymes.

Amino Acids

There is a common plan of construction for the thousands kinds of proteins in living systems. The 20 kinds of naturally occurring amino acid monomers are strung together in unbranched, linear polymer chains of proteins. These are the 20 amino acids specified in the genetic code that is universal to all organisms. Some other kins of amino acids are also found in cells, but they are either degradation products or residues that have been modified form oen of the 20 commonly occurring amino acids after this latter has been inserted into the polymer chain. Hydroxyproline is a major amino acid constituent of collagen in connective tissue, but proline residues are initially included in the protein and are converted to hydroxyproline after polymerization. Hydroxyproline is not one of the encoded acid, but proline is. Many proteins contain fewer than 20 kinds of amino acids. The relative proportions and the absolute number of the amino acid repertory vary from one protein to another, as a reflection of the specific in formation in genes, which are the blue prints for protein construction.

Nucleotides and Nucleic Acids

Nucleotides are involved in at least two major cellular functions: (1) they are monomeric units from which DNA and RNA polymers are constructed, and (2) they act as agents in certain energy-transferring reactions during metabolism. A mononucleotide is made up of one nitrogen-containing organic base, one pentose sugar, and one phosphate residue derivative from phosphoric acid. When there is no phosphate group, the sugar-base combination is called a *nucleoside* (Table 7.5). For this reason, *nucleotides* (phosphate-sugar-base) are also called nucleoside phosphates. Nucleoside mono-, di-, and tri- phosphates contain one, two or three phosphate groups respectively.

Table 7.5. Showing nomenclature of nucleic acids and their constituent units.

Base	*Nucleoside*	*Nucleotide*	*Nucleic Acid*
Purines			
Adenine (A)	Adenosine	Adenylic acid	RNA
	Deoxyadenosine	Deoxyadenylic acid	DNA
Guamine (G)	Guanosine	Guanylic acid	RNA
	Deoxyguanosine	Deoxyadenylic acid	DNA
Pyrimidines			
Cytosine (C)	Cytidine	Cytidylic acid	RNA
	Deoxycytidine	Deoxyadenylic acid	DNA
Thymine (T)	Thymidine	Thymidylic acid	DNA
Uracil (V)	Uridine	Uridylic acid	RNA

The nitrogenous bases commonly found in nucleic acids and their nucleotide building blocks are derivatives of purine and pyrimidine. The commonly occuring purines *adenine* and *guanine* are found in both DNA and RNA, as in the pyrimidine compound *cytosine,* the second kind of pyrimidine in DNA is *thymine,* while its demethylated form, *uracil,* occurs in RNA.

Since each kind of nucleic acid contains one unique pyrimidine, it is convenient to study synthesis and activity of DNA or RNA using isotopically-labelled precursors containing one or the other of these bases. Usually the nucleosides uridine or thymidine, or other nucleotide forms, are added to the biological system under study.

The only difference in the pentose sugars of nucleotides is the presence of a hydroxyl group at carbon atom 2 of D-ribose in RNA, but a hydrogen at carbon-2 of 2-deoxy-D-ribose in DNA monomers and polymers. This seemingly simple difference is partly responsible for profound differences in stabilities, pairing potential, and functions of DNA and RNA.

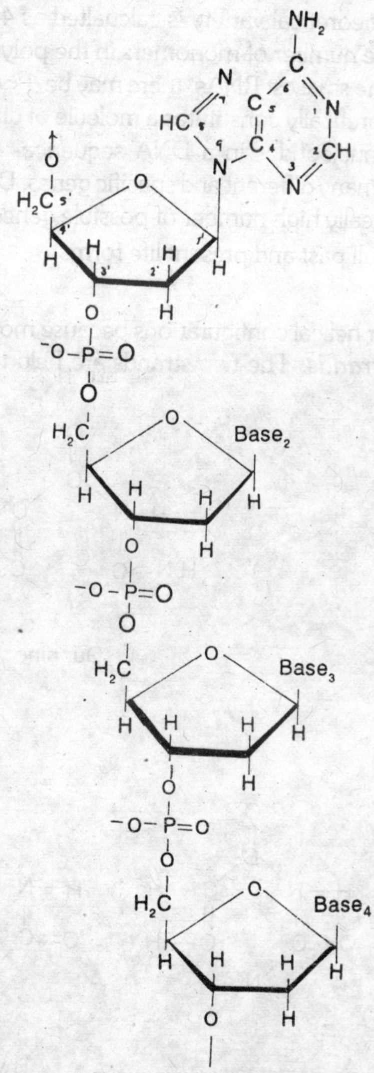

Fig. 7.7. The molecular organization of a backbone of DNA showing phosphodiester and glycosidic bonds.

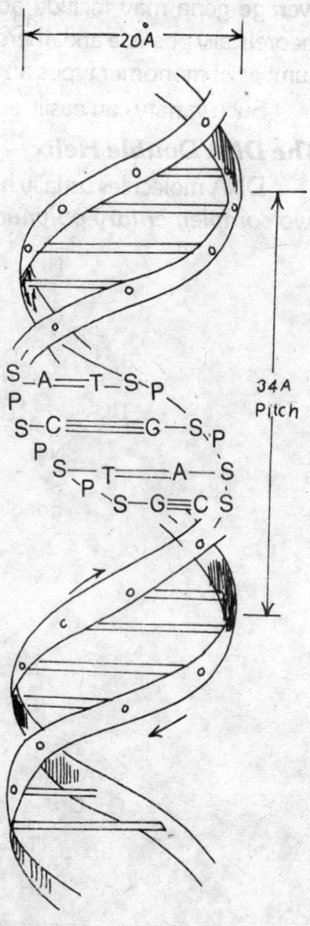

Fig. 7.8. DNA and its paired bases.

Polynucleotides of both DNA and RNA varieties are built from mononucleotides that are linked convalently via phosphodiester bridges between the 3' position of one unit and the 5' position of hte next. Since there is no restriction on the ventrical sequence of adjacent mononucleotides in either DNA or RNA, a considerable variety of molecules is possible even though only 4 kinds of nucleotide monomers (one for each of the four kinds of bases in the combination with sugar and phosphate) are used in polymer construction. The theoretical variety is calcualted as 4”, where 4 is the number of different kinds of nucleotides, and *n* is the number of monomers in the polymer. For a molecule made of only 75 monomeric units, as in some of the smallest RNAs, there may be 4^{75} different arrangements of the constituent units. Each arrangement theoretically constitutes a moleule of different specificity. Where the average gene may include about 500 nucleotides in a DNA sequence, 4^{500} different sequences are theoretically possible and, therefore, that many different and specific genes. Despite the apparently meager number of monomer types an astronomically high number of possible genes can be constructed.

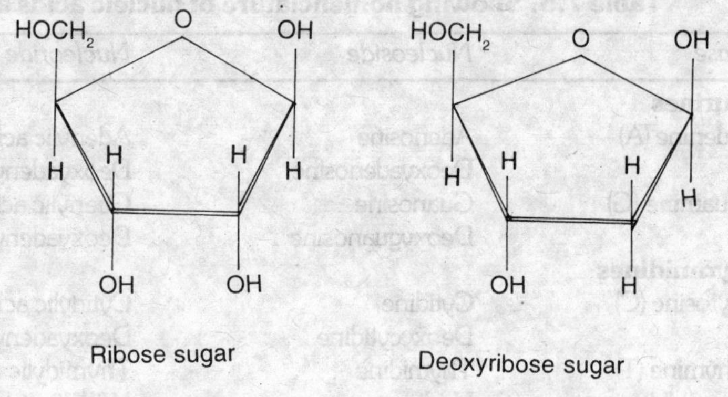

Fig. 7.9. Pentose Sugar.

Such variety can easily account for all past and present life forms.

The DNA Double Helix

DNA molecules usually have regular helical configurations because most DNA molecules consists of two *complementary polynucleotide strands*. The two strands are held together by *hydrogen bonds*

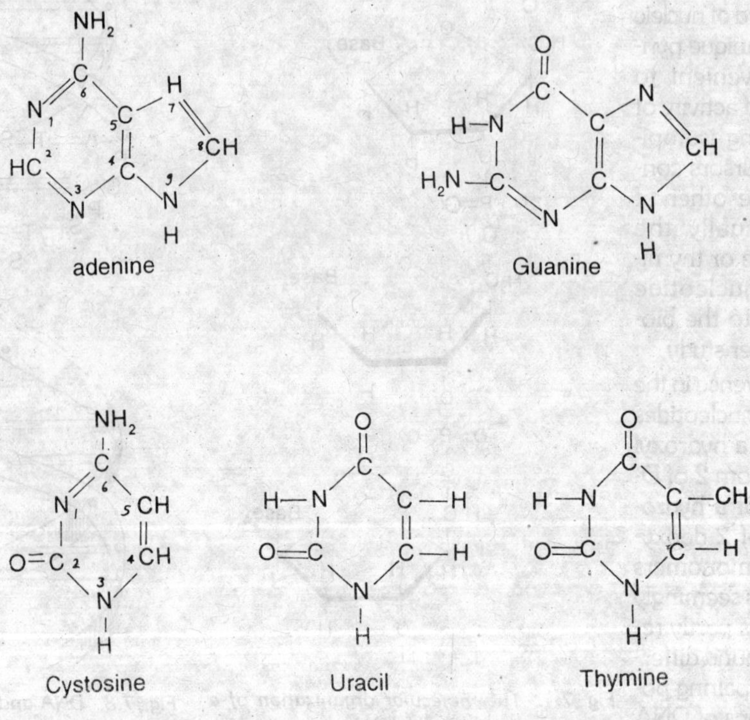

Fig. 7.10. Various heterogenous bases.

between complementary pairs of purines and pyrimidines. Adenine always bind with thymine while guanine always binds with cytosine. This repeated hydrogen bonding within the double helix structure and bonding between virtually all the surface atoms in the sugar and phosphate groups with water molecules serve to stabilize the structure.

Since the purine-pyrimidine pairs are found in the center of the molecule, their flat surface can stack on top of each other and thereby limit their contact with water. In double-helical molecules, a regular structure is possible because the complementary base pairs are exactly the same size. Single polynucleotide chains could not have a regular backbone structure because pyrimidines are smaller than purines, which would cause the angle of helical rotation to vary with the sequence of bases.

DNA double helix molecules are very stable at physiological temperatures because: (1) disruption of the double helix breaks hydrogen bounds and brings hydrophobic purines and pyrimidines into contact with water, which is energetically unsatisfactory; and (2) there are many weak bonds within the DNA molecule, arranged so that most of them cannot break without many others breaking at the same time. Even though some hydrogen bonds may be broken by thermal motion, hydrogen bonds in the rest of the molecule remain intact and the molecule does not fall apart. In fact, when held together by more than ten nucleotide pairs the double helices are quite stable. At room temperature weak bonds is the stability of molecular shape, in proteins as well as in nucleic acids. At abnormally high temperatures there is more frequent breakage of weak bond, which become less stable as temperatures rise above physiological levels. Once a significant number of weak bonds have been broken, a protein or nucleic acid molecule usually loses its original form and changes to an inactive or denatured form.

8

ENZYMES

The cell is a highly sophisticated factory. It derives its raw materials either from some circulatory system or directly from its external environment. It produces a great variety of end production from these raw materials. A cell contains many thousands of different compounds present, ranging from simple inorganic ions to extremely large and complex molecules of DNA. Besides this production capacity, the cell possesses the ability to divide and reproduce itself, to replace and repair its basic structure continuously and ability to respond extremely rapidly to a variety of internal and external stimuli.

All the informations for the performance of these activities are inherited by the cell in the form of deoxyribonucleic acid and each cell makes its own machinery to carry them out. This machinery is the enzymes of the cell and the aim of this chapter is to show how enzymes are able to carry out the function alloted to them.

Term enzyme was first introduced by *W. Khune* in 1878, derived from Greek means 'in yeast'. *Berzelius,* was first who recognized and defined the nature of enzymes. In 1827, he proposed that forments were the catalysts produced by living cells. *J.B. Sumner* (1926) for the first time isolated the enzymes *urease* in crystalline form. He was awarded Noble Prize in 1946 for this work. Since that time about 2000 enzymes have been isolated in highly purified form and more than 150 have been isolated in crystelline forms. The enzymes can be defined as "soluble, collidal, organic, catalysts, which are produced by living cells but are capable of acting independently of the cells." All the enzymes are protein in nature without any exception and exhibit all the properties of proteins.

CLASSIFICATION OF ENZYMES

Enzymes are generally named by adding *ase* to the root indicating the substrate on which the enzyme acts. According to IUB (International Union of Biochemists) the enzymes are classified on the basis of their reaction specificity. There are six classes of the enzyme according to IUB.

1. *Oxidoreductase.* This class includes enzymes which are concerned with biological oxidation and reduction and therefore, concerned with respiration and fermentation processes. About 221 enzymes have been listed in this category.
2. *Transferases.* These enzymes catalyze the transfer of a group from one substance to another. The groups may be alkyl, glycosyl, aldehydes, sulphur or phosphorus.
3. *Hydrolases.* Enzymes, which catalyze the hydrolysis of their substrates by adding constituents of water across the bond they split, are called hydrolases. The substrates may be esters, peptides, glycosides etc.
4. *Lyases.* These enzymes catalyze removal of groups from substrates by mechanism other than hydrolysis leaving double bond in the substrate. The enzymes act on the following bonds: C-C, C-N, C-O, C-S, etc.

5. *Isomerases.* These enzymes catalyze reactions which brings about intramolecular rearrangement of atoms in the substrates.
6. *Ligases.* These are also called *Synthetases.* These enzymes catalyze the linking together of compounds utilizing energy made available by simultaneous breakdown of pyrophosphate bond of ATP or similar high energy bond compounds. These enzymes form C-O, C-S, C-C and C-N bonds.

NATURE OF ENZYMES

All the enzymes are proteins with one or more polypeptide chains or their derivatives. The molecular weight vary from 10,000 upto 50,000 daltons. Each enzyme is made up of amino acids. There are 20 amino acids which participate. These are linked together by peptide bonds. The sequence of amino acids is specific for a particular enzyme and determines the properties of the enzymes.

The enzymes are of two types:

(i) Simple Proteins

Some enzymes such as amylase, pepsin, trypsin and Chymotrypsin and called simple proteins. These on hydrolysis yield only amino acids.

(ii) Conjugated Proteins

Other enzymes are conjugated or derived proteins formed by the combination with other groups (non-proteins) called *cofactors.* Usually the enzyme complex is called *holoenzyme.* Thus the protein part characterized as heat labiie and colloidal is called *apoenzyme* and the heat stable and dialysable part as *co-factor*. Three types of co-factors may be identified, *prosthetic group, coenzyme* and *metal ions.*

(a) Prosthetic group

The prosthetic group is organic compound. These are permanently bonded to the enzymes. If the prosthetic group is removed, the enzyme activity is lost e.g. catalase and peroxidase.

(b) Coenzymes

These are also organic compounds but their association with the apoenzyme is transient, usually occuring only during the course of catalysis. The coenzyme molecule may serve as the cofactor in a number of different enzymes catalyzed reactions. The essential chemical component of many coenzymes are vitamins. The B group vitamins (except biotin and lipoic acid) function as part of coenzyme such as Co-A is a derivative of panthothenic acid (B_3), flavin adenine dinucleotide (FAD) contains riboflavin (B_2), Thiamine pyrophosphate contains riboflavin (B_2), Thiamine pyrophosphate contains thiamine (B_1), Nicotinamide adenine dinucleotide (NAD) and Nicotinamide adenine dinucleotide phosphate (NADP) contains the vitamin niacin (B_5).

(c) Metalic ions

A number of enzymes require metal ions for their activity. These ions form co-ordination bonds with specific side chains at the active site and at the same time form one or more co-ordinate bonds with the substrate. The common ions are Zn^{++}, Mg^{++}, Cu^{++}, Fe^{++}, Co^{++} etc.

General Properties of Enzymes

1. ***Protein Nature.*** All the enzymes are protein in nature. The protein structure provides for weak associations due to side groups of amino acids within it. The enzyme-substrate complex involves the association of the enzyme and substrate by weak associations rather than covalent bondings.
2. ***Colloidal Nautre.*** They are colloidal in nature due to which they present a large surface area for reactions to take place.
3. ***Required in small amount.*** A small amount of enzymes is required by biological systems because the enzyme is reused. It has been estimated that a single enzyme can act upon 5,00,000 substrate molecules per minute. This value is called *turn over number.*
4. ***Remain unaltered at the end.*** At the end of a specific reaction, the enzymes remain unchanged.

5. ***Accelerate the rate of reaction.*** An enzyme changes the rate of reaction only, not the direction of the equilibrium. A reaction will proceed at a faster rate in the presence of an enyzmes, but the equilibrium between substrate and product will remain constant with or without the intervention of the enzyme.
6. ***Reversibility of reactions.*** Like a true catalyst, enzymes have been found to accelerate the chemical reaction in either direction i.e., onwards and the backwards depending upon the availability of suitable energy source.
7. ***Denaturation.*** If the natural conditions are changed. The enzymes undergo denaturation which is accompanied by a parallel loss of enzymic activity.
8. ***Enzyme specificity.*** Enzymes are highly specific in nature i.e., a particular enzymes can catalyze only a particular type of reaction. Thus specificity is an important criterion of enzyme action. The specificity may be of following types:

(a) *Reaction specificity.* Different enzymes bring out different reactions on same substrate.

(b) *Substrate specificity.* This is of following two types

(i) *Absolute specificity.* A particular enzyme acts on a particular substrate.

(ii) *Group specificity.* Several enzyme hydrolyse polypeptide, but each has a preference for peptide linkages involving specific sites of polypeptide chains or bonds involving specific amino acids.

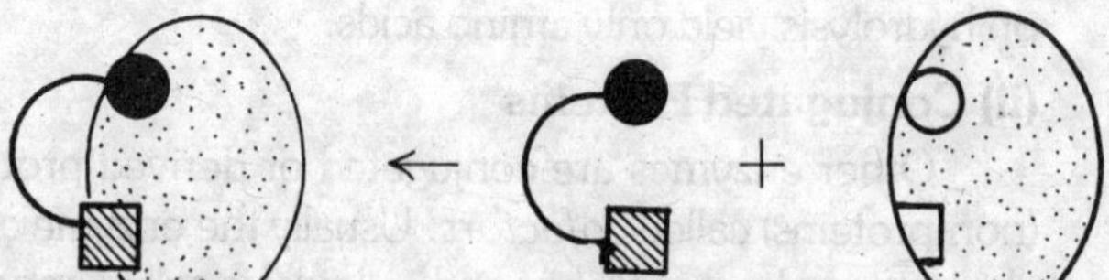

Fig. 8.1. Combination of Enzyme with the substrate in form ES Complex according to Fisher's hypothesis.

(iii) *Stereospecificity or optical specificity.*

An enzyme may have optical specificity for a D or L optical isomers. Thus *L-amino acid oxidase* attacks only L-amino acid and not the D-amino acid.

Mechanism of Enzyme Action

The enzyme speeds up a given reaction, but remains unchanged at the end of the reaction. The most recent view regarding mechanism of enzyme action is that the enzyme combines with the substrate (on which enzyme acts) to form an *enzyme-substrate complex* (*Michaelis* and *Menten,* 1931). For this complex the energy is liberated and the substrate is broken down into the products of reaction.

$$\text{E enzyme (E) + Substrate (S)} \xrightarrow{\text{binding}} \underset{\text{(ES)}}{\text{Enzyme – Substrate}}$$

Catalysis

$$\text{Products (P) + Enzyme (E)} \xleftarrow{\text{release}} \underset{\text{(EP)}}{\text{Enzyme – Products}}$$

The reaction takes place in three steps i.e., binding, catalysis and release. The ES is also known as *Michaelis Complex.*

Most of the catalyzed reactions are reversible in nature i.e., the same enzyme can catalyse a particular reaction in the forward as well as in backward direction. The equilibrium of the reactions may be shifted towards either direction at different times depending upon the concentration of the substrate or products in the reaction medium.

Enzymes Reduce the Energy of Activation

Energy of activation is that minimal amount of energy, which is required by a molecule to take part in a reaction. It is designated by $\Delta E\neq$ or $\Delta G\neq$. For example, decomposition of H_2O_2 (hydrogen peroxide) without a catalyst has an energy of activation of some 18,000 cal/mole. When the enzyme catalyse is added, it is less than 2,000 Cal/mole.

Enzymes accelerate reactions by reducing free-energy of activation, this may be interpretated in term of a *transition state theory* illustrated in the figure. The chemical compound A is in a metastable state and requires energy of activation E to convert the compound in a new activated state of B. The reaction then becomes spontaneous, converting B into C and energy E is recovered. A further amount of energy is released according to the difference of enthalpy (*H) of A and C. Thus enzymes increase the rate of conversion of A into C with a lower energy activation (e).

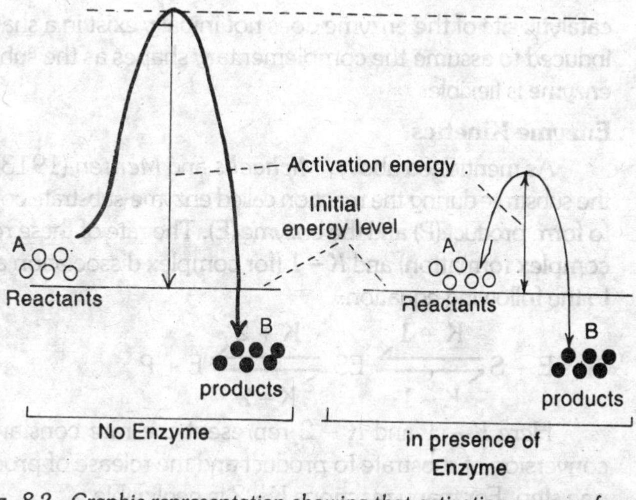

Fig. 8.2. Graphic representation showing that activation energy of an enzyme catalysed reaction is lower than that of an uncatalysed reaction.

The Catalytic Site

Biochemists are of the opinion that some restricted regions the large sized enzyme is concerned with the process of catalysis. This region was termed as the *active site*. Nowadays, scientists refer to the *catalytic* site, since it is now known that other sites of enzymes are also 'active'.

The catalytic sites exert a binding force on the substrate molecule both by hydrophilic and hydrophobic groups. In most of the cases the enzyme substrate formation occurs by the multiple bonding with the substrate. This may involve formation of covalent bonds, hydrogen bonds or electrostatic bonds.

The sites of catalysis and binding of substrate have been reported to lie at different places in the enzyme molecules which remain quite apart from each other in the absence of substrate. However, in the presence of substrate both the sites come closer to each other in such a way that the substrate fits into enzyme in a most convenient manner. If the tertiary and quaternary structure of enzyme protein is destroyed, the enzyme property of protein is lost. The enzyme besides possessing site for the actual substrate, does also possess certain sites for attachment of coenzyme or cofactor (in the enzymes which require these additional factors for their activity) During the course of activity of such enzymes, *enzyme-substrate-cofactor* complexes are formed instead of enzyme-substrate complexes.

Interaction of Enzyme-substrate

The mechanism of enzyme - action has been explained by two theories:

1. Lock and key theory.
2. Induced-fit theory.

1. Lock and Key theory

This theory was proposed by *Emil Fisher* in 1890. According to this theory the enzyme and the substrate have a complementary shape. It was concluded that a very tight fit between enzyme-substrate complex would be required. The theory can be explained easily by the fact that a particular lock can be opened by a particular key. Similarly enzymes have specific sites whose a particular substrate can only by attached.

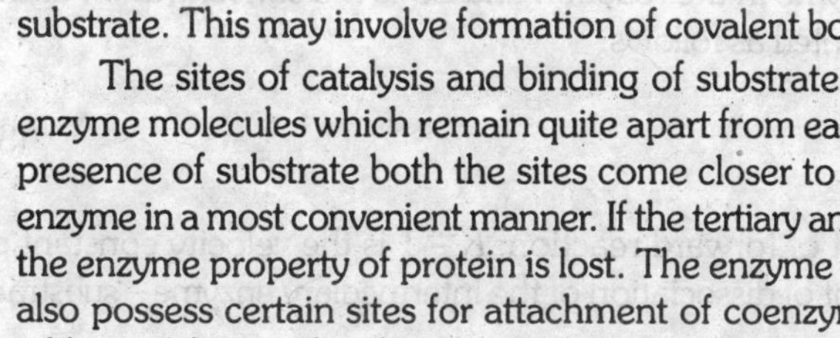

Fig. 8.3. Conformational change in the enzyme as a result of the entry of substrate(s) into the active site; according to Koshland.

2. Induced -fit theory

In recent years the lock and key theory has been given up and a new theory has been proposed. It is called induced-fit theory proposed by *Daniel Koshland* (1960). According to this theory the catalytic site of the enzyme does not initially exist in a shape that is complementary to the substrate but is induced to assume the complementary shapes as the substrate bound. Thus according to this theory the enzyme is flexible.

Enzyme Kinetics

As mentioned above, *Michaelis* and *Menten* (1913) postulated that, enzyme forms a complex with the substrate during the reaction called enzyme-substrate complex. The complex then undergoes dissociation to form product (P) and the enzyme (E). The rate of these reactions are governed by constant *K* + 1 (For the complex formation) and *K* – 1 (for complex dissociation and *K* + 2 (for the product formation) according to the following equation:

$$E + S \underset{K-1}{\overset{K+1}{\rightleftharpoons}} ES \underset{K-2}{\overset{K+2}{\rightleftharpoons}} E + P \qquad (1)$$

Here *K* – 1 and *K* – 2 represent the rate constants of the reverse reactions. For simplicity the conversion of substrate to product and the release of product from the enzyme have been combined into one step. For many reactions *K* – 2 is negligable.

The dissociation constant *Km* of the enzyme-substrate complex is equal to:

$$Km = \frac{(E)\,(S)}{(ES)} \qquad (2)$$

Figures in brackets indicate their molar concentration. If *Eo* is the initial total concentration of the enzyme and *E* is the concentration of the free enzyme in the reaction and *ES* is the concentration of the bound enzyme the above equation may be represented as follows:

$$Km = \frac{(Eo\text{–}ES)\,(S)}{(ES)} \qquad (3)$$

In equation (1), *K* – 1 is the velocity constant of forward reaction, *K* – 1 is the velocity constant of backward reaction and *K* + 2 is the velocity constant of dissociation of the intermediary enzyme – substrate complex into enzyme and products

Hence, rate of forward reaction = (Eo – ES) (S) K + 1

Rate of backward reaction = (ES) K – 1

Role of dissociation of (ES) leading to product formation + [ES] K + 2

when the reaction are in equilibrium

$$(Eo - ES)\,(S)\,K + 1 = (ES)\,K - 1 + (ES)\,K + 2$$
$$= (ES)\,(K - 1 + K + 2)$$

$$\text{or } \frac{(Eo - ES)\,(S)}{ES} = \frac{K - 1 + K + 2}{K + 1} = Km \qquad (4)$$

In actual practice, it is very difficult to get a pure enzyme, hence its molecular weight determination is quite difficult. It is, therefore, better to eleminate (*Eo*) and (*ES*) from the above equation.

The reaction velocity for dissociation (*ES*) 'v' will be proportional to its concentration in the system.

$$\upsilon \propto (ES)$$
$$= K\,(ES)$$

$$\text{or } (ES) = \frac{\upsilon}{K} \qquad (5)$$

By placing (V/K) for (ES) in equation no, 4 we get.

$$Km = \frac{(Eo - ES)(S)}{(ES)} = \frac{(Eo - v/K)(s)}{v/K} \tag{6}$$

If there is large excess of substrate, (S) will be very much greater than (Eo) and whole of the enzyme will be converted into ES and (ES) will be equal to (Eo). Under such conditions, reaction velocity will be maximum and equal to V

$$V \alpha (Eo)$$
$$= K (Eo)$$
$$\text{or } (Eo) = V/K \tag{7}$$

Dividing (5) by (7) we get

$$(ES)(ES) = \frac{\upsilon/K}{V/K} = \upsilon/V \tag{8}$$

Now equation no. 5 can be further simplified as

$$Km = \frac{(Eo - ES)(S)}{(ES)}$$

$$\text{or } Km(ES) = (Eo - ES)(S) = (Eo)(S) - (ES)(S) \tag{9}$$
$$\text{or } Km(ES) + (ES)(S) = (Eo)(S)$$
$$\text{or } (ES)(Km + S) = (S)(Eo)$$

$$\text{or } \frac{(ES)(Km + S)}{(Eo)} = (S) \tag{10}$$

By substituting the value of ES/Eo in equation 10

$$\frac{\upsilon}{V}(Km + S) = (S)$$

$$\text{Therefore, } v\ (Km + S) = V(S)$$
$$\text{or } Km + (S) = (V/v)(S)$$
$$\text{or } Km = (V/v)(S) - (S)$$
$$= (S)(V/v - 1) \tag{11}$$

In equation no. 11, V is the limiting velocity, υ is reaction velocity and S is concentration of the substrate used.

In equation no, 11, when the ratio of limiting and observed velocities (V/v) equals to 2, then this equation may be written as

$$Km = (S)(2 - 1)$$
$$= S \tag{12}$$

Thus, the substrate concentration, which is expressed in moles per litre, required for half attainment of the maximum activity of an enzyme is a characteristic constant of that enzyme and is known as its *Km*. The *Km* is known as *Michaelis constant.* Km Value is very much helpful in evaluating affinity of the enzyme for a substrate. If Km value for a substrate of an enzyme is low, this indicates high affinity or *vice versa*. If also gives an idea regarding the type of inhibition caused by an inhibitor.

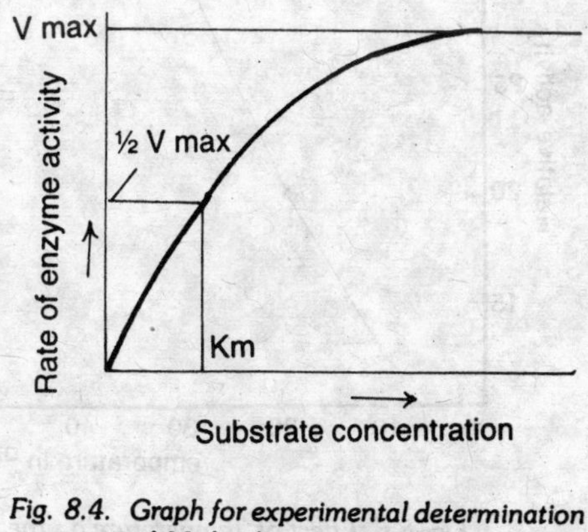

Fig. 8.4. Graph for experimental determination of Michaelis constant (Km).

Michaelis equation no. 11 may be rewritten in the following form

$$v = \frac{V(S)}{Km = (S)} \qquad (13)$$

$$\frac{1}{v} = \frac{Km + (S)}{V(S)}$$

Separating the components of numerator on the right side the equation gives

$$= \frac{1}{V} + \frac{Km}{V} \quad \frac{1}{(S)} + \frac{1}{V(S)}$$

which simplifies to

$$= \frac{1}{V} = \frac{Km}{V} \quad \frac{1}{(S)} + \frac{1}{V(S)} \qquad (14)$$

Equation (14) is a transformation of Michaelis — Mention equation called the *Lineweaver Burke equation*. For enzymes obeying the Michaelis—Mention relationship exactly, a plot of 1/v against 1/(E) yields a straight line. This line will have a slope of Km/v an entercept of 1/V on the 1/v axis and an intercep—1/Km on the 1/(S) axis.

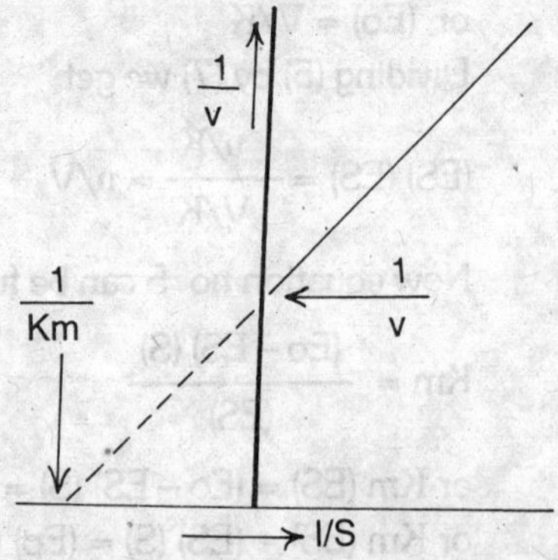

Fig. 8.5. Graphic representation of Lineweaver-Burk plot.

The Michaelis-Mention equation is basic to all aspects of kinetics of enzyme action. If we know Km and V we can calculate the reaction rate of an enzyme at any given concentration of its substrate.

Factor Affecting Enzyme Activity

1. Temperature

Between a range of 5°C to 40°C the activity of an enzyme is increased. When the temperature is raised by 10°C, the reaction speed increase 2 or 3 times. Most enzymes operate at their best in a temperature ranging between 25-40°C after that denaturation take place, as however, some enzymes present in dry tissues such seeds and spores are very resistant to high temperature.

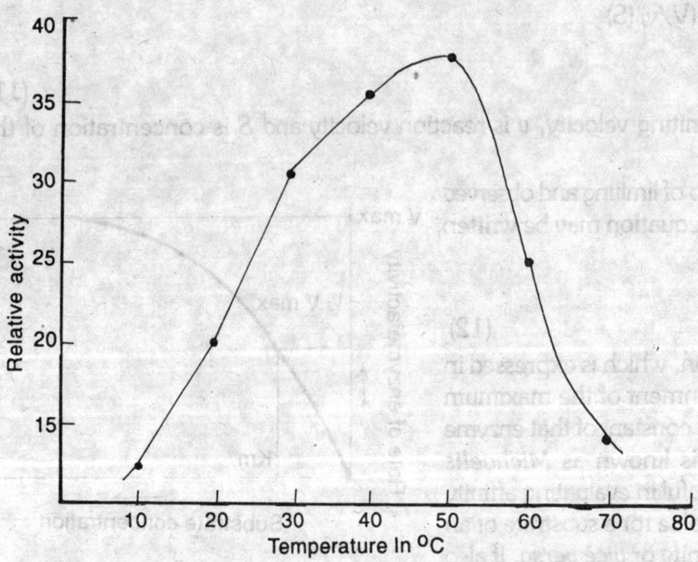

Fig. 8.6. Effect of Temperature on the rate of Enzyme Activity.

2. Hydrogen Ion Concentration (pH)

Because of their protein nature, enzymes are very sensitive to change

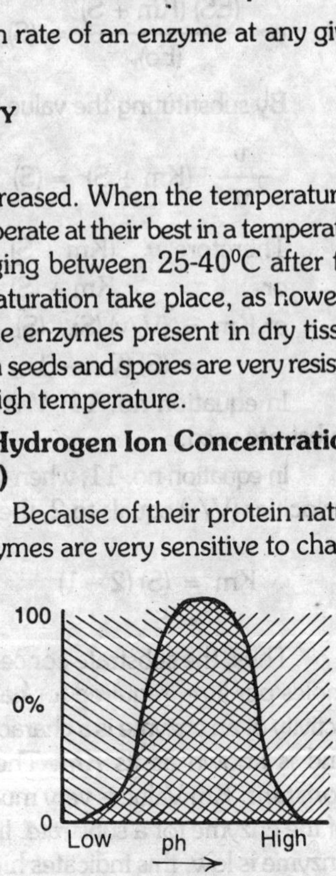

Fig. 8.7. Effect of pH on enzyme activity.

in *pH*. Each enzyme has a characteristic optimum *pH* at which its activity is maximum. The *pH* activity profiles to enzymes reflect the pH at which important proton—accepting or proton—donating groups in the enzyme catalytic sites are in their required state of ionisation. Some enzymes have two optima of pH or on sharp *pH* but a range of *pH*. Generally, the optimum pH lies between 4 and 9.

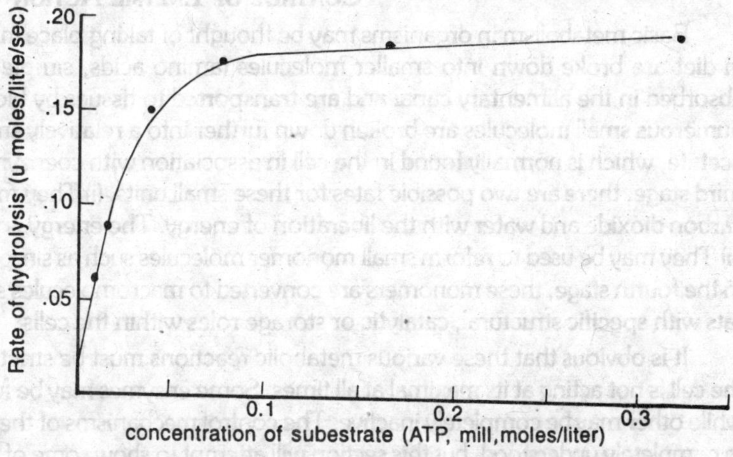

Fig. 8.8. Effect of substrate concentration on the rate of reaction.

3. Substrate Concentration

Enzyme activity increases with increase in concentration of substrate to a maximum and than it levels off. Some enzymes are inhibited by high concentrations of substrate.

Two types of graphs are obtained with substrate concentration and enzyme activity plot. Michaelis type of curve of straight line curve is obtained with the enzymes which are made up of one unit. Sigmoid type of curve is obtained when enzyme which is made up of the several subunits and show interaction amongst subunits.

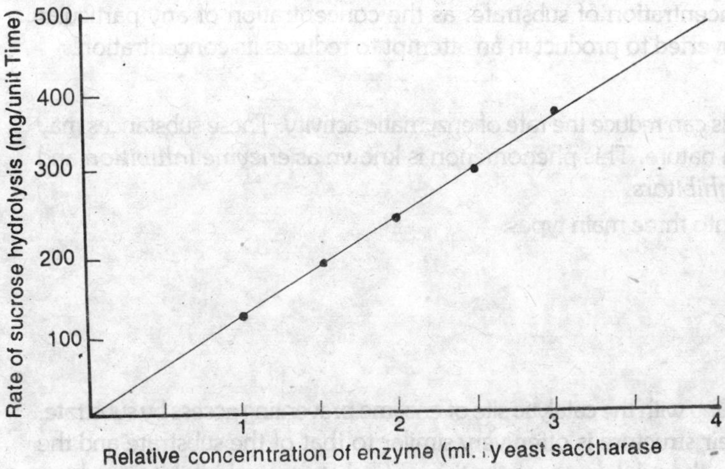

Fig. 8.9. Effect of Enzyme Concentration on the rate of reaction when an excess of substrate is present.

4. Enzyme Concentration

If an excess of substrate is present, doubling the enzyme concentration usually doubles the rate of reaction. This phenomenon usually applies only at the start of the reaction for the end product of the reaction often have an inhibitory effect on the enzyme and decreases its efficiency.

5. Effect of Ions

Many enzymes remain inactive they only become active in the presence of cations such as Mg^{++}, Ca^{++}, Mn^{+}, Zn^{+}, Ma^{+} or K^{+}. In some cases, the cations are loosely bound to the enzyme, while in other cases, they bind to the substrate. Anions are also effective to increase the activity of enzyme. For example, Cl– (chloride) enhances the activity of salivary amylase.

6. Redox Potential

The enzymes are sensitive to redox potential of the cell. Enzymes such as oxidising and reducing are involved in changing the redox components of a cell and hence they are influenced by redox potential. It is due to the presence of readily oxidisable –SH group in their molecule.

7. Effect of Ultra-violet Length

The U.V. light has inactivating effect upon the enzyme activity.

8. Effect of End Products

If the end products of the reactions are allowed to accumulate then these will inhibit the activity of enzyme (negative feed back). It has also been suggested that the end product may combine with the enzyme at a site other than catalytic site and cause a conformational change in the enzyme protein. It has been referred to as *allosteric effect.*

Control of Enzyme Action

Basic metabolism in organisms may be thought of taking place in four stages. In first stage, nutrients in diet are broke down into smaller molecules (amino acids, simple sugars and fatty acids) which are absorbed in the alimentary canal and are transported to tissues by blood stream. In second stage, these numerous small molecules are broken down further into a relatively small number of simple units such as acetate, which is normally found in the cell in association with coenzyme A (CoA) forming acetyl CoA. In third stage, there are two possible fates for these small units. (i) They may either be oxidized completely to carbon dioxide and water with the liberation of energy. The energy is usually stored in the cell as ATP. or (ii) They may be used to reform small monomer molecules such as simple sugar, amino acid and fatty acids. In the fourth stage, these monomers are converted to macromolecules such as proteins, polysaccharides or fats with specific structural, catalytic or storage roles within the cells.

It is obvious that these various metabolic reactions must be strictly controlled. Every enzyme within the cell is not acting at its maximal at all times. Some enzymes may be functioning at greatly reduced levels, while other may be completely inactive. The control mechanisms of the cell are highly complex and, as yet, in completely understood, but this section will attempt to show some of the basic principles underlying such control.

1. Control by Substrate

This is the simplest type of control. Since the rate of an enzyme-catalyzed reaction is proportional at low substrate concentrations to the concentration of substrate, as the concentration of any particular substrate builts up, more of it will be converted to product in an attempt to reduces its concentration.

2. Control by Inhibitors

Some of the elements or compounds can reduce the rate of enzymatic activity. These substances may be of physiological or unphysiological in nature. This phenomenon is known as *enzyme inhibition* and such substances are known as *enzyme inhibitors.*

Enzyme inhibition may be divided into three main types:

(i) Competitive inhibition

(ii) Non-competitive inhibition

(iii) Allosteric inhibition.

(i) Competitive inhibition

Competitive inhibitors act by combining with the catalytic site of enzyme preventing access by substrate. Because of their competitive nature, their structure is often very similar to that of the substrate and the activity of the enzyme will depend upon the relative concentrations of substrate and inhibitor and the relative binding affinities of these molecules for its catalytic site. Enzyme diphosphoglycerate mutase is an enzyme in the metabolism of glucose in RBCs and catalyses the inter conversion of 1, 3-diphosphoglycerate and 2, 3-diphosphoglycerate

$$\begin{matrix} \text{COOP} \\ \text{HO—C—H} \\ \text{CH}_2\text{OP} \end{matrix} \rightleftharpoons \begin{matrix} \text{COO}^- \\ \text{PO—C—H} \\ \text{CH}_2\text{OP} \end{matrix}$$

It is competitively inhibited by low concentration of 2, 3-diphosphoglycerate and hence if this metabolite builds up, the interconversion is slowed.

(ii) Non-Competitive inhibition

The non-competitive inhibitors are equally important in metabolism. Here substrate and inhibitor do not exhibit mutual competition. Most probably, the sites of attachment of substrate and non-competitive inhibitor are different on the enzyme molecules. The inhibitor are most structurally dissimilar to the substrate.

If the inhibitor can be removed from its site of attachment preserving the activity of the enzyme, it is known as *reversible non-competitive inhibition*. If it cannot be removed at all, or only removed after loss of enzymic activity, it is known as *irreversible non-competitive inhibition*. Both reversible and irreversible inhibitions have been found to exhibit similar reaction kinetics.

(iii) Allosteric inhibitions

In mixed inhibitions, the inhibitor molecule attached to the enzyme prevents the breakdown of enzyme-substrate complex, and also at the same time, it interfers with the binding of the substrate of the enzyme. Thus, such inhibitors exhibit partially competitive and partially non-competitive kinetics. The sites other than the enzymatic sites, lying on the different region in the enzyme molecule where an inhibitor molecule may get attached is known as *allosteric site*. Presence of inhibitor on allosteric site changes the conformation around the catalytic site. Thus results into reduced enzyme-substrate complex formation. If the substrate molecules have preoccupied the catalytic site, it may affect conformation around the allosteric site thereby reducing enzyme-inhibitor complex formation and thus reduces the extent of inhibition.

3. Control by Covalent Modification Cascades

Some enzymes exist in the cell as inactive precursors. To activate these enzyme, it is necessary to remove a small peptide from the precursor form called a *zymogen*. For example, trypsinogen in duodenum is released as zymogen and a small amount is activated by another enzyme, enterokinase. Once a small quantity of active trypsin is produced, this in turn activates more trypsinogen and the whole process is autocatalytic. Chymotrypsinogen is another zymogen which is converted to chymotrypsin by trypsin. The net effect in which one enzyme acts to convert a zymogen to its active enzyme is termed as cascade.

ALLOSTERIC ENZYMES

These are also called as *regulatory enzymes*. The kinetic properties of these enzymes cannot be accounted for by the Michaelis—Menten model. When a product is formed after a series of enzymatic reactions and it influences the activity of first enzyme by which the rate of product formation itself is controlled, under such circumstances. The first enzyme is called an *allosteric enzyme*.

$$a \xrightarrow{E_1} b \xrightarrow{E_2} c \xrightarrow{E_3} d \xrightarrow{E_4} e \xrightarrow{E_5} P \qquad \text{(Product)}$$

If accumulation of product P influences the activity of enzyme E_1 then E_1 is known as allosteric enzyme.

ISOENZYMES

It is now known that some of the enzymes exist in many physiologically distinct forms and all the forms of a single enzyme can catalyse the same reaction. The only differences noted in the different forms of a single enzyme are that of electrophoretic mobility and possible of the quaternary structure of the enzyme protein. Such forms of a single enzyme, having same catalytic activity are known as *isoenzymes*. According to International Union of Biochemists (IUB) the term isoenzyme has been described as the multiple forms of an enzyme occuring within a single species. About 100 enzymes at present are known to exist in such multiple forms. For example *lactic dehydrogenase* of human and animal tissues have been shown to exist in five isoenzymatic forms, namely LD_1, LD_2, LD_3, LD_4, and LD_5 and their relative distribution varies considerably from tissue to tissue and from species to species.

Isoenzymes may be *homologous* or *analogous*. Homologous isoenzymes have essentially similar molecular structure and catalytic activity. Cytochrome –C from different animals is the example of homologous isoenzymes. Analogous isoenzymes have similar reactions but different molecular structure and catalytic properties.

ANTIENZYMES

The wall of alimentary canal is not digested by the proteolytic enzymes secreted by itself. This has led to the discovery of *antienzymes*. It has been reported that such tissues possess certain proteins having antagonistic property to the proteolytic enzyme.

9

Endoplasmic Reticulum

Endoplasmic reticulum was first of all observed in 1945 by *Porter, Claude* and *Fullam*. They noted the presence of a network or recticulum of strands associated with vesicle-like bodies in the cytoplasm of the cultural fibroblast or thinly-shaped tissue culture cells. Further electron microscopy by *Porter* and *Thompson* (1947) has revealed that these strands of reticulum are vesicular bodies inter-connected, so as to form a complex network in the inner endoplasmic part of the cytoplasm. As this network is concentrated in the endoplasm of the cell more than in the ectoplasm, therefore, it is known as *endoplasmic reticulum* (ER), or *ergastoplasm* or *vaculolar system* of the cell. The endoplasmic reticulum is not visible in the cytoplasm of a living cell under the phase-contrast microscope but the observations by electron microscope further confirmed the presence of endoplasmic recticulum as reported by *Porter* and his colleagues. Recent studies have further confirmed and accepted the concept of a structural organization of cytoplasm. Recently, under the phase-contrast microscope, *Fawcett* and *Ito* (1958), and *Rose* and *Pomerat* (1960) have studied the structure and distribution of endoplasmic reticulum in the living tissue-culture cells.

Occurrence

The endoplasmic reticulum occurs in all the eukaryotic cells except erythrocytes (R.B.Cs) of mammals. It is absent in prokaryotes. Its development varies considerably in various cell types. It is small and undifferentiated in eggs and in undifferentiated embryonic cells. Only a few vacuoles are present in the spermatocytes and muscle cells. However, it is highly organized in cells synthesizing proteins or in cells that are engaged in lipid metabolism.

Morphology of Endoplasmic Reticulum

The endoplasmic recticulum has been found in all kinds of mature cells except the mature mammalian erythrocyte, which is also devoid of a nucleus. Actually the first description of these structures seemed with electron microscope by *Porter, Claude* and *Fullam* in 1945 in cultured cells. These are membrane bounded sacs in the form of double membranes (cisternae) by *Sjostrand* (1953) or the name *cisternae* was given by *Sjostrand* and the name *tubules* was given by *Kurosumi* (1954). Rounded and irregular *sacs* or *vesicles* were observed by *Weiss* 1953.

Morphologically the endoplasmic reticulum is composed of following three kinds of structures, viz,. 1. cisternae, 2. vesicles and 3. tubules.

Cisternae or Lamellae

They are long, flattened and usually unbranched tubules, which are arranged in parallel arrays. They are of uniform width throughout and their thickness varies from 40-50 μm. This pattern of reticulum is characteristic of basophilic regins of the cytoplasm and of those cells, which are active in protein synthesis. Lamellae or cisternae occur in the liver-cells, plasma-cells, brain-cells and in notochord cells etc.

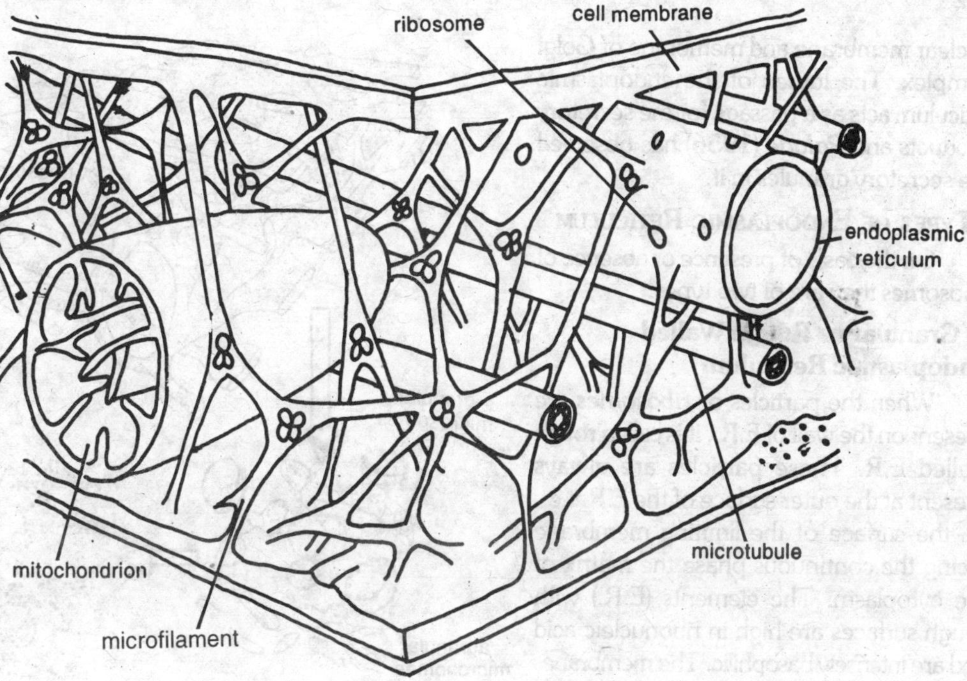

Fig. 9.1. The complex cytoskeleton of an eukaryotic cell.

Tubules

The tubules are small, smooth-walled, branched tubular spaces having a diameter of about 50-190 mμ. These occur in cells that are busy in the synthesis of steroids like cholesterol, glycerides and hormones. These are haphazardly arranged in the cytoplasm of developing spermatids of guinea pig, muscles cells and other non-secretory cells.

Vesicles

The vesicles range in diameter from 25 to 500 mμ and are for the most part rounded in shape. These are abundant in the cells engaged in the protein synthesis as in hepatic and pancreatic cells.

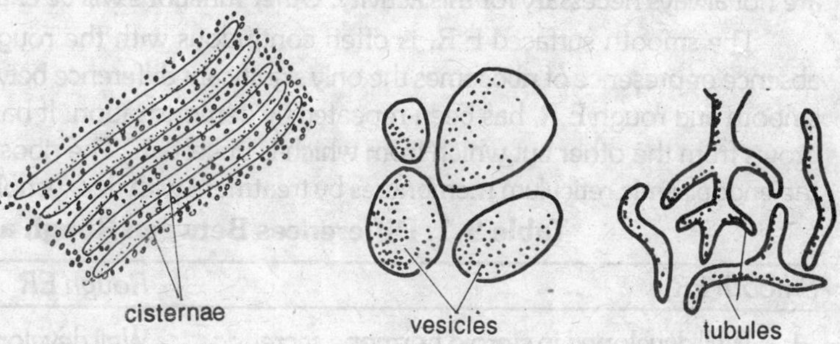

Fig. 9.2. Various components of the endoplasmic reticulum.

All these three patterns of endoplasmic reticulum may occur in the same cell or in different cells. Their arrangement also differs in different cells viz., as parallel rows in the liver-cells of mammals; haphazardly in panceratic cells or in the form of a net-work of tubules in striated muscle-cells. In notochordal cells of *Ambystoma* larva, the pattern of cisternae is of still another type.

Ultrastructure of Endoplasmic Reticulum

All the three structures of the endoplasmic reticulum are bounded by a thin membrane of 50 to 60 Å thick. Like the plasma membrane, nucleus, etc., its membrane is also formed of three layers—the outer and inner dense layers are composed of protien molecules, and the two middle thin and transparent layers are of phospholipids. The endoplasmic reticular membrane is continuous with the plasma membrane, nuclear membrane and membrane of Golgi complex. The lumen of the endoplasmic reticulum acts as a passage for the secretory products and *Palade* (1956) has observed the secretory granules in it.

TYPES OF ENDOPLASMIC RETICULUM

On the basis of presence or absence of ribosomes they are of two types:

(i) Granular or Rough Walled Endoplasmic Reticulum

When the particles or *ribosomes* are present on the wall of E.R., it is called rough walled E.R. These particles are always present at the outer surface of the E.R. i.e., on the surface of the limiting membrane facing the continuous phase the matrix of the cytoplasm. The elements (E.R.) with rough surfaces are high in ribonucleic acid and are intensely basophilic. The membranes themselves are not rough, but associated with their outer surfaces are tiny particulate components 100 to 150Å in diameter. These are called *ribonucleoprotein* (RNP) *particles* or *ribosomes* and contain as the average 40% RNA and 60% protein. The elements having ribosomes are usually of the cisternal type and are found in cells active in protein synthesis. Bichemical studies have indicated that the ribosomes are important in protein synthesis, even though the membranes are not always necessary for this activity. Other functions will be explained later.

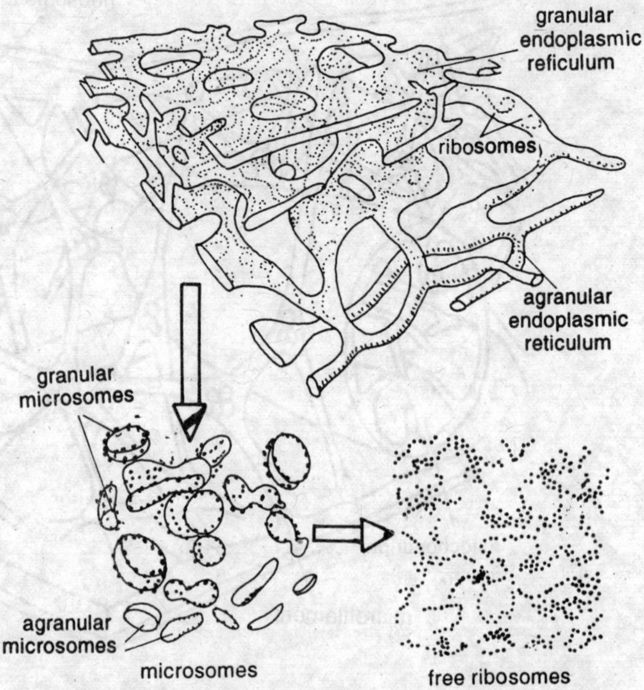

Fig. 9.3. Three-dimensional structure of endoplasmic reticulum showing microsomes and ribosomes.

The smooth surfaced E.R. is often continuous with the rough surfaced E.R. thereby making the absence or presence of ribosomes the only significant difference between the two. The continuity between smooth and rough E.R. has been repeated by demonstration. It has been suggested more over that one grows from the other but which from which is uncertain. The ribosomes can readily be dissociated from the endoplasmic reticulum membranes by treatment with deoxycholate.

Table 9.1. Differences Between Rough and Smooth ER

	Smooth ER	*Rough ER*
1.	Well developed in steroid hormone secreting cells.	Well developed in protein secreting cells.
2.	It tends to be tubular.	It tends to be cisternal.
3.	It is less stable and autolysis readily, after the death of the cell.	It is comparatively more stable. It may persist for sometimes.
4.	It is devoid of ribosomes.	Ribosomes are found associated.

(ii) Smooth Walled Endoplasmic Reticulum

The name smooth walled is given to that portion of endoplasmic reticulum that is devoid of ribosomes. Like the rough walled endoplasmic reticulum smooth form shows a characteristic morphology which is tubular rather than cisternae. The smooth walled endoplasmic reticulum is found in the cells that are active in the synthesis of steroid compounds such as cholesterol, glycerides, and the hormones (testosterone and

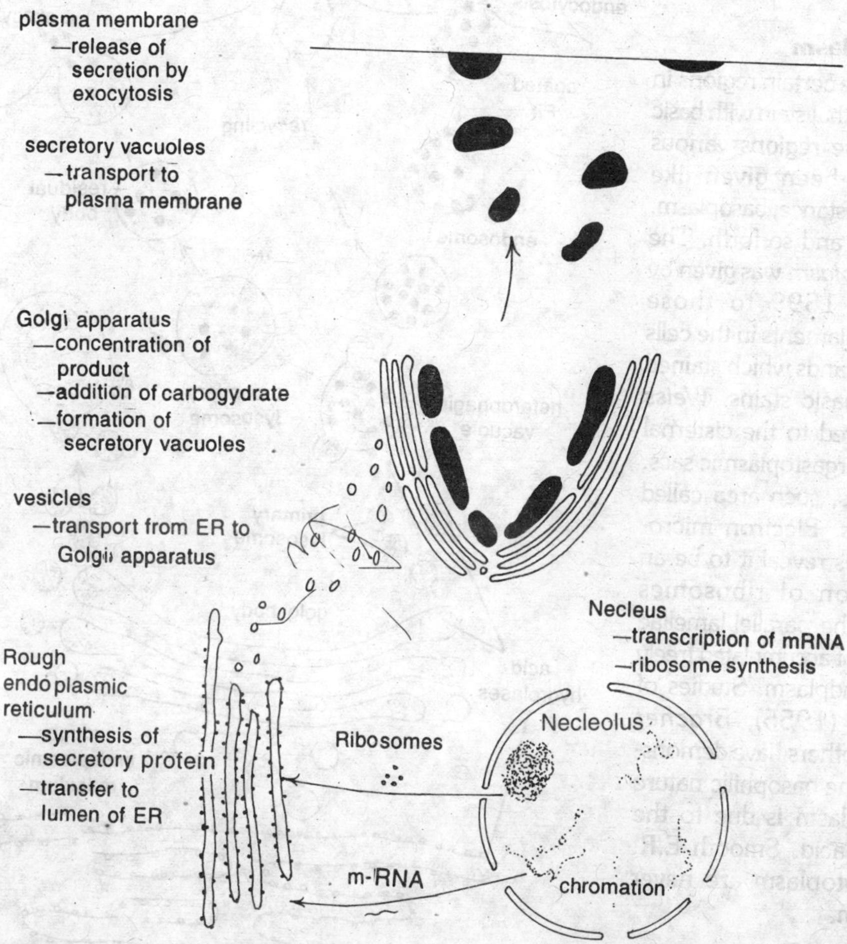

Fig. 9.4. A diagram summarizing the steps in secretion of a protein or a glycoprotein in an exocrine gland cell.

porgesterone). It is studied by *Fawcett* (1960) that they are also present in the pigmented epithelial cells of the retina which are involved in the metabolism of vitamin A in the production of visual pigment. Glycogen storing cells of the liver contain the smooth, tubular elements of the endoplasmic reticulum.

Modifications of Endoplasmic Reticulum

(a) Sarcoplasmic Reticulum

Sarcoplasmic reticulum, found in the skeletal and cardiac muscles is a highly modified form of smooth ER. It was first reported by *Veratti* (1902) as delicate plexuses in skeletal muscles surrounding the myofibrils. Electron microscopy showed it to be composed of a network of membrane like tubules which run longitudinally in the interfibrillar sarcoplasmic space for the length of each sarcomere. At the level of H and I bands, these tubules merge with large cisternal structures. At the H band level this cisterna, called the central cisterna, forms a sieve-like structure round the myofibrils. At the level of I band, these tubules merge with the large terminal cisternae, from whch transverse tubules extend peripherally to the sarcolemma and are continuous with and are deep invaginations of it. It is generally believed that sarcoplasmic reticulum plays a role not only in distributing energy-rich material needed for muscular contraction but also in providing the necessary channels for transmiting impulses along the surface and conveying the action potenital from the surface to the myofibrils within. Moreover, they store calcium ions during relaxation of muscles.

(B) Ergastoplasm

There are certain regions in the cytoplasm that stain with basic dyes. To these regions various names have been given like chromidial substance, basoplasm, ergastoplasm and so forth. The term *ergastoplasm* was given by *Granier* in 1899 to those cytoplasmic filaments in the cells of exocrine glands which stained readily withbasic stains. *Weiss* (1953) referred to the cisternal elements as ergastoplasmic sacs. In nerve cells, such area called *Nissl bodies.* Electron microscopic studies reveal it to be an accumulation of ribosomes situated on the parallel lamellae of ER stalks of accumulated freely in the groundplasm. Studies of *Casperson* (1955), *Brachet* (1957) and others have demonstrated that the basophilic nature of ergastoplasm is due to the ribonucleic acid. Smooth E.R. areas of cytoplasm are never ergastoplasm.

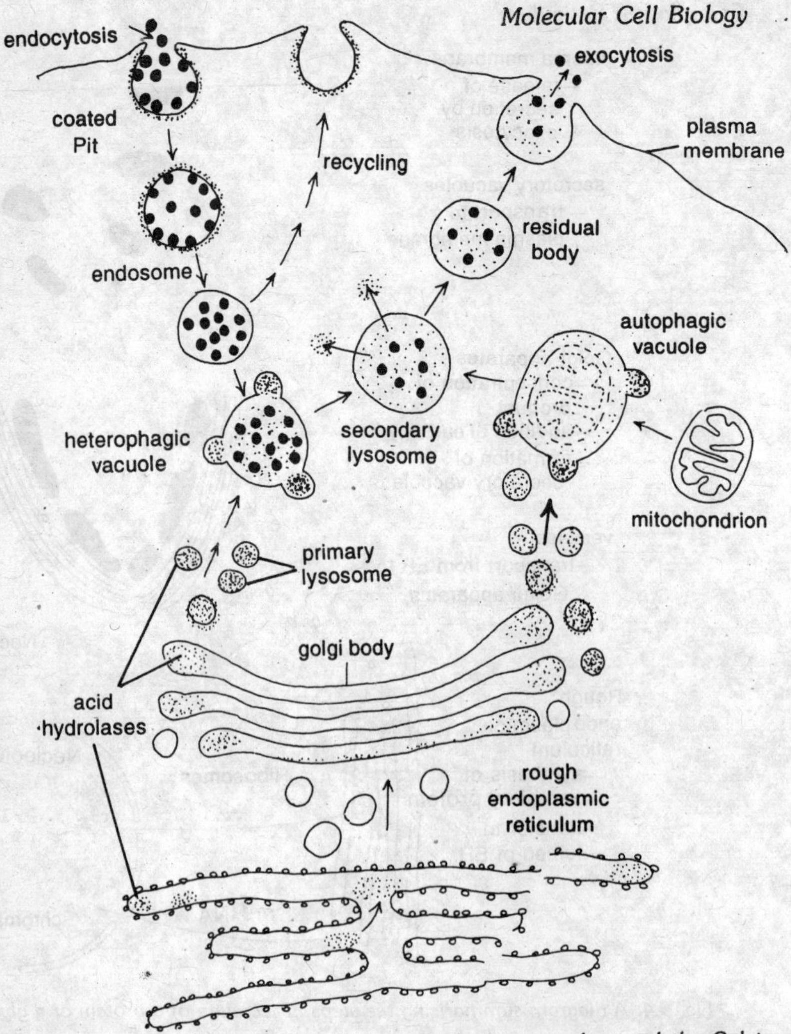

Fig. 9.5. Summary of the role of the endoplasmic reticulum and the Golgi complex in the synthesis of membrane-bond products.

Isolation of Endoplasmic Reticulum

Endoplasmic reticulum can also be isolated mechanically with the help of centrifuge. When tissue or cells are disrupted by homogenization the E.R. is fragmented into many smaller closed vesicles called *microsomes* (100 nm diameter), which are relatively easy to purify.

Microsomes derived from rough E.R. are studied with ribosomes and are called rough microsomes. Many vesicles of size similar to that of rough microsomes, but lacking attached ribosomes, are also found in these homogenates. Such smooth microsomes are derived in part from smooth portions of the ER and in part from vesiculated fragments of plasma membrane, Golgi complex and mitochondria (the ratio depending on the tissue). Thus, while rough microsomes can be equated with rough portions of ER, the origins of smooth microsomes cannot be so easily assigned. An outstanding exception is the liver. Because of the exceedingly large quantities of smooth ER in the hepatocyte, most of the smooth microsomes in liver homogenates are derived from smooth ER.

Ribosomes, which contain large amounts of RNA, make rough microsomes more dense than smooth microsomes. As a result, the rough and smooth microsomes can be separated from each other by sedimenting the mixture to equilibrium in sucrose density gradients. When the separated rough and smooth microsomes of a tissue such as liver are compared with respect to such properties as enzyme activity or polypeptide composition, they are remarkably similar, although not identical. It, therefore, seems that most of the components of the ER membrane can diffuse frelly between rough and smooth regions of the E.R. membrane, as would be expected for fluid, continuous membrane system.

ENZYMES OF THE ER MEMBRANES

The membranes of the endoplasmic reticulum are found to contain many kinds of enzymes which are needed for various important synthetic activities. The most important enzymes are the stearases, NADH-cytochrome—C—reductase, NADH diaphorase, glucose-6-phosphotase-and-Mg^{++} activiated ATPase. Certain enzymes of the endoplasmic reticulum such as nucleotide diphosphatases are involved in the biosynthesis of phospholipid, ascorbic acid, glucuronide, steroids and hexose metabolism. The enzymes of the endoplasmic reticulum perform the following important functions:

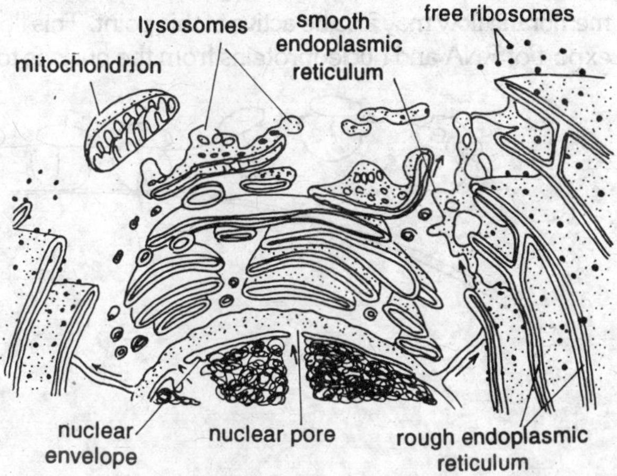

Fig. 9.6. Three-dimensional representation of cytoplasmic membrane systems and their relationship (i.e. ER, Golgi and lysosomes).

1. Synthesis of glycerides, e.g., triglycerides, phospholipids, glycolipids.
2. Metabolism of plasmalogens.
3. Synthesis of fatty acids.
4. Biosynthesis of the steroids, e.g., cholesterol biosynthesis, steroid hydrogenation of unsaturated bonds.
5. $NADPH_2+O_2$—requiring steroid transformations: Aromatic hydroxylations, side-chain oxidation, deamination, thio-ether oxidations desulfuration.
6. L-ascorbic acid synthesis.
7. UDP-uronic acid metabolism.
8. UDP-glucose dephosphorylation.
9. Aryl-and steroid sulfatase.

ROLE OF ENDOPLASMIC RETICULUM

Many functional interpretations of endoplasmic reticulum are based on the polymorphic aspects of its components in a variety of cells and its different stages of activity. More reliable interpretations are based on the same well-known facts together with hypothesis.

1. Mechanical Support

ER contributes to the mechanical support of the cytoplasm by dividing the fluid contents of the cell into compartments. This makes possible the existence of ionic gradient and electrical potentials along ER membranes. This concept has been specially applied to sarcoplasmic reticulum.

2. Exchange of Ions and any other Fluid

The membranes of the endoplasmic reticulum may regulate the exchange between the inner compartment of cavity and the cytoplasmic matrix. The following statistic gives an impressive idea of the surface area available for exchange; I gm of liver contains about 8 to 12 square meter of endoplasmic reticulum. After isolation, microsomes expand or shrink according to the osmotic pressure of the fluid. Diffusion and active transports may take place across the membrane of the endoplasmic reticulum.

3. Intracellular Circulation

The endoplasmic reticulum may act as a kind of circulatory system for intracellular circulation of various substances. Membrane flow may be an important mechanism for carrying particles, molecules and ions into and out of the cells by way of vascular system. The "pinocytosis," or "cellular drinking" also takes place by endoplasmic reticulum.

By this mechanism, particles attached to the surface of the cell or suspended in the fluid medium can be incorporated into the cytoplasm. The similar mechanism but working in a reverse direction can effect the transport of a particle from the interior of the cytoplasm to the outer medium. The continuities observed in some cases between the endoplasmic reticulum and the nuclear envelope suggest that the membrane flow may also be active at this point. This flow would provide one of the several mechanisms for export of RNA and nucleoproteins from the nucleus to the cytoplasm.

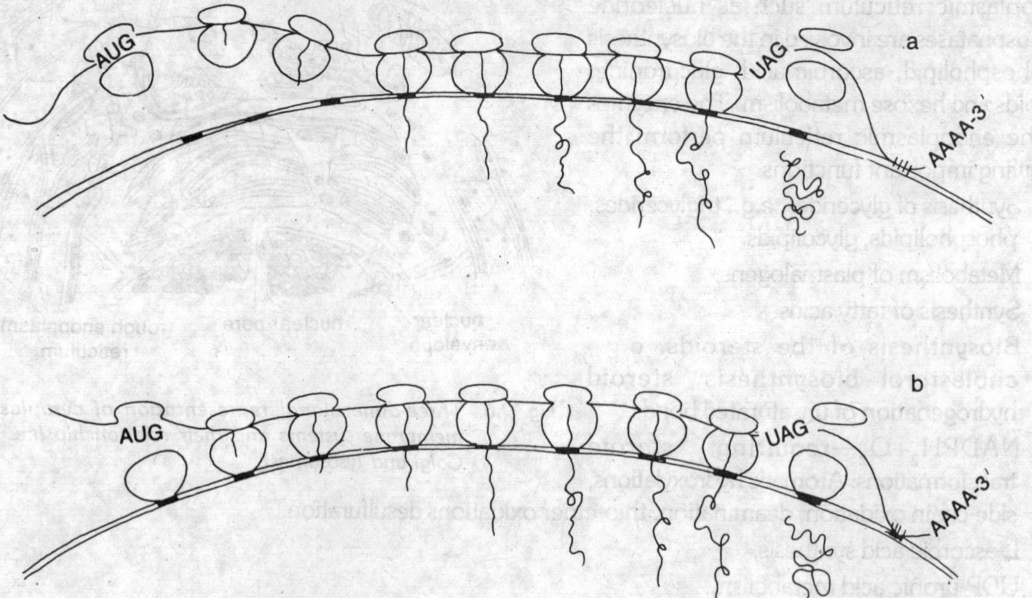

Fig. 9.7. Diagram showing the attachment of ribosomes and messenger RNA to membranes of the endoplasmic reticulum.

4. Protein Synthesis

Proteins may be synthesized to be utilized within the cell or these may have to be exported outside the cell to the site of their utility. It is the later kind of proteins in whose synthesis, endoplasmic reticulum plays an important role. For instance, rough endoplasmic reticulum, which has attached ribosomes carries synthesis of secretory proteins on these ribosomes and export them. Synthesis of tropocollagen, serum proteins and secretory granules are some examples of secretory proteins. The protein molecules synthesized on attached ribosomes are discharged and penetrate into the cavity of ER, where they are stored or exported outside. During the transport of these products, three types of membranes ER—Golgi membrane—plasma membrane should interact and remain interconnected or disconnected due to fusion and fission respectively.

5. Synthesis of Lipids

The cells in which active lipid metabolism takes place are found to contain large amount of the smooth type of endoplasmic reticulum. According to some workers as *Christensen* (1961) and *Claude* (1968) the smooth type of endoplasmic reticulum is related with the synthesis and meabolism of the lipids.

6. Synthesis of Glycogen

The smooth endoplasmic reticulum of the glycogen storing cells of the liver and the cells of certain plants is found to be associated with the synthesis, storage and metabolism of the glycogen. But, *Porter* (1961) and *Peter* (1963) have suggested that smooth type of endoplasmic reticulum is related to glycogenolysis (breakdown of glycogen) and not to glycogenesis (synthesis of the glycogen).

7. Detoxification

Smooth ER is also involved in the detoxification of many endogenous and exogenous compounds. Prolonged administration of certain drugs (phenobarbitol) results in the increased activity of enzymes related to detoxification, as well as other enzymes, and a considerable hypertrophy of the SER (*Claude*, 1970). This is also applicable to administered steroid hormones.

8. Synthesis of Cholesterol and Steroid Hormones

Cholesterol is an important precursor of steroid hormones. The major site of cholesterol synthesis is the ER. In liver cells the SER is believed to be concerned with both the synthesis and storage of cholesterol.

In the testis, ovary and the adrenal cortex the SER has a role in the synthesis of steroid hormones. The enzymes catalysing biosynthesis of androgens have been located in the SER. There is a strong correlation between the amounts of SER in cells and the capacity to synthesize steroid hormones.

9. Amphibian Development

There are evidences to suggest that the ER contributes in several ways to the development of the amphibian embryo.

10. Cell Differentiation

Some specific instance of development have been studied in detail which more or less confirm the contention that the ER is important in the process of cell differentiation. Not only this much, ER also plays role in coordinating the differentiation.

11. Formation of Microbodies

Closely related with the ER are microbodies, which are small granular bodies filled with the electron dense substance and limited by a single membrane. Microbodies are formed as dilations of the ER and frequently show connections with the ER cisternae. They are rich in the enzymes peroxidase (and are hence also called *peroxisomes*), *catalase,* and D-amino acid oxidase. In plant cells the enzymatic content is different and the bodies are called *glyoxysomes* because they, include enzymes of the glyoxylate cycle.

12. Enzyme Activities and Cellular Metabolism

Numerous enzymes mainly those involved in the metabolism of steroids (cholesterol and glycerides), phospholipids and hromones (testosterone and progesterone) are associated with the membranes of smooth endoplasmic reticulum. These membranes provide an increased inner surface for various metabolic reactions and they themselves take an active part in them by means of attached enzymes. This facilitates free union of enzymes with their substrates.

13. Role of Endoplasmic Reticulum in Intra-cellular Impulse Conduction

The existence of endoplasmic reticulum separating the cytoplasm into two compartments makes possible the existence of ionic gradients and electrical potentials across these intra cellular membranes. This idea has been applied to the sarcoplasmic reticulum, a specialized form of smooth surfaced endoplasmic reticulum found in striated muscles which is now being considered as intra-cellular conducting system. On the basis of some evidnece, it has been postulated that the sarcoplasmic reticulum transmits impulses from the surface membrane into deep regions of the muscle fibres.

14. Formation of Plasmodesmata

Electron microscopic studies suggest that the endoplasmic reticulum in plants plays a special role in the interconnection of cells through the cytoplasmic strands called *plasmodesmata.*

15. Role of E.R. During Cell Division

During cell division, some of the elements of reticulum contribute in the formation of the new nuclear membrane after karyogamy. The nuclear membrane break up into fragments in the early part of the division which finally disintigrate into small vesicles (*Moses* 1960). These vesicles move towards the pole of the spindle as the metaphase starts, where they are indistinguishable from the elements of ER. From the polar ends of the cell, elements of ER as well as the fragmented vesicles migrate into the regions around the chromosomes, which are grouping at the poles. Most of these elements of ER join or fuse around each group of daughter chromosomes to form a new nuclear envelope.

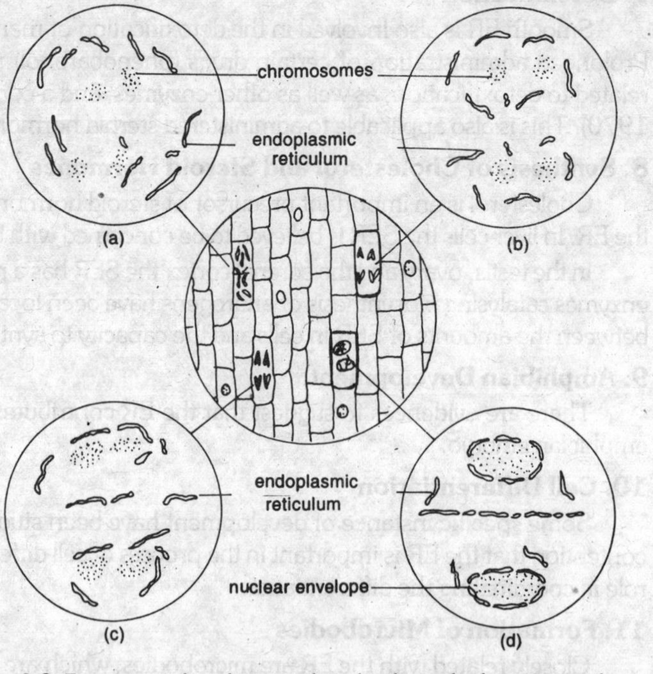

Fig. 9.8. Formation of nuclear envelope by the endoplasmic reticulum.

16. Transportation of Message from Genetic Material

ER provides passage for the genetic material to pass from the nucleus to the various organelles in the cytoplasm, thereby controlling the synthesis of proteins, fats and carbohydrates.

17. ATP Synthesis

ER membranes are the sites of ATP synthesis in the cell. The ATP is used as a source of energy for all the intracellular metabolism and transport of materials.

18. Formation of Cell Organelles

Most of cell organelles like Golgi complex, mitochondria, lysosomes, nuclear membrane and cell plate etc., are usually developed from endoplasmic reticulum.

ORIGIN OF ENDOPLASMIC RETICULUM

(i) Multi step Mechanism

Although the origin of the new endoplasmic reticulum membrane has not been fully understood. The views are there. In fact, one of the possible function attributed to the endoplasmic reticulum is that of membrane biosynthesis. The protein components of endoplasmic reticulum and other membranes may be assembled by activity of the endoplasmic reticulum. There is certainly convincing evidence that Golgi membranes and many cytoplasmic vesicles an be derived from endoplasmic reticulum. Moreover, endoplasmic reticular membranes appear to be continuously synthesized, having a relatively high rate of turn over. At the same time, the several elements of endoplasmic reticulum in the cell are asynchronous in this respect, they are not all replaced at the same time or with the same rate. It has also been suggested that membranes of the endoplasmic reticulum are formed not from pre-existing elements but from the ground substance of the cytoplasm. Thus, the process by which a membrane is modified chemically and structurally is called membrane differentiation.

(ii) From Nuclear Membrane

The vacuoles derived from the evagination of the outer membrane of the nuclear envelope, which separates from its inner partner, leaving cavities between. Shortly after the separation, small vesicles appear near the nuclear envelope, suggesting that parts of the envelope give rise to elements of the endoplasmic reticulum. Thus the endoplasmic reticulum seems to have its origin in the nuclear envelope in undifferentiated cells.

10

GOLGI COMPLEX

In 1898 by means of a silver staining method, *Golgi* discovered a reticular structure in the cytoplasm. The name "Golgi apparatus," generally given to this structure, is confusing because it suggests a definite relationship with the physiologic processes of the cell. Today it seems more appropriate to use the name "*Golgi substance*" or "*Golgi complex,*" to refer to this material that has special staining properties. Because its refractive index is similar to that of the matrix, the Golgi complex is difficult to observe in living cells. The use of the electron microscope has provided a distinct image of this component, and its submicroscopic structure has been revealed.

For years the Golgi complex was thought to be an artifact of various fixation and staining precedures. In other words, many scientists believed that the structure observed during numerous microscopy procedures and termed the Golgi did not actually exist in the living cell. *Guilliermond* (1923), *Parat* (1927) *Walker* and *Allen* (1921), doubts arose with regard to the existence of Golgi complex.

The different lines of objections against the existence of Golgi complex are as follows:

1. Nothing corresponding to the Golgi elements exists in the living cells, what appears Golgi bodies in the processed cells are myeline figures, produced by the action of cytological reagents or lipids *Walker* and *Allen* (1921); *Palade* and *Claude* (1949), however, believe in the existence of Golgi bodies in the germ cells but no such things in the somatic cells.
2. The Golgi apparatus of the fixed material is produced by the deformation and metallic impregnation of several unrealed cytoplasmic bodies (*Guilliermond* 1923, *Parat* and *Painleve,* 1924, 1926 and *Baker* 1949).
3. The dictyosome or acroblasts of male germ cells homologised with the Golgi apparatus of nerve cells. *Gatenby, Bowen* and *Hirschler, Parat* (1928) believed that they are nothing but the modified mitochondria.
4. The Golgi apparatus is a stacks of the smooth endoplasmic reticulum. (*Palade* 1956).

Answer to the Objections

1. In a number of cells these bodies have been studied in *vivo* or in *vitro* with phase contrast microscopy (*Hirsch* 1948, *Schneider et. al.* 1953, *Dortin,* 1949; *Yangiwasa* 1953; *Sosa* 1953).
2. The Golgi bodies have been separated by centrifugal fractionation of homogenized cell (*Worley* 1951; *Schneider et.al.* 1953; *Dalton* and *Felix* 1954; *Schneider* and *Kuff,* 1954).
3. The objection of artifact is discarded on the basis that artifact cannot show the pattern of the same type as shown by these bodies.
4. Their ultra structure has been recently studied with the electron microscope by *Sjostrand* and *Hanzon* (1954) and *Dalton* and *Felix* (1954).

5. Dictyosomes of male germ cells which are impregnated with silver and osmium, have same type of chemical composition as the Golgi apparatus of somatic cells (*Dalton* and *Felix* 1956).
6. According th *Srivastava* (1953, 1954) dictysomes in male germ cells, are favourable materials for the study of Golgi bodies even under the light microscope.
7. The idea that the dictyosomes are modifed mitochondria has been discarded on the basis of biochemical tests.

(A) The dictyosomes of male germ cells contain appreciable quantity of polysaccharides. It is also present in the somatic Golgi bodies but not found in mitochondria. Similarly alkaline phosphatase is associated with Golgi bodies, but not found in the mitochondria.

(B) *Tiwari* and *Broune* (1962, 1963) have found that the distribution of adenosine triphosphatase is localized in mitochondria but not in Golgi bodies. In the same way succinic dehydrogenase and cytochrome oxidase are noted from the mitochondria, but not in the Golgi bodies.

8. The statement that the Golgi bodies of neuron are nothing but endoplasmic reticulum, is discarded on the basis of electron microscopic structure and the width of the Golgi membrane is different from that of the endoplasmic reticulum.

In concluding the controversy of the Golgi apparatus, it should be pointed out that *Baker* (1963) and his students were the last to accept the reality of the Golgi apparatus when *Baker* wrote, "The author accepts, after long hesitation, the view that the Golgi apparatus in the neurons of vertebrates with the organelle of the same name, in other cells."

Terminology of Golgi Complex

Holmgren referred to the Golgi complex as *trophospongium* (*Cajal* referred it as the *Golgi-Holmgren canals*). *Baker* used the term *lipochondria* because of the presumed lipid content. The term *Dalton Complex* was given after the name of its observer *Dalton* in 1952. *Sjostrand* proposed the term cytomembranes for the Golgi system.

Sosa has suggested the following nomenclature for the Golgi Complex.

1. *Golgiokinesis.* Division of the Golgi apparatus during nuclear division.
2. *Golgiosomes.* Corpuscles produced by the Golgiogenesis are called as Golgiosomes which are described as Gogli material in invetebrates.
3. *Golgiolysis.* Process of dissolution of the Golgi apparatus.
4. *Golgiorrhexis.* Fragmentation of the Golgi apparatus.
5. *Golgiogenesis.* Formation and differentiation of the Golgi body during embryonic development.
6. *Golgi-cytoarchitecture.* Study of structure of cell in relation to Golgi apparatus.

Occurrence

The Golgi complex occurs in all cells except the prokaryotic cells (viz., Mycoplasma, bacteria and blue green algae) and eukaryotic cells of certain fungi, sperm cells of bryophytes and pteridophytes, cells of mature sieve tubes of plants and mature sperm and red blood cells of animals. Their number per plant cell can vary from several hundred as in tissues of corn root and algal rhizoids (i.e., more than 25,000 in algal rhizoids, *Sievers,* 1965), to single organelle in some algae. In animal cells, there usually occurs a single Golgi complex, however, developing oocytes of chordates may contain many Golgi complex.

Morphology of Golgi Complex

The morphology of the Golgi complex varies from cell to cell depending uupon the type of cell in which they are found. Two forms of Golgi complex have been observed.

1. *Localized form.* In polarised cells of vertebrates (which have base and apex). Golgi complex occurs singly and occupies a fixed position. It lies between the nucleus and secretory pole. This can be best seen in thyroid cells, in exocrine cells of pancreas and the mucous cells of intestine.

2. *Diffused form.* In some specialized cells of vertebrates (nerve cells and liver cells), in most plant cells and in the cells of invertebrates several units of Golgi complex are found scattered along with the elements of endoplasmic reticulum. Each unit is called a *dictyosome.* In liver cells these occur as many as 50 dictyosomes per cell and in certain plant cells their number may reach upto hundreds.

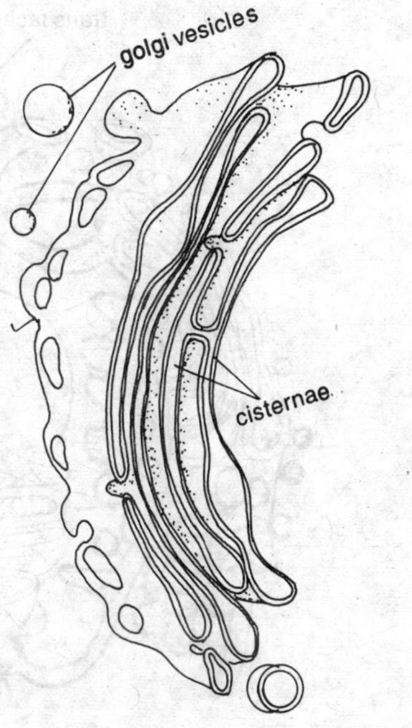

Fig. 10.1. ***Diagram showing the structural organization of a Golgi body.***

Shape

The shape of the Golgi complex is quite variable in different somatic cell types of animals. Even in the same cell there are variations in different functional stages. The shape is, however, constant with each cell type. It varies in form from a compact mass to a disperse filamentous network.

Number

The number of Golgi stacks per cell varies enormouly, depending on the cell type—from as few as one to the hundreds. There is a single large in some cells while in the case of *Paramoeba* there are two. In *Stereomyxa* (a species of *Amoeba*) there are many Golgi complexes. Nerve cells, liver cells and most plant cells also have multiple Golgi complexes, there being about 50 in liver cells. The Golgi complex can even account for a large fraction of the cell volume in some specialized cells. One example is the goblet cell of the intestinal epithelium, which secretes mucus into the gut; the glycoproteins in mucus are glycosylated principally in the Golgi complex.

Size

The size is likewise very variable. It is large in the nerve and gland cells and small in the muscle cells. In general the Golgi complex is well-developed while the cell is in active state. When the cell grows old, the complex progressively diminishes in size and disappears.

Position

The position of the Golgi complex is relatively fixed for each cell type. In the cells of ectodermal origin, the Golgi complex is polarized from the time of the embryonic state between the nucleus and the periphery (*Cajal,* 1914). In the secretory exocrine cells that have in general a typical polarization the Golgi complex is found between the nucleus and the secretory pole. In the endocrine glands the polarity of this organoid is variable, except in the thyroid, where it is oriented towards the centre of the follicle. In the yonger cells and often in the older ones it lies most commonly at one side of nucleus but in certain cases may completely surround it. In ganglionic cells of the mouse the position in perinuclear.

Detailed Structure of Golgi Complex

Dalton and *Felix* (1954) described the Golgi complex in the epididymis of the rat after taking the first electron micrographs. The following description of the Golgi complex is a composite one based on the work of several authors.

1. Cisternae

The cisternae or saccules are similar to the smooth surface ER, and appear in section as stacks of closely spaced membrane-delimited sacs. The number of saccules varies from about 4 to 8 in most animal and plant cell types. In *Euglena,* the number may go upto 20. The membrane of the saccule is roughly 60 to 70Å in thickness which encloses a cavity about 150Å wide whose edges are often dilated. According to most authors, there are two well-defined faces of the cisternae, i.e., convex and the concave; the later is

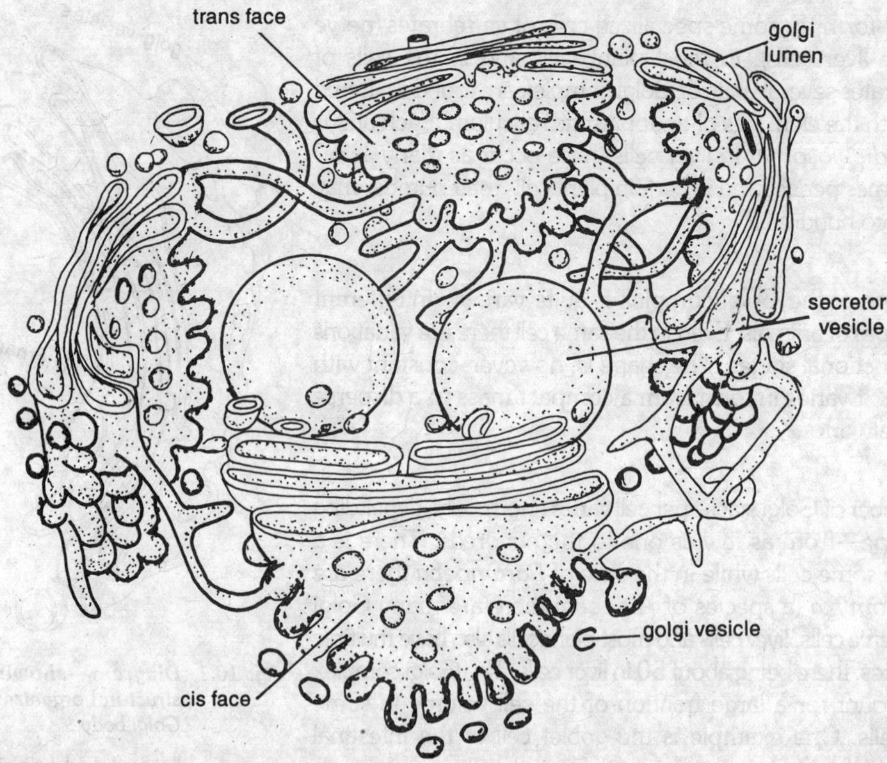

Fig. 10.2. Three-dimensional drawing of a Golgi apparatus; it is drawn from electron micrographs of a secretory animal cell.

generally referred to as the mature or forming or distal face and the convex side is assumed to be the immaure or excit or proximal face, the cisternae lie in parallel array are separated from each other by a space of about 200 to 300Å. What holds them together is not yet known but in few cells a thin layer of electron opaque, sometimes dense material is seen between the saccules which at certain regions are more prominent to which *Amos* and *Grimstone* (1968) applied the term nodes. *Mollenhauer et al.*, (1973) explored in some detail intercisternal elements and plaques in certain plant Golgi complex.

2. Tubules

From the peripheral area of cisternae arise a complex, anastomosing flat network of tubules of 300 to 500Å diameter. *Clowes* and *Juniper* (1969) have compared this tubular network to disc of lace.

3. Vesicles

The vesicles are small droplet-like sacs which remain attached to tubules at the periphery of the cisternae. They are of following two types:

(a) Smooth vesicles

The smooth vesicles are of 20 to 80 mμ diameter. They conain secretory material (so often are called secretory vesicles) and are budded off from the ends of cisternal tubules within the net. Often more than one tubule connect to, and presumably fill, a single forming vesicle.

(b) Coated vesicles

The coated vesicles are spherical protuberances, about 50 mm in diameter and with a rough surface. They are found at the periphery of the organelle, usually at the ends of single tubules, and are morphologically quite distinct from the secretory vesicles. Their function is unknown.

4. Golgian Vacuoles

These are large rounded sacs present on the maturing face of Golgi. These are formed either by the expanded cisternae or by the fusion of secretory vesicles. The vacuoles are filled with some amorphous or granular substance.

The Golgi Complex is Structurally and Biochemically Polarized

The Golgi complex has two distinct faces: a *cis* or *forming face* and a *trans,* or *maturing face.* The cis face is closely associated with a smooth transitional portion of the rough ER. In secretory cells, the trans face is the face closest to the plasma membrane; here, the large secretory vesicles are found exclusively in association with the trans face of a Golgi stack, and the membrane of a forming secretory vesicle is often continuous with that of the trans face of the last ("trans-most") cisterna. In contrast, the small Golgi vesicles are localized more evenly along the stack, proteins are commonly thought to enter a Golgi stack from the ER on the cis side and to exit for multiple destinations on the trans side; however, neither their exact path through the Golgi complex nor how they travel from cisterna along each stack are known.

The two faces of the Golgi complex are biochemically distinct. For example, a variation in the thickness of the Golgi membranes can be detected across the stack in certain cases, with those at the cis side being thinner (ER-like) and those at the trans side being thicker (plasma membrane like). More striking are the results obtained when certain histochemical tests are used in conjunction with electron microscopy to localize particular proteins within the Golgi complex. Some of these tests reveal membrane-bound enzyme activities that show a distinct polarity in their localization within the Golgi stack. A particularly intriguing biochemical finding was the discovery that lysosomal enzymes, such as acid phosphatase, are concentrated with the trans-most cisterna of the Golgi stack and within some of the coated vesicles nearby. This suggests that specific vesicles leaving for lysosomes are assembled in this region.

Secretory proteins are found by histochemical methods in all of the stacked cisterna, even though the large secretory vesicles in which these products are concentrated and associated only with the trans-most Golgi cisterna.

Chemical Composition

Regarding the chemical composition of Golgi complex, it has been demonstrated that the following substances are present:

1. *Phospholipids.* Phospholipids composition of Golgi membranes is intermediate between those of endoplasmic membranes and plasma membranes.
2. *Proteins and Enzymes.* Golgi complex from different plant and animal cells show great variations in the protein and enzyme contents. Some of the enzymes are ADPase, ATPase, NADPH cytochrome-C-reductase, glycosyl transferases, galactosyl transferase, thiamine pyrophosphate etc.
3. *Carbohydrates.* Both plant and animal cells have some common carbohydrate components, like glucosarine, galactose, glucose, mannose, and fructose. Plant Golgi lack sialic acid, but

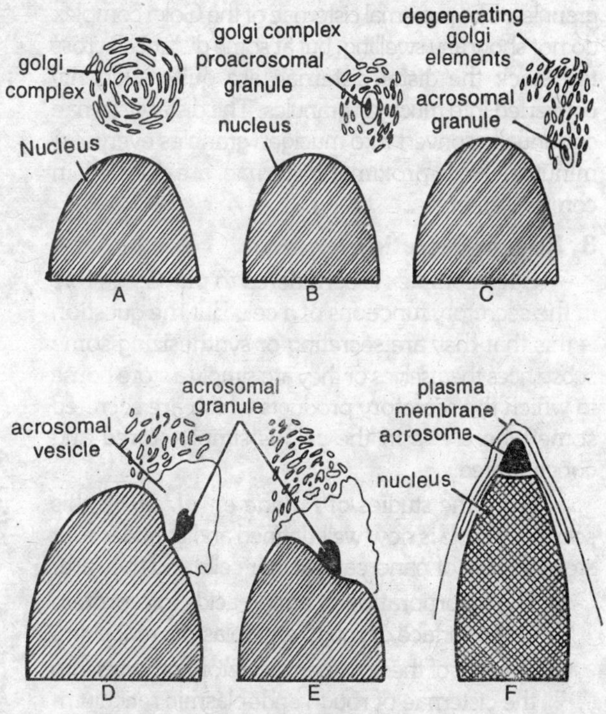

Fig. 10.3. Diagram illustrating gradual modification in Golgi complex to form acrosome during spermiogenesis.

it occurs in high concentration in rat liver. Some carbohydrates like xylase and arbinose are present in plant cells only.

4. *Vitamin C.* The function of vitamin C stored in the Golgi complex has been shown by *Tomitte.* According to him Golgi complex stores vitamin C and liberates it slowly into the cytoplasm in sufficient amount of prevent-oxidation of the cell products.

FUNCTION OF GOLGI COMPLEX

1. Formation of Acrosome during Spermiogenesis

During the maturation of sperm the Golgi complex plays a role in the formation of acrosome (*Burgos* and *Fawcett,* 1955).

In early stages, the Golgi appears as a spherical body, comprising cisternae arranged in parallel stacks and numerous small vesicles. The latter always pinched off from the cisternae. As development proceeds, the Golgi complex becomes irregular in shape and large vacuoles are formed by dilations of cisternal sac. In the centre of these large vacuole or vacuoles is present a dense granule, the *proacrosomal granule.* This granule which is derived from Golgi complex continues to grow within the vacuole by a process known as *accretion.* This vacuole and granule approaches the anterior pole of the nuclear membrane, constituting *acrosomal granule.* With the elongation of the spermatid, the *acrosomal vesicle* spreads over the nuclear surface and finally collapses with the nuclear membrane, forming the *cap material*. The acrosomal granule becomes the *acrosome* which lies at the apex of the nucleus and apparently comprises certain enzymes involved in the process of fertilization.

2. Synthesis and Secretion of Polysaccharides

Studies on goblet cells by autoradiography and electron microscopy have established the inter-relationship between protein synthesis, carbohydrate addition and sulphation. The goblet cells of the colon produce mucigen. This secretory material contains a large proportion of carbohydrate. The Golgi complex is found just above the nucleus. Towards the free surface of the cell are gradually enlarging mucigen granules. The proximal cisternae of the Golgi complex do not show any swelling, but at some distance across the stack the distal cisternae are quite suddenly converted into mucigen granules. The distal cisternae continually convert into mucigen granules every 2-4 minutes. New proximal cisternae are formed in compensation.

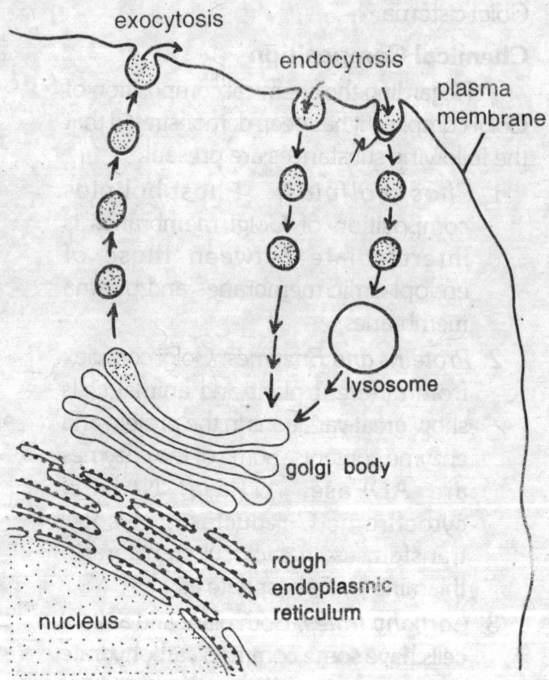

Fig. 10.4. Role of Golgi complex in the recycling and reuse of plasma membrane components.

3. Role in Secretion

Golgi complex is considered to play some role in the secretory functions of a cell. But the question is this that they are secreting or synthesizing some substances themselves or they are simply a store house in which the secretory products which are secreted some where else in the cell, is simply stored and concentrated.

From the studies of *Palade et. al.* 1962 this secretory cycle is now well-defined and includes four steps in case of pancreatic acinar cells and they are:

(i) The incorporation of amino acids into protein at the surface of rough endoplasmic reticulum.

(ii) Transfer of these nascent secretory proteins into the cisternae of rough endoplasmic reticulum.

(iii) The intracellular transport of these proteins to the Golgi complex.

(iv) The migration of *zymogen granules* towards the apex of the cell where they discharged into lumens.

4. Role of Golgi Body in Oogenesis

Srivastava (1965) has given a brief review on the Golgi complex during oogenesis.

According to *Afzelius* (1956), the Golgi complex of a sea-urchin egg, as seen under electron microscope, consists of stacks of lamellae forming walls of flat pouches, which may occasionally be swollen. There is some indications of transverse divisions of these bodies. *Sotelo* (1959) and *Sotelo* and *Porter* (1959) have described the Golgi complex in rat-ovum as seen under the electron microscope and found

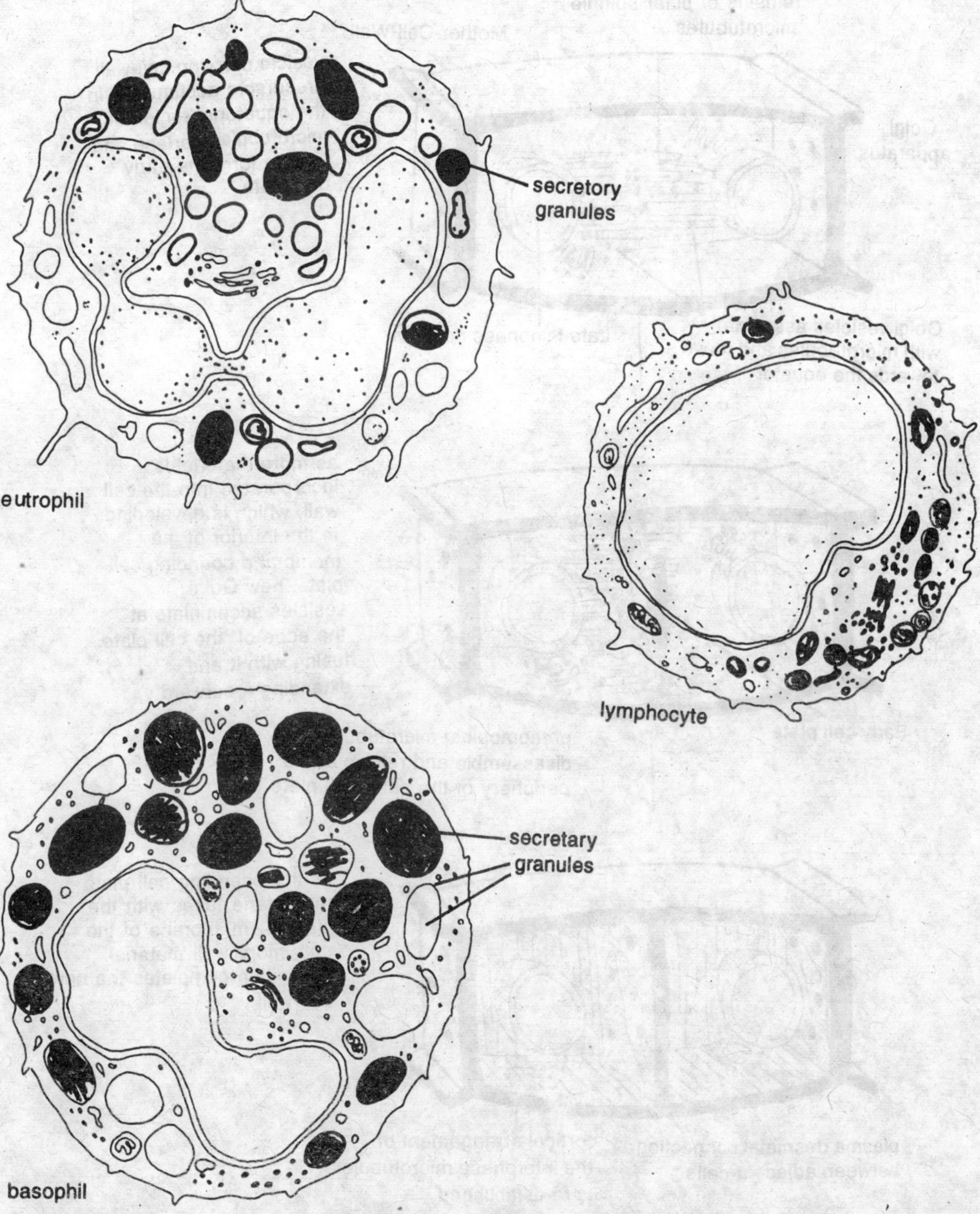

Fig. 10.5. Secretory granules in some leucocytes.

juxtra nuclear localization of this organelle in early oocytes. In the next stage, these resolve into fragments and in the third stage, these move towards the cortex. In all these cases, their structures remain to be of closely packed arrays of slender, double profiles (flattened sacs) and spherical vesicles. In the early oocytes the complex is compactly organized. In later stages, discrete bundles of profiles, surrounded by small vesicles are found scattered in the cortical zone. In the early oocytes, the Golgi complex and centrosome are closely associated.

5. Absorption of Compounds

Hirsch et. al. have discovered that when iron sugar is fed to an animal, iron becomes absorbed on Golgi complex (*Kedrowsky*) *Van Teel* has shown that Golgi systems also absorb compounds of copper and

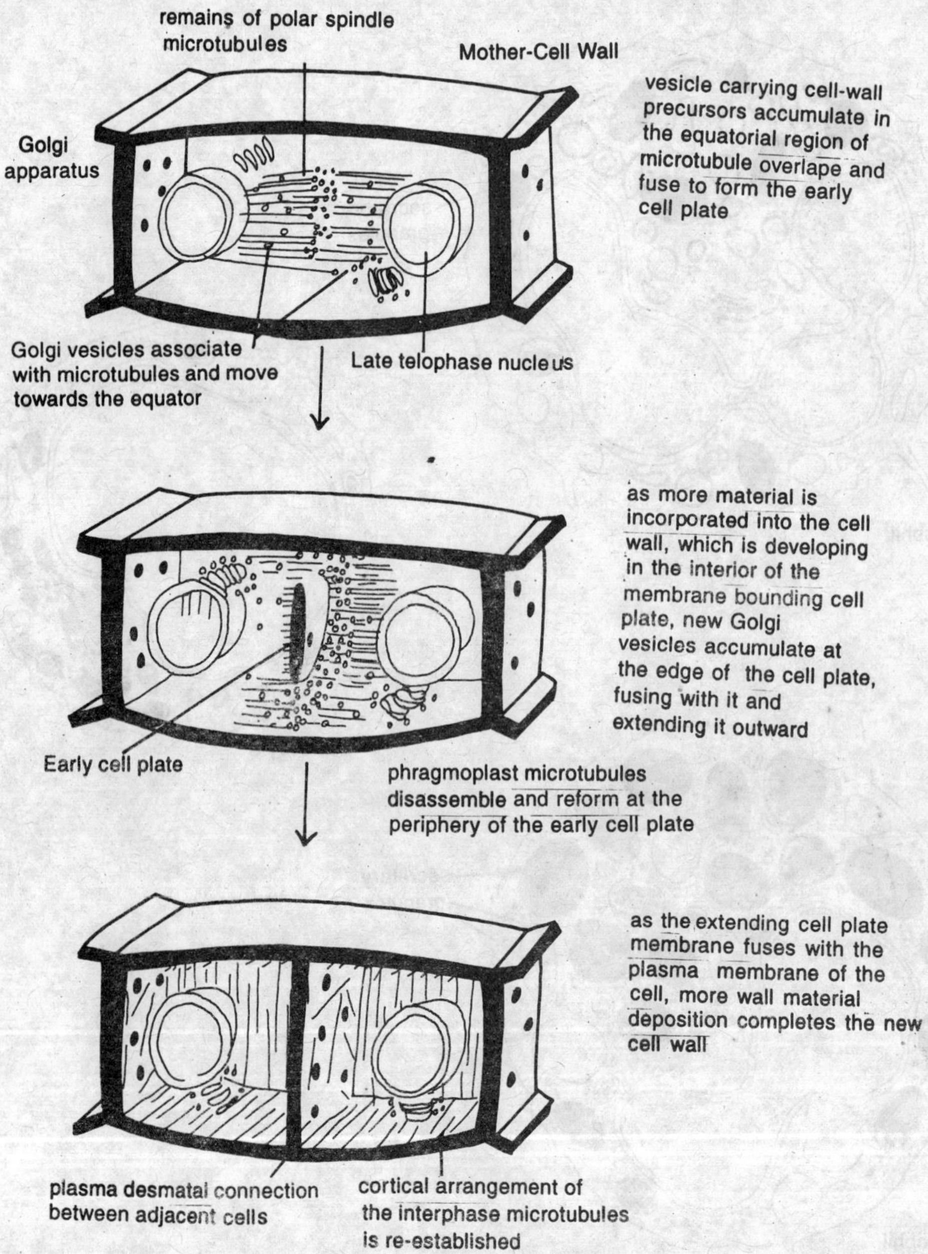

Fig. 10.6. Cell plate formation.

gold. *Kedrowsky* has shown that Golgi complex of *Opalina* can absorb bismutose (compound of albumin and bismuth) and protargol (compound of albumin and silver). Thus, *Kirkman* and *Severinghaus* state that Golgi complex acts as a condensation membrane for the concentration of products into droplets or granules.

6. Plant Cell Wall Formation

The cell walls of plants are made up of fibrils which predominantly contain polysaccharides, along with some lipids and proteins. During cytokinesis a *cell plate* is formed between the two daughter nuclei, and has around it a membrane which later becomes the plasma membrane of the daughter cells. There is clear evidence that the polysaccharides are formed in the Golgi complex and transferred to the new cell wall which is laid down while the cells are still growing.

Substances like pectins and hemicelluloses, which form the matrix of the cell plate separating the plasma membranes, are also contributed by the Golgi complex.

7. Formation of Intracellular Crystals

In the marine isopod, *Limnoria ligmorum,* which is a burrowing form there are present midglands whose cells consist of crystals. These range up to 30Å in length and 15Å thick. It has been proved that these crystals are formed by Golgi complex and are known to contain protein and iron. They are without enclosing membrane and usually spheroidal in shape. They are concerned with the secretory activity.

8. Milk Protein Droplet Formation

In the lactating mammary gland of mice are produced protein droplets which are related with Golgi complex. These droplets usually open on to the cell surface by the fusion of their enclosing membrane with the plasma membrane.

9. Formation of Lysosomes and Vacuoles

Primary lysosomes are formed from the Golgi membranes the same way as the secretory vesicles. There is good evidence that dictyosomes accumulate hydrolytic enzymes in their more mature regions. Some vacuoles in plant cells have been found to certain small amounts of hydrolytic enzymes and these are presumed to have been derived from Golgi complex.

10. Pigment Formation

In many mammalian tumour and cancer cells the Golgi complex has been described as the site of origin of pigment granules (melanin).

11. Regulation of Fluid Balance

A homology has been suggested between the Golgi complex and the contractile vacuole of lower Metazoa are Protozoa. The contractile vacuole expel surplus water from the cell. In certain Protozoa the Golgi complex is also concerned with regulation of fluid balance.

ORIGIN OF GOLGI COMPLEX

The different sources have been proposed from which new Golgi complex may arise:

1. From Endoplasmic Reticulum

Essner and *Novikoff* (1962) and *Beams* and *Kessel* (1968) have proposed that the Golgi cisternae arise from the ER. The rough endoplasmic reticulum after synthesizing specific proteins loses ribosomes and changes into smooth ER. Small transitory vesicles pinch off from smooth ER. These migrate to dictyosome.

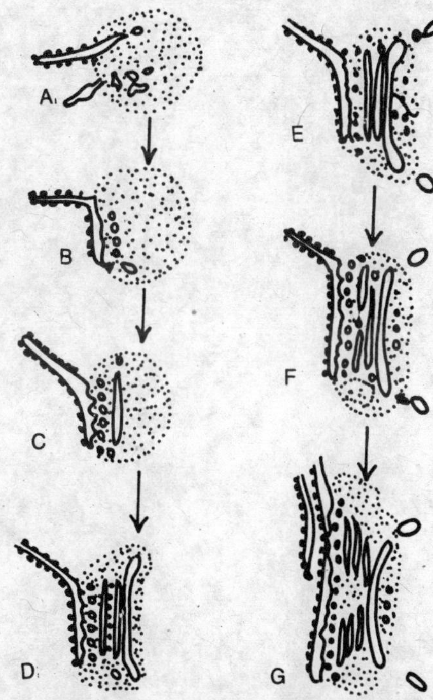

Fig. 10.7. Origin of dictyosome from endoplasmic reticulum from A to D and division of dictyosome—E, F, G.

On reaching the forming face of dictysome these fuse to form new cisternae and thus contribute to its growth. By the fusion of these vesicles new cisternae are formed continuously on the forming face and on the maturing face the old cisternae break down into secretory vesicles. Thus Golgi exhibit a phenomenon of 'membranous flow.'

2. From Nuclear Membrane

Bouch (1965) described the origin of Golgi from outer membrane of nuclear envelope in brown algae. Vesicles are pinched off from outer nuclear membrane which fuse to form cisternae on the forming face of dictyosome.

Presence of zones of exclusion in relation with smooth ER or nuclear membrane, the occurence of zones of exclusion in dormant seeds of higher plants and the formation of dictyosome from these zones in germinating seeds provide evidence in support of the above two theories about the origin of dictyosome.

3. By the Division of Pre-existing Dictyosome

It has been observed that during cell division in both plants and animals, the number of dictyosomes increases and the number of dictyosomes in each daughter cell just after division is almost equal to the number in the parent cell prior to division. From this and other direct observations on the dividing cells it has been presumed that dictyosomes also divide during cell division.

11

MICROSOMES

LYSOSOMES

The concept of the lysosome originated from the development of cell fractionation techniques by which different subcellular components are isolated. By 1949 a class of particles having centrifugal properties some what intermediate between those of mitochondria and microsomes was isolated by *de Duve* and found to have a high content of acid phosphatase and other hydrolytic enzymes. Because of their enzymatic properties they were named lysosomes (Gr. *Lysis* = dissolution, *soma* = body). According to *Gohan* (1972) membrane bounded storage granules containing digestive enzymes are considered lysosomes of plant Alls. Hence *spherosomes, aleurone granules* and *vacuoles* of plant cells are supposed to have lysosome like functions.

Occurrence

With the exception of mammalian R.B.C., the lysosomes have been reported practically from all the animal cells. The presence of lysosomal particles have also be suspected (and in some cases established) in protista (protozoan, slime-moulds, fungi, algae and prokaryotic protista). In plant cells, considering the evidences as a whole, there now seems little doubt about their presence. Further, they have strong affinities with the lysosomes of animals and protista (*Pitt* and *Gaplin* 1973). In plants, further they should not be confused with spherosomes in function. According to *Pitt* lysosomes and spherosomes are two different organelles and the later are comparable to lipid droplets of animals. *Yatsu* and *Jack* (1972) have clearly shown that spherosomes are morphologically distinct organelles. *Gahm* (1973) reviewed the occurrence and histochemistry of plant lysosomes.

MORPHOLOGY

Shape and Size

The shape and size of lysosome is variable. Morphologically they can be compared with *Amoeba* and white blood cells (W.B.C.). Due to their changing habit they cannot be accurately identified as the basis of their shape. Normally lysosomes vary in size from 0.4 to 0.8μ, but they may be as large as 5μ in mammalian kidney cells and are exceedingly large in phagocytosis.

Structure of lysosomes

Like other cytoplasmic complexes, lysosomes are like round tiny bags filled with dense material and digestive enzymes. They consist of two parts (1) limiting membrane and (2) inner dense mass.

1. *Limiting membrane.* This membrane is single, unlike that of mitochondria and composed of lipoprotein. The chemical structure is homologous with the unit membrane of plasmalemma consisting of bimolecular layer as suggested by *Robertson*.

2. *Inner dense mass.* This enclosed mass may be solid or of very dense contents. Some lysosomes have a very dense outer zone and less dense inner zone. Some others have cavities of vacuoles within the granular material. Usually they are supposed to possess denser contents than mitochondria.

They show their polymorphic nature and their contents vary with the stage of digestion, as they help in intracellular digestion.

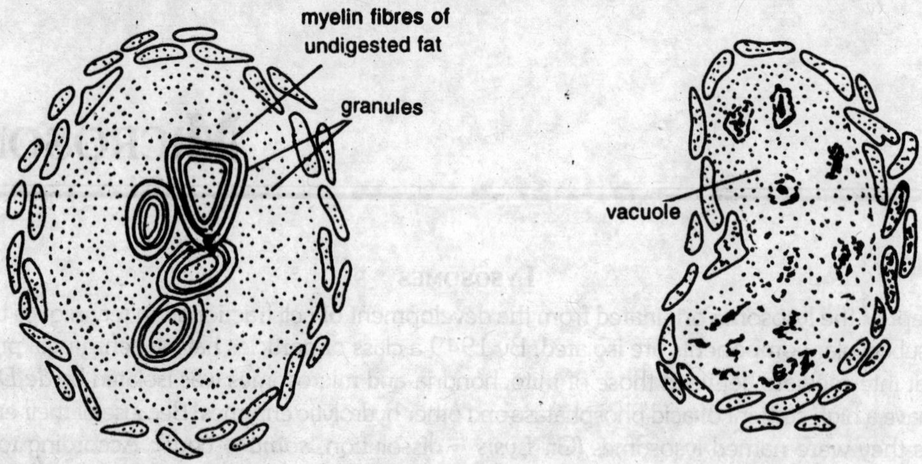

Fig. 11.1. Two types of Lysosomes from the kidney cells of rat.

Permeability of Lysosomal Membrane

The lysosomal membrane is impermeable to substrate of the enzymes contained in the lysosome. Certain substances, called *labilizers,* cause instability of the lysosomal membrane, leading to release of enzymes from the lysosome. Other substances, called *stabilizers,* have a stabilizing action on the membrane. A list of some labilizers and stabilizers is given in table 11.1.

A labilizer might increase the permeability of the lysosomal membrane to small solutes like source. The osmotic swelling which results might completely disrupt the membrane.

The limited permeability of lysosomal membrane explains why lysosomal hydrolases do not have a direct access to cellular components. This prevents uncontrolled digestion of the cell contents by the lysosomal enzymes.

Table 11.1. Lysosomal Labilizers and Stabilizers.

Labilizers	*Stabilizers*
Vitamin A	Cholesterol
Vitamin B	Cortisone
Vitamin K	Cortisol
Vitamin E (high dose)	Vitamin E (low conc.)
Progesterone	Chloroquine
Testosterone	Phenothiazines
β-estradiol	Antithistamines
Ubiquinone	Heparin
Digitonin	
Polyene antibiotics	

Polymorphism

Lysosomes are polymorphic in nature. The polymorphic nature is due to variation in contents of lysosomes with different stages of digestion. Generally lysosomes can be traced in four forms given below:

1. *Primary lysosomes.* These are also called the *true, pure* or *original lysosomes*, having a single unit membrane containing enzymes in the inactive forms.

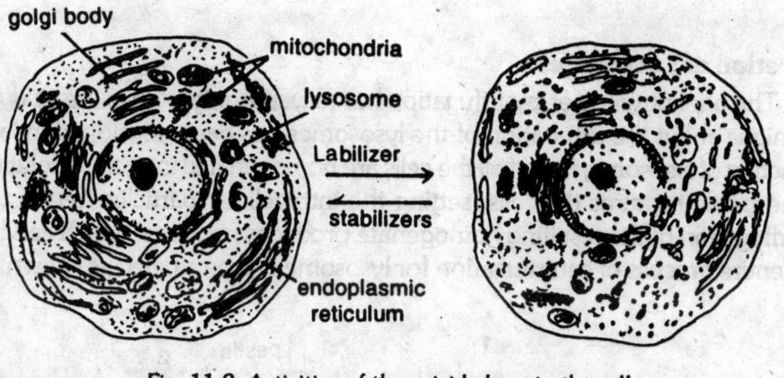

Fig. 11.2. Activities of the suicide bags in the cell.

2. *Secondary lysosomes.* These are also called the *phagosomes* as they contain the engulfed material and enzymes. The fused mass is called the secondary lysosomes. The enzyme present in such lysosomes gradually digests the engulfed material.
3. *Residual or Post lysosomes.* Lysosomal membrane characterized by the presence of undigested material like myelin figures in called residual body.
4. *Autophagic Vacuoles.* The autophagic vacuoles are also known as *autophagosomes* or *cytolysosomes.* The autophagic vacuoles are formed when the cell feeds on its intracellular organelles such as the mitochondria and endoplasmic reticulum by the process of autophagy. In such caeses, the primary lysosomes are concentrated around the intracellular organelles and digest them ultimately. The autophagic vacuoles are formed in special pathological and physiological conditions. *C. de Duve* (1967) and *Allison* (1967) have observed that during starvation of the organisms many autophagic vacuoles developed in the liver cells which feed on the cellular components.

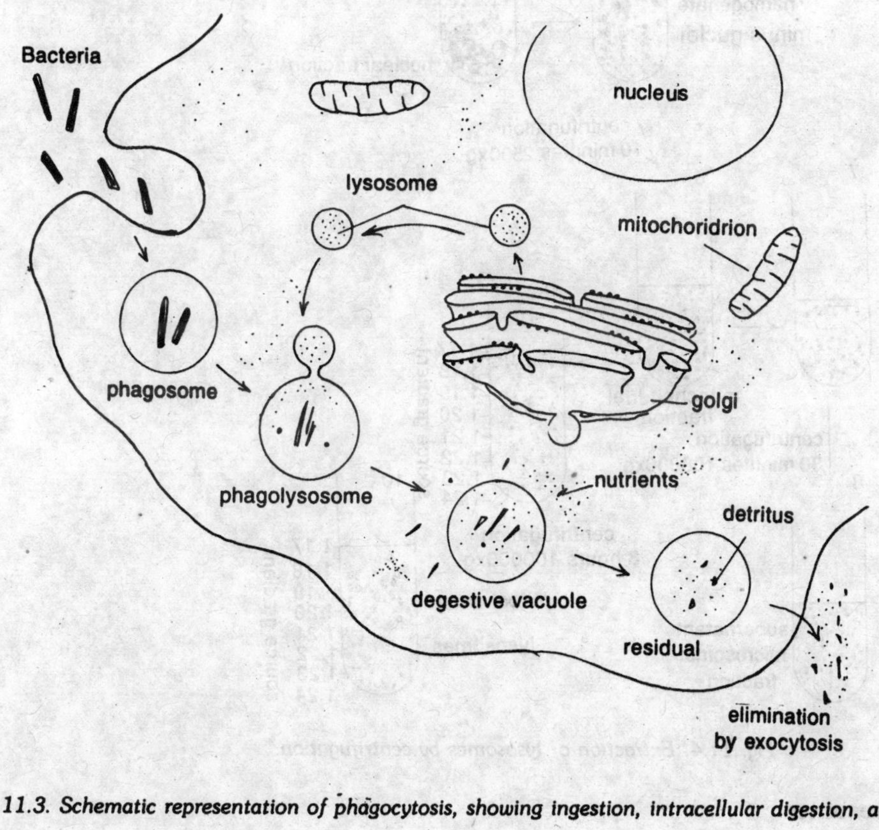

Fig. 11.3. Schematic representation of phagocytosis, showing ingestion, intracellular digestion, and exocytosis.

Excretion of Lysosomes

The phenomenon of centrifugation has played a great role in the study of cytoplasmic inclusion. The technique about the separation of the lysosomes has been developed in laboratory of *de Duve*. For the extraction of lysosome, first of all the cells are homogenated in sucrose solution. Rapid mechanical rotation of the pestle ruptures the cells, setting the intracellular particles free in the medium. Then successive centrifugation of the resulting homogenate produces fractions, which can be separated by micro-needle. The entire process of centrifugation for lysosome can be studied as shown in the Figure 11.4.

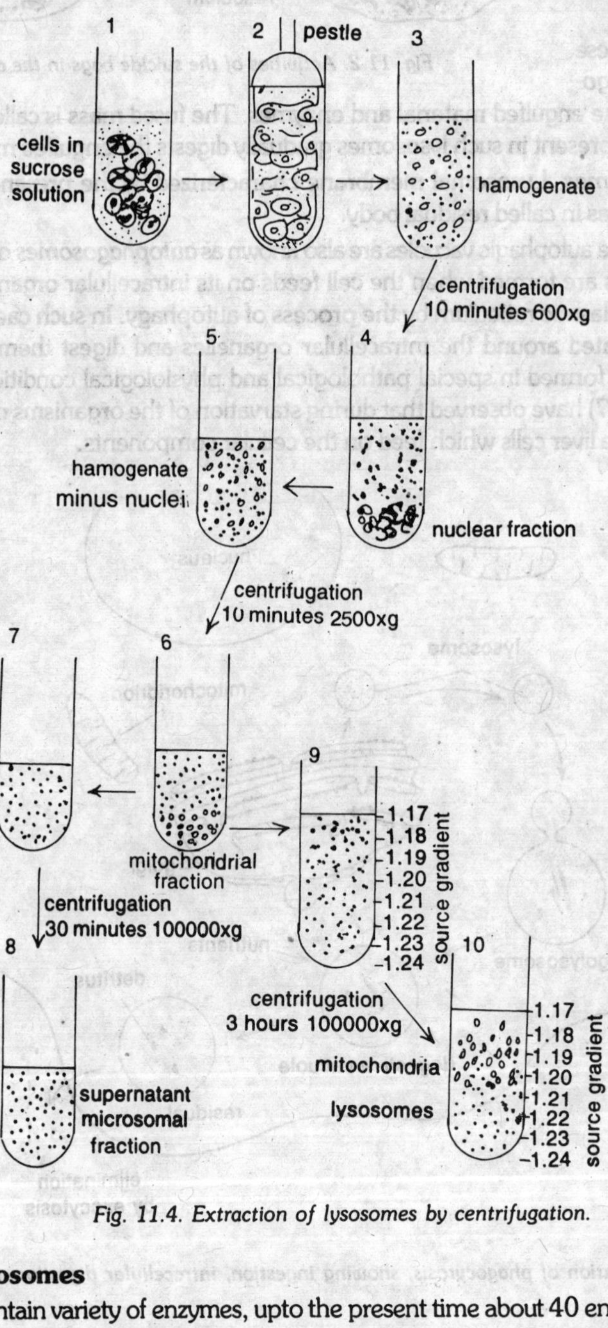

Fig. 11.4. Extraction of lysosomes by centrifugation.

Chemistry of Lysosomes

Lysosomes contain variety of enzymes, upto the present time about 40 enzymes have been isolated in variety of tissue type. Some common enzymes, and β-galactosidase, β-glucuronidase, β-N-acetyl-glucosaminidase, α-glucosidase, α-mannosidase; cathepsin A (Acid protease), cathepsin B (Acid protease), aryl sulphatase A, aryl sulphatase B, acid ribonuclease, acid deoxyribo-nuclease, acid phosphatase, acid lipase, phospholipase A, phosphotidic acid phosphatase hyaluronidase, phosphoprotein phosphatase, amino peptidase A, dextranase, sacchariase, lysozyme (muramidase), Mg++ activated ATPase, indoxylacetate, esterase and plasminogen activator. All these enzymes of the lysosomes are enclosed within the single lipoprotein membrane. Most of these enzymes function more efficiently under slightly acidic medium, *pH* optima around 5.0, as such they are collectively called as acid hydrolases.

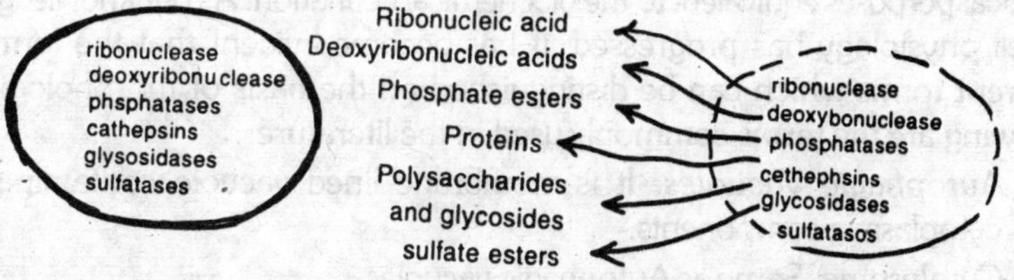

Fig. 11.5. The enzymes of lysosomes and their substances.

Table 11.2. Lysosomal Enzymes, the Substrate on Which these Work and the End Products.

Enzymes	Substrate	End Product
1. Nucleases		
(i) Acid ribonuclease	Polynucleotides of RNA sugar	Nitrogenous bases + Phosphate + ribose
(ii) Acid deoxyribonuclease	Nitrogenous base + DNA	phosphate + deoxyribose sugar
2. Phosphatases		
(i) Acid phosphatase	Phosphomonoesters	Monophosphates
(ii) Phosphodiesterase	Oligonucleotides, phosphodiesters	Monophosphates
3. Proteases and Peptidases		
(i) Cathepsins (A, B) and peptides	Various proteins	Amino acids
(ii) Collagenase	Collagen	Amino acids
(iii) Peptidase	Peptides	Amino acids
4. Glycosidase		
(i) β-galactosidase	β-galactosides	
(ii) α-glucosidase	Glycogens	
(iii) β-glucosidase	β-glycosides	
(iv) α-Mannosidase	α-mannosides	Monosaccharides
(v) β-glucuronidase	Polysaccharides and mucopolysaccharides	
5. Sulphatases	Sulphate esters	Fragments of lipids
6. Lipases	Lipids	Fragments of lipids
7. Esterases	Fatty acid esters	Fragments thereof
8. Lysosome	Mucopolysaccharides, bacterial cell alls	Fragments thereof
9. Sphingomyelinase	Sphingomyelin	Fragments thereof

Different Nomenclature

The cytochemical defination of lysosomes based on the presence of a single unit membrane and a positive staining reaction for acid phosphatase and some related enzymes, may be considered for most practical purposes equivalent to the biochemical defination. As our knowledge of the singificance of lysosomes in cell physiology has progressed, it has become evident that the term lysosomes covers a variety of different forms which can be distinguished on the basis of morphological and functional criteria. The following are the terms commonly used in the literature:

(i) *Autophagic Vacuoles.* It is membrane lined vacuole containing morphologically recognizable cytoplasmic components.

(ii) *Cytolysome.* Same as Autophagic vacuoles.

(iii) *Cytosome.* Particles referred to cytosomes are usually lysosomes. Some workers include the unrelated micro-bodies under this term.

(iv) *Cytoergosome.* Same as Autophagic vacuoles.

(v) *Microbody.* A particle found in liver and kidney, bounded by a single unit membrane and containing a finely granulated material. According to *de Duve,* they are definitely not lysosomes.

(vi) *Multivesicular Bodies.* Structures lined by a single membrane and containing inner vesicles resembling Golgi complex and are considered to be lysosomes.

(vii) *Residual bodies.* Membrane lined inclusions characterized by undigested residues comprises telolysosomes and hypothetical post-lysosomes.

FUNCTIONS

1. Lysosomal Digestion of External Particles

Large molecules are taken into the cell by the process called *phagocytosis.* This is a perfectly adequate and accurate term that implies a cell that eats. But recently the new term *endocytosis* has won favour. The first clear indication of a relationship between lysosomes and engulfment of extracellular material was provided by *Stians* (1952, 54, 55). As shown in the figure, the cell engulfs the particles and then forms and invagination that becomes pinched off from the cell membrane to become an internal sac or body. It is referred to as a *phagosome.* A phagosome then moves towards the lysosome. Exposure of the material to the lysosomal hydrolases occurs through fusion of the phagosome with a lysosome. This results in the formation of a *secondary lysosome* or *digestive vacuole.* The lysosome involved in the process may be primary or secondary, depending on the relative sizes of the two partners. The process may appear to an observer as a lysosome discharging enzymes into a phagosome or as a phagosome shedding its contents into a lysosome, as may be the case in hepatic parenchymal cells; or simply as a mutual sharing of the contents of the two vacuoles, if they are of a comparable sizes. Now the enzymes from the lysosome can come into contact with the molecules brought into the cell in the phagosome and digestion occurs. Once the molecules are digested, the digested products can diffuse out of the so-called digestive vacuole into the cytoplasm of the cell leaving the residue in the digestive vacuole. The digestive vacuole now moves on to the cell membrane where the so-called reverse phagocytosis or defecation occurs.

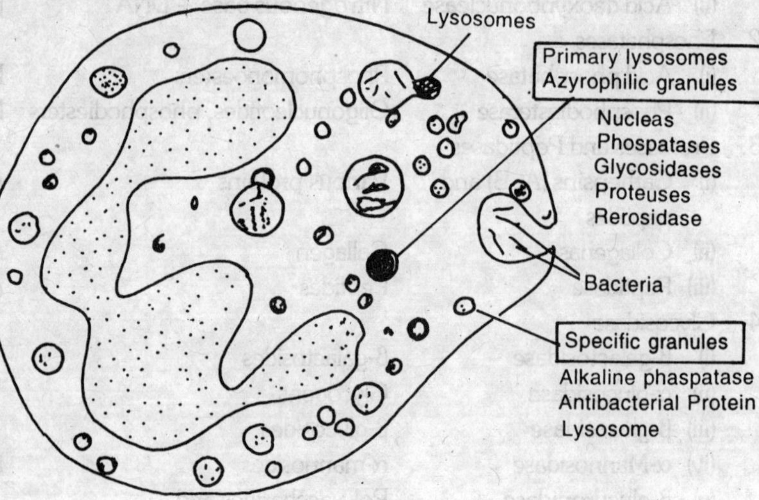

Fig. 11.6. A professional phagocyte.

2. Digestion of Intracellular Substance

In certain cases portions of the cell somehow, find their way inside the cell's own lysosomes and are broken down. This process is termed as *cellular autophagy.* How do they get in, is not clear, and the role which autophagy plays in cell function can only be summarized. Proteins, fats and polysaccharides can all be synthesized, and stored in the cell. During the starvation of cell these stored food materials are digested by lysosomes to give energy. What stimulates autophagy to take place and how do the large molecules get into the lysosome is not clear.

3. Cellular Digestion

When a cell dies, the lysosomal membrane ruptures. The liberated enzymes become free in the cell, which then quickly digested the entire cell. The hypothesis has been advanced that this is a built in mechanism for removing dead cells. In multicellular animals, many cells are constantly being formed, live for a short period of time, and then die. The self-digestion may occur as a pathological mechanism just for example, it a cell is cut of from its oxygen supply or poisoned, the lysosomal membrane may rupture, thereby permitting the enzymes to dissolve the cell. Therefore, they are also considered as *suicide bags* of the cells.

4. Extracellular Digestion

A cell can discharge lysosomal enzymes to destroy the surrounding structures. This function is performed by reverse phagocytosis. A pocket of enzymes from a lysosome is released outside the cell where it then digests contigious structures. This is thought to explain, how sperms penetrate the protective coating of the ovum during fertilization. It may also explain how oesteoblast cells that destroy bones, function. This may also be explanation for the well-known ability of white blood cells to pass quickly out of the blood vessels and into thetissues spaces at the site of an infection.

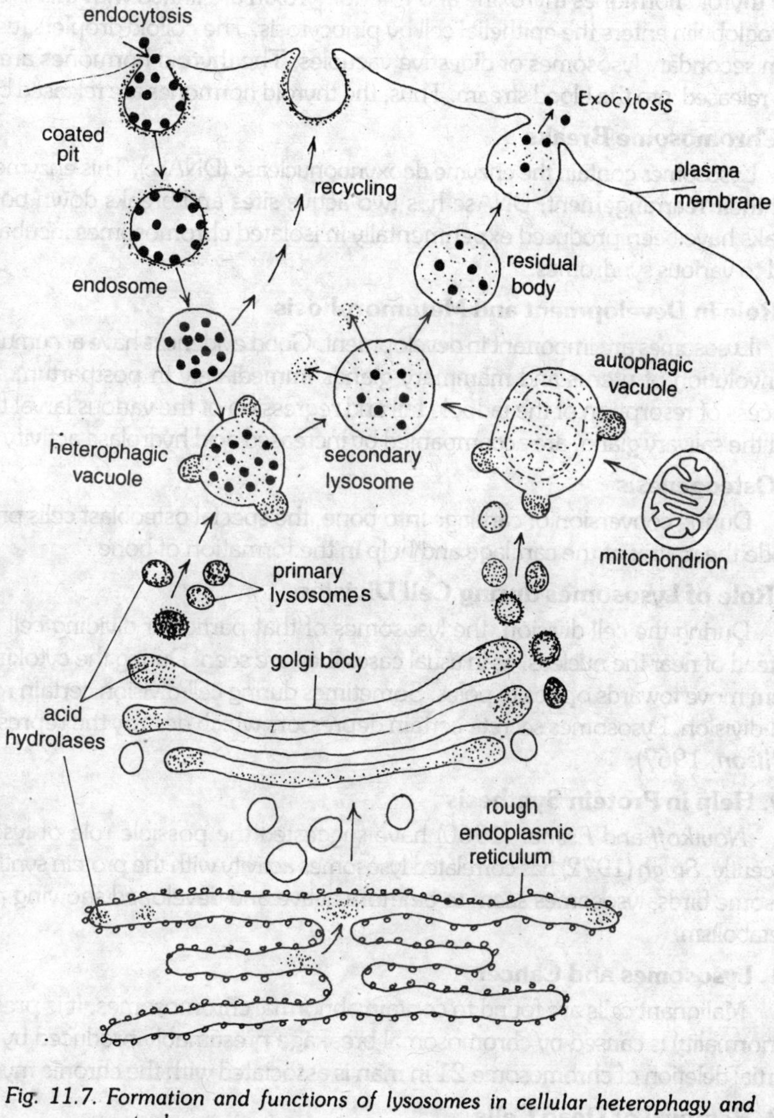

Fig. 11.7. Formation and functions of lysosomes in cellular heterophagy and autophagy.

5. Role in Secretion

In recent years evidences have started acumulating suggesting the role of lysosomes in the formation of secretory products in secretory cells. The phenomenon of lysosomes-mediated thyroid hormones secretion is the best known example of direct lysosomes involvement in secretory process. Lysosomes also play possible role in the regulation of hormone secretion. It is believed that mammotrophic hormones of the anterior pituitary is synthesized on the ribosomes of the RER and is packaged into secretory granules by passage through the Golgi.

The eptihelial cells of the thyroid also contain lysosomes rich in lysosomal enzymes. The follicles of the thyroid gland contain high molecular weight protein thyroglobulin, which is stored as colloid in the lumen. The thyroid hormones thyroxine and triiodothyroxin are linked with this protein. The colloid containing thyroglobulin enters the epithelial cell by pinocytosis. The colloid droplets fuse with primary lysosomes to form secondary lysosomes or digestive vacuoles. The thyroid hormones are split from the thyroglobulin and released into the blood stream. Thus, the thyroid hormones are released by hydrolysis of thyroglobulin.

6. Chromosome Breaks

Lysosomes contain the enzyme deoxyribonuclease (DNAse). This enzyme causes chromosomal breaks and their rearrangement. DNAse has two active sites and breaks down both the strands of DNA. The breaks have been produced experimentally in isolated chromosomes incubated in DNAse. These breaks lead to various syndromes.

7. Role in Development and Metamorphosis

Lysosomes are important in development. Good evidences have accumulated on the role of lysosomes in involution of uterus and mammary glands immediately in postpartum. During metamorphosis, the process of resorption of the tadpole tail and regression of the various larval tissues, including the fat body and the salivary gland, are accompanied by increased acid hydrolase activity (Weber).

8. Osteogenesis

During conversion of cartilage into bone, the special osteoblast cells produce lytic substances which erode the matrix of the cartilage and help in the formation of bone.

9. Role of Lysosomes during Cell Division

During the cell division, the lysosomes of that particular dividing cell move towards the periphery instead of near the nucleus, as in usual cases they are seen. During the cytokinesis roughly equal number of them move towards opposite poles. Sometimes during cell-division certain repressors in cytoplasm inhibit cell-division. Lysosomes secrete certain depressors which destroy the repressor and results in cell division (*Allison*, 1967).

10. Help in Protein Synthesis

Novikoff and *Essner* (1960) have suggested the possible role of lysosomes in protein synthesis. Recently, *Singh* (1972) has correlated lysosomes activity with the protein synthesis. In the liver and pancreas of some birds, lysosomes seem to be more active and developed showing possible relationship with cell metabolism.

11. Lysosomes and Cancer

Malignant cells are found to contain abnormal chromosomes. It is presumed that the chromosomal abnormality is caused by chromosomal breakage presumably produced by the lysosomal enzymes. The partial deletion of chromosome 21 in man is associated with the chronic myeloid leukemia (blood cancer).

12. Removal of Dead Cells

Hirsch and *Cohn* (1954) suggested that lysosomes help in the removal of dead cells in tissue. The lysosomal membrane ruptures in these cells, releasing the enzyme into body of cell, so that whole cell may be digested. This process of tissue degeneration (necrosis) is due to this lysosomal activity.

13. Fertilization

During fertilization the sperm releases hydrolytic enzymes from the acrosome vesicles. These enzymes help in the penetration of the sperm through the envelopes of the egg. Fluorescence microscopy studies of acridine-stained spermatozoa of the guinea pig show that the acrosome vesicles contain several enzymes, including hyaluronidase and proteases, which are also found in lysosomes. In fact, the acrosome vesicle has been looked upon as a *giant lysosome*. The acrosome vesicle enzymes also apparently activate the egg by breaking down its cortical granules.

14. Lysosomes and Disease

Inhalation of foreign particles like silica, asbestos etc., leads to inflammation and deposition of fibrous tissue in the lungs. The particles of silica or asbestos increase the permeability of lysosomal membranes and rupture of lysosomes. This leads to the lysis of lung cells resulting in their inflammation.

A metabolic disorder, the *gout*, is caused by the accumulation of sodium urate crystals in the joints. These are picked up by the phagocytes resulting in their lysosome rupture. This leads to acute inflammation and increased collagen synthesis. Pome's disease is another example which is caused due to absence of an lysosomal enzyme which hydrolize the glycogen. Thus, liver cells are engorged with glycogen.

Table 11.3. Some of the Disease Related to Genetic Abnromalities of Lysosomes are listed below (*Allison*, 1974).

S.No.	*Disease*	*Substance accumulated*	*Enzyme defect*
1.	Ceramide lactoside	Ceramid lactoside	β-glactosidase
2.	Gancher's disease	Glucocerebroside	β-glactosidase
3.	Generalized gandliosidosis	Ganglioside GM_1	β-glactosidase
4.	Krabbe's disease	Galactocerebroside	β-glactosidase
5.	Meta chromatic leucodystrophy	Ceramide glactose-3 sulphate	Sulfatidase
6.	Niemann-Pick disease	Sphingomyelin	Sphingomyelinase
7.	Tay-Sach's disease	Ganglioside GM_2	Hexosaminidose-A
8.	Type II glycogenosis	Glycogen	α-Glycosidase.
9.	Fabry's disease	Ceramide trihexodise	α-Glactosidase

Origin of Lysosomes

They have multiple origin depending upon the tissue in which they are located or on their function in a specific cell.

(i) *Extracellular origin*. Lysosome may be the vacuole absorbed into cell by the process, pinocytosis. The pinocytic vacuole may later become cytoplasmic particle and thereon enzymatic activity becomes developed.

(ii) *Origin from Golgi complex*. There are evidences that lysosomes originate from the Golgi complex and represent zymogen granules. Their similar function and structure with Golgi complex support this view. Recent studies have shown that accumulation of secretory products within Golgi vacuoles leads to the formation of lysosomes, and membranes surrounding the products are derived from Golgi membrane.

(iii) Origin from ER. *Novikoff* (1965) reported that the lysosomes originate directly from granular endoplasmic reticulum, by a process of blebbing.

MICROBODIES (PEROXISOMES AND GLYOXISOMES)

Under electron microscopic examination, in 1954 in the cells of kidney tubules, a membrane-bound with a dense granular structure appeared. It measured about 0.5-1.0 mm in diameter. It was named as *microbody*. In 1956, an another similar structure was described in the liver tissue. In these cells some type of crystallines core was seen within the vesicles. When cells are homogenized and subjected to differential centrifugation, microbodies sediment with small particle fraction containing the lysosomes and mitochondria. These microbodies contain different enzymes and they are involved in particular types of oxidative reactions. It has been shown that the absence of microbodies in liver and kidney tissue is associated with a fatal disease of infancy, *cerebrohepatorenal syndrome*.

Peroxisomes

The term peroxisomes is coined by *de Duve* (1969) for those microbodies which are rich in enzymes *peroxidases* and *catalases,* and produce hydrogen peroxide (H_2O_2) during their degenerative activity.

Structure

There are oval or spherical structures limited by a single membrane. They contain fine granular substances that may condense to form an opaque core or *nucleoid.* If the nucleoids are absent such peroxisomes are called *microperoxisomes.*

Following enzymes have been isolated from peroxisomes. These are uric acid oxidase, D-amino acid oxidase, α-hydrolytic acid oxidase, NADH-glyoxylate reductase, NADP—isocitrate dehydrogenase catalase and glycolate oxides.

Function

In green plants peroxisomes are involved in *photorespiration*. The photorespiration occurs in two steps. In the first step-oxidation of one of a variety of substrates using molecular. Oxygen as the oxidizing agent, resulting in the formation of hydrogen peroxide (H_2O_2). These reactions are catalysed by various oxidases. Since H_2O_2 is a very reactive and toxic oxidizing agent, some means must be available in the organelles to destroy this compound so that it cannot accumulate. In the second step H_2O_2 is degraded into water and oxygen. This reaction is catalyzed by the enzyme catalase.

O_2

RH_2

oxidases

R_2

H_2O_2

$R'H_2$

Catalase

R' O_2

$2H_2O$

(here RH_2 may be amino acid or uric acid).

Depending upon the plant and its environmental conditions, such as light, oxygen availability etc., chloroplasts have the potential to produce a 2-carbon compound celled *glycolate* the glycolate is formed as:

Ribulose - 1, 5-diphosphate $\xrightarrow{O_2}$ phosphoglycolate + 3, phosphoglycerate.

Phosphogycolate ⟶ glycolate.

As soon as glycolate is formed in chloroplast, it diffuses out into peroxisomes

Glycolate + O_2 $\xrightarrow{\text{Glycolate oxidase}}$ Glyoxylàte + H_2O_2

The glyoxlate of peroxisome can be metabolized as

Glyoxylate + Glutamic acid ⟶ Glycine + α-ketogluteric acid.

Glycine is transferred to the mitochondrion, where it is converted into serine.

Glycine ⟶ Serine + CO_2 + NH_3.

Origin of peroxisomes

The close morphological association between peroxisome and endoplasmic reticulum suggests that they are originated from endoplasmic reticulum and their enzymes are synthesized by the ribosomes.

Glyoxisomes

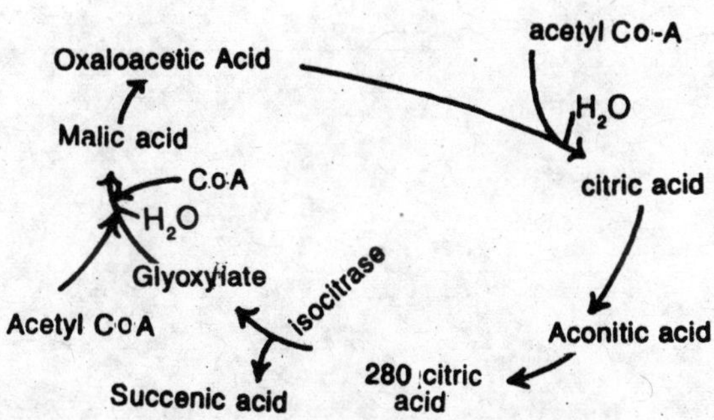

Along with peroxjsomes, plants contain another organelles which, in addition to glycolate oxidase and catalase, also contains a number of other enzymes not found in animal cells. These are called *Glyoxisomes* and found only in the cells of endosperm or plant seedlings.

Glyoxysomes were described by *Briedenbach* (1967). Their structure is similar to that of peroxisomes except in place of granules, crystals are bounded by a single membrane. Two enzymes *isocitrase* and *malate synthetase* are present besides several essential enzymes of Krebs' cycle.

Function

One of the primary activities of glyoxisomes in the germinating seedlings is the conversion of stored fatty acids into carbohydrate i.e., gluconeogenesis. Since the product of fatty acid hydrolysis is acetyl Co-A, one might expect that glucose formation could occur by shutting the acetyl Co-A through the Kreb's cycle.

An alternate pathway for acetyl Co-A in plants cells is afforded by the *glyoxylate cycle,* which takes place in glycoxisomes. In this cycle, the first step is the condensation of acetyl Co-A with oxalo-acetic acid (as seen in Kreb's cycle), results in the formation of citric acid. Two decarboxylation steps of Kreb's cycle are by passed and succinic acid is formed along with glyoxylate. The succinic acid can be utilized in gluconeogenic reactions. Glyoxylate can serve to regenerate oxalo-acetic acid for an additional round of the cycle. Presence of catalise and glycolate oxidase provide the means for other compounds to be converted to glyoxylate, which can then enter the cycle.

12

RIBOSOMES

Ribosomes are the basophilic granules present in the cytoplasm of the cell. Since this material had an affinity far basic stains similar to that of chromatin granules of the nucleus, so it was for a time called *chromidial* or *chromophil substance.* The ribosomes were first noted in plant cells by *Robinson* and *Brown* in 1953 while studying been roots with the electron microscope and shortly afterward *Palade* (1955) observed them in animal cells. He isolated the ribosomes and detected the RNA in them that is why also called RNP or *ribonucleoprotein particles* or *Palade granules. Claude* named them as *microsome* but the name *ribosome* was designated by *Robert* in 1958.

Occurrence

They are universally distributed throughout the animal kingdom and plant kingdom. In prokaryotes too, they are found. The only cell types devoid of ribosomes is the mammalian RBC. The density of ribosomes per unit area being rather constant for any given type. It is high in the cells which are active in protein synthesis and low in cells where protein synthesis is low.

Distribution

In prokaryotic cells the ribosomes often occur freely in the cytoplasm. In eukaryotic cells the ribosomes either occur freely in the cytoplasm or remain attached to the outer surface of the membrane of endoplasmic reticulum. When they are not attached to the ER they are called free ribosomes. Free ribosomes serve as sites for the synthesis of proteins that are required to maintain the enzyme constitution of the cytoplasmic matrix.

Method of Isolation

The ribosomes are usually isolated from the cell by the differential centrifugation method in which analytical centrifuge is employed. The sedimentation coefficient of the ribosomes is determined by the various optical and electronic techniques. The sedimentation coefficient is expressed in the *Svedberg unit,* e.g., 'S' unit. The S is related with the size and molecular weight of the ribosomal particles.

Number and Concentration of Ribosomes

In all cells that contain endoplasmic reticulum, a good number of ribosomes can be observed. For example, at the base of gland cells, in plasma and liver cells, in all rapidly growing plant and animal cells and in bacteria, quantity of RNA can be related with the concentration of ribosome. In reticulocytes of rabbit nearly 100 ribosomes are found per $\mu3$, which corresponds to 1×10^5 particles per reticulocyte and about .5% of total volume of the cell mass, or about 20,000 to 30,000 per cell. However, it the rate of protein synthesis is slowed down by unfavourable nutritional conditions, the number of ribosomes can drop considerably in protein synthesizing cells and in bacteria.

Structure of Ribosomes

Ribosomes are remarkable for their uniformity in size and composition through the wide range of cells in which they have been studied. Ribosomes of higher plants and animals are oblate, spheroids and their diameter is approximately 250Å. However, the ribosomes of bacteria are some what smaller, because they contain less amount of proteins than the higher animal ribosomes. However, in quantity of RNA per particle, bacterial ribosomes resemble all the other presently studied ribosomes.

In the electron microscopic studies negative staining reveals a cleft that divides the ribosomes into a larger sub-unit and a smaller sub-unit. In *E. coli* the larger particle is somewhat 'cup' shaped or dome shaped (140 to 160Å) and smaller one forms a 'cup' (90 to 110Å) that is applied to the flat surface of the other (*Huxley* and *Zubey* 1960). In higher animals and plants it was shown that the ribosomes are attached to endoplasmic reticulum by the large sub-units.

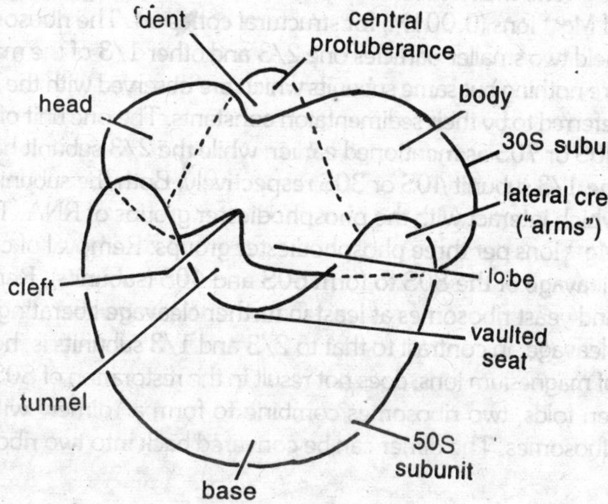

Fig. 12.1. Stoffler and Wittman's model of 70S ribosome.

The fine structure of the ribosome is very complex and not yet fully elucidated. Since the ribosomes are highly porous and hydrated, the RNA and protein are probably interwined within the two subunits. In sections, stained with uranyl ions (RNA selective stain), each ribosome appear as a star shaped body with four to six arms implanted on a dense axis. The isolated 50S subunit of *Bacillus subtilis* appears as a compact particle of 160 to 180Å having a pentagonal face, in the center of which is a round area of 40 to 60Å (*Nanninga,* 1967). An electron transparent core, which is negative stained ribosomes correspond to an electron opaque region, has been described in the large subunits. The 40S subunit is not regular and tends to be subdivided into two portions that are inter-connected by a strand 30 to 60Å thick.

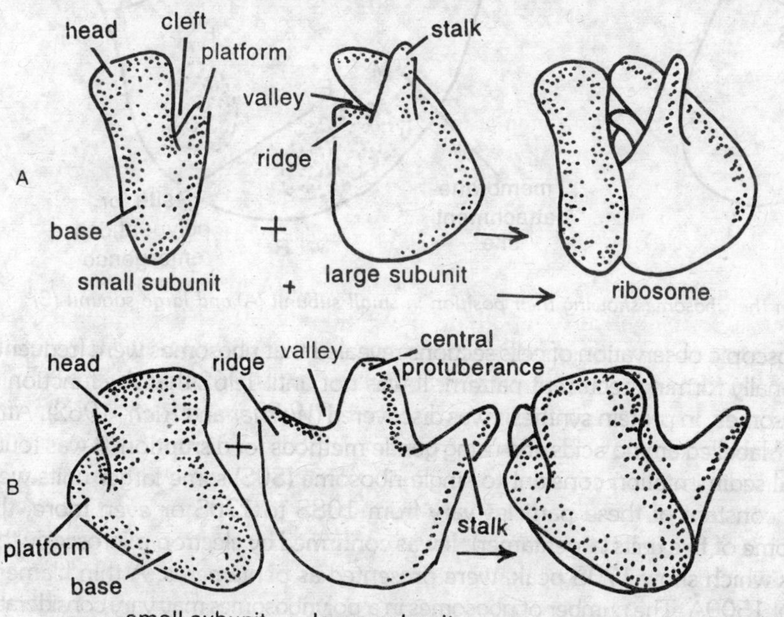

Fig. 12.2. Lake's model of the 70S ribosome. Note the three dimensional structure of the ribosome in two different orientations.

Ribosomal Subunit

Another property which is common to all ribosomes in their subunit structure and their sedimentation constant by ultracentrifugation. On the basis of sedimentation constant there are two main types of ribosomes. Those form bacteria usually have a coefficient of 70S (S-Svedberg unit) corresponding to a molecular weight of 2.7 x 10^6.

The others are the ribosomes of eukaryotic (nucleated cells, either plant or animal) such ribosomes posses sedimentation constant of 80S with a molecular weight of about 4 x 10^6 daltons.

The two, above mentioned (cup and cap) structural subunits of ribosomes require low concentration of Mg^{++} ions (0.001M) for structural cohesion. The ribosomes may be cleaned by removal of Mg^{++} ions to yield two smaller particles one 2/3 and other 1/3 of the mass of the original intact ribosome. Actually they are nothing but same subunits which are observed with the help of electron microscope. These subunits are referred to by their sedimentaion constants. The one unit of ribosome possesses a sedimentaion constant of 80S or 70S as mentioned earlier, while the 2/3 subunit has a sedimentation constant of 60S or 50S and the 1/3 subunit 40S or 30S respectively. Both the subunits are bound together through magnesium ions which interact with the phosphodiester groups of RNA. The fully magnesium saturated particle contains Mg^{++} ions per three phosphodiester groups. Removal of calculated 1/3 of these magnesium ions result in cleavage of the 80S to form 60S and 40S (subunits). Removal of further Mg^{++} results in the case of pea and yeast ribosomes at least in further cleavage liberating what are apparently 1/6 subunit. This further cleavage, in contrast to that to 2/3 and 1/3 subunits is, however, irreversible in the sense that restoration of magnesium ions, does not result in the restoration of 80S particles. If the Mg^{++} concentration is increased ten folds, two ribosomes combine to form a "dimer" with twice the molecular weight of the individual ribosomes. The dimer can be convered back into two ribosomes by lowering the Mg^{++} concentration.

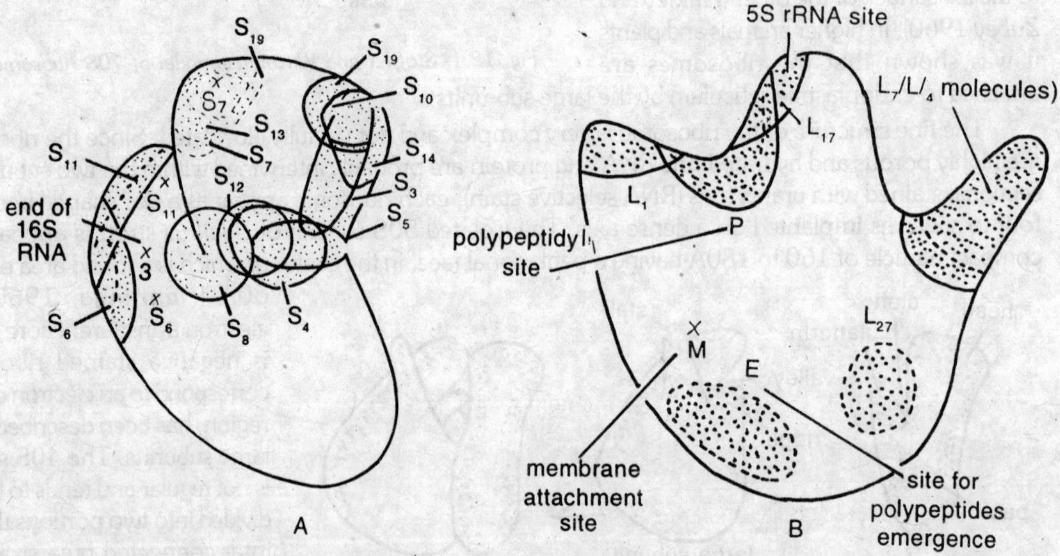

Fig. 12.3. Map of proteins in the ribosome showing their position in small subunit (A) and large subunit (B).

The early electron microscopic observation of cells sections revealed that ribosomes were frequently associated in groups occasionally forming recurrent pattern. It was not until 1962 that the function of these polyribosomes, or polysomes, in protein synthesis was discovered (*Warner* and *Rich*, 1962). After treating reticulocytes with C^{14} labelled amino acids and using gentle methods for disruption, it was found that in addition to the typical sedimentation constant to single ribosome (80S) some larger units were present. The sedimentation constant of these particles vary from 108S to 170S or even more; this corresponded to a polyribosome of five units (a pentamer). It was confirmed by electron microscopy that about 75% of the ribosomes which show 170S peak, were presented as pentameres. A thin filament, interpreted as in mRNA, about 1500Å. The number of ribosomes in a polyribosomes may vary considerably and seems to be related to the length of the mRNA that should be 'read' in the translation process.

In *E. coli* and in cells from chick embryo, polyribosomes composed of about 50 units have been observed (*Rich* 1967). Polyribosomes may be free in the cytoplasm or bound to the membranes of the endoplasmic reticulum. For free polyribosomes, a helical configuration has been postulated with the small subunits arranged around the central axis and the large subunits disposed at the periphery. In sections of various cells polyribosomes have been observed in a helical array (*Weiss* and *Grover*, 1968). It is assumed that in the polysome the mRNA is situated in between the two subunits of the ribosome.

Types of Ribosomes

The ribosomes are of two basic types, 70S and 80S ribosomes. The S refers to Svedberg units. Actually it is the sedimentation coefficient which shows how fast a cell organelle sediments in an ultracentrifuge. Sedimentation coefficients are not additive. 80S ribosomes are found in eukaryotes (organisms whose cells have true nuclei bounded by nuclear envelopes) e.g., algae, fungi, higher plants and animals. The 80S ribosome of animals consists of a large 60S subunit and a small 40S subunit.

70S ribosomes are relatively smaller and are found in prokaryotes (organisms whose DNA is not bounded by a nuclear envelope) e.g., bacteria. The 70S ribosome consists of a large 50S subunit and a small 30S subunit.

Table 12.1 Characteristics of Different Ribosomes Types.

Ribosome Type	*Found in*	*Large subunit*		*Small subunit*	
		Size	*RNAs*	*Size*	*RNAs*
80S	Eukaryotes (animal)	60S	28-29S+5S+5.8S	40S	18S
80S	Eukaryotes (plants)	60S	25S+5S+5.8S	40S	16-18S
70S	Prokaryotes	50S	23S+5S	30S	16S
55S	Mitochondria	40S	16-17S+5S	30S	12-13S

Table 12.2. Differences between 70S and 80S Ribosomes.

Presence	*In Prokaryotes (bacteria)*	*In eukaroytes (algae, fungi, higher plants and animals)*
Sedimentation Coefficient	64S-72S (average 69S)	79-85S in fungi 80S in mammals.
Size	Relatively smaller	Relatively larger
Molecular wt.	3 x 10^6	4 - 5 x 10^6
Subunits	Small 30S and large 50S	Small 40S and large 60S.
RNA	3 molecules of RNA: 16S RNA in 30S subunit, 23S and 5S RNA in 50S subunit	4 molecules of RNA 16S-18S RNA in 40S subunit; 25-29S, 5.8S and 5S RNA in 60S subunit.
M.Wt. of RNA	16S RNA-550,000 23S RNA-1,100,000 5S RNA-40,000	18S RNA-700,00 28S RNA-1,700,000 5.8S RNA-51,000 5S RNA-39,000
No. of proteins	21 (S1-S21) in small subunit 34 (S1-L34) in large subunit Total in prokaryotes: 50-60 proteins.	33 in small subunit 49 in large subunit Total in eukaryotest 70-80 proteins
Average M.Wt of protein	18,000	21,000
No of amino acids	8,000	16,000
RNA-Protein ratio	2.1	1:1

Ribosomes found in mitochondria and chloroplasts of eukaryotes are closer to prokaryote ribosomes rather than the 80S eukaryote ribosomes. Vertebrate mitochondria, for example, contain 55S ribosomes, each with a large 40S subunit and a small 30S subunit. The sedimentation coefficients 80S, 70S and 55S are rounded off values. Actual S values in different organisms may be slightly higher or lower.

Chemical Composition

The main constituents of ribosomes are RNA and proteins. The lipids are entirely absent present in traces. The ribosomes of *E. coli* passes nearly 60-65% of RNA and 35-40% protein of their weight. Table 12.3 is showing approximate quantity of RNA and protein in different types of ribosomes.

Ribosomal RNA differ in size and base contents from tRNA and from other RNA classes of most cells. Two types of RNAs are found in all ribosomes. They are an integral component, and cannot be removed easily. RNA of rat liver particles contain chiefly the common bases adenine, guanine, cytosine, and uracil with small amount of pseudouridine, the usual bases which occur in the soluble RNA are found in particles RNA only in very small amount.

Table 12.3. Showing the Chemical Composition of Ribosomes

Ribosomes	*Sedimentation constant*	*Subunits*	*S.U. of % RNA*	*RNA in complete ribosomes*	*% protein in complete Ribosomes*
Liver	80S	60S	28S	45%	55%
Yeast	80S				
Neurospora	80S	40S	18S		
Mitochondria	70S	50S	23S		
E.coli	70S			65%	35%
Chloroplasts	70S	30S	16S		

Amino acids

The amino acid composition of trichloracetic acid insoluble protein of rat liver RNP has been determined by *Crompton* and *Petermann* (1959). Aromatic and sulphur containing amino acids were present in very small amount while leusine and arginine were high over 10%. The composition of both rabbit reticulocytes and pea seedling RNPs are similar remarkably.

A hydrochloric extract of partially purified rat live RNP has yielded arginine (*Butter,* 1960), but whether these aminoacids were derived from the proteins itself has not yet been determined.

Protein

Protein contains linear chain of amino acids. The similar in ribosome composition of aminc acids of ribosomal protein is remarkably of different origin and may be quite different from that of protein formed by the ribosome. Therefore, it is common to distinguish between ribosomal structural protein and growing peptide chain of enzyme under going fabrication by the ribosome that protein is associated with ribosomal RNA by hydrogen bonding is clear from the fact that dissociation of the two is rapidly and readily accomplished by reagents which attack hydrogen bonds as for example by guanidium bromide. Experiments have been done by *Yin* and *Bock* (1960), who obtained a stable yeast ribosomal protein. *Watson* (1960) got the protein of *E.coli* and by analysis of such a group indicates the molecular weight of each of sub-units to be about 30,000.

Ribosomal enzymatic proteins

Most of the ribosomal proteins act as enzymes thereby catalyzing protein synthesis. Initiation proteins IF1, IF2, and IF3 initiate the process of protein synthesis whereas transfer proteins (G-factor, Ts-factor) help in translocation of ribosomes over mRNA and transfer to t-RNA residue from one site of the ribosome to the other site. Another enzymes peptidyl transferase helps in the transformation of peptide chains to aminoacyl 1-tRNA and other enzymes—termination of completed polypeptide chain.

As a result of washing of ribosomes with NH_4Cl and subjected them to column chromatography, *Ochoa* and *coworkers* (1960) isolated above factors in protein synthesis. Among them, three initiation factors—IF1, IF2, and IF3—are loosely associated with 30S subunit. IF1 factor is a basic protein having molecular weight of 9200 daltons. It is involved in the binding of F-met-tRNA. IF2 factor is also a protein of mol. wt. 8000 daltons and contains-SH groups which help in binding with GTP. Third protein factor—IF3, does not require GTP and is involved in binding mRNA to 30s subunits. It is a basic proteins having molecular weight of 30,000 daltons. IF3 may also act as dissociation factor for 70s ribosomes. *Ochoa* et. al. (1972) have further reported interference factor (i) in bacteria *E. coli*. These factors bind to IF3 factor, change its specificity and thus regulate the translation of genetic message at the beginning.

Elongation factors are essential for the elongation of polypeptide chain. There are EFG (also called G factor or translocase) and EFT factor. As described earlier EFG or G factor is involved in the translocation of mRNA. In *E. coli*, it consists of single polypeptide chain having mol. wt. of 72,00 daltons. EFG + GTP promotes the translocation of newly elongated peptidyl 1-tRNA. Another EFT factors has two kinds of proteins namely Tu (temperature-unstable) and Ts (temperature stable). The EFTu factor+GTP forms complex aminoacyl-tRNA before its binding to the acceptor site of ribosome being catalyzed by Ts factor. Moreover, 50s ribosomal subunit has an enzyme-peptide synthetase or peptiydl transferase associated in the formation of peptide bond. Termination factors R1 and R2 are releasing proteins helping in liberating the polypeptide chain.

Biogenesis of Ribosomes

The biogenesis of ribosomes in bacterial prokaryotic cells takes place inside the cytoplasm because of the absence of nucleous. The rRNAs originate from the specific codons of the genome or the ribosomal DNA (rDNA).

In eukaryotic cells the process of biogenesis of ribosomes is complicated and occurs inside the nucleolus. One of the chromes of a set has a specific nucleolus organizer region, which contains nucleolar RNA molecule which is a precursor of both 28S and 18S rRNA. The process of conversion of 45S-RNA into 28S and 18S rRNA is illustrated below:

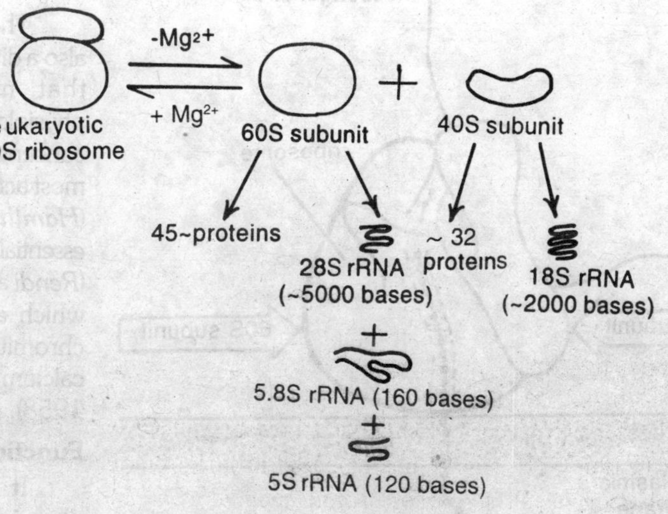

Fig. 12.4. Different RNA and protein components of eukaryotic ribosomes.

45S nucleolar RNA molecules are methylated (-CH_3 group is added). These methylated molecules of 45S nucleolar RNA become associated with the necessary proteins present in the nucleolus forming 80S ribonucleoprotein molecule particles (RNP). These 80S RNP with 45S molecular RNA split up into 32S and 18S rRNA through several intermediate steps, by which non-methylated portion of the molecules are lost. 18S molecule together with its protein molecules immediately transported to the cytoplasm. The 32S rRNA remains inside the nucleolus for sometime and then gets cleaved into 28S rRNA.

The 5S rRNA is synthesized outside the nucleolus and the genes lie adjacent to nucleolar organizer region of the chromosome.

The 18S rRNA together with its proteins leaves the nucleus through nucleopore and comes out into the cytoplasm, where in association with proteins this assembles into small subunit (40S) of the ribosome.

The 28S rRNA also leaves the nucleus and gets incorporated with 5S rRNA and the proteins forming 60S subunit.

The synthesis of ribosomal occurs partly inside the nucleolus and partly inside the cytoplasm. The proteins synthesized in the cytoplasm assemble in the nucleolus to be utilized in ribonucleoprotein molecule particles (RNP).

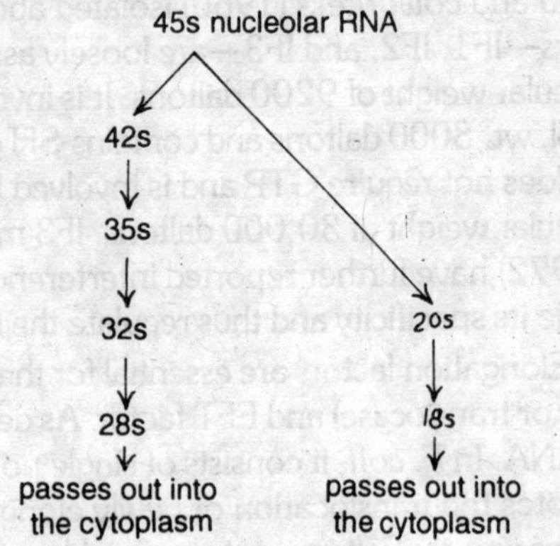

Fig. 12.5. Diagrammatic representation of the steps involved in the conversion of 45S nucleolar RNA into 28S and 18S rRNA.

Ribosomal RNA

The 70S ribosomes contain three types of rRNA, viz., 23S rRNA, 16S rRNA and 5S RNA. The 23S and 5S rRNA occur in the larger 50S ribosomal subunit, while the 16S rRNA occurs in the smaller 30S ribosomal subunit. The 23S rRNA consists of 3200 nucleotides, 16S rRNA contains 1600 nucleotides and 5S rRNA includes 120 nucleotides in it (*Brownlee*, 1968, *Fellner*, 1972).

The 80S ribosomes also contain three types of rRNA, viz., 28S rRNA, 18S rRNA and 5S rRNA. The 28S and 5S rRNA occur in the larger 60S ribosomal unit, while the 18S rRNA occur in the smaller 40S ribosomal subunit.

The 28S rRNA has the molecular weight 1.6×10^6 daltons and its molecule is double stranded and having nitrogen bases in pairs. The 18S rRNA has the molecular weight 0.6×10^6 daltons. The molecule of 5S rRNA has a clover leaf shape and a length equal to 120 nucleotides (*Forget* and *Weissmann*, 1968).

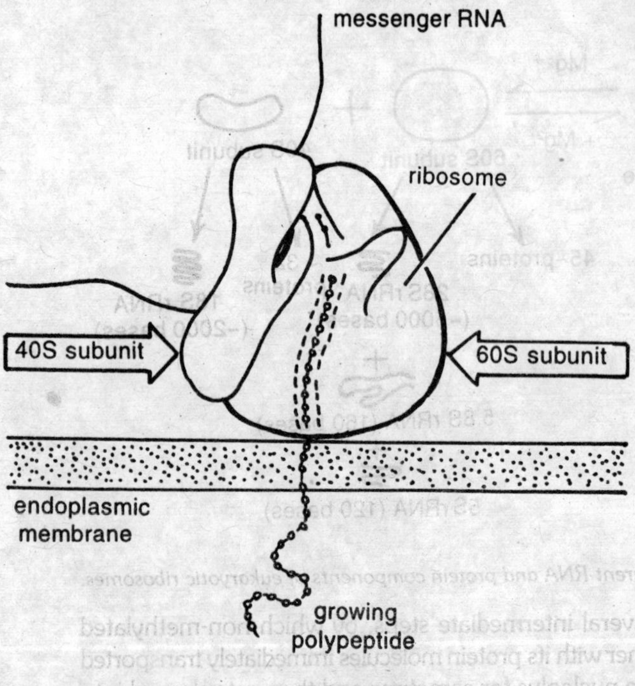

Fig. 12.6. A three dimensional model of eukaryotic cytoplasmic ribosome.

Other constituents

The metal contents or ribosomes are also a difficult question. There is little doubt that magnesium is the chief ion physiologically, since it is present in high concentration in rat liver ribosomes. It is the most active in maintaining the 80S structure (*Hamilton* and *Petermann*, 1959) as it is essential for amino acid incorporation in *vitro* (*Rendi* and *Hultin*, 1960). The other metals which are present in the ribosomes are chromium, manganese, nickel, iron and calcium (*Walker* and *Valle*, 1958 and *Tso* 1958).

Function of Ribosomes

It is a well-known fact that the ribosomes are factories of manufacturing of protein in a cell but actually one ribosome cannot take part in the protein synthesis. It was known from 1962 when a report was published which showed that active units are no individual ribosome, but a group of these units, which is called as polyribosome. A detailed role of polyribosome was published in Scientific American Dec. 1963 by *Gric* and *Hall*. There is much evidence that ribosomes are alike and are at least interchangable.

In case of haemoglobin with a chain of 150 amino acids the most usual number of ribosomes which were linked together was 5. There may be large number of ribosomes and upto 40 ribosomes have been described. So the name polysome can also be applied at the place of polyribosome.

Protein consists of linear chain of amino acids. The chain may be short or long in both the cases it is called polypeptide chain. Polypeptide can be folded in a specific manner often combine to form a complex protein.

The ribosomes in a polysome are separated by gap of 50-150Å. Positive staing by Uracil-acetate reveals that the ribosomes are connected by a thin thread 10-15Å in diameter, which is about the thickness of a single strand of RNA. From the size of the gap between ribosomes, the total measurement of the gap containing five ribosomes, is about 1500Å, the five ribosomes have inter ribosomal gap of 50-150Å each.

13

MITOCHONDRIA

Kolliker (1880) was the first who observed the granules (mitochondria) in muscle cell of insects. *Flemming* (1882) named the mitochondria as *fila. Altmann* in 1894 observed them and they were called *Altmann's granules bioplasts.* The term *mitochondria* was applied by *Benda* (1897-98) to these granules which were described as *cytomicrosomes* by *Velette St. George. Benda* Stained mitochondria with alizarin and crystal violet. *Kingsbury* (1912) related them to cellular respiration and *Warburg* (1913) observed the presence of respiratory enzymes. In 1934 *Bensley* and *Horr* isolated mitochondria from liver cells and *Porter* and *Palade* described their electron microscopic structure.

Distribution

Ordinarily mitochondria are distributed in the cytoplasm. They may, however, be localized in certain regions. In proximal convoluted tubules of the kidney they are found in the basal region of the cell, opposite the renal capillaries. In skeletal muscles they lie between the myofibrils. In insect flight muscle several large mitochondria are in contact with each fibril. In cardiac muscle the mitochondria are situated in clefts between the myofibrils, numerous lipid droplets are associated with the mitochondria. In many sperms the mitochondria fuse into one or two structures which lie in the middle piece of the tail, surrounding the axil filament. In columnar or prismatic cells they are oriented parallel to the long axis of the cell. In leucocytes they are radially arranged.

Orientation

Mitochondria may have a more or less definite orientation. For example in cylindrical cells they are generally oriented in the basal-apical direction, parallel to the main axis. In leukocytes, mitochondria are arranged radially with respect to the centrioles. It has been suggested that these orientation depend upon the direction of the diffusion currents within cells and are related to the submicroscopic organization of the cytoplasmic matrix and vaculolar system.

Plasticity of Mitochondria in a Cell

Lewis and *Lewis* (1914-15) concluded that the mitochondria are extremely variable bodies, which are continually moving and changing shape in cytoplasm. There are no definite types of mitochondria, as any one type may change into another. They appear to arise in cytoplasm and to be used up by cellular activity. The shape may change fifteen to twenty times in ten minutes; it can be changed by heat, hypertonic and hypotonic media or by acids, fat solvents, potassium permaganate and osmotic changes. *Frederic* (1958), *Littre* (1954), *Tobioka* and *Biesels* (1956) studied the effect of large number of chemical and physical agents on mitochondrial behaviour. Some material, such as detergents, show some effect *in vivo* as on mitochondria isolated from homogenates.

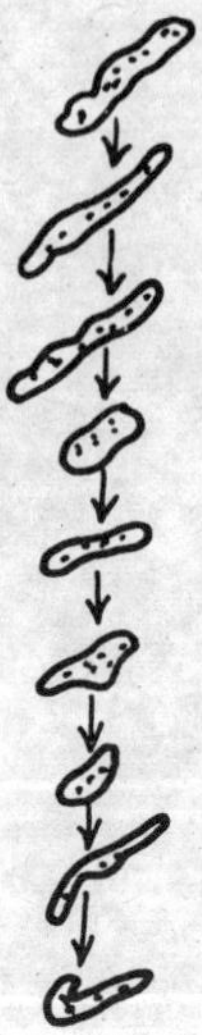

Fig. 13.1. Rapid changes of shape in living cells.

MORPHOLOGY

Shape

The shape is variable but is characteristic for a cell or tissue type, this too is dependable upon environment or physiological conditions. In general they are *filamentous* or *granular.* They may swell out one end to become *club-shaped* or hollow out at one end to assume a shape of *tennis-racket.* They may become *vesicular* by the appearance of central clear zone. *Rod-shaped* mitochondria are also observable.

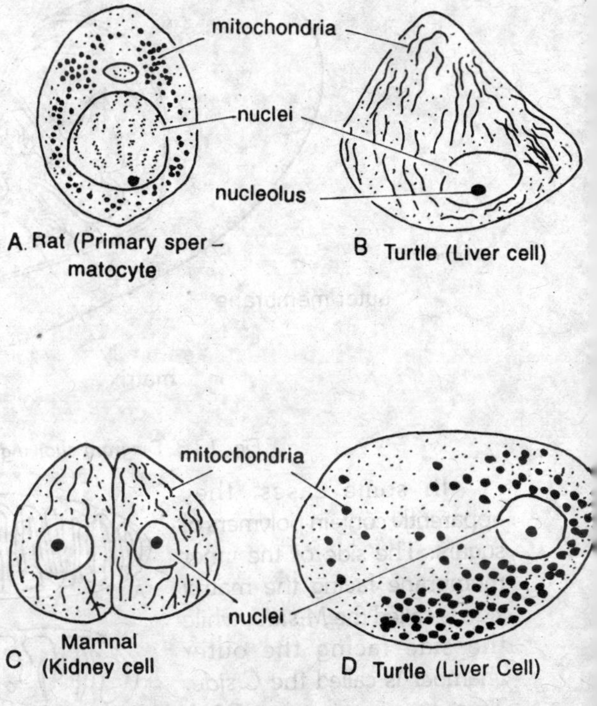

Fig. 13.2. Mitochondria of the flight muscle of a dragon fly, showing profuse cristae.

Size

The size of mitochondria also varies. In majority of the cells, width is relatively constant, about 0.5μ, but the length varies and, sometimes, reaches a maximum of 7μ. The size of cell also depends on the functional stage of the cell. Very thin mitochondria, about 0.2μ, or thick rods 2μ are also seen. The size and shape of the fixed mitochondria are determined by the osmotic pressure and pH of the fixative. In acid, pH mitochondria are fragmented and become vesicular. Mitochondria, in the rat-liver, are usually 3.3μ in length; in mammalian exocrine pancreas, they are about 10μ in length and in oocytes of Amphibia, they are approximately of the length of 20 to 40μ.

Number

Mitochondria are found in the cytoplasm of all aerobically respiring cells with the exception of bacteria in which the respiratory enzymes are located in the plasma membrane. The mitochondria content of a cell is difficult to determine, but in general, it varies with the cell type and functional stage. It is estimated that in liver mitochondria constitute 30 to 35 percent of the total protein content of the cell and in kidney, 20 percent. In lymphoid tissue the value is much lower. In mouse liver homogenates there are about 8.7×10^{10} mitochondria per gram of fresh tissue. A normal liver cell contains about 100 to 1600 mitochondria, but this number diminishes during regeneration and also in cancerous tissue. This last observation may be related to decreased oxidation that accompanies the increase to anaerobic glycolysis in cancer. Another interesting finding is that there is an increase in the number of mitochondria in the muscle after repeated administration of the thyroid hormone, *thyroxin.* An increased number of mitochondria has also been found in human hyperthyroidism.

Thus cell with high metabolic activity have a high number of mitochondria, while those with low metabolic activity have a lower number. Large sea urchin eggs have 13,000-14,000, while renal tubules have 300-400. In the sperm there are as few as 20-24 mitochondria while in some oocytes there are about 300,000. In the protozoan *Chaos chaos* there are about 500,000 mitochondria. Some algal cells contain only one mitochondria.

Structure of Mitochondria

A typical mitochondria in sausage shaped with an average diameter of about 0.5μ. When it is properly fixed in osmimum containing fluid and studied under electron microscope which reveals that there is hardly any difference between plant and animal mitochondria. In both the cases the mitochondria is bounded by two membranes, the *outer membrane* and the *inner membrane.* The space between the two membranes is called the *outer chamber* or *inter-membranous space.* It is filled with a watery fluid, and is 40-70Å in width. The space bounded by the inner chamber is called the *inner chamber* or *inner membrane space.* The inner membrane space is filled with a *matrix* which contains dense granules (300-500Å), ribosomes and mitochondrial DNA. The granules consist of insoluble inorganic salts and are believed to be the binding sites of divalent ions like Mg^{++} and Ca^{++}.

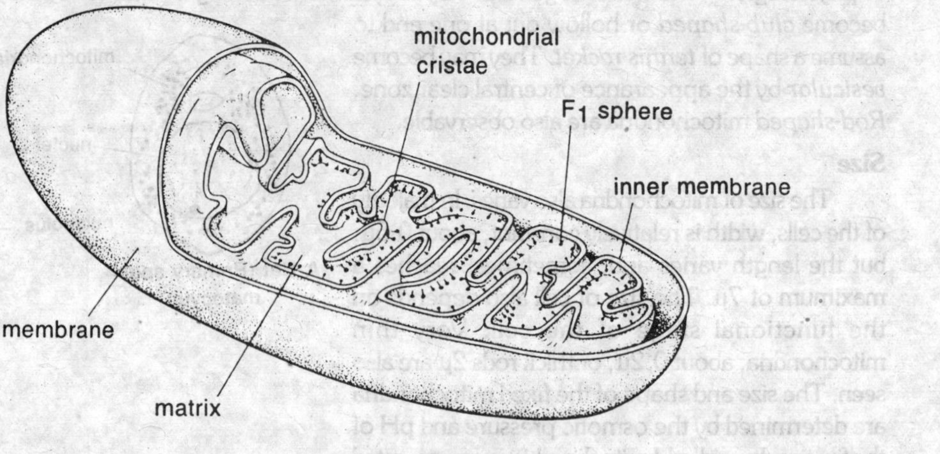

Fig. 13.3. Diagram showing the fine structure of a mitochondrion.

In some cases, they apparently contain polymers of sugars. The side of the inner membrane facing the matrix side is called the *M-side*, while the side facing the outer chamber is called the *C-side.* Two to six circular DNA molecules have been identified with mitochondria. These ring may either be in the open or in the twisted configuration. They may be present free in the matrix or may be attached to the membrane. The enzymes of the Krebs' cycle are located in the matrix.

The inner membrane is thrown up into a series of folds, called *cristae mitochondriales,* which project into the inner chamber. The cavity of the cristae is called the *intercristae space,* and is continuous with the inner-membrane space.

The space and arrangement of crests is variable and may be of the following types:

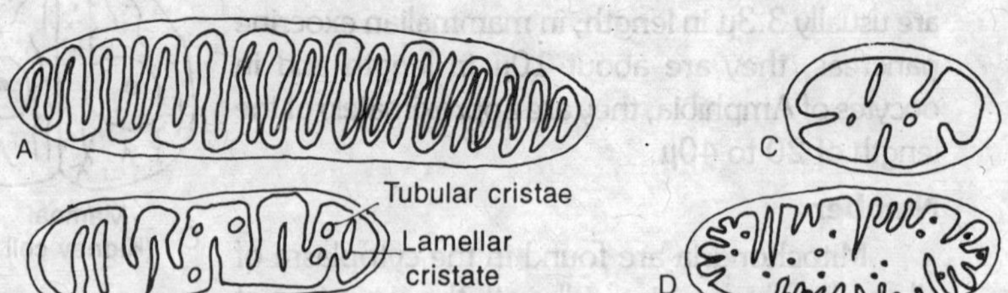

Fig. 13.4. Structural variations in mitochondrial cristae. A—Tubular, B—Lamellar cristae, C—Cristae in salamander, D—Plate like cristae, E—Parallel cristae, F—Cristal with sharp angulation, G—Honey corubs cristae, H—Usual cristae, I—Longitudinal cristae, J—Transverse and longitudinal cristae.

(i) Parallel to the long axis of mitochondria as in the neurons and the striated muscles cells.

(ii) Concentrically arranged as in the matrix of certain spermatids.

(iii) Interlaced to form villi as in *Amoeba*.

(iv) Cristae in the form of vesicles which form a network of interconnected chambers as in the cells of parathyroid gland and W.B.C. of man.

(v) Arranged in a tubular fashion but perpendicular to mitochondrial axis as in the cells of adrenal gland.

(vi) Haphazardly distributed as in the cells of kidney of insects and hepatic cells.

(vii) Cristae externally small and irregular as in the interstitial cell of *Opossum*.

(viii) Rarely the mitochondrial wall is smooth with no cristae. The number and size of cristae in a mitochondrion directly affects its efficiency. The greater and larger are the cristae, the faster is the speed of oxidation reaction.

(ix) Perpendicular to the long axis of mitochondria.

Mitochondrial particles

According to earlier description (*Green* and *Perdue*, 1966) the outer surface of the outer membrane and the inner surface of the inner membrane were supposed to be covered with thousands of small particles. Those on the outer membrane were described as being stalk less and were called the *subunits of Parson*. There may be as may be as many as 10,000 to 100,000 particles per mitochondrion. However recent studies have shown that stalkless particles are absent. The stalked inner membrane particles were called the *subunits of Fernander-Moran, Elementary particles, F_1 particles* or the *oxiosomes* or *ETP* or *electron transport particles*. These particles are about 84Å in diameter and are regularly spaced at intervals of 10mm on the inner membrane. There may be as many as 10^4 to 10^5 elementary particles per mitochondrion.

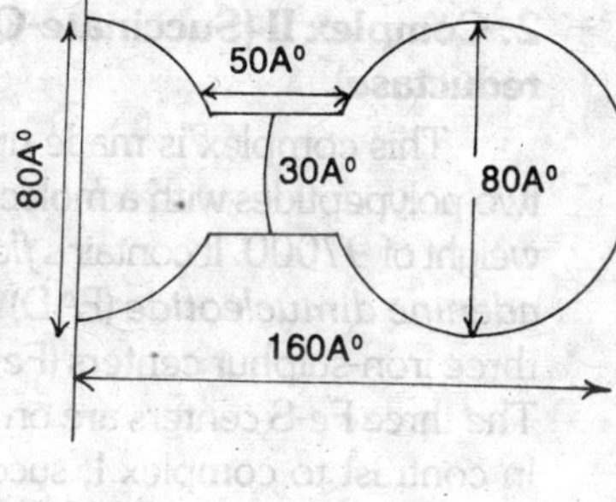

Fig. 13.5. Structure and measurements of the mitochondrial particles.

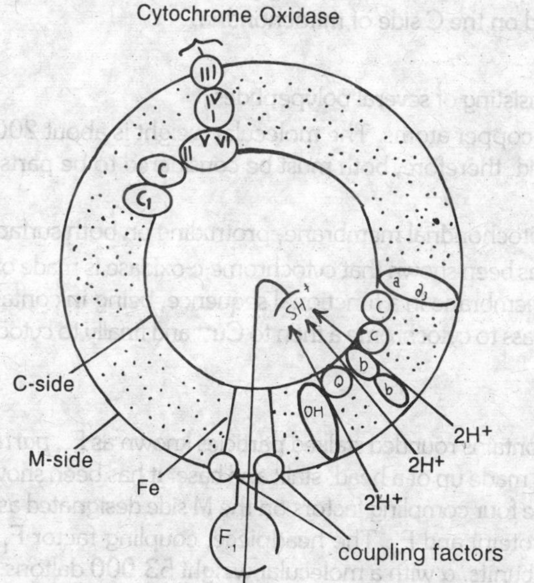

Fig. 13.6. Inner membrane of mitochondrion showing distribution of respiratory chain complex.

Isolation of Mitochondria

Mitochondria can be isolated from the cell in the living form for their physiological studies. The cell first treated with deoxycholate for their break down. Then they are passed in sucrose solution. The homogenate should be centrifuged for 10 minutes at the speed of 6000 X g. From this homogenate upper substance is centrifuged at the speed of 8500 X g for 10 minutes. After this centrifugation, the upper microsomal fraction is discarded while the lower fraction consists of mitochondria and other particles like lysosomes. This fraction is passed through sucrose gradient as shown in the figure number 10.4. The mitochondrial fraction then centrifuged at the speed of 10,000 X g upto 3 hrs. The upper part of this centrifuged material have mitochondria and lower part of lysosomes.

The Respirating Chain Complexes

Green et. al. have recognized five main complexes which, if mixed in correct ratios, can reconstitute to form ETC. These complexes are:

1. Complex I (NADH-Q-reductase)

This is largest complex, with a molecular weight about 500,000 and structure consisting of 15 subunits. It contains as prosthetic groups *flavin mononucleotide* (FMN) and six iron-sulphur centers. The NADH reaction site lies at M side of mitochondrion. Contract between NADH and CoQ is apparently made in the middle of the membrane. Complex I spans the inner mitochondrial membrane made and is able to translocate the protons across it, from M side to C side of mitochondrion.

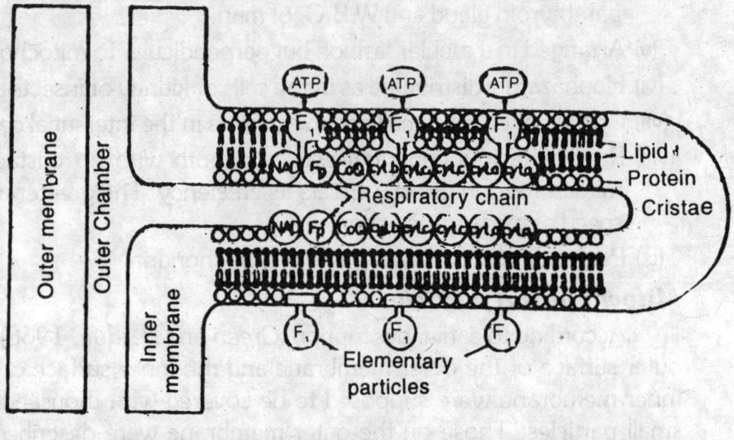

Fig. 13.7. The arrangement of the respiratory chain in cristae.

2. Complex II (Succinate-Q-reductase)

This complex is made up of two polypeptides with a molecular weight of 97000. It contains *flavin adenine dimucleotide* (FAD) and three iron-sulphur centers (Fe-SS_1, Fe-SS_2 and Fe-SS_3). The succinate binding site occurs on the M side. The three Fe-S centers are on the M side and there is a close interaction between Fe-S centre 3 and CoQ. In contrast to complex I, succinate-Q-reductase apparently is unable to translocate protons across the membrane.

3. Complex III (Q H_2-cytochrome-c-reductase)

This complex contains a number of subunits with a molecular weight 280,000. It contains cytochromes b, cytochrome C_1 and iron sulphur protein. There are two types of cytochrome b, these are bT (transducing cytochrome b) and b_k (Keilin-type cytochrome b). The heme containing part of cytochrome C_1, which transfer the electrons to cytochrome-c, is located on the C side of mitochondria.

4. Complex IV (Cytochrome-C-exidase)

Cytochrome oxidase is a large complex consisting of several polypeptides.

It has two cytochromes, a and a_3, and two copper atoms. The molecular weight is about 200,000. The a and a_3 however, never been separated and, therefore, both must be considered to be parts of the same complex.

Complex IV is thought to transverse the mitochondrial membrane, protruding on both surface.

In experiments with yeast mitochondria it has been shown that cytochrome-c-oxidase is made of seven subunits. The seven subunits are arranged the membrane in a functional sequence, being in contact with cytochrome c on the C-side. The electrons then pass to cytochrome a then to Cu^{++} and finally to cytochrome a_3 and oxygen at the M side.

5. Complex V (ATPase complex)

Towards the M side, the inner membrane contains rounded stalked particles known as F_1, *particles* or *Fernandez-Moran particles*. Each F_1 particles is made up of a head, stalk and base. It has been shown that complex V is identicle with F_1 particles. These are four compling factors on the M side designated as F_1, F_2, F_5 or OSCP (oligomycin-sensitive confering protein) and F_6. The headpiece, coupling factor F_1, is the ATPase proper. The F_1 contains five types of subunits. α with a molecular weight 53,000 daltons, β with a molecular weight 50,000 daltons, γ with a molecular weight 33,000 daltons, δ with a molecular weight 17,000 daltons and ε with a molecular weight 7,000 daltons. In addition to these subunits there is ATPase inhibitor (I) with a molecular weight 10,000 dalton. This inhibitor can be removed while treating with trypsin.

The stalk, made up of protein, connects the head-piece to the base. This portion corresponds to the OSCP (oligomycin—sensitive conferring protein) and F_6. These are needed to bind F_1 to the membrane. Treatment with ammonia releases OSCP. While treatment with silicotungustate removes F_6.

The basepiece lies within the inner mitochondrial membrane. It contains the proton translocating mechanism. This corresponds to Fo in the figure.

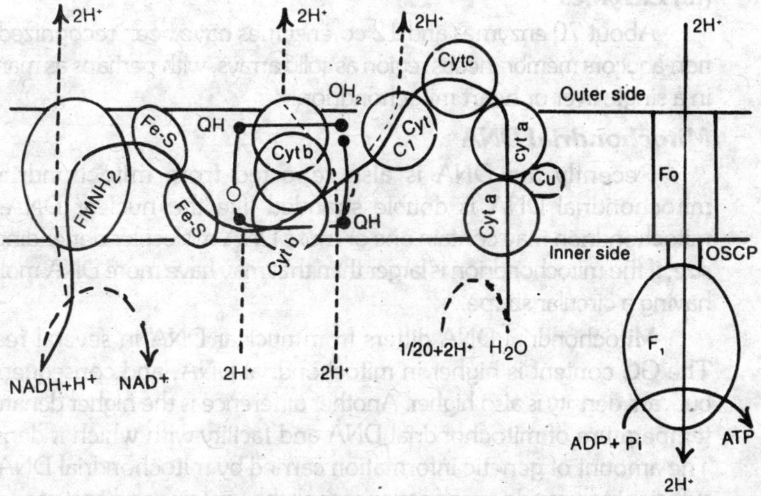

Fig. 13.8. Current view of the location of the enzymes and components of the respiratory chain in the inner membrane of the mitochondrion.

BIOCHEMISTRY OF MITOCHONDRIA

Lindberg and *Ernster* (1954) have given the data of chemical composition of mitochondria as follows: proteins 70 to 75%, Lipids 25-30%, and RNA 5% of the dry weight. But the recent biochemical analaysis shows the following components:

(i) Proteins

The proteins are the main constituent which is insoluble in water. The outer limiting membrane of mitochondria contains less than 10 percent of the total protein. There are about 14 different proteins having molecular weight from 12,000 to 22,000. The inner membrane contains about 60% protein having molecular weight varying from 10,000 to 90,000. The protein composition of mitochondrial membranes is not fully known.

Table 13.1. Localization of Enzymes, Obtained in Fractionation Studies.

	Mitochondrial fraction	*Enzymes located*
1.	Outer membrane.	Monoamine oxidase, "Rotenone-insensitive" NADH-cytochrome—C—reductase, Kynurenine hydroxylase, Fatty acid CoA ligase, Glycerolphosphate acyl transferase, Nucleoside diphosphokinase.
2.	Inter-membrane space.	Adenylate kinase, Nucleoside diphosphokinase, Nucleoside monophosphokinase,
3.	Inner membrane	Respiratory chain enzymes, β-Hydroxybutyrate, dehydrogenase, Ferrochelatase, Carnitine palmityl-transferase, Fatty acid oxidation system, Xylitol dehydrogenase.
4.	Matrix	Malate, isocitrate and glutamate dehydrogenase, Fumarse, Aconitase, Citrate synthetase, Ornithine-Carbonyl transferase, Fatty acid oxidation system, Pyruvate carboxylase,

(ii) Lipid

The lipid forms about 1/5th of the weight of the membranes. It is present almost entirely in the form of the molecules known *phospholipid.* It has been reported by *Meluick* and *Packer* in 1971 that the outer membrane fraction has a 40% lipid contents as compared to 20% in the inner membrane.

(iii) Enzymes

About 70 enzymes and 12 co-enzymes have been recognized in the mitochondria. Enzymes lie in a non-aqucors membraneous region as solid arrays, with perhaps as many as 5000 to 20,000 such assemblages in a single liver or heart mitochondrion.

Mitochondrial DNA

Recently the DNA is also reported from mitochondria. The mitochondrial DNA is double stranded like the nuclear DNA. Each mitochondrion may contain one or more DNA molecules depending on its size, if the mitochondrion is larger than that may have more DNA molecules, having a circular shape.

Mitochondrial DNA differs from nuclear DNA in several respects. The GC content is higher in mitochondrial DNA, and consequently the buoyant density is also higher. Another difference is the higher denaturation temperature of mitochondrial DNA and facility with which it denatures. The amount of genetic information carried by mitochondrial DNA is not sufficient to provide specifications for all the proteins and enzymes present in this organoid. The most likely possibility is that mitochondrial DNA codes for some structural proteins.

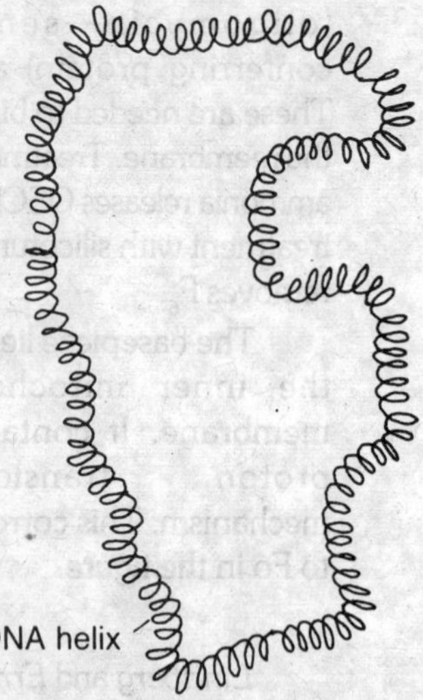

Fig. 13.9. Mitochondrial DNA.

Yeast mitochondria have been shown to contain DNA polymerase (*Winter Berger,* 1966) and more recently *Kalf* 1968 has succeeded in isolating the enzymes from rat liver mitochondria. Mitochondrial DNA polymerase appears to be involved in DNA replication rather than repair (*Karol* and *Simpson* 1968) and possesses properties which are different from those to the nuclear enzymes. These include a differing requirement for metal ion (*Meyr* and *Simpson* 1968). *Yeast* mitochondrial DNA polymerase appears to be smaller than its nuclear counter part, and is active at different stages of the cell cycle (*Iwshima* and *Rabinowitz,* 1969). Visual evidence showing what appear to be rat liver mitochondrial DNA in the process of replication has been presented by *Kirschner, Wolsten Holme* and *Gross* (1968).

Mitochondrial DNA does not appear to have histones associated with it as has nuclear DNA of higher organisms. In this respect mitochondrial DNA resembles with the bacterial DNA.

Mitochondrial RNA

South and *Mehlear* (1968) suggested that the amount of mt-RNA is about 10 to 20 times that of mt-DNA. All sorts of RNA have been identified in mitochondria. The present evidences point out conclusively that mitochondria contain complete set of *t*-RNA (*Wintersberger* and *Tuppy,* 1965), aminoacyl RNA synthetases (*Barnett, Braun and Epler,* 1967), as well as ribosomal RNA (*Rogers, Preston, Titchener* and *Linnane,* 1967). All these components differ from their respective counterparts in the groundplasm. The presence of m-RNA, transcribed from mitochondrial DNA is still uncertain. However, there are authorties who suggests its presence. The ribosomal RNA's are coded for by mitochondrial DNA and are thus apparently synthesized within the mitochondria by a mitochondrial DNA-dependent RNA polymerase system (*Wintersberger,* 1964).

Mitochondrial Ribosomes

Mitochondria appear to contain ribosomes which are smaller in diameter than cytoplasmic ribosomes (*Swift,* 1965) and *Yeast* mitochondria contain RNA species of 23^s and 16^s (*WinterBerger* 1966) which would correspond to a 70^s ribosome of bacterial type rather than the 80S ribosome of the cytoplasm. Ribosome like particles with sedimentation values of 81S and 55S have also been reported, and the extent of degradation suffered by the particles during isolation is not yet clear. Polysome like aggregation of ribosomes have been observed in sections of yeast mitochondria by *Vignais, Huet* and *Andre* in 1969. High molecular weight and RNA species associated with mitochondria which differ in sedimentation value from cytoplasmic ribosomal RNA have been reported in *Yeast, Neurospora* and *He-La cells.* Mitochondrial ribosomes require a higher concentration of Mg^{++} ions to maintain their integrity than do cytoplasmic ribosomes.

Protein Synthesis

In general, mitochondria can code and synthesize protein, but the DNA present therein is unsufficient to code for all the proteins. It is suggested that mitochondria can synthesize the proteins of structural nature (cytochrome oxidase), but much of the proteins if not all, of the soluble proteins of the matrix as well as the proteins of the outer membrane and a number of proteins located in the cristae (*Borst*, 1972) are under the control of nuclear DNA. Of the proteins coded by nuclear DNA, it is generally agreed that the m-RNA derived from nucleus are translated in the cytoplasm, and the resulting proteins are then transported into the mitochondria. How then protein enter the mitochondria? Two methods have been proposed:

1. The precursors enter the mitochondrion and inside are changed into the end products, thereby affecting a undirectional flow of material into the mitochondrion.
2. There is the synthesis of lipoprotein vesicles which they merge and combine with the growing mitochondrion.

FUNCTIONS

1. Role of Mitochondria in Yolk Formation

There has been a good number of investigations, whose account reveals that, mitochondria help in the formation of yolk in a developing ovum. The first study in this field was made by *Loyez* (1911) and latest probably by *M.D.L. Srivastava* (1965), with the help of light microscope. The evidence adduced depends upon the topographical, and size relationship, and staining reactions of mitochondria and early protein yolk.

In the modern cytology with the investigation of electron microscope a new era has started and the studies of yolk formation do not remain away from electron microscope. With the help of electron microscope *Farvard* and *Carasso* (1958) come to the conclusion that mitochondrial transformed into yolk granules in the egg of *Planorbis coneus*.

The main structural changes which they observed in the mitochondria are as follows:

(i) The cristae become disorganised in a few membranes, remaining concentric to the other membrane before dropping completely.

(ii) In the matrix appeared a few minute granules which have scattered first, but become aggregated eventually in masses in regular pattern.

2. During Cell division and Spermiogenesis

Early cytologists, *Benda, Dulberg* and *Meves* were of opinion that mitochondria also divide equally during the cytoplasmic division and perhaps playing a part in inheritance.

Wilson (1928) commented that not the slightest proof has been produced of a fusion between the paternal and maternal chrondrisomes. *Frederic* (1958) has briefly summarized various changes in mitochondria during cell division. The first phase shows a decrease in the total volume of mitochondrial material; gradually ceases its movements, pronounced thinning, fragmentation into small spheres, loss of optical density and finally assimilation into the cytoplasm.

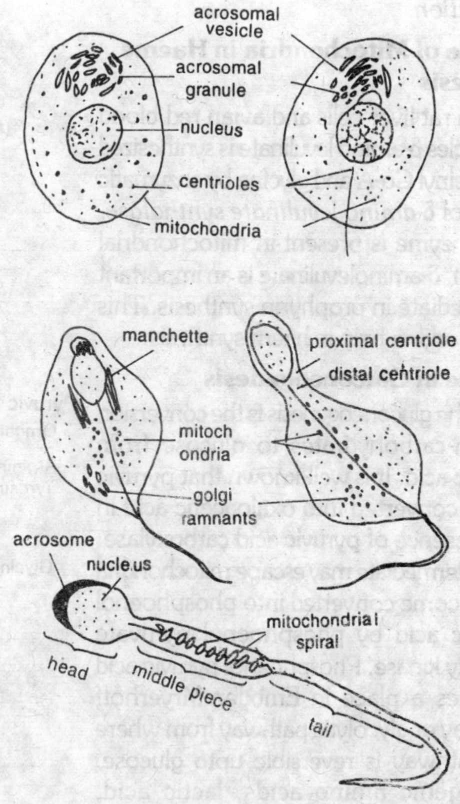

Fig. 13.10. Role of mitochondria during spermiogenesis.

In the second phase, when the cell divides into two, the modified mitochondria are separated passively, into the daughter cells. In the third phase, the modified mitochondria are reconstituted by addition of elements assimilated in the cytoplasm. *Wilson* found in *Opisthacanthus*, during spermatogenesis, the number of mitochondria gradually reduces. *Pollister* (1930) describes in *Gerris* that mitochondria arranged themselves into a well-defined ring, but without fusion. Modern microscopic studies provide firm conclusions regarding mitochondrial division during mitosis. *Pyne* (1952) noted in fowl adrenal cortex the mitochondria frequently appeared as pairs. This suggested that division, rather than fusion, was occurring. In the transformation of spermatids into spermatozoa many mitochondrial changes are observed. *Franzen* (1956) observed in those sperms, which are shed directly in water, that mitochondria are present in form generally of four or five spheres below the sperm head, and in the case of sperms discharged in viscous medium these spheres transform into two enlongated ribbion like filamentous mitochondria. Sometimes these develop into '*nebenkern spheres*' which may elongate and twist around the axial filament to form the mitochondrial sheath. *Yasuzumi* (1958) found an electron opaque body within 'nebenkern' of spermatids of *Drosophila;* it is described as indistinguishable from a lipid droplet.'

3. Role of Mitochondria in the Production of Energy

Mitochondria play a very important role in cellular respiration or the production of energy. Energy is produced inside the cell, partly outside the mitochondria and mainly inside the mitochondria. The ATP molecules that are produced in the non-mitochondrial cytoplasm are generated by a process referred to as *anaerobic respiration.*

4. Role of Mitochondria in Haeme Synthesis

In rat liver cells and avian red blood corpuscles δ-amino levulinate is synthesized for succinyl Co-A and glycine by enzymatic action of *δ-amino levulinate synthatase.* This enzyme is present in mitochondrial fraction. δ-aminolevulinate is an important intermediate in prophyrin synthesis. This mitochondria help in haem synthesis.

5. Role in Gluconeogenesis

The gluconeogenesis is the conversion of non-carbohydrates to glucose from pyruvic acid. It is well-known that pyruvic acid is converted into oxaloacetic acid in the presence of pyruvic acid carboxylase. This intermediate may escape mitochondria and become converted into phosphoenol pyruvic acid by phosphoenol pyruvate corboxy kinase. Phosphoenol pyruvic acid occupies a place in Embden-Mryerhoff pathway or glycolytic pathway from where the pathway is reversible upto glucose. Glucogenic amino acids, lactic acid, glycerol and in some cases propionate can,

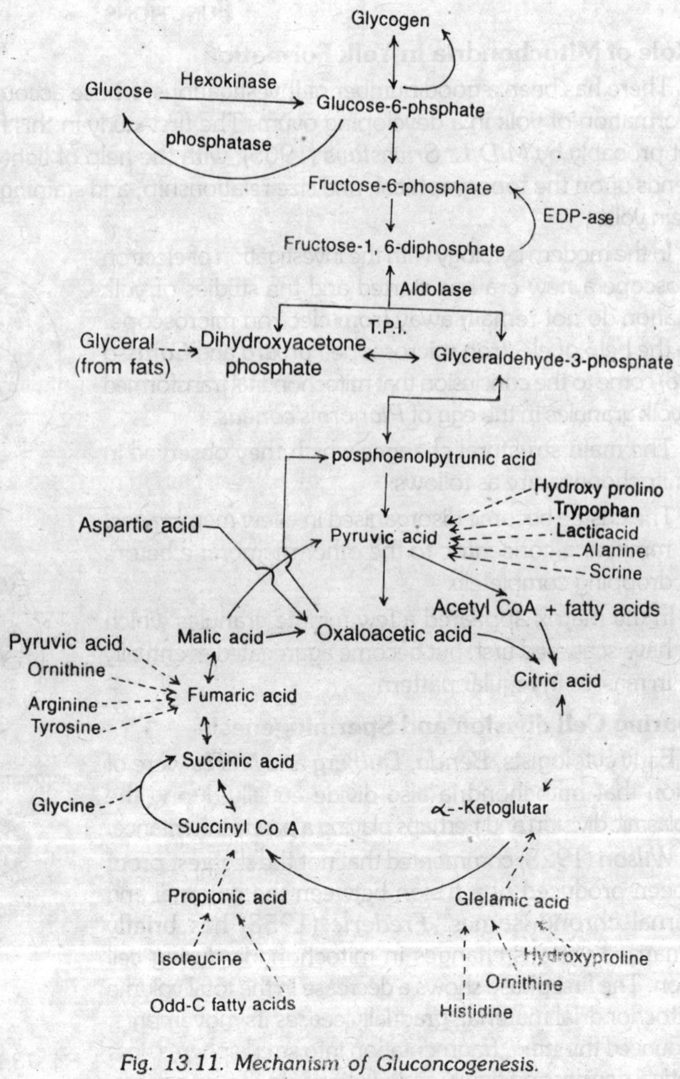

Fig. 13.11. Mechanism of Gluconcogenesis.

after suitable modification, be fed at one or the other point in Krebs' cycle. Oxaloacetic acid and malic can come out of mitochondria to be finally converted into glucose.

6. Role of Mitochondria in Amino Acid Metabolism

The enzymes for oxidative deamination of amino acids are present in mitochondria. These are glutamate dehydrogenase, proline dehydrogenase, δ-amino levulinate synthetase etc.

7. Role of Mitochondria in Lipid Metabolism

They are able to oxidize fatty acids. Oxidation of fattyacid requires complete oxidation of acetyl CoA in the Krebs' cycle so that free CoA may be generates. Reversal of fatty acid oxidation leads to fatty acid synthesis. During starvation, the mitochondria utilize fat to produce energy.

ORIGIN OF MITOCHONDRIA

Our understanding of the process where by mitochondria are produced, is still very incomplete. *Lehninger* (1964) classified the various theories of possible routes of mitochondrial genesis into three main groups:

1. Formation from other membranous structures in the cell.
2. Growth and division of pre-existing mitochondria.
3. *De novo* synthesis from submicroscopic precursors.

1. Formation from other Membranous Structures in the Cell

The formation of mitochondria by "pinching off" or budding from pre-existing cell structures has been suggested for a range of cells membranes including those of the plasmalemma (*Robertson* 1959), endoplasmic reticulum, nuclear envelope, and Golgi complex (*Novikoff,* 1961). But the support for such evidence, in the absence of supporting biochemical data, cannot be wholly conclusive. Part of the problem undoubtedly lies in our fragmentary knowledge of the structure and composition of, and differences between cell membranes in general. Indeed the similarities discussed in the literature between the mitochondrial membrane and the endoplasmic reticulum, could lend weight to the idea that probably mitochondria are formed when cytoplasm pushes into a cavity surrounded by an internal membrane, which they, pinches off and separates from the continuous system.

2. Growth and Division of Pre-existing Mitochondria

Electron microscopic evidence for mitochondrial division by fission, although plentiful, is difficult to assess the danger of producing artifacts is very real because of the harsh chemical and physical agents brought to bear on the test material during processing. Interpretation is not made easier by the ability of mitochondria of undergo extreme changes in shape *in vivo* which may or may not be associated with mitochondrial fission. There are numerous reports of mitochondria connected to each other by narrow bridges of membranes, especially in rapidly melopolizing tissue, and it is thought that such figures may represent mitochondria in an early stage of fission. By observing serial sections of rat liver *Stempak* (1967) was able to show that "dumb bell-shaped" mitochondria can be the sections of cupshaped bodies. Such bodies have also been observed in rapidly growing tissues of fern and may represent the beginning stages of division.

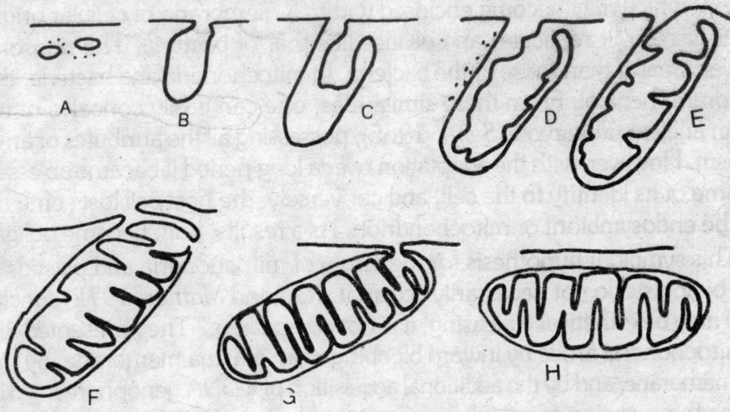

Fig. 13.12. A possible mechanism for the formation of mitochondria from the plasma membrane.

An early stage in mitochondrial division may involve the separation of mitochondrial contents into two or more compartments. The presence of mitochondria with internal "partitions" has been formed in several cell types (*Tandler et. al.*, 1969) although the possibility that they are manifestations of mitochondrial fusion cannot easily be ruled out. *Lafontaine* and *Allard* (1964) have presented electron micrographs of rat liver mitochondria which exhibit what appear to be partitions dividing the inner membrane complex into two masses, the whole being surrounded by continuous outer membrane. *Tandler et. al.* (1969) has demonstrated the partitions of mitochondria in liver which was recovering from riboflavin deficiency.

3. De novo Synthesis

The possibility of *de-novo* synthesis of mitochondria arose with experiments in the early part of the century, when mitochondria containing larvae were seen to develop from sea urchin egg cytoplasm which had apparently been freed of mitochondria by centrifugation (*Novikoff* 1961). Using the greater resolving power of the electron microscope, it was later shown, that mitochondria could not be dislodged by centrifugation of the egg (*Lansing, Hiller, Rosenthall* 1952). In the earlier experiments, mitochondria had probably been present in the "centripetal end" of the egg cell after all, and these mitochondria could have served as precursors in subsequent mitochondrial production.

In the above description, a number of views were described, regarding the generation of mitochondria in the cytoplasm of different types of cells. But it is perhaps unwise to group the evidence to suit one or other of a limited number of clear cut methods by which mitochondria could replicate. The actual situation is probably complex, and it may well be that different methods of replication take place in different tissues, and at different stages in development. One could imagine the early mitochondria being formed from membrane structures in the developing embryo concentration of mitochondria around the nuclear membrane have been noted in embryonic tissues from several phyla (*North* and *Pollak*, 1961) and formation of mitochondria from this membrane could involve the transfer of nuclear genetic information essential for subsequent mitochondrial growth and multiplication by division. The multiplication of mitochondria could then proceed by the incorporation of large pre-fabricated molecules and association of molecules, with division by fission, when the mitochondria reached a critical stage.

5. Prokaryotic Origin of Mitochondria

This fact that mitochondria can grow, divide and are capable of mutations support a long-held view that mitochondria originated with their host. Bacteria would have originated the mitochondria and blue green algae, the chloroplasts.

There are numerous homologies between mitochondria and bacteria. In bacteria the electron transport system is localized in the plasma membrane which can be compared with the inner membrane of mitochondria. Some bacteria even have membranous projections extending from the plasma membrane (*Fitz-James*, 1960) which are comparable to mitochondrial crests since both contain the respiratory chain (*Salton* and *Chapman*, 1962). The inner membrane and matrix, it has been postulated may represent the original symbiont which may become enclosed within a membrane of cellular origin (ER). Further mitochondrial DNA is circular, it replicates and divides like that of bacteria. The ribosomes are also found which are, however, smaller than those of the bacteria. In mitochondria and bacteria, the protein synthesis is inhibited by chloramphenicol. From these similarities, one can easily conceive of mitochondria as being evolved from an ancient prokaryote (*Swift*, 1965), possessing all the attributes of an independent, probably aerobic organism. However, with the adaptation over a long period it became an essential and dependent symbiont, lost some of its identity to the cell, and conversely, the host cell lost some of its functions, deriving it now from the endosymbiont or mitochondrion. As a results, both became obligatory symbionts to each other.

This symbiont hypothesis for the origin of mitochondria and plastids has acheived wide popularity, but all biologists do not necessarily accept it. *Raff* and *Mahler* (1972) concluded that "while the symbiotic theory may be esthetically pleasing, it is not compelling." They presented a lot of evidences and proposed that mitochondria arose by inward blebbing from plasma membrane, by the acquisition some how of an outer membrane, and by the additional acquisition of a DNA genophore from the DNA of the protoeukaryote in which the evolution of mitochondria occurred. *Borst* (1972) proposed an *episome theory* and supposed that the DNA of mitochondrion left the "nuclear' DNA by a sort of amplification to become mapped within a membrane containing the respiratory chain.

14

CELLULAR RESPIRATION

Respiration is one of the most fundamental phenomenon of life and indeed life itself is inseparable from respiration in some forms. Respiration involves the oxidative breakdown of certain organic substrates present in the cells. Respiration, therefore, is not the concern of special organs or tissues, it is an integral part of the life of every cell. Cellular respiration is the aerobic oxidation of carbohydrates in the cell with the help of atmospheric oxygen, in the course of which energy is liberated and carbon dioxide is given out. In glycolysis and fermentation, glucose is oxidized in the absence of oxygen, i.e. *anaerobically*. Organisms that live in the presence of oxygen are aerobes, those that live in its absence are anaerobes.

The first series of reactions in oxidative break-down of glucose is called the *Embden-Meyerhof (E-M) pathway*. This was described by two German biochemists *Embden* and *Mayerhof*. The term *glycolysis* is sometimes used as a synonym for the E-M pathway, but biochemists usually apply the term glycolysis to the E-M pathway plus a form of fermentation, which together lead from glycolysis to lactic acid. The E-M pathway is a several-step process that occurs in the cytoplasmic matrix of cell. It can proceed in either the presence or absence of oxygen but makes available only a small portion of the potential energy that could be expected from glucose.

I. The E-M Pathway

From six-carbon glucose molecule the E-M pathway starts and the formation of 2-three-carbon molecules of pyruvic acid is the final step of this pathway. Following steps take place in the entire pathway.

1. *Phosphorylation*. In the first step glucose reacts with adenosine triphosphate (ATP) to form glucose-6-phosphate in the presence of *hexokinase* and Mg^{++} ions.

$$\text{Glucose + ATP} \xrightarrow[\text{Mg}^{++}]{\text{hexokinase}} \text{Glucose-6-phosphate + ADP.}$$

This reaction is exergonic in nature so that it is irreversible.

2. *Isomerization*. Glucose-6-PO_4, in the presence of the enzyme *phospho-hexose-isomerase*, is isomerized into fructose-6-PO_4.

$$\text{Glucose-6-phosphate} \xrightarrow{\text{Phosphohexose-isomerase}} \text{Fructose-6-phosphate.}$$

3. *Phosphorylation*. Fructose-6-PO_4 reacts with one molecule of ATP in the presence of Mg^{++} ions and *phosphofructokinase* to form fructose-1, 6-diphosphate. This reaction is also irreversible due to its strongly exergonic nature.

$$\text{Fructose-6-phosphate + ATP} \xrightarrow[\text{Mg}^{++}]{\text{Phosphofructokinase}} \text{Fructose-1-6-disphosphate.}$$

4. *Cleavage.* Fructose-1, 6-diphosphosphate is cleaved under the influence of an enzyme *aldolase* with the formation of dihydroxy acetone phosphate or phosphodihydroxy acetone (DHAP) and 3-phosphoglyceraldehyde or glyceraldehyde-3-phosphate (PGAL). Both the products are triose phosphates. The two trioses can be reversibly changed into one another by enzyme *Triosphosphate isomerase.*

$$\text{Fructose-1-6-diphosphate} \xrightarrow{\text{Aldolase}} \left[\begin{array}{l}\text{Dihydroxyacetone phosphate (DHAP)}\\ \text{Triose phosphate isomerase.}\\ \text{3-phosphoglyceraldehyde (PGAL)}\end{array}\right.$$

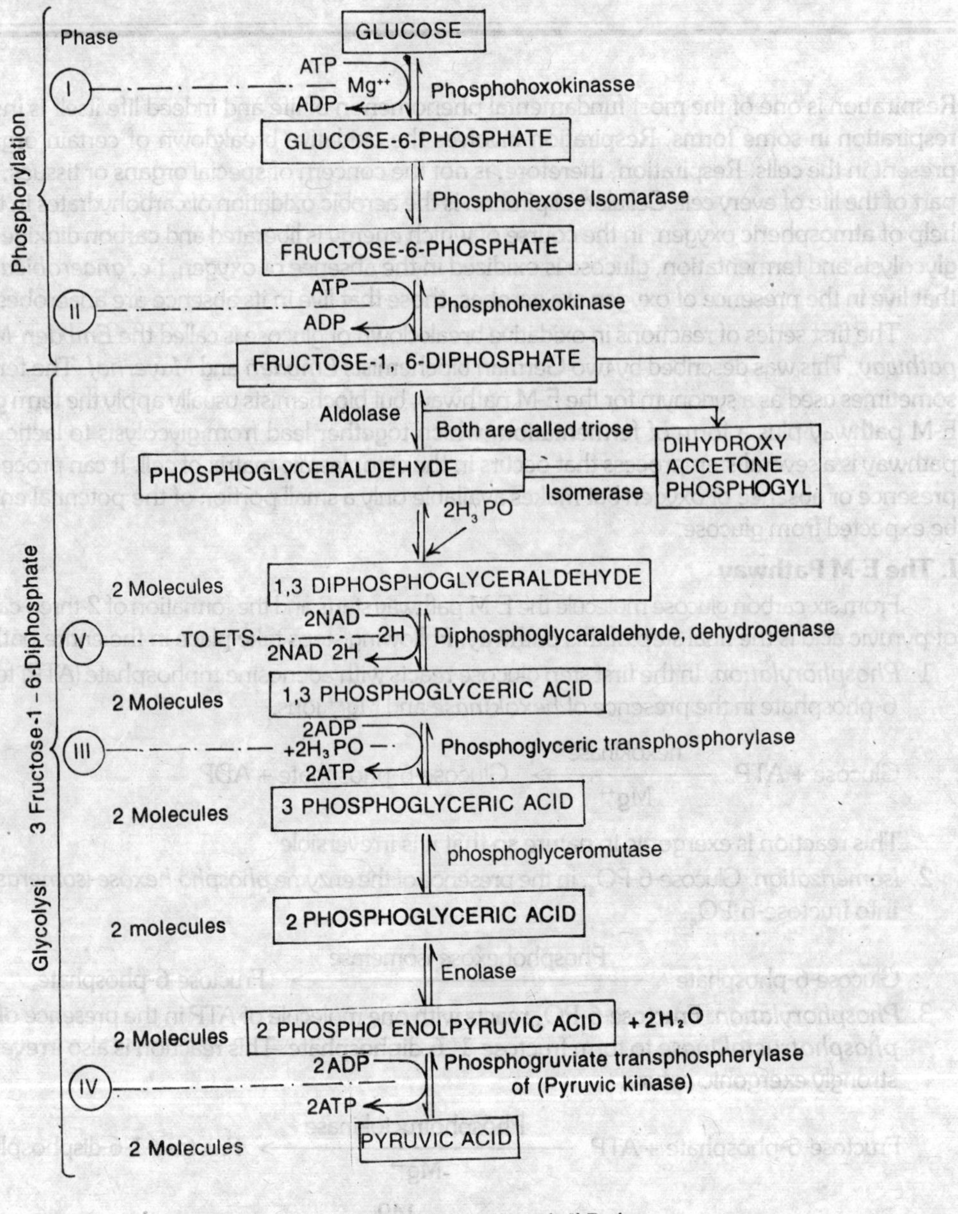

Fig. 14.1. Glycolysis of Embeden Meyerhof Pathway.

5. *Phosphorylation and oxidative dehydrogenation.* This is the first reaction in the energy-yielding phase of the E-M pathway. 3-phosphoglyceraldehyde (PGAL) reacts with phosphoric acid (H_3PO_4) to form 1-3-diphosphoglyceraldehyde in the presence of enzyme *glyceraldehyde-3-phosphate dehydrogenase.*

$$\underset{\text{(2 molecule)}}{\text{3-phosphoglyceraldehyde}} + \underset{\text{(2 mole.)}}{H_3PO_4} \longrightarrow \underset{\text{(2 mole.)}}{\text{1, 3-diphosphoglyceraldehyde.}}$$

Coenzyme nicotinamide adenine dinucleotide (NAD) reacts with 1-3-diphosphoglyceraldehyde removing 2 electrons and 2 protons (the equivalent of 2 H atoms) forming 1, 3-diphosphoglyceric acid.

$$\underset{\text{(2 mole)}}{\text{1, 3-diphosphoglyceraldehyde}} + 2NAD^+ \xrightarrow{\text{Glyceraldehyde-3-phosphate dehydrogenase}} \underset{\text{(2 mole.)}}{\text{1, 3-diphosphoglyceric acid} + 2\ NADH + H}$$

6. *ATP generation.* The high-energy phosphate group in 1-3-diphosphoglyceric acid is transferred to ADP resulting in the formation of 3-phosphoglyceric acid (3 PGA) and ATP. This reaction takes place in the presence of enzyme *phosphoglycerate kinase* and Mg^{++} ions.

$$\underset{\text{(2 mole.)}}{\text{1, 3-diphosphoglyceric acid}} + 2\ ADP \xrightarrow[Mg^{++}]{\text{phosphoglycerate kinase}} \underset{\text{(2 mole)}}{\text{3-phosphoglyceric acid} + ATP}$$

7. *Isomerization.* 3-phosphoglyceric acid (3-PGA) is isomerized into 2-phosphoglyceric acid (2-PGA) with the help of enzyme *phosphoglyceromutase.* This enzyme catalyse the transfer of a phosphate group from one carbon to another carbon of the same compound.

$$\underset{\text{(2 mole.)}}{\text{3-phosphoglyceric acid}} \xrightarrow[\text{acid}]{\text{Phosphoglyceromutase}} \underset{\text{(2 mole.)}}{\text{2-phosphoglyceric}}$$

8. *Dehydration of 2-PGA.* The 2-phosphoglyceric acid molecule is dehydrated by the action of enzyme *enolase* to yield phosphoenol pyruvic acid (PEPA). Presence of Mg^{++} ions is also necessary.

$$\underset{\text{(2 mole.)}}{\text{2-phosphoglyceric acid}} \xrightarrow[Mg^{++}]{\text{Enolase}} \underset{\text{(2 mole.)}}{\text{2-phosphoenol pyruvic acid}} + 2H2O$$

9. *ATP generation.* In the final step of E-M pathway the high energy phosphate group in 2-phosphoenol-pyruvic acid (PEPA) is transferred to ADP in the presence of *pyruvate kinase* and Mg^{++} ions with the formation of ATP and pyurvic acid.

$$\underset{\text{(2 mole.)}}{\text{2-phosphoenolypyruvic acid}} + 2ADP \xrightarrow[Mg^{++}]{\text{Phruvative kinase}} \underset{\text{(2 mole.)}}{\text{Pyruvic acid} + 2ATP}$$

Energy Yield of the E-M Reactions

How much energy or ATP has the çells gained after all these reactions? That depends upon whether the reactions have taken place under anaeròbic or aerobic conditions. In general, the steps of energy production in E-M pathway are (1) during conversion of 1,3-diphosphoglyceric acid into 3-phosphoglyceric acid and (2) during conversion of 2-phosphoenol pyruvic acid into pyruvic acid. If ADP is present in these steps, energy of energyrich compounds is captured with the formation of ATP. One molecule of ATP at each of the above two steps in synthesized per molecule of 1, 3-diphosphoglyceric acid or 2-phosphoenolpyruvic acid utilized for forward reaction. Thus, in all, 4 molecules of ATP are formed at these two steps per molecules of glucose used under anaerobic conditions.

Besides these two steps, these exists one more site of energy production which comes into play under aerobic conditions. This step takes place during conversion of 3-phosphoglyceraldehyde into 1, 3-

diphosphoglyceric acid when one molecule of reduced NAD, is oxidized with the formation of 1, 3-diphosphoglyceric acid.

In order to evaluate the net gain of energy the energy expenditure has also got to be taken into account and must be substracted from total energy production. Under anaerobic conditions 2ATP molecules are used up during reactions and there is a net gain of only 2ATP molecules.

II. Oxidative Decarboxylation

Pyruvic acid, when enters the mitochondria does not itself enter Kreb's cycle reactions. Entry into the Kreb's cycle requires preparation of a 2-carbon fragment called *acetyl group*. This takes place in several steps. In the first step there is a decarboxylation (a carbon dioxide molecule is released), that shortens the carbon chain by one carbon. A molecule of NAD is also reduced. The remaining 2-carbon fragment i.e., acetyl group is attached to a carrier molecule, *coenzyme A*. This complex, made up of two carbon fragment and coenzyme A is called *acetyl CoA*. The acetyl CoA is in an active form ready to be fed into Kreb's cycle.

$$\underset{\text{(Pyruvic acid)}}{CH_3\,COCOOH} + \underset{\text{(CoA)}}{CoASH} + NAD \xrightarrow[\substack{\text{Thymine pyrophosphate}\\ \text{Lipoic acid amide}}]{\substack{\text{Pyruvic dehydrogenase}\\ \text{multienzyme complex}}} \underset{\text{(acetyl-S-CoA)}}{CH_3COS\,CoA} + NADH + H^+ + CO_2$$

III. Krebs cycle or Tricarboxylic Acid Cycle (TCA)

The krebs cycle completes the oxidation of carbohydrates and supplies electrons to be passed through electron transport system. Enzymes catalyzing Kreb's cycle reactions are located in the mitochondrial matrix.

The cyclical nature of the reactions by which further oxidations beyond pyruvic acid occur was established after a series of experimental analyses by *Albert Szent-Gyorgyi, Hans Krebs* and others. *Krebs* was awarded Nobel Prize in 1953 for this work.

The Kreb's cycle is a *cycle* because it is series of reaction events, each catalyzed by a specific enzyme, that ends up back at its original point. The reactions involve a series of organic acids with different number of carbon atoms.

(i) *Formation of citric acid.* Acetyl Co-A, formed from pyruvic acid condenses with oxaloacetic acid forming citric acid. This reaction is catalyzed by enzyme *citrate synthatase* or condensing enzyme.

$$CH_3CO\text{-}S\text{-}CoA + Oxalicacid + H_2O \xrightarrow[\text{CoASH}]{\text{Citrate synthetase}} \text{Citric acid} + CoASH$$

(ii) *Dehydration.* Citric acid is converted into *cis*-aconitic acid by the enzyme *aconitase* with release of one molecule of water.

$$\text{Citric acid} \xrightarrow{\text{Aconitase}} \textit{Cis}\text{-aconitic acid} + H_2O$$

(iii) *Hydration I.* By hydration of *Cis*-aconitic acid, isocitric acid is formed. This reaction is catalyzed by enzyme *aconitase.*

$$\textit{Cis}\text{-aconitic acid} + H_2O \xrightarrow{\text{Aconitase}} \text{Isocitric acid.}$$

(iv) *Dehydrogenation I.* Isocitric acid is dehydrogenated to form oxalo succinic acid in the presence of *isocitric dehydrogenase* and one pair of hydrogen atoms removed is accepted by NAD+ or NADP+ which forms NADH + H^+ or NADPH + H^+ (Mitochondria have been found to contain two types of isocitric dehydrogenases, one is NAD specific and other i NADP specific).

$$\text{Isocitric acid} + NAD+ \text{ or } NADP \xrightarrow{\text{Isocitric dehydrogenase}} \text{Oxalosuccinic acid} + NADH + H^+ \text{ or } NADPH + H^+$$

(v) *Decarboxylation I.* In the presence of enzyme *carboxylase,* the oxalosuccinic acid undergoes decarboxylation to form α-Ketoglutaric acid. One CO_2 is removed. Thus α-Ketoglutaric acid is 5-carbon molecule.

$$\text{Oxalosuccinic acid} \xrightarrow{\text{Carboxylase}} \alpha\text{-Ketoglutaric acid} + CO_2$$

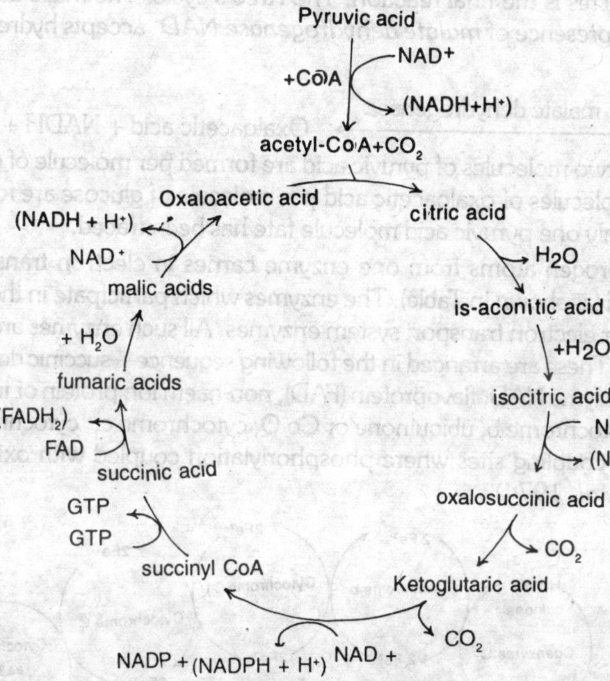

Fig. 14.2. The Krebs cycle.

(vi) *Dehydration II and decarboxylation II.* α-Ketoglutaric acid undergoes simultaneous decarboxylation and dehydration and joins with coenzyme A to form succinyl Co-A, a 4-carbon atom derivative of coenzyme A. This reaction is catalyzed by enzyme *α-Ketoglutaric dehydrogenase* in the presence of thiamine pyrophosphate (TPP) and lopoic acid. The hydrogen atoms released are accepted by NAD^+ to form $NADH + H^+$

$$\alpha\text{-Ketoglutaric acid+Co-ASH+NAD}^+ \xrightarrow[\text{dehydrogenase}]{\alpha\text{-Ketoglutaric}} \text{Succinyl CoA} + CO_2 + NADH + H^+$$

(vii) *Phosphorylation*. Succinyl CoA on hydrolysis yields to the formation of succinic acid and CoA. Succinyl CoA possesses a high-energy thiol group. In the presence of guanosine diphosphate (GDP) and inorganic phosphate the energy of thiol group is transferred to GDP with the formation of guanosine triphosphate (GTP). This reaction is catalyzed by enzyme *succinic thiokinase*.

$$\text{Succinyl- CoA} + GDP + H_3PO_4 \xrightarrow{\text{Succinic thiokinase}} \text{Succinic acid} + GTP + \text{CoASH.}$$

(vii) *Dehydrogenation III.* Succinic acid undergoes dehydrogenation under the catalytic action of *succinic dehydrogenase* to form fumaric acid. This dehydrogenase contains flavin adenine dinucleotide (FAD^+) as its prosthetic group which accepts the pair of hydrogen atoms removed from succinic acid.

$$\text{Succinic acid} + FAD^+ \xrightarrow{\text{Succinic dehydrogenase}} \text{Fumaric acid} + FADH + H^+$$

(ix) *Hydration II.* Fumaric acid is hydrated to form malic acid. This reaction is catalyzed by the enzyme *fumarase.*

$$\text{Fumaric acid} + H_2O \xrightarrow{\text{Fumarase}} \text{Malic acid.}$$

(x) *Dehydrogenation IV.* This is the final reaction. The Kreb's cycle. The malic acid is transferred to oxaloacetic acid in the presence of *malate dehydrogenase NAD⁺* accepts hydrogen atoms to form NADH + H⁺.

$$\text{Malic acid} + NAD^+ \xrightarrow{\text{malate dehydrogenase}} \text{Oxaloacetic acid} + NADH + H^+$$

It should be noted that two molecules of pyruvic acid are formed per molecule of glucose during E-M pathway. Therefore, two molecules of oxaloacetic acid per molecule of glucose are formed at the end of Kreb's cycle. In the figure only one pyruvic acid molecule fate has been traced.

During transfer of hydrogen atoms from one enzyme carries in electron transport system, large amount of energy is released (as shown in Table). The enzymes which participate in the process are called respiratory chain enzymes or electron transport system enzymes. All such enzymes are found in the inner membrane of mitochondria. These are arranged in the following sequence—succinic dehydrogenase which remains associated with coenzyme NAD, flavoprotein (FAD), non-haem ion protein or iron-sulphur protein, flavoprotein (FAD or FP), cytochrome b, ubiquinone or Co Q, cytochrome c_1, cytochrome c, cytochrome a cytochrome a_3 and three coupling sites where phosphorlylation coupled with oxidation leads to the production of ATP (*Hall et. al.* 1974).

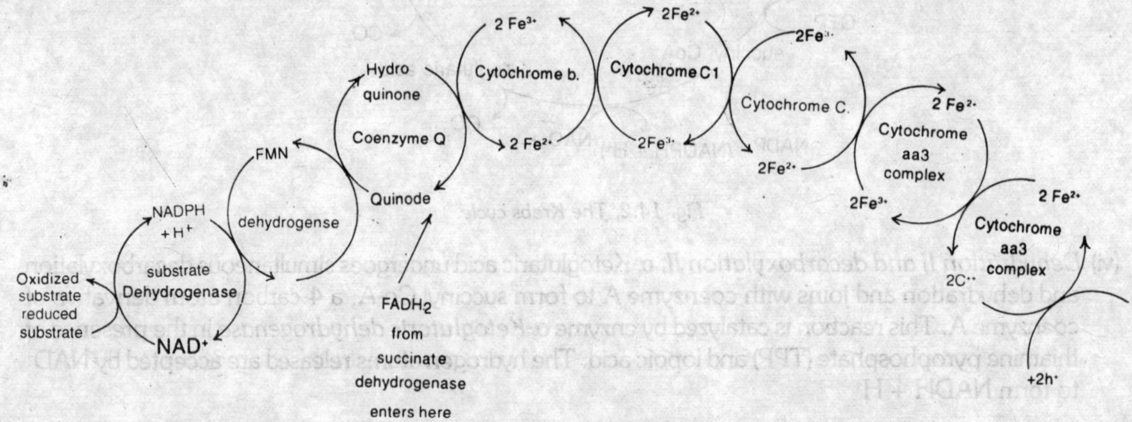

Fig. 14.3. Electron Transport system enzymes.

The respiratory chain system manufactures the ATP molecules during oxidation of reduced coenzymes in the following steps:

1. *Dehyderogenase* enzymes accept hydrogen from different substrates of Krebs' cycle. The hydrogen atom becomes ionized into protons (2H⁺) and electrons (2e).

   Protons or hydrogen ions reduce the oxidized NAD

   $$NAD^+ + 2H^+ \longrightarrow NADH + H^+$$

   The NADH molecule passes through M-side of inner mitochondrial membrane and enters the respiratory chain system.
2. The NADH is oxidized into NAD⁺ by transferring its proton (H⁺) to flavoprotein FAD. The FAD acts as a hydrogen carrier. From the flavoproteins each proton is discharged into perimitochondrial space through the C-side of inner mitochondrial membrane and electrons are passed onto non-haem iron protein.

3. In normal course, reduced FADH is reoxidized by ubiquinone (Co Q) which is the next member of ETS. The ubiquinone is reduced to dihydroubiquinone which is oxidized by ferric (Fe^{++}) cytochrome—C—(Fe^{+++}) which is reduced to ferrocytochrome. At this level of oxidation, hydrogen atoms undergo ionization to form $2H^+$ and $2e^-$. The protons cannot be further carried by ETS, however, the electrons can be carried easily by the cytochromes. One molecule of each cytochrome is capable of carrying only one electron at a time. Thus, for each molecule of hydrogen which releases two electrons, two molecules of each cytochrome are required. At the level of cytochrome-b, two protons per molecule of hydrogen are released in the mitochondrial system. Ferrocytochrome b is oxidized in the presence of ferric cytochrome c_1. From ferrocytochrome c_1 these electrons are taken by cytochrome a and than to a_3 (Both cyto-a and a_3 together called cytochrome oxidase.
4. These electrons from cytochrome oxidase react with one atom of oxygen making it reactive.
5. This activated oxygen react with $2H^+$ (protons) forming molecule of water.

$$O + 2e \longrightarrow O^- \xrightarrow{2H^+} H_2O$$

With the release of 2 electrons and 2 H atoms from every molecule of NADH + H^+ and their reaction with oxygen, 3ATP molecules are synthesized. The position of ATP molecules. Energy released at these positions reach and stored in ATP molecules. The process is known as oxidative phosphorylation because it occurs in respiration in the presence of oxygen hence it is called oxidative phosphorylation.

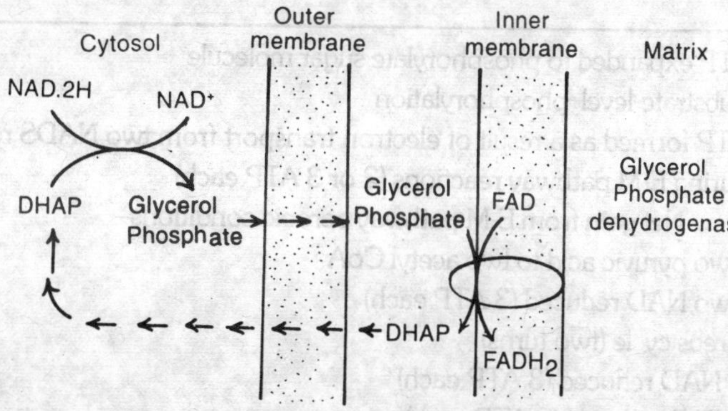

Fig. 14.4. Glycerol-phosphate shuttle.

Now we can summarize the energy retum from the complete oxidation of a glucose molecule under aerobic conditions. Under aerobic conditions, the NADH produced in the E-M pathway can transfer electons to the ETS inside mitochondria, these by salvaging more usable energy than is possible under anaerobic conditions.

However, reduced NAD cannot enter a mitochondrion directly, so it cannot pass its electrons directly to an ETS in the inner mitochondrial membrane. Instead, the electrons are passed to a shuttal carrier that takes them into the mitochondrion and donates them to an electron transport system.

These are several such electron shuttles for example malate-asperate shuttle, glycerol phosphate shuttle etc., but the common one is *glycerol-phosphate shuttle,* which picks up electrons from reduced NAD, carries them into mitochondrion and donates them not to NAD, but to ubiquinone. In this case the electrons do not start at the 'top' of the ETS, so the possibility for one phosphorylation is bypassed. If mitochondria use this particular mechanism, a reduced NAD in the cytoplasm ends up being worth only 2ATP molecules. In this case complete oxidation of glucose will result in a gain of 36 ATP molecules.

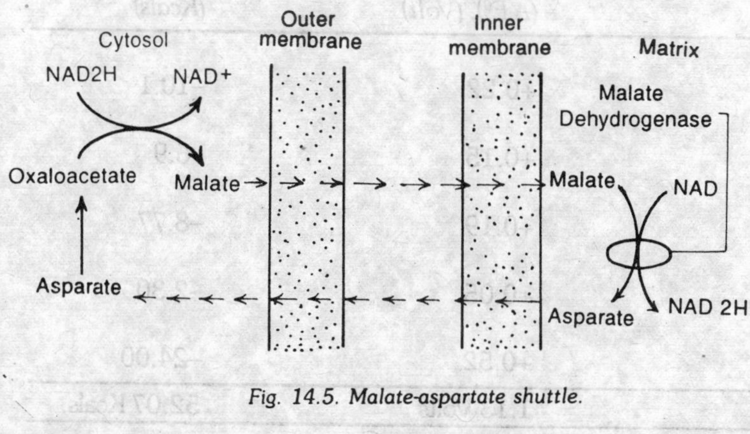

Fig. 14.5. Malate-aspartate shuttle.

In *malate-asperate shuttle,* the electrons are transferred from NADH in cytoplasm to malate which traverses the inner mitochondrial membrane and reoxidized to form NADH + H^+ thus resulting in the formation of oxaloacetate. Oxaloacetate does not readily cross the inner mitochondrial membrane 20 a transmission reaction is needed to form asperate which does traverse this barrier. 3ATP molecules are generated for each pair of electrons. In this case complete oxidation of glucose will result in a gain of 38 ATP molecules.

Oxidative Phosphorylation Mechanism

Oxidative phosphorylation was discovered in 1939 but the mechanism of coupling between ETS and ATP synthesis is not yet known. However, three principal hypothesis have been proposed for the coupling of oxidation and phosphorylation. These are:

1. Chemical Coupling hypothesis.
2. Cheiniosomotic Coupling hypothesis.
3. Conformational hypothesis.

Table 14.1. Showing ATP Formation During Oxidation of Glucose.

Embden-Meyerhof Pathway	*ATP calculations*	*Sum*
ATP expanded to phosphorylate sugar molecule	– 2	
Substrate-level phosphorylation	4	
ATP formed as a result of electron transport from two NADS reduced during E-M pathway reactions (2 or 3 ATP each)	4 or 6	
Net gain from E-M pathway aerobic conditions		6 or 8
Two pyruvic acid to two acetyl CoA	6	
Two NAD reduced (3 ATP each)		6
Krebs cycle (two turns)		
6 NAD reduced (3 ATP each)	18	
2 FAD reduced (2 ATP each)	4	
Two substrate-level phosphorylation (1 ATP each)	2	
Total from Kreb's cycle (2 turns)		24
Total ATP formation per molecule of glucose		36 or 38

Table 13.2. Free-Energy Change at Different Steps in the Electron Transport Chain.

Steps of Transport	*Difference in Redox-potential (ΔE^0), (Volt)*	*Free energy change (Kcals)*
NAD $\xrightarrow[2e^-]{2H^+}$ FAD	+0.22	–10.1
FAD $\xrightarrow[2e^-]{}$ cyt.b	+0.15	–6.9
2Cyt.b $\xrightarrow[2e^-]{}$ 2 cyt c	+0.19	–8.77
2 Cyt.c $\xrightarrow[2e^-]{}$ 2 cyt.a	+0.05	–2.30
2 Cyt.a $\longrightarrow$ Oxygen	+0.52	–24.00
Total	1.13 Volts	52.07 Kcals.

1. Chemical Coupling Hypothesis

According to this hypothesis the ET is coupled to ATP synthesis by a sequence of consecutive reactions in which a high-energy covalent intermediate is formed by electron transport and subsequently is cleaved and donates its energy to ADP to form ATP. For example, when an oxidoreduction reaction occurs between A^{red} and b^{oxi}, the factor I is incorporated into the formation of an energy rich $A^{oxi} \sim I$ complex (here ~ represents a linkage having an energy-rich nature.)

$$A^{red} + I + B^{oxi} \rightleftharpoons A^{oxi} \sim I + B^{red}$$

In the following reaction an enzyme (E) replaces A^{oxi} from the above complex forming an energy rich E ~ I complex. Later on inorganic phosphate (P_i) reacts with E ~ I complex, forming phosphoenzyme complex (E ~ P). The phosphoenzyme complex reacts with ADP to form ATP liberating enzyme free.

$$A^{oxi} \sim I + E \rightleftharpoons A^{oxi} + E \sim I$$

$$E \sim I + P_i \rightleftharpoons E \sim P + I$$

$$E \sim P + ADP \rightleftharpoons ATP + E$$

An extensive search has been made for the nature of E ~ I and E ~ P, but non have been found in mitochondria. The chemical coupling hypothesis is only of historical importance today.

2. Chemiosmotic Coupling Hypothesis

This hypothesis was proposed by *Peter Mitchell* in 1961. He proposed that electron transport and ATP synthesis are coupled by a proton gradient, rather, than by a covalent high-energy intermediate or an activated protein. This theory suggests that a transmembrane electrochemical gradient is critically important to the phosphorylation process.

Chemiosmotic coupling hypothesis suggests that during electron transport, a gradient of H^+ ions is produced across the inner mitochondrial membrane. The mobile electron carrier that gains H^+ ion is reduced one side of the membrane and loses proton as it is oxidized near the other side. These reactions lead to the transport of a pair of H^+ across the inner mitochondrial membrane from the matrix to the space between the mitochondrial membrane. Thus there is a H^+ ion gradient across the membrane. There is a strong tendency for these H^+ ions to move back toward the matrix because of the excess of +ve charge is the space between the two membrane.

Mitchell further suggests that there are special channels through which the H^+ ions can move in response to the forces acting on them (electrical as well as concentration differences). In same way, the movement of H^+ ions back through the membrane is coupled in ATP synthetase complex to the phosphorylation of ADP to form ATP.

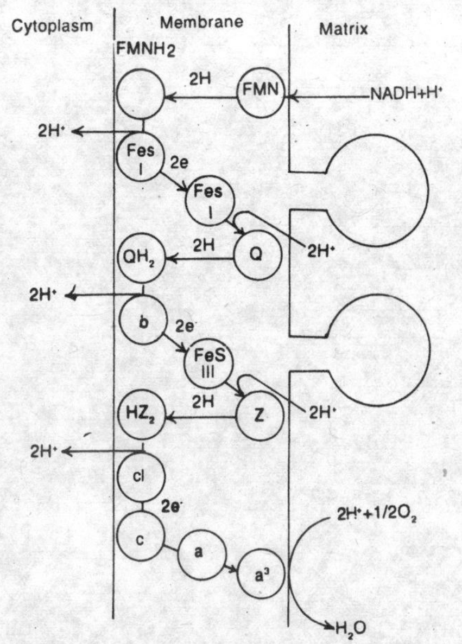

Fig. 14.6. Showing Michell's chemiosmolic coupling hypothesis.

Oxidation-reduction loop

Mitchell has proposed that reducing equivalents are transferred as H atoms by some of the electron carriers (ubiquinone) and as electrons by other (Fe-S center and cytochromes). According to him hydrogen carrying and electron carrying proteins alternate in the respiratory chain to form three functional loops called *oxidation-reduction loops* (O/γ loops).

A single loop consists of a hydrogen carrier and an electron carrier. In each loop two H^+ are carried outward through the inner membrane and delivers two H^+ to the C side, the corresponding pair of electron is then carried back from outer to the inner surface of the membrane. Thus, each pair of reducing equivalents passing through such loop carries $2H^+$ from the matrix to outside. Each loop is thought to provide the osmotic energy to form one molecule of ATP.

Mitchell's hypothesis is supported by a wealth of evidences.

3. Conformation Coupling Hypothesis

According to this proposed by *Boyer* (1965), the electron transport and phosphorylating systems are in molecular contact and that information is transmitted by short-arange interactions which may, in part, be electrostatic in nature. It is postulated that upon acceptance of the two electrons, each electron transfer protein undergoes a conformational change which is then transmitted to the ATPase by short range interactions. However, the mode of conformational changes that take place in the inner mitochondrial membrane is not yet clearly understood.

According to some recent views (*Ernster,* 1977) the conformational and chemiosmotic mechanisms of electron transport linkage to phosphorylation do not represent mutually exclusive alternative but may indeed be two aspects of the same mechanism.

15

BIOENERGETICS AND CELLULAR HOMEOSTASIS

Bioenergetics is the field of biochemistry deals with the transformation and use of energy by living cells. Cells constantly manufacture new substances, they perform the mechanical work of motion, they transport substances and they generate heat. All these activities are controlled by the enzymes. Enzymes speed up the chemical reactions to certain extent which is enough, but not too much, of the products of the reaction. The ability of enzymes to regulate chemical reactions provide the cells with a homeostatic system i.e., one that maintains proper functioning of cell inspite of changing environment and need.

Concept of Energy

The term *energy* refers to the capacity for acting or to perform work. It is one of the fundamental components of any system. Physicists and chemists recognize many kinds of energy such as *thermal energy,* in which a hot system can be made to do work as in a steam-engine; *gravitational energy,* in which work is obtained from an object as it falls from high place to a low one; *electric energy, light energy* and so on. These different forms of energy are interconvertable and during conversion of one form of energy to another, there is always some loss. Moreover, when a work is done, there is also a loss of useful energy. The kind of energy in which we are most interested here, however, is called *Gibbs free energy* or *free energy* in short.

Thermodynamic Principles

Many quantitative observations made by physicists and chemists on the interconversion of different forms of energy have led to two fundamental laws of thermodynamics. These laws helps in understanding the direction of a reaction, the accomplishment of work and the source of energy.

I. The first law of thermodynamics or Principla of Conservation of Energy

According to this law, "in any physical or chemical change, the total amount of energy in the universe remains constant." Thus according to this law the energy cannot be created nor destroyed, whenever energy is used to do work or is transformed from one kind to another the total amount of energy is unchanged.

This law can be mathematically represented as:

$$\Delta E = E_B - E_A = Q - W \quad (1)$$

where ΔE is change in internal energy, ΔE_A is the energy of a system at the start of work, ΔE_B is energy of a system at the end of work, Q is heat absorbed by the system and W is work done by the system.

II. The second Law

According to this law, "All physical and chemical changes tend to proceed in such a direction that useful energy undergoes irreversible degradation into a randomized form called *entropy.* They come to a stop at an equilibrium point, at which the entropy formed is the maximum possible under the existing conditions."

This simple, somewhat abstract defination requires some explanation of free energy, heat energy and entropy.

The *free energy* is the kind of energy that can do work at constant temperature and pressure.

The *heat energy* can do work only through a change of temperature.

Entropy is the energy in a state of randomness or disorder. It is unavailable useless energy.

A quantitative rigorous defination of entropy requires mathematical statements of randomness.

There is another aspect of second law which is necessary to understand its operation particularly in biological system. Firstly, it is necessary to identify the reaction system, the collection of matter that is undergoing a particular physical or chemical process (this may be a call, an animal or two reacting components). Secondly, the surrounding with which the reacting systems can exchange energy must be defined. The system plus the surroundings together constitute the universe. Universe literally includes the entire earth rather even the outer space. Some physical or chemical process can be made to take place in isolated systems, which are unable to exchange energy with their surroundings. But in the biological world, reacting system do exchange energy, and often they exchange matter with their surroundings.

Changes in free energy, heat energy and entropy in chemical reactions at constant temperature and pressure, the conditions existing in biological systems, are related to each other quantitatively by following equation

$$\Delta G = \Delta H - T\Delta S \tag{2}$$

here ΔG is the change in free energy of a reacting system, ΔH is the change in its heat content or *enthalpy* of this system, T is the absolute temperature at which the process is taking place and ΔS is change in the entropy of this system or universe.

When any chemical reaction proceeds to its equilibrium point, the entropy of universe always increases and ΔS, therefore, always has a positive sign in the biological world. The minus sign indicates release of useful energy. The Δ is equation by definition means final state minus initial state. For example, $\Delta H = H_2 - H_1$, in the transition form a heat content of H_1 to a heat content of H_2. Hence, a system that releases heat will have ΔH i.e. $\Delta O < O$. Such reactions are called *exothermic*. The exothermic reactions are characterized by a warming of their environment at heat is liberated from reacting system as heat leaves the reacting system. On the other hand, reactions that cool their surrounding do so because they are absorbing heat energy. Such reactions are called *endothermic* ($\Delta H > O$). If the ΔG is zero, the reaction system is at equilibrium and no net change takes place.

With regards to the free energy change, ΔG of a reacting system, the points need to be emphasized are:

(i) The ΔG of a reaction system depends only on the free energy of the products minus that of the reactants. Obviously, the mechanism of a reaction has no effect on ΔG. For example, the value of ΔG is the same for the oxidation of glucose to carbon dioxide and water whether it takes place by enzymatic-catalytic reactions or combination.

(ii) The value of ΔH does not provide any information about the rate of a reaction. A–ΔG indicates that a reaction can occurs spontaneously. As already pointed out, that the rate of reaction rather depends on the free energy of activation ($\Delta G \neq$), which is unrelated to ΔG.

We must emphasize that the formation of entropy or disorder, is not a totally useless activity. The increase in entropy of the universe during biological process provides their driving force and gives direction to all biological activities. Living systems constantly produce entropy in their surrounding as a necessary cost of maintaining their own internal order.

For example

$$C_6H_{12}O_6 + 6O_2 \rightarrow 6CO_2 + 6H_2O$$

if above reaction taking place at temperature 25°C or 298 Kelvin and at a pressure 760 mm Hg (1.0 atm.) which are standard conditions in thermodynamic calculations, the following energy changes take place per molecule of glucose oxidized.

$\Delta G = -686{,}000$ cal/mol (Free energy of the system)

$\Delta H = -673{,}000$ cal/mol (the reacting molecules have released heat).

$$\Delta S = \frac{\Delta H - \Delta G}{T} \quad (3)$$

$$= \frac{-673000-(-686000)}{298}$$

= + 44 cal/deg (The entropy of the universe has increased).

Relationship between standard free energy change and equilibrium constant.

Each chemical reaction has a characteristic standard free energy change ΔG^0. The ΔG^0 is not the same as ΔG given in equation. When a reacting system has achieved its state of lowest free energy, the system is said to be in *equilibrium.* At that pont free energy change for the reaction has declined to zero (ΔG = O). In other words, if the reaction is started with all participants already at their equilibrium concentration, the reaction goes no-where and cannot be made to do work. To apply this concept to chemical reactions, consider the isomerization.

$$A \rightleftharpoons B$$

α-D glucose $\rightleftharpoons$ β-D glucose.

Let us give the name state A to a situation in which the chemical bond is intact and state B is the situation where the same bond is broken. It we examine this system when it has reached equilibrium at any absolute temperature (T) the ratio of broke to intact bond will be given by the equation.

$$\Delta G^0 = -2.303\ RT \log \frac{(B)_{eg}}{(A)_{eg}}$$

This expression is a form of *Boltzmann equation.* Here R is universal gas constant (8.314) Joules/ mol/degree kelvin), 2.303 allows us to express the equation in term of common (base 10) logarithms instead of natural (base-e) logarithms and, eg is remind us the equation is valied only at equilibrium. ΔG^0 is standard state free energy change which indicates where the equilibrium position of reaction will be.

The ratio $(B)_{eg}/(A)_{eg}$ is an equilibrium constant designated as K_{eg}. The above reaction can be now written as

$$\Delta G^0 = -2.303\ RT \log K_{eg}$$

In replacing 2.303 RT with 5936, the temperature was assumed to be that of the human body (37°C or 310 K). ΔG^0 in above equation is expressed in the term of Joules per molecule.

It is clear that by knowing ΔG^0 should permit one to K_{eg} or *vice-versa.* The importance of ΔG^0 is two-fold, however, for it also permits one to calculate the amount of useful work (Gibbs free energy change) that can be obtained from a system as it moves toward the equilibrium.

$$\Delta G + \Delta G^0 + 2.303\ RT \log \frac{(B)}{(A)}$$

ΔG^0 has Characteristic Values for Different Chemical Reactions

$$\text{Glucose-1-phosphate} \xrightarrow{\text{Phosphoglucomutase}} \text{Glucose-6-phosphate.}$$

The chemical analysis shows that if the reaction starts with, say 0.0200 M glucose-1-phosphate and allow to go in forward direction or if we start with 0.0200 M glucose-6-phosphate to go in the reverse direction, the final equilibrium mixture in either case will contain 0.00100 M glucose-1-phosphate and 0.019M glucose-6-phosphate at 25°C and *pH* 7.0. From these datas we can than calculate the equilibrium constant to three significant figures

$$Keg = \frac{\text{Glucose-6-phosphate} = 0.0190}{\text{glucose-1-phosphate } 0.00100} = 19.0$$

from this value of K_{eg} we can calculate the standard free energy change

$$\Delta G^0 = -2.303 \text{ RT} \log K_{eg}$$
$$= -2.30\,(1.987)\,(298) \log 19.0$$
$$= -1360\,(1.28)$$
$$= -1740 \text{ cal/mol.}$$

ΔG of adenosine triphosphate (ATP) hydrolysis can be calculated as

$$\Delta G = \Delta G^0 + 2.303 \text{ RT} \log \frac{(ADP)\,(P_1)}{(ATP)}$$

$$DG = -7300 + 1360 \log \frac{(2.50 \times 10^{-4})\,(1.65 \times 10^{-3})}{2.25 \times 10^{-3}}$$

$$= -7300 + 1360 \log 1.83 \times 10^{-4}$$
$$= -7300 + 1360\,(-3.74)$$
$$= -7300 - 5100$$
$$= -12400 \text{ cal/mole}$$

or 12.4 kcal/mole.

In the above calculation, seen in human RBC, the concentration of ATP, ADP and *Pi* are 0.25, 0.25, and 1.65 mM respectively. The $\Delta G^0 = -7300$.

Thus the actual free energy (ΔG) is – 12.4 kcal/mol.

Difference Between ΔG and ΔG⁰

The actual free energy change (ΔG) of a chemical reaction is a function of the condition of *pH*, concentration and temperature under which the reaction is taking place. Moreover, it is always negative in an outgoing chemical reaction. It becomes less negative as the reaction proceeds and becomes zero at the point of equilibrium. This indicates that no more work can be done by the reaction. Thus, in every spontaneous chemical or physical, process, the free energy (ΔG) of the reaction system is always decreases i.e., it is negative.

On the other hand, the value of standard free energy change (ΔG^0), for a chemical reaction is characteristic and unchanging and may be positive, negative or zero depending on the equilibrium constant of the reaction.

Energy Rich Compounds

In order to maintain biological activities, the cells must get free energy from their environment. The autotrophs get supply of free energy by coupling their metabolism to some exergonic process in their surrounding. For example, green plants utilize the energy of sun, on the other hand, animals get free energy by coupling their metabolism to the breakdown of complex organic molecules in their environment.

A large number of compounds are found in biological system. On hydrolysis they yield free energy changes of the order of –2 to –5 kcal per molecule under close to physiological conditions. These compounds ae catagorized as *low energy bond compounds*. These compounds are very stable. Glucose-6-phosphate, glucose-1-phosphate, 3-phosphoglyceric acid, α-glycerophosphate, ethylacetate are some examples of this category. There is an another category of compounds, which on hydrolysis yield -6 to 13 kcal per molecule under physiological conditions. This category of compounds is called *high energy bond compounds*. The number of such compounds is relatively small. A dinosine triphosphate, adinosine diphosphate, phosphenol pyruvic acid, acetyl Co-A creatine phosphate, arginine phosphate etc., are the examples of this category.

In the energy rich compounds, the high energy is trapped in a particular bind. This bond is known as *energy rich bond*. Such bonds are represented by a sigmoid (~) in order to differentiate from the ordinary bonds, such bonds are represented by *covalent bonds* (–).

CLASSIFICATION

Energy rich compounds may be broadly classified into two categories-namely phosphate group and thioester group.

1. *Phosphate group*. This group includes:
   (i) *Phosphate anhydrites*. In this catagory adinosine triphosphate, adinosine diphosphate and 1-3 diphosphoglyceric acid are included.
   (ii) *Enolic phoshpate*. It include phosphoenol pyruvic acid.
   (iii) *Guanidinium phosphate*. It includes creatine phosphate and orginine phosphate.
2. *Thioester group*. This group has acetyl Co-A, 5-adensyl-methionine etc.

1. Phosphate Group Compounds

(i) Adinosine triphosphate

Fiske and *Subbarow* (1929) first of all discovered ATP in the muscle extract. At the same time *Lohamann* (1929) also discovered it independently. *Lipmann* (1944) presented a detailed hypothesis of its function as an energy rich compound.

At neutral *pH*, the two terminal phosphate groups are in an ionized form and form a complex with the divalent magnesium ions (Mg^{++}). In most of the enzymatic reactions, where ATP participate as phosphate donor, its active form is magnesium—ATP Complex.

On hydrolysis of each of the terminal phosphate group—8.8 kcal/mole of free energy (ΔG^0) is obtained. In this reaction ATP is converted into ADP with the liberation of inorganic phosphate. On further hydrolysis of second phosphate group, — 8.8 kcal/mole energy is liberated. The ADP is converted into AMP with inorganic phosphate liberation. The hydrolysis of the first phosphate which is linked by an ester bond to ribose sugar, releases only-3 kcal/mole energy. To distinguish the terminal two pyrophosphate bonds from the first ester bond, the terminal bonds are represented by (~) high energy bond and the ester bond is represented by (~) straight bond.

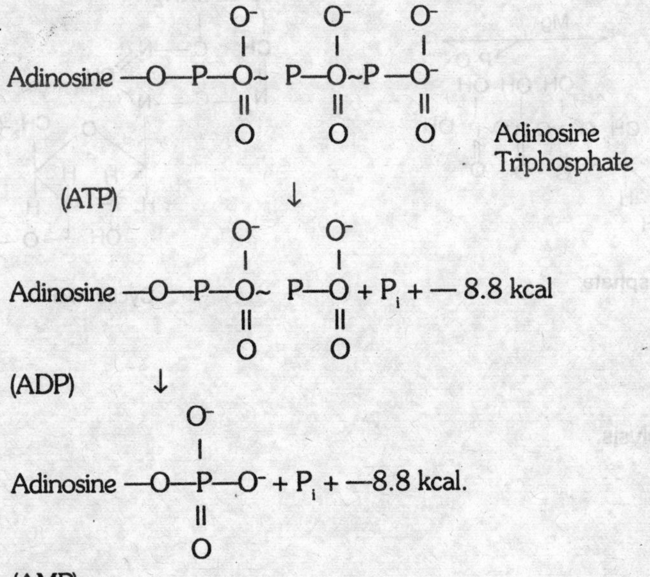

In the cell ATP and ADP both occur. ATP is utilized, but soon regenerated through a coupled reaction where ADP accepts high energy phosphate to form ATP. In reality, an ATP/ADP cycle connects these process which generate ~ P to these processes which utilize ΔP.

(ii) Phosphocreatine

It is the phosphagen found in the muscles of vertebrates and some invertebrates. It is the storage form of energy in muscles.

Although ATP is also present in muscles, but it is present in small amount and is used up in muscle contraction to form ADP. It is replanished by transfer of energy rich phosphate from creatine phosphate to ADP.

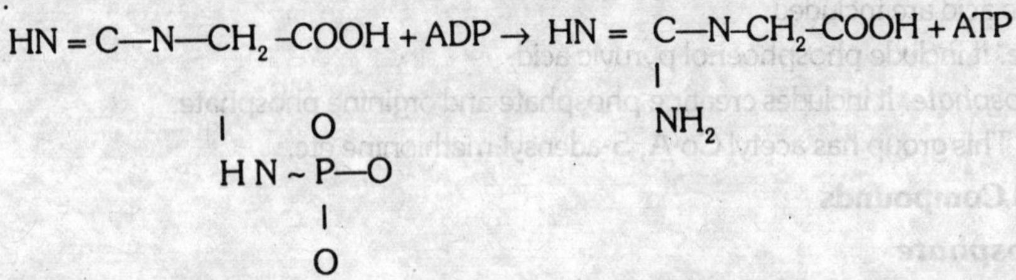

Creatine phosphate creatine

In the invertebrates the energy rich phosphagen is arginine phosphate.

(iii) 1, 3-diphosphoglyceric acid

It is formed during Embden—Mayerhoff pathway or glycolysis. The phosphate group is attached to the carboxylic group by high energy bond.

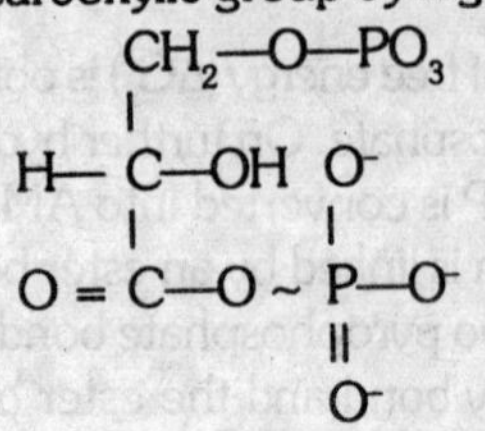

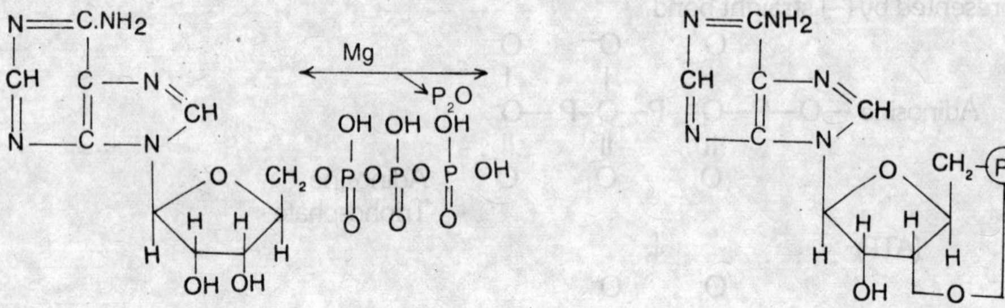

Adenosine triphosphate (ATP) Cyclic AMP

(iv) Phosphoenol pyruvic acid

This is also formed during glycolysis.

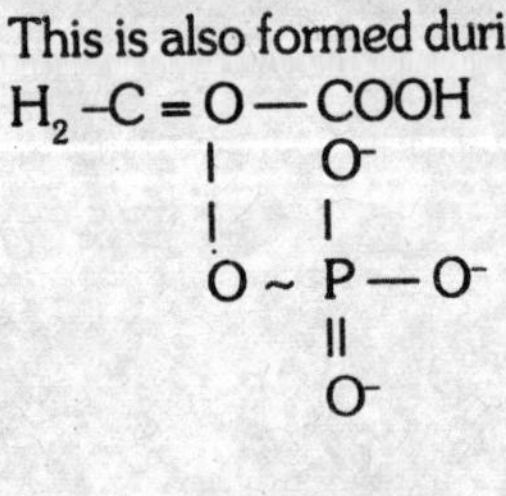

Besides above high energy rich compounds, two more such compounds are present in every cell of organisms. These are cyclic nucleotides namely — adinosine 3', 5'-cyclic monophosphate (cAMP) and other is guanosine 3', 5'-cyclic monophosphate (cGMP). The first one is formed from adinosine triphosphate and other from guanosine triphosphate. The reactions are catalized by adenylate cyclase and guanylate cyclase.

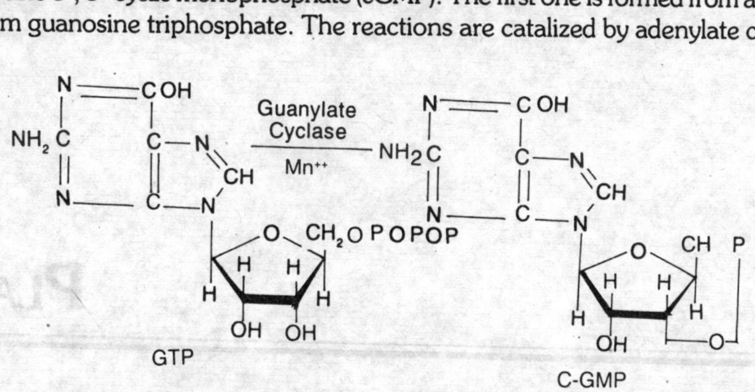

Fig. 15.2. Structure of cGMP.

On hydrolysis cyclic nucleotides yield—11.9 kcal/mole free energy. The significance of this compounds lies not in their high energy character, but in the regulatory roles in metabolism.

2. Thioester Group Compound

S-adenosyl methionine yields —10 kcal/mole free energy on hydrolysis.

Table 15.1. Standard Free Energy (ΔG^0) of Hydrolysis Some Energy Rich Compounds at pH 7.

Compounds	ΔG^0 *kcal/mole.*
Acetyl adenylate	— 13.3
Phospho enol pyruvic acid	— 13.0
cAMP	— 11.9
Acetyl phosphate	— 10.5
Acetoacetyl CoA	— 10.5
S-adenosyl methionine	— 10.0
Phosphocreatine	— 9.0
ATP (Terminal bond A)	— 8.8
ATP (Inner bond B)	— 8.8
Acetyl Co A	— 7.7
Palmityl Carnitine	— 7.7
Uridine diphosphate glucose	— 7.6
Phosphodiesters	— 6.0
Glucose -1-phosphate	— 5.0
Alanyl glycine	— 4.0
Glutamine	— 3.4
Glyceral-3-phosphate	— 3.0

16

PLASTIDS

The plastids are characteristic large cytoplasmic bodies of the cell intimately related to the metabolic processes and which impart colour to the tissues. They are found mostly in plants with possible exception of the very lowest group of plants. However, plastids are also found in few animals, especially the protozoan flagellates. Plastids are found scattered in majority of the plant-cells except those of the lowest groups of plants. They are not found in bacteria, certain algae, myxomycetes and fungi. On the basis of presence of the plastids, the living world has been distinguished into plant and animal kingdoms. Old literature indicates that *Pringsheim* first to investigate the plastids. The term plastid was used in 1885 by *A.F.W. Schimper* to include the cell organelles whch are primarily involved in the formation and storage of soluble and insoluble carbohydrates. *A. Meyer* in 1883, *F. Schmitz* in 1884 and *A.F.W. Schimper* in 1885 made detailed cytological studies of these cell organelles and showed that green plastids now known as chloroplasts could not be formed from cytoplasm, but only from existing chloroplasts. The plastids vary in size, number, shape and chemical organization with the particular tissue and organism concerned. In contrast to the smaller cytoplasmic granules such as the mitochondria, important physical and chemical differences in the plastids can be easily recognized, such as change in size or shape, the presence or absence of the chloroplastic chlorophyll, carotenoids and starch. Mature plastids can be shown to arise from more or less undifferentiated plastid primordia which occur in the meristematic cells, or those of the developing embryo.

The plastids are self-duplicating bodies and their life history, especially of the chloroplast, may be divided into three phases. In the first phase self-duplication and enzymatic activities with some concern to RNA takes place. The second phase is mainly concerned with differentiation to form the enzymes required for photosynthesis. In the third phase the activities of the enzymes that predominates are those concerned with photosynthesis.

VARIOUS TYPES OF PLASTIDS

Plastids may be divided into two general groups; the *chromoplasts* (coloured plastids); and the *leucoplasts* (colourless plastids). The chromoplasts may be further divided into two subgroups; those that are photosynthetically active and those that are not.

(A) Photosynthetically Active Chromoplasts

The photosynthetically active chromoplasts are chloroplasts, phacoplasts, rhodoplasts, bluegreen chromoplasts and chromatophores of photosynthetic bacteria.

1. *Chloroplasts.* These are green plastids of green algae and higher plants and contain chlorophyll a and b in the ratio of 2:3:1. These have different appearance and varying sizes.
2. *Pheoplasts (Brown or yellow plastids).* This type of plastid is commonly present in the brown algae, diatomos and dinoflagellates. In these pheoplasts, brown carotenoids mask the chlorophyll.

As reported by *Strain* (1951) the yellow or brown colour of the pheoplasts appears to result from the physical state or condition of the pigment from their geometrical arrangement in the plastids. Brown algae contain fucoxanthin which can absorb light and transfer energy to chlorophyll.

3. *Rhodoplasts*. Various members of Rhodophyceae contain red coloured plastids. The red colour is due to a pigment phycoerythrin (*Strain* 1949). The colour of plastids may be the result of light absorption. There may be phycocyanin too in traces. *Haxo*, *Blink* and *Young* (1952) found that light is absorbed by phycobilin and is transmitted to chlorophyll *a*.

4. *Blue-green chromatophore*. Two accessory protein phycocyanin and phycoerythrin impart blue-green colour to algae. These pigments and chlorophyll *a* alongwith carotenoid form lamellar structure, the *chromatophore*.

Wilson and *Morrison* (1967) mentioned that all the organisms capable of performing photosynthesis do not necessarily contain plastid structures; for example, blue-green algae and photosynthesis bacteria. In blue-greens, the light-sensitive pigments are chlorophyll and phycobilin pigments. These pigments are associated with a system of membranes and lamellae in the cytoplasm of the cell. In photosynthetic bacteria also wide diversity of structure are associated with the photosynthesis. These structure are as follows:

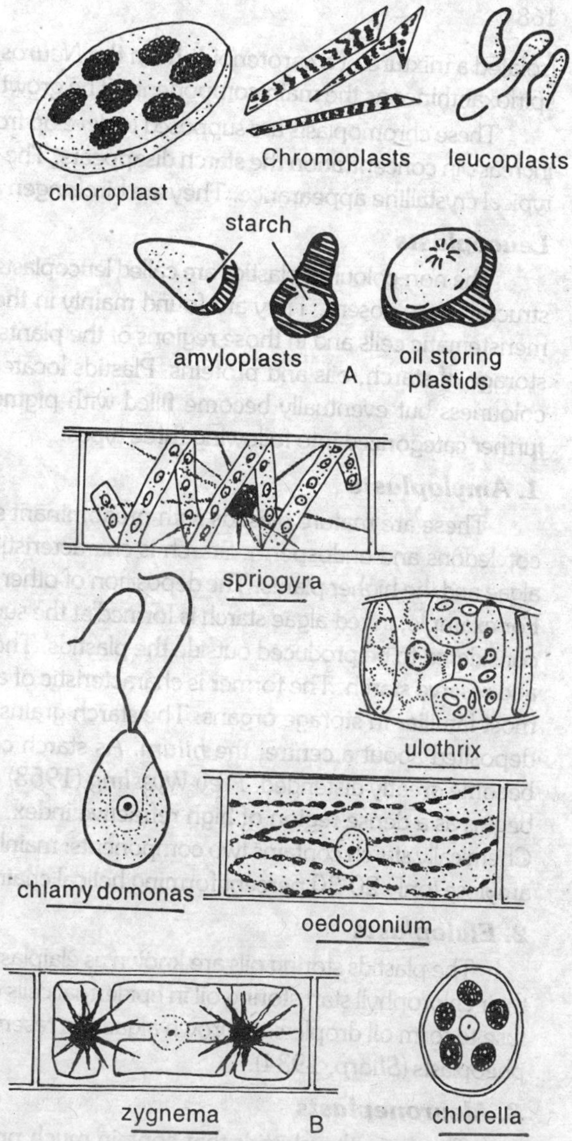

Fig. 16.1. A—Different kinds of plastids; B—Various types of choloroplasts found in Algae.

1. Lamellae in some purple bacteria; for example, *Rhodomicrobium,* (2) Vesicular membrane-bound structures, known as chromatophores found in non-sulphur and purple sulphur bacteria; for example, *Rhodospirillium, Chromatium,* and (3) Ribosome like granules found in green sulfur bacterium, *Chlorobiumthio sultatophilum*. All these photosynthetic bacteria possess a pigment system consisting of a specialized form of chlorophyll, known as bacteriochlorophyll and various carotenoid pigments.

(B) Chromoplasts without Photosynthetic Activity

These types of chromoplasts contain carotenoids only. They are red to yellow in colour and are much more variable in shape. They impart specific colourto various plant tissues, fruits and flowers. The new colour that arise often due to the formation of special carotenoids, for example, the *leucopene* of tomato fruits, *prolycopene* of berries, and *capsanthin* of red pepper. Xanthonhyll is also present in adequate proportions. Coloured carotenoids are also important constituents of the coloured parts of plant. These may be found in fungi and bacteria. Acidic carotenoids of fungi are similar to astaxanthin. *Haxo* (1949) isolated a mixture of a carotenoids from the *Neurospora crassa* which includes lycopene, carotene, and spirloxanthines as the main components. The growth of carotenoid becomes double in presence of light.

These chromoplasts are supposed to develop from the leucoplasts that contain starch. As the carotene increase in concentration the starch disappears. The older chromoplasts appear thin plates as needles with typical crystalline appearance. They are birefringen and dichroic (*Frey Wyssling,* 1935).

Leucoplasts

The non-coloured plastids are called leucoplasts. They do not contain pigments and the membranous structures are absent. They are found mainly in the embryonic and sexual cells. They are also found in meristematic cells and in those regions of the plants which do not receive light. They serve mainly for the storage of starch, oils and proteins. Plastids located in cotyledons and the primordium of stem are first colourless but eventually become filled with pigments and become coloured plastids. Leucoplasts are further categorized into following three types:

1. Amyloplasts

These are mature plastids with predominant starch contents and generally found in storage tubers, cotyledons and endosperm. Starch is characteristic of the coloured and the colourless plastids of green algae and the higher plants. The deposition of other carbohydrates may not necessarily be with the plastids. For example, in red-algae starch is formed at the surface of the plastids. In *Euglena,* paramylum grains are considered to be produced outside the plastids. The starch grains may be of two types: assimilative starch and reserve starch. The former is characteristic of actively photosynthesizing chloroplasts and the latter is most familiar in storage organs. The starch grains are made of a series of concentric layers successively deposited about a centre, the *hilum.* As starch continues to be deposited, the walls of the amyloplast become greatly distended. *Frey Wyssling* (1953) observed that the formation of a new concentric layer begins as a dense region of high refractive index, then it becomes a dense layer due to more hydration. Chemically starch contains two components: mainly α-amylose (which stain red with iodine) and minorly β-amylose (only 20-30 percent forming helical chain structure and stains blue with iodine), *Hassid* (1954).

2. Elaioplasts

The plastids storing oils are known as elaiplasts. In most monocotyledons the chloroplasts after losing their chlorophyll start storing oil in epidermal cells of Orchidaceae and Liliaceae. The disorganised plastids fuse to form oil droplets. In brown algae the reserve storage is oil and this oil appears to be formed in the pheoplasts (*Sharp,* 1934).

3. Aleuroneplasts

There are the plastids that contain much protein and often termed as the *proteinoplasts.* Protein crystals and granules are produced in the seeds of numerous plants especially in those seeds that also form large amount of oil, e.g., the seeds of *Ricinus. Mottier* (1921) referred that protein and oil formation involves the activity of permanent plastid primordia. Protein crystals may devleop independently in the cytoplasm without any aid of the aleuroneplasts, as for example in cells poor in starch in peripheral layers of potato. Proteins crystals in epidermal cells may also represent virus material or may be the result of virus action (*Granide*).

Amyloplasts are the most important of all the different types of leucoplasts, the amyloplasts resemble proplastids and differ from them only in size. They contain fewer lamellae, but in presence of light they can build up a thylacoid structure found in chloroplasts. However, chloroplasts can also synthesize starch and store it in stromal region. The starch in chloroplasts disappears quickly and is, therefore, known as transitory starch. The starch present in leucoplasts on the other hand can be stored for longer periods and is, therefore, known as reserve starch.

The starch development begins with the formation of a particle in the stroma. This particle has several tubuli and is surrounded by additional rings of starch. It grows till the amyloplasts is filled with starch. As a result, the tubuli are pressed against the wall and ultimately starch granule exceeds the size of amyloplast. The wall of the plastid then ruptures and finally withers away to that only the starch granules can be seen.

CHLOROPLASTS

Chloroplasts are the plastids that are most common and of greatest biological importance, since by *photosynthesis* they produce most of the chemical energy used by living organisms, without chloroplasts there would be no animals, because animals feed on the food stuffs produced by plants.

Distribution of Orientation

Chloroplasts are sometimes distributed homogeneously within the cytoplasm, but are frequently packed near the nucleus, or close to the cell wall. The distribution and orientation of the chloroplasts within the cell may vary with the amount of light energy. Chloroplasts apparently multiply by division—elongation of the plastid and constriction of the central portion. They have a higher density than the cytoplasm and migrate to the centrifugal pole of the cell when submitted to the action of centrifugal force.

Table 16.1. Showing Occurrence, Functions and Pigments of Various Plastids

Kinds of Plastids	*Occurrence*	*Function*	*Pigment*
I. Chromoplasts.			
A. *Photosynthetically active chromoplasts*			
1. Chloroplasts	Higher plants and green algae	Photosynthesis	Chlorophyll-a and Chlorophyll-b.
2. Pheoplasts	Brown algae, diatoms and dinoflagellates	Light absorption	Fucoxanthin
3. Rhodoplasts	Red algae (Rhodophyceae)	Light absorption	Phycoerythrin
4. Blue-green	Blue-green algae	Photosynthesis	Phycocyanin Phycoerythrin Chlorophyll-a and Carotenoids
5. Chromoplasts of Photosynthetic bacteria	Purple and non-purple sulfur bacteria	Absorption of infra-red portion of light	Bacteriochlorophyll
B. *Chromoplasts devoid of photosynthetic activity*			
6. Carotenoids	Tomato, red pepper, flower parts fungi, orriamental bacteria and fruits		Lycopene and Capsanthin
II. Leucoplasts			
7. Amyloplasts	Storage tubers, cotyledons and endosperm	Starch-storage	None
8. Elaioplasts	Epidermal cells of Orchidaceae and Liliaceae	Oil-storage	None
9. Aleuroneplasts	Epidermal cells of *Helleborus* and seeds of *Ricinus* and Brazil nut	Protein-storage	None

MORPHOLOGY

Shape

Higher plant chloroplasts are generally bioconvex or planoconvex. However, in different plants cells, chloroplast may have various shapes, viz., *Filamentous, saucer-shaped, spheroid, ovoid* or *discoid shaped.* They are vesicular and have a colourless centre. Some chloroplasts are *club shaped,* having a thin middle zone and bulging ends filled with chlorophyll. Algae often possess a single huge chloroplast that appears as a network, a *spiral band,* or a stellate plate.

Size

The size of chloroplasts varies from species to species but is constant for a given cell type. The average size of chloroplasts in higher plants measure 4-6μ in diameter and 1-3μ in thickness. The chloroplast of polyploid plant cells are larger than those in the corresponding diploid cells. In general, chloroplasts of plant grown in the shade are larger and contain more chlorophyll than of those plants grown in sunlight.

Number

The number of chloroplasts is relatively constant in the different plants. In higher plants there are 20 to 40 chloroplasts per cell. It has been calculated that the leaf of *Ricinus communis* contains about 400,000 chloroplasts per square millimeter of surface area. When the number of chloroplasts is insufficient, it is increased by division; when excessive, it is reduced by degeneration.

Conformational Changes and Ion Movements

Chloroplasts undergo striking conformational changes which are dependent on the action of light. In the dark they undergo swelling and separation of the membranes of the thylakoid. This phenomenon is reversible; in the presence of light the membranes again return to close apposition. It was mentioned in Chapter 9 that mitochondria eject H^+ ions during the electron transport and that the interior becomes more alkaline. In the case of chloroplasts the opposite reactions occur, under the action of light H^+ ions are absorbed, and the interior becomes acidified. At the same time K^+ and Mg^{++} are ejected into the medium. In the dark, H^+ is ejected and K^+ and Mg^{++} are reabsorbed by the chloroplasts. Thus, the chloroplast membrane has a vectoral organization that is the reverse of the mitochondrial membrane. In physical terms it is understood that dark reactions,being of a thermochemical nature, do not need an ordered structural arrangement (i.e., they are carried out by soluble enzymes). On the contrary, photophosphorylation requires the separation of positive and negative charges into specific pathways of electron flow.

Structure

Plastids are dynamic organelles whose structure changes in environment. A mature plastid is having following parts:

1. Limiting membrane
2. Stroma
3. Grana

1. Limiting membrane

Each mature plastid is covered by a double membrane as in mitochondria, each unit is about 40 to 60Å in thickness. The limiting membrane functions as a semipermeable membrane as well as a boundary.

2. Stroma

The inner space which is formed by a limiting membrane is filled with a protienaceous matrix called the stroma. A number of starch grains and sometimes osmiophilic droplets are embedded in the stroma. When the plastid becomes inactive or ages, the number of osmiophilic droplets may increase.

3. Grana or lamellar system

These are flat bodies shaped like platelets or discs, which in a lateral view appear as dense bands perpendicular to the limiting membane of plastid. A number of grana are embedded in

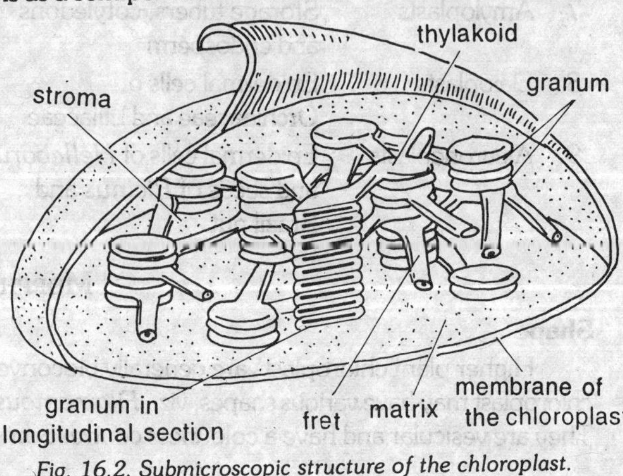

Fig. 16.2. Submicroscopic structure of the chloroplast.

the stroma or matrix of plastid. The size of the grana varies between 0.3-1.7μ depending on the species. Each granum consists of a stack of double membrane of lamellae. The lamellar membranes are composed of lipids and proteins. Protein contents are about 45% while lipids contents are 55% in amount. Each lamella is 30-35-Å thick and the space between two such lamellae is of 65-70Å. The lamellae can be placed under two groups:

(i) Stroma lamellae
(ii) Grana lamellae

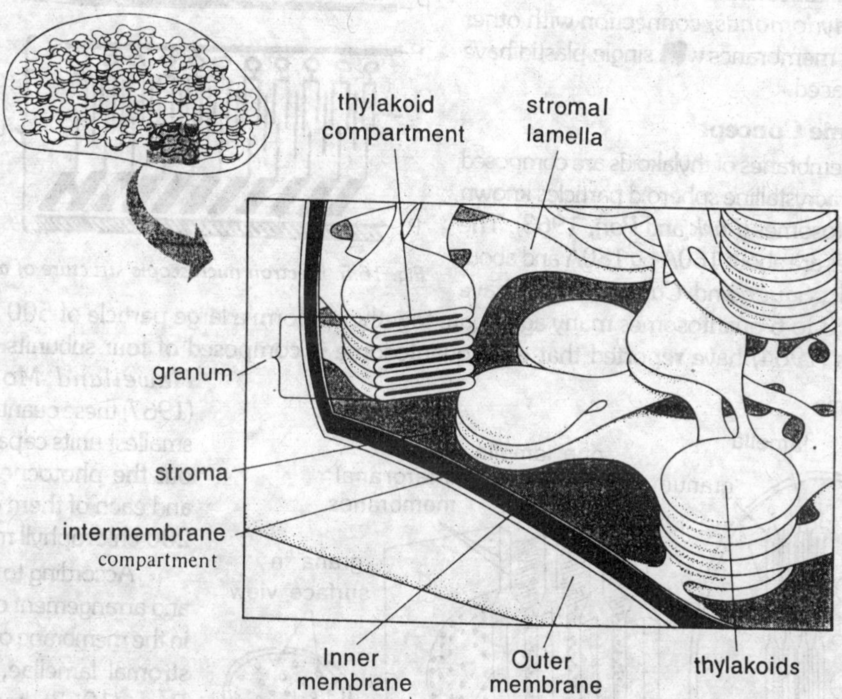

Fig. 16.3. Structural organization of a chloroplast.

Those lamellae, extending through the entire length of the plastids are called the stroma lamellae. These are not directly connected with the cell wall. The grana lamellae are always densely packed. Any way, the system of lamellae represents a closed unit, and thus both stroma and grana lamellae are closely related to one another, i.e., each granum lamella represents a local adhesion of two or more stroma lamellae. Both stroma and grana lamellae may stay coupled for a long distance and then separate again from each other. Another term *thylakoid* (*Greek;* sac like) has been used for the double membranous lamellae already explained. The stroma lamellae are larger thylakoids while the smaller, stacked thylakoids are the grana lamellae. One more term has been added by *Weier* and others, which is *fret* to describe stroma lamellae as a flattened tubular channel connecting adjacent grana. Recent studies have provided evidence that the membranes within the chloroplasts form a continum and the cavities of the thylakoids (locuili) communicate via fret work connection.

Fig. 16.4. Grana membranes consisting of a string of the particles.

The total number of grana in plastids of certain plants have been calculated. There is a considerable variation in the number of grana in a single plastid, depending upon the species. In *Euglena* only one granum is present in one plastid while in Spinach, where about 40-60 grana are present in each plastid.

In *chlamydomonas,* the plastid envelope is sometimes continuous with elements of the endoplasmic reticulum. In *Ochromonas* as well as in *Chalmydomonas,* connection with other cytoplasmic membranes w[illegible] single plastid have also been traced.

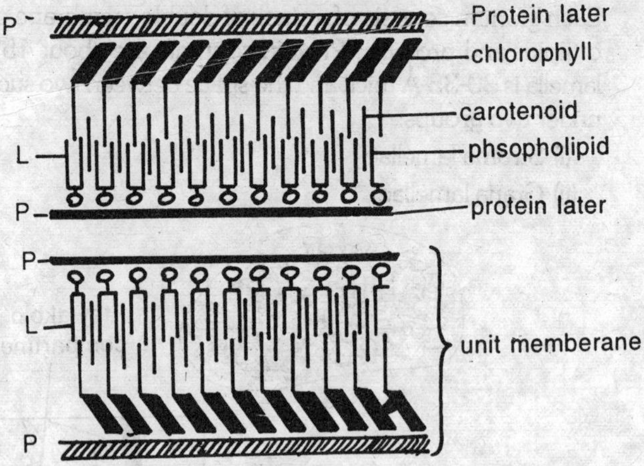

Fig. 16.5. Electron microscopic structure of a Chloroplast.

Quantosome Concept

The membranes of thylakoids are composed of many paracrystalline spheroid particles known as the quantosomes (*Park* and *Pon,* 1963). The quantosomes are about 160Å by 180Å and about 100Å thick. *Sauer* and *Calvin* (1962) have shown that 3 to 6 quantosomes many aggregate together to form a large particle of 500 x 100Å. *Park* and *Begins* (1964) have reported that each quantosome is composed of four subunits. According to *Hawelland Moundrianakis* (1967) these quantosomes are the smallest units capable of carrying out the photochemical reaction and each of them contains about 200 chlorophyll molecules.

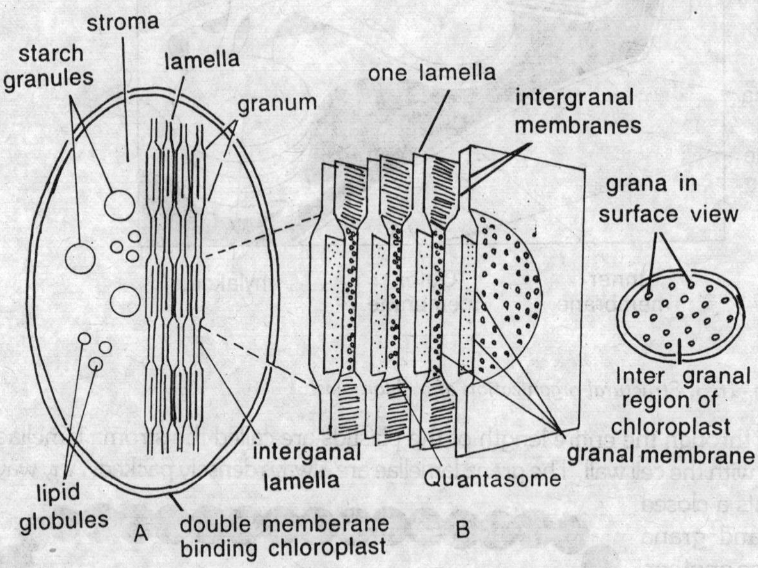

Fig. 16.6. Diagram to show detailed structure of a chloroplast and its internal membranous system.

According to the size, nature and arrangement of quantosomes in the membrane of thylakoids and stromal lamellae, *Branton* and *Park* (1967) have recognised three types of membranes which are as follows:

1. Membrane with quantosomes.
2. Membrane with smaller particles (110Å).
3. Membrane with rough surface and with a few or no particles.

The chloroplasts, thus, act as elaborate biochemical machines and transform the energy at the molecular level.

Chemical Composition

Chloroplasts of spinach leaf cells contain about 56% protein, 32% lipid and 8% chlorophyll. The lipids include xanthophylls, carotenoids, triglycerides, steroids, waxes and phospholipids. *Chloroplasts* of higher plants contain approximately 35-55% proteins, 20-30% lipids, 9% chlorophyll, 4-5% carotenoids and 2-3% nucleic acids. In cytoplasm the protein-lipid ratio is generally 90:1. In plastids this ratio is about 40:30. Thus the grana have a high concentration of lipid. The nucleic acid DNA and RNA have been demonstrated in the chloroplasts of higher plants.

Chlorophyll has a tetrapyrrol structure and consists of a hydrophilic head and a hydrophobic tail. The head is made up of a magnesium atoms surrounded by four pyrrolic nuclei (Tetrapyrrole structure). The long tail consists of a phytol chain. Green plants have two types of chlorophyll, *chlorophyll* a and *chlorophyll b*. In chlorophyll *b* one of the methyl ($-CH_3$) groups of chlorophyll *a* is replaced by a formyl (–CHO) group.

Functions of the Chloroplast

Photosynthesis. Photosynthesis involves two major sets of reactions, each set consisting of many steps. One set, the *light reactions,* transform ("transduce") energy acquired through the absorption of light into usable chemical forms, namely ATP and the reduced form (NADPH) of the coenzyme NADP. This is accomplished via oxidation-reduction sequences that include the removal of electrons from water, releasing oxygen and protons (H^+). These protons and others transferred across the thylakoid membranes during the oxidation-reduction reactions contribute to an ATP-producing proton gradient like the one described for mitochondria in the last chapter.

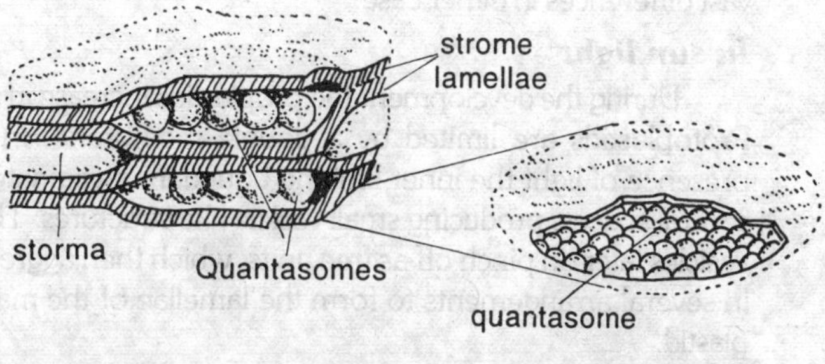

Fig. 16.7. Young chloroplast of germinating corn embryo showing equal division.

The other reactions in photosynthesis, the *dark reactions,* occur both in the light and in the dark. They use the ATP and NADPH to build carbohydrates from CO_2. The most common pathway by which this is one is the "carbon reduction cycle" (the *Calvin cycle*), which starts with the reaction of CO_2 with the 5-carbon sugar ribulose 1, 5-diphosphate. This generates two 3-carbon molecules that undergo the cycle schematized. This initial CO_2 fixation is analyzed by the enzyme ribulose 1, 5-diphosphate ("bisphosphate") carboxylase, often called RuDP carboxylase or carboxydismutase.

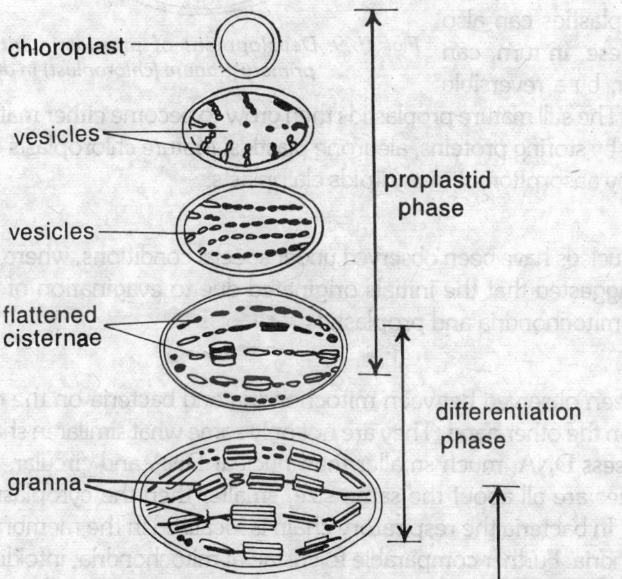

Fig. 16.8. Development of proplastids into chloroplast in the presence of light

Origin of the Plastids

Plastids are generated only by other pre-existing plastids or proplastid body. They are self-reproducing cell organelles like the chromosomes and the centrioles. The proplastid may be regarded as the "stem" plastid, giving rise to metabolically active leucoplasts or chloroplasts which may later develop into other types of chromoplasts through a kind of degeneration. Plastids are transmitted directly to daughter cells during division, providing a genetic continuity in plastid lineage. Although the origin of intials that develop into proplastids is not entirely clear, some observations in fern suggest the nuclear envelope as a possible source.

In lower plants

In algae and lower plants, the plastids are quite large and often of different shapes and the process of their division is easy to follow. Here the division begins with an infolding of the inner membrane of the limiting chloroplast envelope. Later on the outer membrane invaginates until the plastid is divided into two. The daughter plastids thus formed, may or may not be equal in size.

In higher or superior plants

In higher plants the process of formation of plastids is more complicated. The process can be effected even by sun light, as the formation of plastids in sun light and dark show vast differences in either case.

In sun light

During the development the proplastids increase in size. Protoplastids are limited by a double membrane. In the presence of light the inner limiting boundary invaginates at several spaces, producing small vesicle like structures. These vesicles later on pinch off as free units, which then aggregate in several arrangements to form the lamellae of the mature plastid.

In dark

If the proplastids are left in dark, the process of formation upto vesicle stage is the same, as in presence of light. But the arrangement of vesicles is different. In dark the clusters of vesicles or discs constitute the lamellar system or a series of concentric rings are developed.

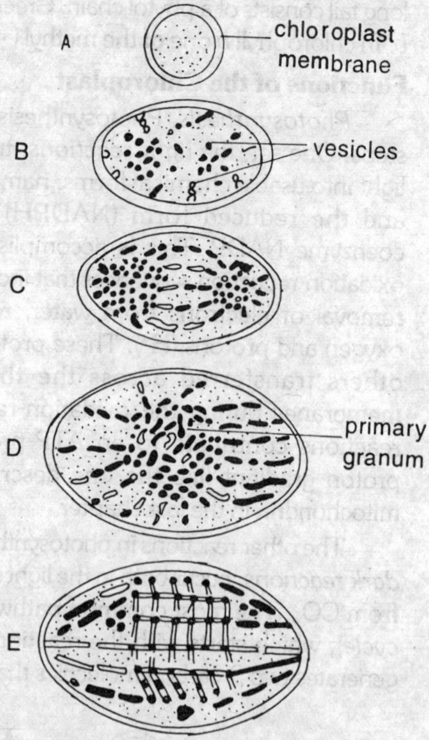

Fig. 16.9. Development of proplastid into a primary granum (chloroplast) in dark.

Formation of leucoplasts, chromoplasts and chloroplasts from proplastids. The non-lamellated proplastids can also transform themselves into leucoplasts. These, in turn, can develop into immature lamellar plastids or, by a reversible process, into starch containing amyloplasts. The still mature proplastids then grow to become either mature chloroplasts or other chromoplasts or even, by storing proteins, aleurone plastids. Mature chloroplasts can become chlorophyll free chromoplasts or, by absorption of many lipids elaioplasts.

Nuclear origin

A large number of initials around the nucleus have been observed under special conditions, where the cell organelles are breaking-down. This suggested that the initials originated due to evagination of the double nuclear membrane and gave rise to mitochondria and proplastids.

Symbiotic origin

To much surprise resemblance has been observed between mitochondria and bacteria on the one hand and chloroplast and blue-green algae on the other hand. They are not only some what similar in shape and structure but also in function. All possess DNA, much smaller than nuclear DNA and circular, not organised in chromosomes; their ribosomes are all about the same size, smaller then the cytoplasmic matrix ribosomes of animal and plant cells. In bacteria the respiratory chain is localised in the membrane comparable to inner membrane of mitochondria. Further comparable to cristae of mitochondria, infoldings of plasma membrane (mesosomes) have been reported in certain bacteria. In mitochondria and bacteria both, the protein synthesis is inhibited by chloramphenicol, where as the extracellular protein synthesis of higher cells is not effected. These similarities have led some to consider the mitochondria and chloroplasts of today as generally modified descendents of an ancestral free living bacteria and blue-green-algae respectively (autotrophic micro organisms) which at one time came together with the ancestral eukaryotic cell (heterotrophic host cell) to form a symbiotic relationship. This hypothesis is not doubt highly attractive, but still it is evident that the electrontransport system in chloroplasts and the enzymes requires for making the photosynthetic pigments are controlled by the nuclear genes (*Kirk,* 1960).

17

PHOTOSYNTHESIS

Photosynthesis (Gr. *Photos*-light; *synthesis*-putting together) is an anabolic process of manufacture of organic compounds inside the cell, containing chlorophyll, from carbon dioxide and water with the help of sunlight as an energy source. A simple equation of photosynthesis can be represented as

$$6CO_2 + 12\,H_2O \xrightarrow[\text{Radient energy}]{\text{Chlorophyll}} C_5H_{12}O_6 + 6H_2O + 6O_2$$

Basically, photosynthesis is the absorption of light energy and its conversion into chemical potential.

Histological Background

1. *Aristotle and Theophrastus* (320 B.C.). Thought that plants derive all their nourishment directly from soil.
2. *Hales* (1727). Concluded that plants are nourished, in part, from atmosphere.
3. *Joseph Pristley* (1772). Discovered the evolution of oxygen by plants. He also showed that oxygen is consumed by animals.
4. *Janingen-Housz* (1779-96). Purification of air is carried out by green plants only and that light is essential to the process. He associated these reactions with the green parts of plants.
5. *Jean Senebier* (1782). The rate of purification of air by green plants depends upon the concentration of foul air.
6. *Nicolas de Saussure* (1804). Noted that water is involved in photosynthesis, plants pick up carbon dioxide and release oxygen.
7. *Dutrochet* (1837) Established connection between photosynthesis and chlorophyll.
8. *Liebig* (1845). Organic matter synthesized during photosynthesis is derived from carbon dioxide and water.
9. *Von Mayer* (1845). Green plants convert solar energy into chemical energy during photosynthesis.
10. *Sachs* (1862). Proved that starch is synthesized by plants in light dependent reaction i.e., photosynthesis.
11. *Englemann* (1882). Showed that red light is the most efficient stimulator of photosynthesis.
12. *Meher* (1883). First described details of chloroplast structure.
13. *Blackman* (1905). Propounded the principle of limiting factors. He also proposed the occurrence of a dark phase in photosynthesis.
14. *Wilstaller and Stoll* (1913). Isolated chlorophyll and later determine its structure.
15. *Warburg* (1919). Found that the efficiency of photosynthesis is higher with intermittent light.
16. *Warburg and Negelein* (1922). First measured the efficiency of photosynthesis.

17. *Thunberg* (1923). Recognized that carbon dioxide is reduced and water oxidized during photosynthesis.
18. *Emerson and Arnold* (1932). Performed flashing light experiments at different temperature and found out the existence of light and dark phases of photosynthesis.
19. *Wood and Werkman* (1936). Distinguished between the light reaction and dark reaction.
20. *Hill* (1938). Evolution of oxygen occurs in light reaction performed by isolated illuminated chloroplasts in the presence of suitable electron acceptors and the absence of carbon dioxide.
21. *Ruben, Randall, Kamen and Hyde* (1941). Showed that the oxygen liberated during photosynthesis comes from water and not from carbon dioxide.
22. *Calvin and Benson* (1948). Showed that phosphoglycerate is an early product of carbon dioxide fixation.
23. *Calvin* (1954-55). Traced the pathway of carbon dioxide fixation by using $^{14}CO_2$ and gave the C_3 cycle, now known as *Calvin Cycle. Calvin* was awarded Nobel Prize in 1960.
24. *Hatch and Slack* (1965). Discovered supplementary mechanism of carbon dioxide fixation called C_4 pathway in certain tropical plants.
25. *Huber et. al.* (1985). Crystallized and analysed the photocentre of *Rhodobactor* through X-ray differaction technique. They were awarded Nobel Prize in 1988.

RAW MATERIALS

1. Carbon Dioxide

In terrestrial plants, the carbon dioxide is obtained from the atmosphere through the stomata. A small amount of carbonates are also absorbed from the soil through the roots. Aquatic plants obtain carbon dioxide supply from the aquatic environment as bicarbonates. These bicarbonates are absorbed by the hydrophytes through their general surface.

2. Water

In 1924, *von Niel* proposed a mechanism accounting for the dual nature of photosynthetic reaction based on his observation of bacterial photosynthesis. The bacteria required hydrogen sulphide for their carbon fixation. There was no evolution of oxygen. Sulphur globules accumulated as waste product.

$$6CO_2 + 12H_2S \longrightarrow C_6H_{12}O_6 + 6H_2O + 12S$$

Van Niel suggested that water was a substance from which H^+ ions and electrons could be derived for the reduction of carbon dioxide in oxygenic photosynthesis of all organisms.

$$2\,H_2O \xrightarrow{\text{light}} O_2 + 4(H)$$

$$4(H) + CO_2 \longrightarrow CH_2O + H_2O$$

The nature of the reductant was not known, and it was clear that the second reaction was complex and probably involved several steps.

Ruben and *Komen* (1941) were able to use isotropically enriched oxygen ($^{18}O_2$) to established that all of the oxygen produced in photosynthesis was derived from water and none from carbon dioxide.

Hill (1937) illuminated the isolated chloroplasts of *Stellaria media* in the presence of leaf extract or hydrogen acceptors in the absence of carbon dioxide. The chloroplast evolved oxygen. The reaction involving the production of oxygen by the illuminated chloroplasts in the absence of carbon dioxide fixation is called *Hill reaction*. The hydrogen acceptor of Hill reaction is nicotinamide adenine dinucleotide phosphate (NADP).

$$2NADP + 2H_2O \xrightarrow{\text{Chlorophyll}} 2\,NADPH + 2H^+ + O_2$$

3. Light

Since the primary reactions of photosynthesis involve the absorption of light by pigments. It is necessary to examine some of its basic properties. Light is electromagnetic energy propagated in discrete corpuscles called *photons* or *quanta*. As the energetics of chemical reactions are usually described in terms of kilocalories per molecule of the chemicals (1 mole = 6.02×10^{23} quanta).

The colour of the light is determined by the wavelength (λ) of light radiation. At any given wavelength, all the quanta have the same energy. The energy (E) of a quantum is inversely proportional to its wavelength.

Depending upon the wavelength electromagnetic spectrum consists of eight types of radiations cosmic rays, gamma rays, x-rays, ultra-violet rays, light spectrum, infrared rays, electric rays and radio rays. Visible light consists of radiations having a wave length between 390-760 nm. It can be resolved into light of different colour—*violet* (390-430 nm), *blue* (430-470 nm), *bluegreen* (470-500 nm), *green* (500-580 nm), *yellow* (580-600 nm), *orange* (600-650 nm), *orange-red* (650-660 nm) and *red* (660-760 nm). Red light above 700 nm is called *farred*. *Infrared rays* having wavelength of 760-10,000 nm. Radiations shorter than those of violet are called *ultra violet rays* having a wavelength of 100-390 nm. Solar radiations or sunlight reaching the earth have wavelength between 300-2600 nm. The proportion of light energy that is available in photosynthesis has long been a matter of controversy. Much attention was focused in the period 1930-1950 on the quantum requirement of photosynthesis, a measure of the efficiency of the process. The initial thought was that since 1 quantum can conduct only one molecular reaction, the number of quanta absorbed per molecule of oxygen produced might indicate the number of reaction steps involved.

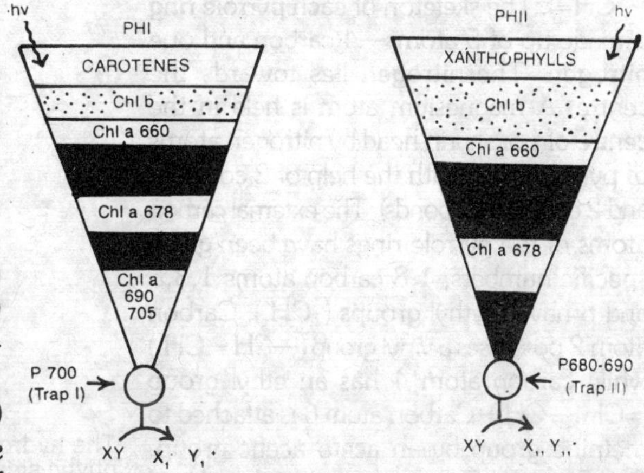

Fig. 17.1. Hypothesis for the approximate distribution of the various pigments in the two photosystems in green plants. The numbers indicate the approximate absorption maxima in the far red end of the spectrum. In diatoms and brown algae, chlorophyll c replaces chlorophyll b. In red and blue-green algae. phycobilins replaces chlorophyll b.

In complex experiments *Warburg* found quantum requirements of 4 quanta per molecule of oxygen produced. This means 1 quantum per (H) derived from H_2O for the reduction of carbon dioxide. The energy required to derive one reducing equivalent from H_2O is close to 30 K cal/mole, and the energy available in quanta of red light, which *Wasburg* used in his experiments, is only about 40 kcal/einstein.

4. Photosynthetic Pigments

The photosynthetic pigments are those which occur inside the chloroplasts and can take part in absorption of light energy for the purpose of photosynthesis. There are two catagories of these pigments—*chlorophyll* and *carotenoids*.

(i) Chlorophyll

So far the only light—absorbing pigment we have mentioned is chlorophyll. Initially the green pigment of plants was recognized as the substance responsible for light absorption in photosynthesis, absorbing red and blue light but not green. There are five types of chlorophyll occurring in plants. These are *bacteriochlorophyll* and *bacterioviridin*. In highter plants only chlorophyll *a* and *b* are present. Chlorophyll *a* occurs in all the photosynthetic plants. It is bluish green in the pure state. Its empirical formula is $C_{35}H_{72}O_5N_4Mg$. Chlorophyll *b* is olive green in pure state and its empirical formula is $C_{55}H_{70}O_5N_4Mg$. Both the chlorophylls are soluble in fat solvents such as ether, chloroform, carbon tetrachloride, alcohols etc. Chlorophyll *a* is more soluble in ether whereas chlorphyll *b* in methyl alcohol.

In 1912, the structure of chlorophyll was studied by *Wilstatter*, *Stoll* and *Fischer*. The chlorophyll is a tetrapyrrole and bears a close resemblance to the chemical structure of heme and the cytochromes. Chlorophyll differs from these iron enzymes in that it contains an atom of magnesium, which does not appear to participate directly in electron transfer reactions as does the iron of cytochromes.

The chlorophyll has a tadpole-like structure, with a head called *porphyrin* and a tail (made up of long chain alcohol called *phytol*). The head is made up of four *pyrrole rings* which are linked by methine bridges (—CH=). The skeleton of each pyrrole ring is made up of 5 atoms—4 carbon and one nitrogen. The nitrogen lies towards the centre. A magnesium atom is held in the centre of porphyrin head by nitrogen atoms of pyrrole rings (with the help of 2 covalent and 2 co-ordinate bonds). The external carbon atoms of the pyrrole rings have been given specific numbers, 1-8 carbon atoms 1,3,5 and 8 have methyl groups (–CH_3). Carbon atom 2 possesses a vinyl group (—CH = CH_2) while carbon atom 4 has an ethyl group (=CH_2—CH_3). Carbon atom 6 is attached to metnine group by an aceto acetic group. Carbon atom 7 is connected to the phytol tail through a propionic acid residue.

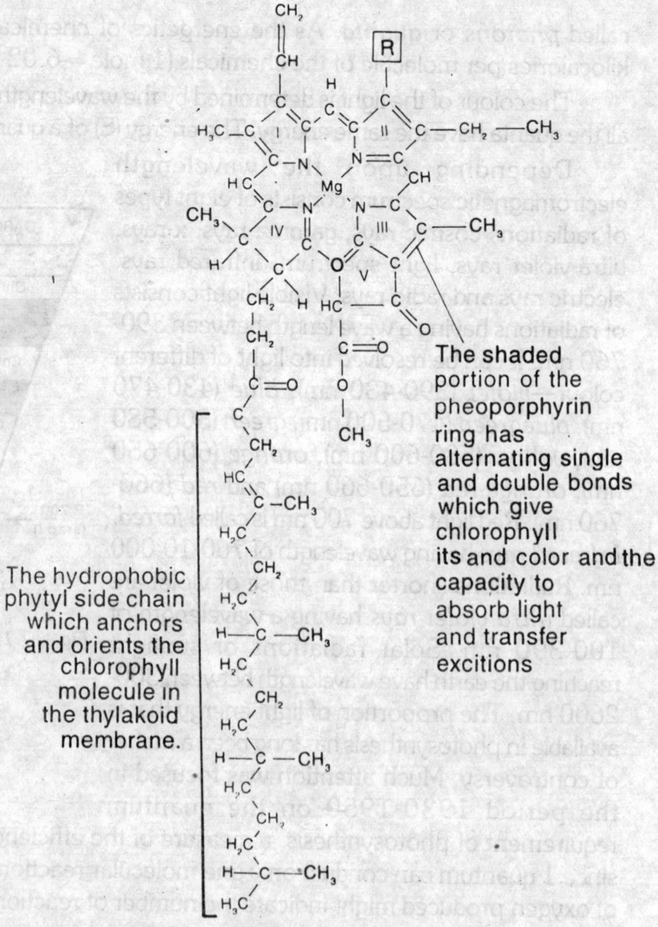

Fig. 17.2. The structure of chlorophyll a.

Chlorophyll *b* differ from chlorophyll *a* in having formly group (–CHO) instead of methyl (–CH_3) group at carbon atom 3. Chlorophyll *b* is present in most green plants, but its place is taken by *phycocyanin* blue-green algae, *fucoxanthin* in brown algae and *phycoerythrin* in red-algae. These are called *accessory phycobilin pigments*. The chemical structure of these accessory phycobilin pigments is similar to that of chlorophyll in that they are tetrapyrrole compounds; however, they differ in being linear instead of cyclic in structure, and they do not have a metal component.

(ii) Carotenoids

The carotenoids are a group of yellow, brownish is reddish pigments which are associated with the chlorophyll inside the chloroplast. These remain alone inside the chromoplasts. Like chlorophyll *b* these are also called accessory photosynthetic pigments. They hand over the energy to chlorophyll *a* absorbed by them. The carotenoids are of two types:

(a) *Carotenes.* These are closely related to the vitamin A and are basically long chain lipophilic molecules with a more active terminal group at each end. The empirical formula is $C_{40}H_{56}$.

(b) *Xanthophylls.* They are oxygen containing derivatives of carotenes. The empirical formula is $C_{40}H_{56}O$ i.e., cryptoxanthin and $C_{40}H_{56}O_2$ i.e., zeaxanthin.

The carotenoids are soluble in fat solvents like ether, chloroform, acetone, carbon disulphide.

Fluorescence

The fluroescence is the property of emission of radiations absorbed by a substance and the substance which can emit back the absorbed radiations is called *fluroscent substance*. It has been noted that all the photosynthetic pigments are fluroscent substances. However, most of the fluroscence emitted by photosynthetic organs is due to chlorophyll *a* because other pigments hand over their absorbed energy to it through resonance.

The normal state of the atoms or molecules is called *ground state* or *singlet state*, when an electron of a molecule or atom absorbs a quantum of light, it is raised to a higher energy level which is called *excited second singlet state*. The excited state is unstable and lasts for about 10^{-9} second. The excited electron may return to ground state in two ways (a) either losing its remaining energy in the form of heat or (b), by losing extra energy in the form of radiation energy. This later process is called fluorescence. The fluroscent substances emit fluroscent light only during the period they are exposed light. The fluroscent light is of higher wave length than the incident light.

Absorption spectrum of chlorophyll

If a substance is extracted and illuminated with light of various wavelengths, it is possible to determines its absorption spectrum, the amount of light of different wavelengths that the substance absorbs or does not absorb. Chlorophyll's absorption spectrum has peaks in the blue and red regions, which means that light of these wavelengths in absorbed. The major peak at 675 nm and minor peak at 450 nm. The unabsorbed or reflected light appears green, thus making chlorophyll, the chloroplasts and the leaves that contain it, appear green to our eye.

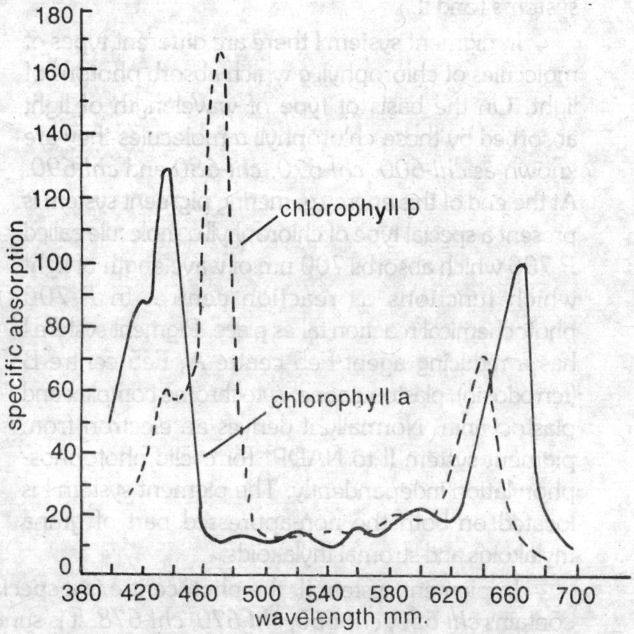

Fig. 17.3. Absorption spectra of ether extracts of chlorophyll a and b.

Action spectrum

The curve showing the relative rates of photosynthesis at different wavelengths of light is called action spectrum. The curve showns that maximum photosynthesis occurs in red part and blue-violet part of light. Thus the action spectrum of photosynthesis corresponds closely to absorption spectra of chlorophyll *a* and *b* showing that chlorophyll *a* as well as *b* are the main photosynthetic pigments, however, photosynthesis also occurs in the mid part of light spectrum where carotenoid are active.

Photosynthetic Units (PSU)—Quantosome

As already stated that light travels in the form of waves of different wavelengths. In certain respects light behaves as if it consists of tiny particles called *photons* or *quanta*. The quanta is indivisible units. Using the technique of inter-mittent light in their experiments, *Emerson* and *Arnold* (1932) found that the evolution of one molecule of oxygen by *Chlorella* required the functioning of approximately 2500 molecules of chlorophyll. It was also estimated that in reduction of one molecule of carbon dioxide and the liberation of one molecule of oxygen eight quanta of light energy are consumed. Putting these observations together it is easy to deduce that a group of 300 pigment molecules is the smallest unit that is capable of utilising one quantum of light energy. This is called the *photosynthetic unit*. Under electron microscope made by *Park* and *Biggins* (1964) showed that minute structures about 100Å in diameter are attached to the chloroplast lamellae. These strips of lamellae contained about 230 chlorophyll molecules each and represent photosynthetic units to which they gave the name of *quantasome*.

The quantosome has a *photocentre* or *reaction centre* which is fed by about 200 harvasting pigment molecules. The photocentre consists of a special chlorophyll *a* molecule. It absorbs light energy at longer wave lengths. The harvasting molecules are of two types, the *antenna molecules* and *core molecules*. The antenna molecules get excited by absorbing light of various wavelengths but shorter than that of trap centre. The antenna molecules come to ground state by hand over their energy to the core molecules. The energy picked up by core molecules is supplied to the reaction centre. On absorption of this energy, the reaction centre gets excited and extrudes an electron after which it comes to ground state to repeat the cycle.

Photosystems or Pigment Systems

Photosynthetic units occur in the form of two distinct groups called *photosystems* I and II or *pigment systems* I and II.

In pigment system I there are different types of molecules of chlorophyll *a* which absorb photons of light. On the basis of type of wavelength of light absorbed by these chlorophyll *a* molecules they are known as *chl-600, chl-670, chl-680* and *chl-690*. At the end of this energy gathering pigment system is present a special type of chlorophyll *a* molecule called *P-700* which absorbs 700 nm of wavelength of light which functions as reaction centre. In *P-700* photochemical reaction takes place. Pigment system I has a reducing agent Fe5 centre A, Fe5 centre B (ferrodoxin), plastoquinone, cytochrome complex and plastocyanin. Normally it derives an electron from pigment system II to $NADP^+$ for cyclic photophosphorylation independently. The pigment system I is located on both the non-appressed part of grana thylakoids and stromal thylakoids.

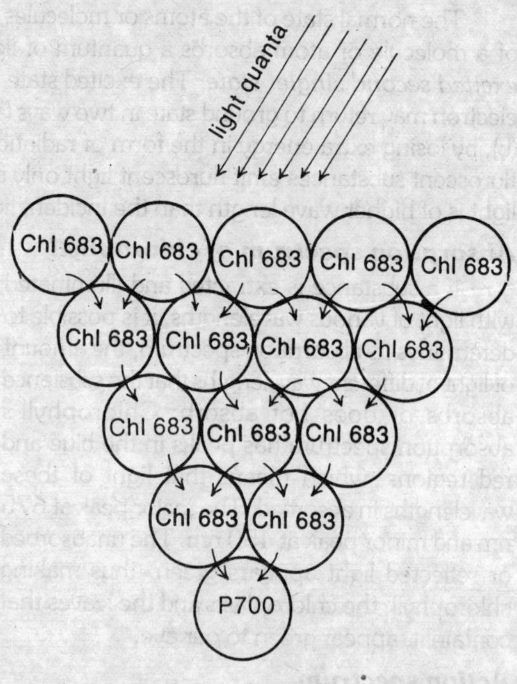

Fig. 17.4. Schematic representation of the harvest of light energy by the quantasome. The long-wave pigment P_{700} acts as energy trap receiving energy initially absorbed by chl. 683.

In pigment system II, the photocentre is a special chlorophyll *a* molecule called P-680. Besides it contains *chl-650, chl-660, chl-670, chl-678*. It is surrounded by chlorophyll *b* and carotenoid molecules. Pigment system II also contains Mg^{++}, Cl^-, Quencher molecule Q, plastoquinone, cytochrome complex and plastocyanin. It picks up electrons released during photolysis of water. In this system photons of lower wavelength of light are absorbed. It is inactive in farred light. This system is located in the appressed part of grana thylakoids only.

Mechanism of Photosynthesis

Photosynthesis is an oxidation-reduction process, in which water is oxidized to release oxygen and carbon dioxide is reduced to form carbohydrates. Photosynthesis occurs in two phases, *photochemical reactions* (also called *light* or *Hill reaction*) and *biochemical reactions* (also called *dark* or *Blackman's* reaction).

I. The Light or Photochemical Reaction in Photosynthesis Evidence for Light Reaction

Following evidences indicate that the over all reaction in photosynthesis takes place in two steps: one is light mediated and the other is light independent.

1. Temperature coefficient studies

The rate of photosynthesis of two groups of plants were compared. Both were supplied with an adequate concentration of carbon dioxide, but one group was kept under light of high intensity and other group in light of low intensty. When the rates of photosynthesis were measured at different temperatures, it was found that *high light* group had a Q_{10} = 2 but the *low light* group Q_{10}=1. Strictly, chemical reactions characteristically have a Q_{10} from 2 to 3. This fact indicates that at least one of the reaction involved in photosynthesis is of a purely chemical type. Since it was first pointed out by *Blackman*, this reaction is called as dark reaction. The other reaction of Q_{10} indicates that one of the reactions proceeds only at the expanse of absorbed light. It is called dark reaction. The Q_{10} of light reaction is 1.

2. Flashing light experiments

Photosynthesis involves photochemical and biochemical reactions is also shown by the results of investigations in which plants are exposed to intermittent light. *Warburg* (1919) obtained higher rates of photosynthesis in *Chlorella* when it was exposed to rapid and alternate periods of light and darkness instead of continuous illumination. *Emerson* and *Arnold* (1932) found that at 25°C, the maximum photosynthesis took place when light and dark periods were respectively 10^{-5} second and 0.055 second. At 1.1°C maximum photosynthesis could be obtained with the same light period but the dark period has to be increased to 0.4 second. It means that temperature influences reactions of the dark period and reactions of light phase are photochemical.

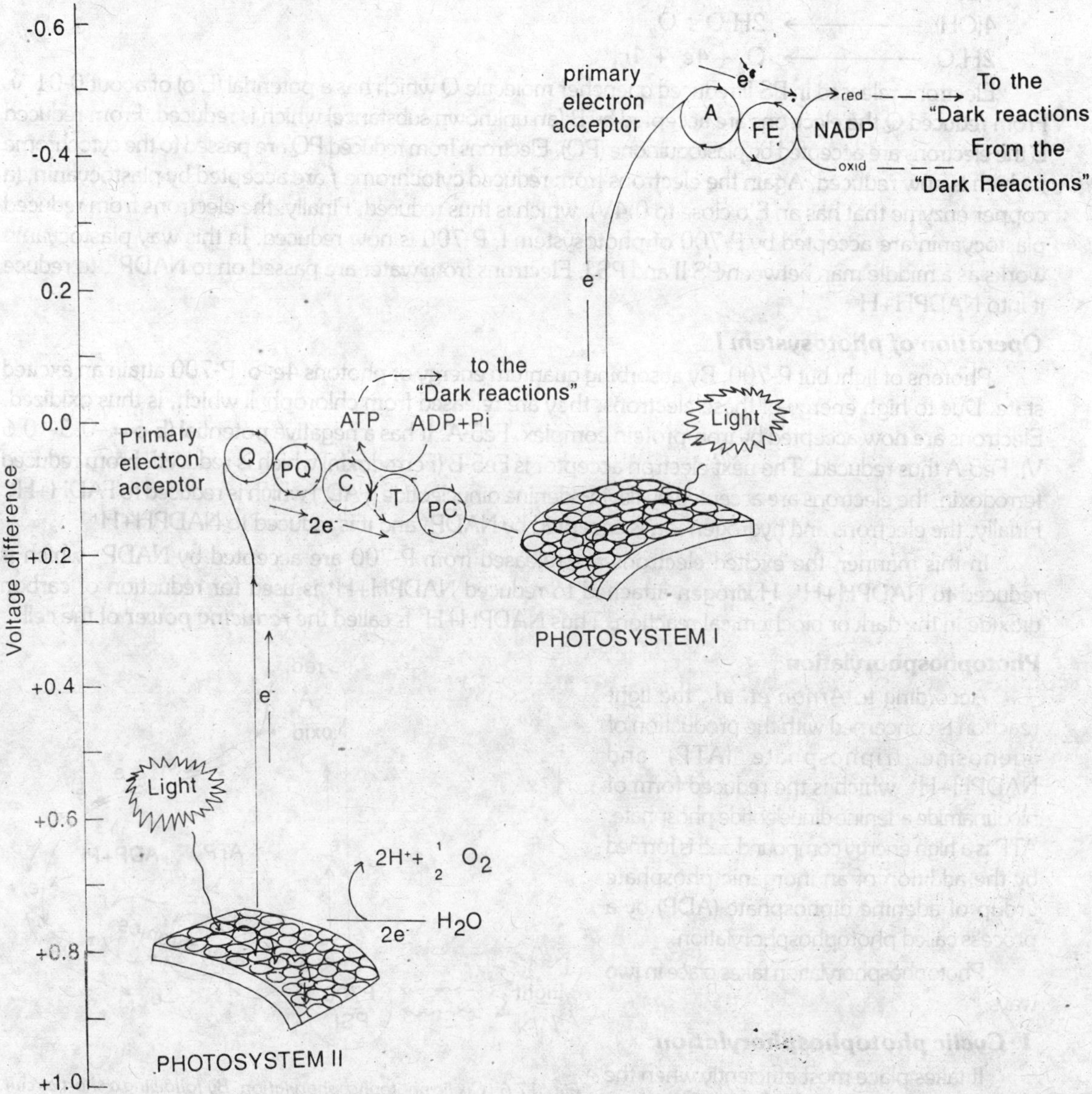

Fig. 17.5. Participation of the two photosystems in energy transfer. Electrons are raised to the energy level required to reduce NADP in the two photosystems. A and Q are the incompletely known primary electron acceptors of the photosystems. Electron carries shown are: plastoquinone (PQ), cytochromes (CYT), plastocyanin (PC), and ferredoxin (FE).

Operation of Photosystem II

The main function of PS II is the photolysis of water to release oxygen. In this act, when pigments of PS II absorb light, electrons are released from this system, in this manner chlorophyll molecules get oxidized. Oxidized pigments that take electrons from OH^- ions and are again reduced. The OH (hydroxyl radical) is left from OH^- and 4(OH) radicals give rise to two molecules of water and one molecule of oxygen. In the above process H^+ atoms are released which are used for the reduction of $NADP^+$. Manganese ions play an important role in this process.

$$4H_2O \longrightarrow 4H^+ + 4OH^-$$
$$4OH^- \longrightarrow 4(OH) + 4e^-$$
$$4(OH) \longrightarrow 2H_2O + O_2$$
$$2H_2O \longrightarrow O_2 + 4e^- + 4H^+$$

Electrons released in PS II reduced quencher molecule Q which has a potential (E'o) of about 0-01. v. From reduced Q the electrons are accepted by B (an unknown substance) which is reduced. From reduced B the electrons are accepted by plastoquinone (PQ). Electrons from reduced PQ are passed to the cytochrome *f* which is now reduced. Again the electrons from reduced cytochrome *f* are accepted by plastocyanin, (a copper enzyme that has an E'o close to 0.4V), which is thus reduced. Finally, the electrons from reduced plastocyanin are accepted by P-700 of photosystem I, P-700 is now reduced. In this way plastocyanin works as a middle man between PS II and PS I. Electrons from water are passed on to $NADP^+$, to reduce it into $NADPH+H^+$.

Operation of photosystem I

Photons of light but P-700, By absorbing quantum energy of photons $4e^-$ of P-700 attain an excited state. Due to high energy of these electrons, they are released from chlorophyll which, is thus oxidized. Electrons are now accepted by iron-protein complex, Fe5-A. It has a negative potential (E'o = –0.5 – 0.6 V). Fe5-A thus reduced. The next electron acceptor is Fe5-B (Ferrodoxin) which is reduced. From reduced ferrodoxin, the electrons are accepted by flavin-adenine dinucleotide (FAD^+) which is reduced to $FADH+H^+$. Finally, the electrons and hydrogen ion is accepted by $NADP^+$ and it is reduced to $NADPH+H^+$.

In this manner, the excited electrons are released from P-700 are accepted by $NADP^+$ which is reduced to $NADPH+H^+$. Hydrogen attached to reduced $NADPH+H^+$ is used for reduction of carbon dioxide in the dark or biochemical reaction. Thus $NADPH+H^+$ is called the *reducing power* of the cell.

Photophosphorylation

According to *Arnon et. al.*, the light reaction is concerned with the production of adenosine triphosphate (ATP) and $NADPH+H^+$, which is the reduced form of nicotinamide adenine dinucleotide phosphate. ATP is a high energy compound and is formed by the addition of an inorganic phosphate group of adenine diphosphate (ADP) by a process called photophosphorylation.

Photophosphorylation takes place in two ways:

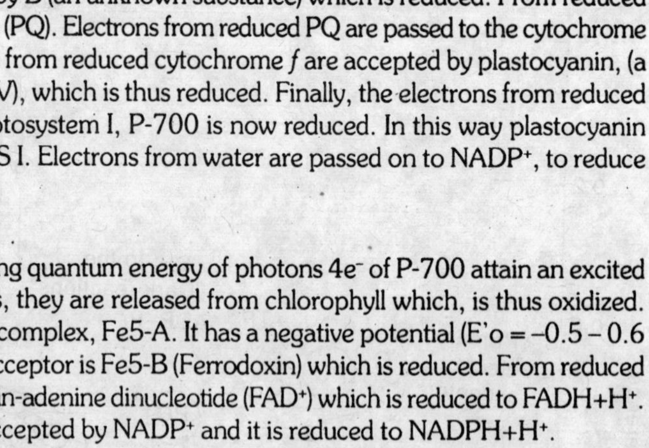

Fig. 17.6. Cyclic photophosphorylation. By following a short circuit, electrons excited by light can generate ATP even when no $NADP_{oxid}$ is available to accept them. The cytochrome pathway is used as in ordinary photophosphorylation. The cycle is valuable because ATP has many other uses besides driving the dark reaction.

1. Cyclic photophosphorylation

It takes place most efficiently when the Hill reaction is prevented, i.e., the medium is devoid of the hydrogen acceptor NADP. Here the electrons released from P-700 move through Fe5-A to cytochrome b_6, to plastoquinone to cytochrome-f to plastocyanin and again return to P-700. While passing between ferrodoxin and plastoquinone, the electrons lose sufficient energy to form ATP from ADP and inorganic phosphate (P_i).

$$ADP + P_i \xrightarrow[\text{Chlorophyll}]{\text{Light}} ATP$$

well-known poisons of photosynthesis 3-(3,4 dichloro phenyl) dimethyl urea (DCMU) do not inhibit the cyclic photophosphorylation.

2. Non-cyclic photophosphorylation

It is the normal process of photophosphorylation and is unidirectional. It is carried out in collaboration of both photosystems I and II. Photoexcited chlorophyll molecules in photosystem II donate electrons to acceptors. The extruded electron has an energy equivalent to 23 kcal/mole. The electrons pass through Q, B, PQ, cytochrome-f, plastocyanin and enter of P-700. While passing over plastoquinone to cytochrome-f, the electron loses sufficient energy for the synthesis of ATP.

The overall photochemical reaction results in the production of ATP and reduced NADP and the oxygen is released to the atmosphere. The products of photochemical reaction are utilized in the reduction of carbon dioxide to the carbohydrate level.

II. Dark Reaction or Biosynthetic Phase or Blackman's Reaction

The dark reaction does not require light, although this takes place in light also, that is why this is called dark reaction. This process takes place in the stroma of chloroplasts. These are two main pathways for the dark reaction.

I. Calvin—Benson cycle

II. Hatch—Slack pathway

I. Calvin—Benson cycle

During the last four decades considerable progress has been made in the understanding of the dark reaction in photosynthesis with the help of tracer techniques and paper chromatography. The path of

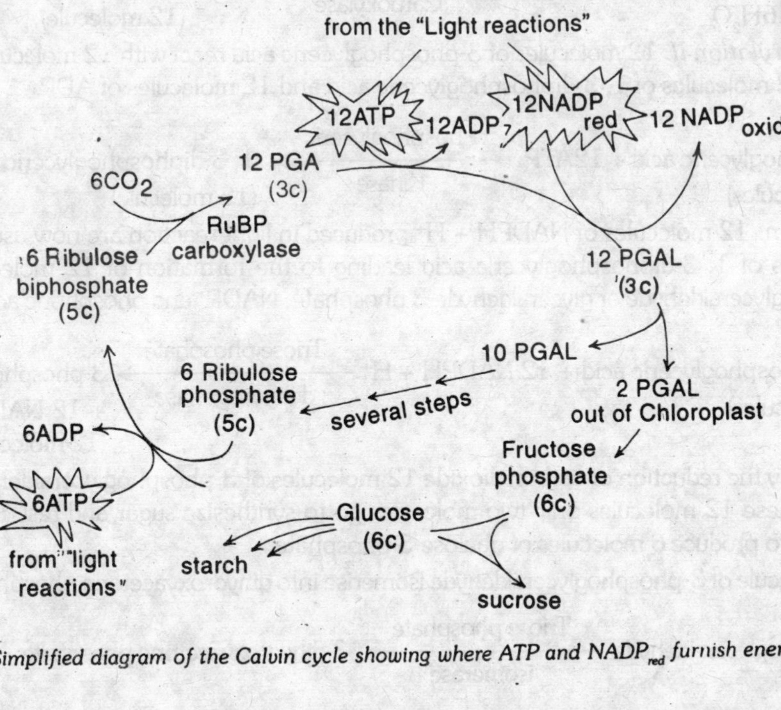

Fig. 17.7. Simplified diagram of the Calvin cycle showing where ATP and $NADP_{red}$ furnish energy to the cycle.

carbon was studied with the help of radioactive tracer technique using $C^{14}O_2$ in a unicellular alga, *Chlorella*. Radioactive carbon C^{14} has a half life of 5200 years. Therefore, the path of CO_2 fixation can be easily traced with its help. Here the radioactive carbon is provided in a solution of bicarbonate to photosynthesizing cells. At regular interval of time, the cells are killed and analyzed quantitatively by chromatography for various compounds containing C^{14}. In cells killed immediately after one second expose to the light C^{14} is found mainly in phosphoglyceric acid (PGA). In cells killed after five seconds expose to the light, C^{14} appears in traces of PGA, alanine, aspartic acid and malic acid. In cells killed after 15 seconds of illumination, in addition to the above substances, faint traces of sucrose containing C^{14}. It may also be detected in some pentose, glucose phosphate and fructose phosphate. After one minute of illumination, in the killed cells abundant C^{14} occurs in glucose phosphate and fructose phosphate. In lipids and proteins also, the C^{14} appears after 5 minutes of illumination. Thus radioactive C^{14} obtained from $C^{14}O_2$ is rapidly spread in all the compounds found in plant cell (*Calvin*, 1950-51, 62; *Bassham* and *Calvin*, 1962 and *Bassham et. al.*, 1955).

After many panstaking calculations, *Calvin* worked out the pathway of carbon dioxide fixation. The acceptor molecules was found to be ribulose-1,5 biphosphate. The followings are the steps taking place in dark reaction of photosynthesis.

1. *Phosphorylation I.* Carbon dioxide reduction starts with 5- carbon sugar i.e., ribulose monophosphate. Six molecules of ribulose monophosphate react with 6 molecules of ATP (produced during light reaction) to form 6 molecules of *ribulose 1, 5 biphosphate* and 6 molecules of ADP.

$$\text{6 Ribulose-5-phosphate + 6 ATP} \xrightarrow{\text{Phosphopentokinase}} \text{6 Ribulose-1,5 biphosphate 6 ADP}$$

2. *Carboxylation.* Ribulose 1,5 biphosphate (RuBP) acts as CO_2 accepter, 6 molecules of RuBP react with 6 molecules of CO_2 and 6 molecules of water giving rise 12 molecules of 3-phosphoglyceric acid.

$$\begin{array}{l} \text{6 Ribulose- 1, 5 biphosphate} \\ +6\ CO_2 + 6H_2O \end{array} \xrightarrow[\text{Carboxylase}]{\text{Ribulose 1, 5 biphosphate}} \begin{array}{l} \text{3-phosphoglyceric acid.} \\ \text{(12 molecule).} \end{array}$$

3. *Phosphorylation II.* 12 molecules of 3-phosphoglyceric acid react with 12 molecules of ATP giving rise to 12 molecules of 1, 3-diphosphoglyceric acid and 12 molecules of ADP.

$$\begin{array}{l} \text{3-phosphoglyceric acid + 12ATP} \\ \text{(12 molecules)} \end{array} \xrightarrow[\text{Kinase}]{\text{Phosphoglyceric}} \begin{array}{l} \text{1, 3-diphosphoglyceric acid + 12 ADP} \\ \text{(12 molecule)} \end{array}$$

4. *Reduction.* 12 molecules of NADPH + H⁺ produced in light reaction are now used to reduces 12 molecules of 1, 3-diphosphoglyceric acid leading to the formation of 12 molecules each of 3-phosphoglyceraldehyde or glyceraldehyde-3 phosphate, $NADP^+$ and phosphoric acid.

$$\begin{array}{l} \text{1, 3 diphosphoglyceric acid + 12 NADPH + } H^+ \\ \text{(12 molecules)} \end{array} \xrightarrow[\text{dehydrogenase}]{\text{Triose phosphate}} \begin{array}{l} \text{3-phosphoglyceraldehyde} \\ +\ 12\ NADP^+\ 12\ H_3PO_4 \\ \text{(12 molecules)} \end{array}$$

In this way the reduction of carbon dioxide 12 molecules of 3-phosphoglyceraldehyde are formed. Out of these 12 molecules only two molecules go to synthesize sugar and rest 10 molecules are recycled to produce 6 molecules of ribulose-5-phosphate.

5. One molecule of 3-phosphoglyceraldehyde isomerise into dihydroxyacetone phosphate.

$$\text{3-phosphoglyceraldehyde} \xrightarrow[\text{isomerase}]{\text{Triose phosphate}} \text{3-dihydroxyacetone phosphate}$$

6. One molecule of 3-phosphoglyceraldehyde unite with 3-dihydroxyacetone phosphate in the presence of aldolase to form fructose— 1, 6-diphosphate.

$$\text{3-phosphoghyceraldehyde + 3-dihydroxyacetone phosphate} \xrightarrow{\text{Aldolase}} \text{Fructose-1, 6-diphosphate.}$$

7. Three molecules of phosphoric acid are released from one molecule of fructose-1, 6-diphosphate to form one molecule fructose-6-phosphate in the presence of phosphatase.

$$\text{Fructose-1, 6-diphosphate} + H_2O \xrightarrow{\text{Phosphatase}} \text{Fructose 6-phosphate} + 3H_2PO_4$$

8. Fructose -6-phosphate is converted to fructose-1-phosphate which forms glucose-1-phosphate.

$$\text{Fructose-6-phosphate} \longrightarrow \text{Fructose-1-phosphate}$$

$$\text{Fructose-1-phosphate} \longrightarrow \text{Glucose-1-phosphate.}$$

9. Fructose-1-phosphate and glucose-1-phosphate condense to form sucrose.

$$\text{Fructose-6-phosphate + Glucose-1-phosphate} \xrightarrow{\text{Phosphatease}} \text{sucrose.}$$

10. *Regeneration of ribulose-5-phosphate.* Remaining 10 molecules of 3-phosphoglyceraldehyde regenerate 6 molecules of ribulose-6-phosphate after a series of complex reactions as follows.

(a) 3-phosphoglyceraldehyde reacts with fructose-6-phosphate to yield a pentose (5-carbon) xylulose-5-phosphate and a triose (4-carbon) erythrose-4-phosphate in the presence of transketolase.

$$\text{Fructose-6-phosphate + 3-phosphoglyceraldehyde} \xrightarrow{\text{Transketolase}} \text{Xylulose-5-phosphate + Erythrose-4-phosphate.}$$

Xylulose-5-phosphate readily isomerizes to ribulose-5-phosphate.

(b) Erythrose-4-phosphate combines with dihydroxyacetone phosphate in the presence of aldolase to form sedoheptulose 1, 7-diphosphate, a 7- carbon sugar.

$$\text{Erythrose-4-phosphate + Dihydroxyacetone phosphate} \xrightarrow{\text{Aldolase}} \text{Sedoheptulose-1, 7-diphosphate}$$

(c) Sedoheptulose -1, 7-diphosphate looses one phosphate group in the presence of phosphotase to form sedoheptulose-7-phosphate.

$$\text{Sedoheptulose-1, 7-diphosphate} \xrightarrow{\text{Phosphatese}} \text{Sedoheptulose-7-phosphate}$$

(d) Sedoheptulose-7-phosphate reacts with 3-phospho-glyceraldehyde in the presence of transketolase to form xylulose-5-phosphate and ribose-5-phosphate.

$$\text{Sedohepatulose-7-phosphate + 3-phosphoglyceraldehyde} \xrightarrow{\text{transketolase}} \text{Xylulose-5-phosphate + ribose-5-phosphate.}$$

(e) Xylulose-5-phosphate is converted into ribulose-5-phosphate (5-carbon sugar) in the presence of phosphoketopentose epimerase.

$$\text{Xylolose-5-phosphate} \xrightarrow[\text{epimerase}]{\text{Phosphoketopentose}} \text{Ribulose-5-phosphate}$$

(f) Ribose-5-phosphate is also converted into ribulose-5-phosphate in the presence of phosphopentose isomerase.

$$\text{Ribose-5-phosphate} \xrightarrow[\text{isomerase}]{\text{Phosphopentose}} \text{Ribulose-5-phosphate}$$

(g) Ribulose-5-phosphate is finally converted into ribulose-1, 5-diphosphate in the presence of phosphopentose kinase and ATP.

$$\text{Ribulose-5-phosphate + ATP} \xrightarrow[\text{Kinase}]{\text{Phosphopentose}} \text{Ribulose-1, 5-diphosphate + ADP.}$$

In this manner the Calvin's cycle is completed. The net reaction of C_3 dark fixation of CO_2 is

$$6RuBP + 6CO_2 + 18\ ATP + 12\ NADP_2H \longrightarrow 6RuBP + 12\ NADP^+ + C_6H_{12}O_6 + 18\ ADP + 18\ P.$$

II. The Hatch-Slack Cycle or C_4 Pathway

It was worked out by *Hatch* and *Slack* (1965, 1967), *Kortschak et. al.* (1965) first of all found that labelled carbon dioxide ($C^{14}O_2$) assimilated by sugarcane leaves first appeared in 4-carbon compound i.e. oxaloacetic acid. *Hatch* and *Slack* found it a regular mode of carbon dioxide fixation in a number of tropical plants such as maize, sorghum, *Salsola, Atriplex, Amaranthus* etc. These plants are called C^4 plants because of the first stable photosynthetic product being a 4-carbon compound. These plants live in hot, arid and saline habitats. These plants have kranz anatomy as the mesophyll is undifferentiated and its

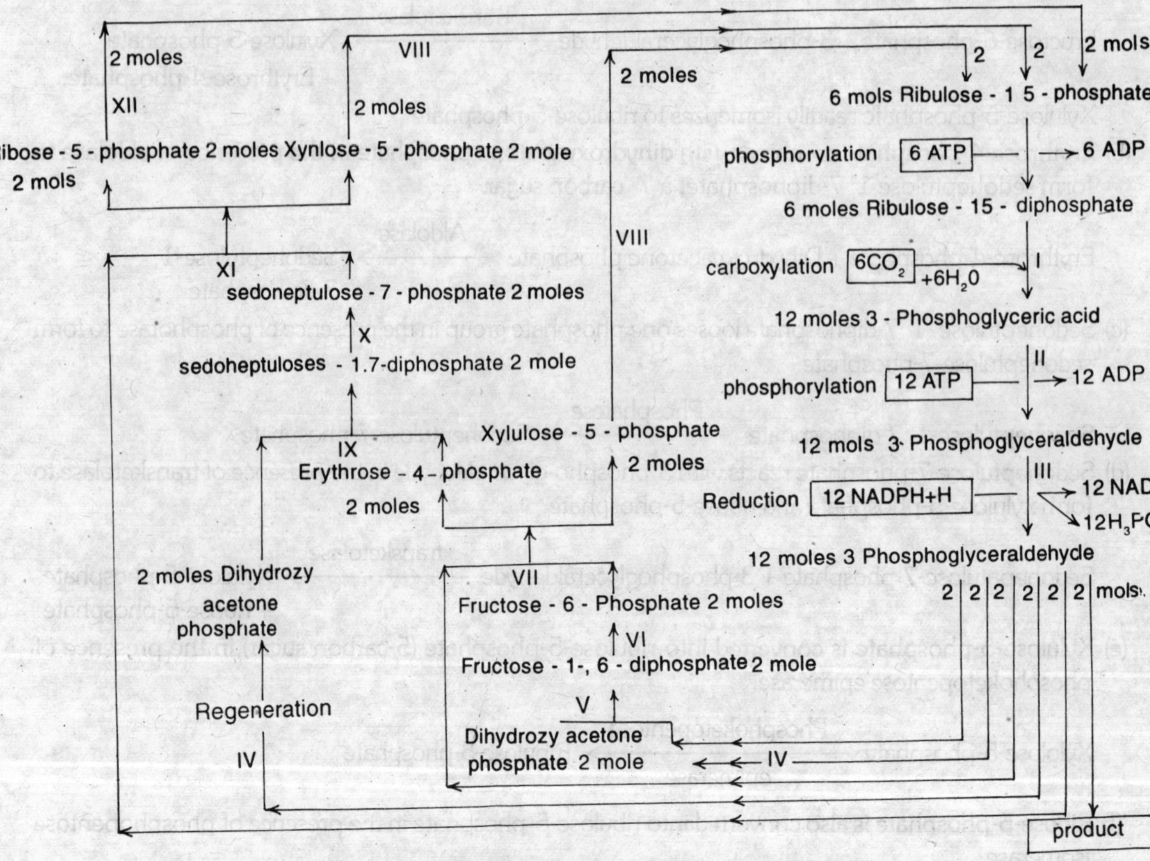

Fig. 17.8. The Calvin or Dark reaction of Photosynthesis.

cells occur in concentric layers around vascular bundles. In C[4] plants, initial fixation CO_2 occurs in mesophyll cells. In this cycle, the primary acceptor of CO_2 is phosphenol pyruvic acid (PEP).

1-XII Enymes: I, Carboxydismutase, II, 2-phosphoglycerate kinase, III. NADP + glyceraldehyde-3-phosphate dehydrogenase, IV. Triose phosphate isomerase, V. Vldolase, VI. Fructose-1, 6-diphosphate, VII. Transketolase, VIII. Ribulose-5-phosphate-3-epimerase, IX. Aldolase, X. Sedoheptulose-1, 7-disphosphate 5, XI. Transketolase XII. Ribosephosphate isomerase, XIII. Phosphoribulokinase.

1. The first step involves the carboxylation in phosphenol pyruvic acid in the chloroplasts of mesophyll cells to form oxaloacetic acid and phosphoric acid in the presence of phosphenol pyruvic acid carboxylase.

$$\text{PEP} + CO_2 + H_2O \xrightarrow[\text{acid carboxylase}]{\text{Phosphoernol pyruvic}} \text{Oxaloacetic acid} + H_2PO_4$$

2. Oxaloacetic acid is reduced to malic acid or aminated to form aspartic acid in the presence of transaminase and malate dehydrogenase respectively.

$$\text{Oxaloacetic acid} + NADPH_2 \xrightarrow{\text{malate dehydrogenese}} \text{Malic acid} + NADP^+.$$

$$\text{Oxalsacetic acid} + NH_3 + NADPH_2 \xrightarrow{\text{transaminase}} \text{Aspartic acid} + NADP+ + H_2O.$$

3. From chloroplasts of mesophyll cells, the malic acid or aspartic acid is translocated to the chloroplasts of bundle sheath cells where it is decarboxylated as in malic acid or deaminated as in aspartic acid to form pyruvic acid and carbon dioxide.

$$\text{Malic acid} + NADP+ \xrightarrow{\text{decarboxylase}} \text{Pyruvic acid} + CO_2 + NADPH_2.$$

4. Now second carboxylation occurs in the chloroplasts of bundle sheath cells through Calvin cycle. RuBP of Calvin cycle is called *final* or *secondary acceptor* of CO_2 in C_4 plants. The pyruvic acid produces above is sent back to mesophyll cells. Here it is phorphorylated to regenerate phosphoenol pyruvic acid. Energy is required for this. It is provided by ATP.

$$\text{Pyruvic acid} + ATP \; (+ P_i) \xrightarrow[\text{Kinase}]{\text{Pyruvate}} \text{phosphoenol pyruvic acid} + AMP + PPi \text{ (Pyrophosphate).}$$

Ribulose-diphosphate accepts CO_2 produced in step 3 in the presence of carboxydimutase system and ultimately yields 3-phosphoglyceric acid as mentioned above in Calvin cycle. Some of the 3-phosphoglyceric acid is utilized in the formation of glucose-6-phosphate and sucrose while rest regenerates ribulose - 1, 5 biphosphate in the system.

In the Calvin' cycle 3 ATP molecules are required for the C[4] plants, 5 ATP molecules are required for the fixation of CO_2. For the formation of a glucose molecules C_4 plants require 30 ATP while C_3 plants utilize only 18 ATP.

$$\text{6-phosphoenol pyruvic acid} + 6\ RuBP + 30\ ATP + 12\ NADPH_2 \longrightarrow \text{6 phosphoenol pyruvic acid} + 6\ RuBP + C_6H_{12}O_6 + 30\ ADP + 12\ NADP^+ + 30\ H_3PO_4.$$

Significance of Hatch — Slack Pathway

(i) This pathway is a modification of Calvin cycle and is advantageous to plants growing in dense, stands of tropical vegetation where the carbon dioxide concentration may be very much reduced.

(ii) Concentric arrangement of mesophyll cell produce a smaller area in relation to volume for better utilization of available water and reduce the intensity of solar radiations.

(iii) Normal oxygen concentration is not inhibitory for the growth in contrast to C_3 plants.

Glycolic Acid or C_2 Pathway

Calvin et. al (1950) reported that glycolic acid in labelled early during photosynthesis with $C^{14}O_2$. When the carbon dioxide concentration of the air is lowered below 1% glycolic acid is formed in large amount. Glycolic acid may be formed from ribulose-5-phosphate and, therefore, may be a secondary product of the photosynthetic carbon-reduction cycle, but, on the other hand, it may be direct product of photosynthesis.

At high concentrations of carbon dioxide in the atmosphere, the synthesis of glycolic acid is inhibited and the phosphoglyceric acid is the favoured product. The mechanism of glycolic acid pathway is unknown.

According to *Gibbs* (1971), biosynthesis of glycolic acid during photosynthesis coupling of photosynthetic carbon-reduction cycle to the light reactions cycle to the light reactions may be associated either with PS-I or PS-II.

18

Cytoskeleton

Like bony skeleton of body, the cytoskeleton or the skeleton of cell also has two primary functions: to provide a supporting framework to the cytoplasmic matrix and to provide the machinery by which the movement of cells and their organelles can be accomplished. Besides providing support and facilitating movement, the cytoskeleton is thought to play an important role in organising the contents of a cell.

The cytoskeleton is composed of following three components:

1. Microtubules
2. The microfilaments
3. Intermediate filaments

In addition, there is a less well defined system of fibrous cytoplasmic elements, the *microtrabeculae*. The microtrabeculae appears to be associated with the cytoskeleton and may serve to integrate the entire complex of materials into co-operative functional unit.

Microtubules

Microtubules are the structures universally present in the cytoplasm of eukaryotic cells. As the name implies, microtubules are small, hollow, cylindrical structures. They were first seen by *DeRobertis* and *Franchi* (1953) in the axoplasm extruded from myelinated fibres. These are called *neurotubules*. They appears as elongated, unbranched cylindrical elements, 20-30 nm in diameter. Cytoplasmic microtubules typically has an outer diameter of approximately 250 Å, a wall diameter of 50Å, leaving an internal diameter of 150 Å. careful ultrastructural analysis has revealed that the wall is a polymer composed of globular subunits. In a cross section the wall of microtubules is made up of 13 subunits making up the complete circumference of the wall. When the surface of isolated microtubules is examined the subunits are seen to be arranged in longitudinal rows. These rows are called *protofilaments*. When the microtubules are caused to split open and flatten out, the 13 protofilaments remain intact. This indicates that the association of subunits with one another remains tight in the protofilaments.

Microtubules are constricted chiefly of proteins called *tubulins*. Some other proteins, the *microtubule-associated* proteins (MAPs) are also present but, have yet to be studied intensively. One subclass of MAPs is known as *Tau proteins*. Each subunit consists of two molecules of proteins, one of α-tubulin and one of β-tubulin organised within the microtubule. The subunits are asymmetrical and all oriented in the same direction along the axis. Each protofilament will have an α-tubulin molecule at one end and β-tubulin molecule at the other end. The structural polarity of the microtubule is thought to be an important factor in allowing these structures to participate in the direction of mechanical movements. Tubulins have molecular weights of approximately, 55,000 daltons purified tubulin preparations from several sources consist of mixtures of equal amounts of two proteins (α-and β-tubulins), differing slightly in size and composition. The present consensus is that in the cell, the two tubulins are associated as *heterodimers* of two tubulin molecules, one of each type.

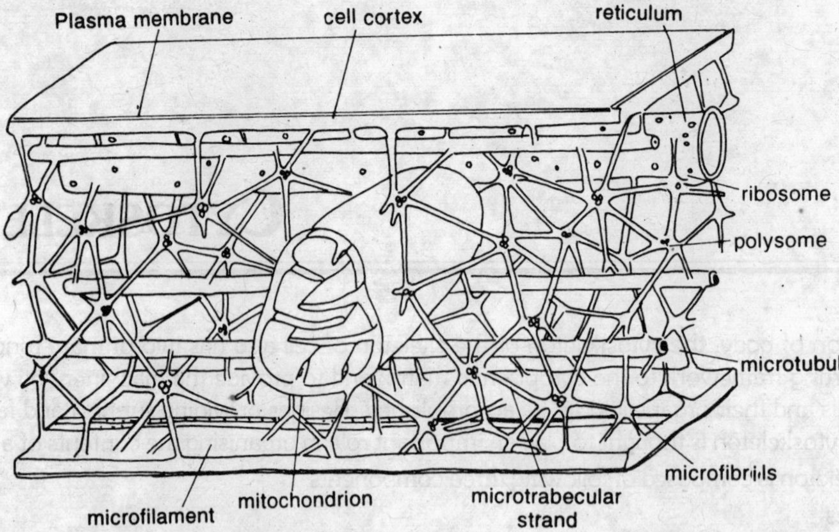

Fig. 18.1. Microtrabecular lattice which serves to anchor many cell's organelles in position.

There are hints that variations in composition or structural details sometimes occur among microtubules of different cells. An interesting possibility, still unknown, in that differences in microtubules function in different cells or regions of a cell is due to difference in the MAPs. The MAPs are believed to include proteins that extend laterally from the wall of the tubule, accounting for the existence of a clear zone of 5-10 nm from which other large structures are excluded.

Functions

The microtubules perform following functions

Mechanical function

The microtubules are considered to be a frame work which gives a definite shape to the cell and redistributed its contents. The integrity of microtubules is necessary for the individual shape of many cells and the rigidity of elongated processes like dendrites and axons.

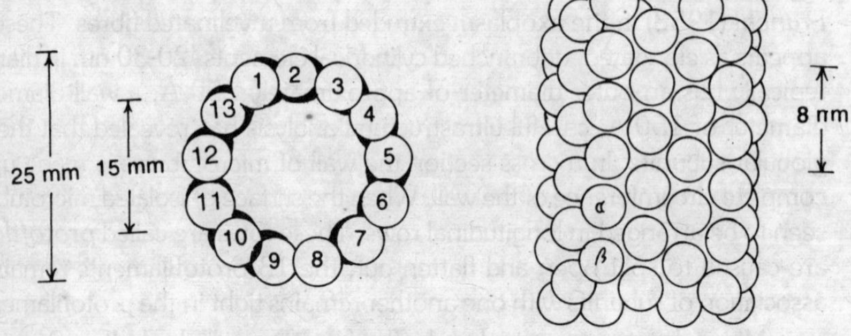

Fig. 18.2. Microtubular construction. In cross section the 13 sub-units making up the hollow cylinder are illustrated. The α and β proteins and their helical arrangements are shown in longitudinal section.

Morphogenesis

Microtubules play a role in the shaping of the cell during cell differentiation. The morphogenetic changes that occur during spermiogenesis provides an example. The enormous elongation that takes place in the nucleus of spermatid is accompanied by the production of an orderly array of microtubules. The microtubules are wrapped around the nucleus in a direction perpendicular to the nuclear axis and that have a double helical arrangement.

Cellular polarity and motality

Following forms of movement have been observed when treated culture cells with colcemid. These are (i) membrane ruffling (ii) endocytosis (iii) attachment to cell surface and (iv) extension of microvilli. Only a stationary movement of particles was found after destruction of the microtubules.

In *vitro,* if the cultural cells are treated with dibutryl cyclic AMP, they become elongated and the number of microtubules increases. These results suggest that microtubules are responsible for maintenance of the shape of the cell, the polarity of movement and distribution of certain organelles.

Circulation and transport

Microtubules may be involved in the transport of macromolecules in the interior of cell. One example can be quoted here. In the melanocytes melanin granules move centrifugally and centripetally with different stimuli. The granules have been observed moving between channels created by the microtubules in the cytoplasmic matrix.

Sensory transduction

In sensory receptors, the microtubules are arranged in bundles. It is possible that they help in transduction of different incident energy.

Assembly and Disassembly of Microtubules

Self-assembly is a reaction having a positive enthalpy i.e., it absorbs heat. Enthalpy change (ΔE) can be determined in the laboratory by noting the small decrease in temperature as the reaction proceeds. In order for such a reaction to occur spontaneously i.e., in order for ΔG to be negative in the equation.

$$\Delta G = \Delta H - T\Delta S$$

The entropy term, ΔS, must be positive. However, since the subunits themselves are getting more highly ordered as polymerization proceeds, some other substances must be getting more disordered. That substance is water which turns out released during polymerization. Thus self-assembly is generally an entropy driven, hydrophobic interaction among subunits.

Basically, microtubules are composed of protein subunits called α-tubulin and β-tubulin, each with a molecular weight of 55,000 deltons.

Tubulin polymerizes spontaneously at 37°C or body temperature to form the 13 protofilament structure seen *in vitro*. Its depolymerization occurs again at 0°C into tubulin dimers. A small amount of MAPs also obtained. When the MAPs are removed, purified tubulin still polymerizes at 37°C but only very slowly. Addition of MAPs results in rapid polymerization. Other factors required for assembly of microtubules are guanosine triphosphate (GTP), Mg^{++} and Ca^{++} ions. One molecule of GTP and one Mg^{++} bind tightly to each dimer. In so doing appear to cause a comformational change in the protein that favours polymerization. GTP is hydrolized to GDP in the process of polymerization. The role of Ca^{++} is unknown. Observation have revealed that concentration above 1 mM (10^{-3}M) cause depolymerization instead of polymerization. It appears that the concentration of free Ca^{++} regulates assembly *in vitro*.

The function of a microtubule within a living cell is dependent on its location and orientation within the cytoplasm. Evidences from the studies of *in vitro* assembly suggest that formation of microtubules occur in two phases—*initiation* and *elongation*.

All the various structures that have the capacity of promote micotubule formation are classified as *microtubule organizing centres* (MTOCs). The basal bodies, centrioles, pericentriolar material and the kinetochore of the mitotic chromosomes are MTOCs. Once microtubule assembly begins in the region of MTOC, continued incorporation of microtubule subunits appear to occur at distal end of growing structure i.e., the microtubule grows away from its site of origin. How these centres turn on or off so that assembly occur only at the right time in the life of the cell is unknown.

MICROFILAMENTS

In cells, there are a variety of different types of filaments in a cell be visualized under electron microscope, the term *microfilaments* generally applied for a specific type of fibrous element having a diameter of 60Å and is composed of protein *Actin*. Actin is a globular protein having a molecular weight about 42,000 daltons. Actin, when present in the monomeric form, is termed *G-actin*. G-actin is capable of assembling *in vivo* or *in vitro* into a double-stranded filament called *F-actin*. Actin filaments are capable of interacting with another protein called *myosin*. With myosin it forms actomyosin.

Observation made on a wide variety of calls have indicated that microfilaments form the major component of contractile machinery of the cell. In addition to their role in muscle contraction, microfilaments also help in cytokinesis, cell movement, nerve outgrowth, tubular gland formation, movement of the intestinal microvilli, cytoplasmic streaming and gastrulation.

Tilney et. al. have shown that in certain invertebrate sperms, polymerization of actin occurs at a very fast rate. For example, when sperm of *Thyone,* comes in contact with the egg, a long acrosomal process shoots out from the sperm head. The acrosomal process in this system is 15 times as the sperm head and forms in about 10 seconds.

INTERMEDIATE FILAMENTS

The intermediate filaments are long, unbranched filaments having a diameter of approximately 100Å. The diameter is intermediate between that of microtubules and that of microfilaments. Unlike microfilaments and microtubule, the intermediate filaments are a chemically heterogenous group of structures whose protein subunits can be divided five major classes. This group of proteins does not share common structural properties that enable them to form similar types of filaments. The intermediate filaments are having the property to resist denaturation and insolubility when compared with other components of the cell. Ionic detergents are required for their extraction and depolymerization. After depolymerization, they are able to undergo reassembly *in vitro.*

Like other elements of cytoskeleton, the intermediate filaments are thought to play a structural role in maintaining the shape of cell. They also promote the mechanical activities of the cytoplasm. In some cells it has been seen that the intermediate filaments are connected to the microfilaments and microtubules by cross bridges. Like other cytoskeletal elements. The intermediate filaments form extended arrays that are able to undergo dynamic changes in organization during mitosis and locomotion.

There are five major classes of intermediate filaments:

1. Keratin Containing Intermediate Filaments

A large variety of epithelial cells, such as myoepithelial cells, hepatocytes, pancreatic acinar cells and epidermal cells, have been found to contain a network of filaments. These filaments have a diameter of 80Å. They are composed of a family of proteins molecular weight 42,000-65,000 daltons. The proteins are very similar to α-kertin. On this basis the filaments are called *cytokeratins*. They are often collected in bundles that make up the tonofibrils of the desmosomes. In the epidermis, the cytokeratins become increasingly cross-linked as the cell differentiate and eventually contribute to the outer layer of dead skin cells forming stratum corneum.

2. Desmin Containing Filaments

This class of intermediate filaments has a diameter of 100Å. It occurs in all the three types of muscle fibres i.e., striated, non-striated and cardiac. The specific protein is called *desmin.* Its molecular weight is 53,000 daltons. Regarding the function, it has been suggested that desmin serves to spatially integrate the various components of a muscle cell into a functional unit capable of shortening. Desmin in found in the Z-line of skeletal muscles.

3. Vimentin Containing Filaments

These intermediate filaments contain protein *vimentin*. The molecular weight is 57,000 daltons. These are the most widespread filaments of all the other intermediate filaments. Depending on the type of cell, vimentin filaments can coexist in the cytoplasm with keratin or desmin or glial filaments. Usually the filaments remain interconnected with the nuclear membrane and plasma membrane. This suggests that intermediate filaments plays a role in supporting the nucleus and maintaining its position within the cell.

4. Glial Filaments

The cytoplasm of neuroglial cells contains a scattered array of bundles of 80Å filaments. The protein present in this filament is *glial fibrillary acidic protein* (GFA) and its molecular weight is 55,000 daltons.

5. Neurofilaments

In the cytoplasm of neurons, along with the microtubules, loosely packed bundles of 100Å filaments are present. These are called neurofilaments. In mammals these are composed of three distinct proteins having molecular weight 68,000, 145,000 and 220,000 daltons respectively. All the three proteins copolymerize to form the intact filament.

Neurofilaments and microtubules are connected to each other by side—arm appendages. It is believed that both form the three dimensional framework for the extended axonal and dendritic processes of nerve cells.

19

Cilia & Flagella and Microtubular Organelles

The cilia and flagella are filamentous processes of cytoplasm which move the entire cell or organism. They also help in capturing food and perform many mechanical functions for the cell. In some cases cilia are immobile cytoplasmic extensions as seen in epididymis and sensory patches of internal ear. In such cases they are called *stereocilia, Kinocilia* are the true cilia found in *Paramecium, Opalina,* trachea etc. They arise from the cytoplasm and are also connected with the plasma membrane. Morphologically and physiologically both cilia and flagella are identical but even then they show following differences:

Cilia	*Flagella*
1. The cilia are more numerous. In *paramecium* there are 10000 cilia.	1. The flagella are few in number. They may be one or two in number.
2. The cilia may cover the entire surface of the cell or organism.	2. The flagella occur at one end of the cell.
3. Cilia measure 5μ to 10μ in length	3. The flagella may be as long as 150μ in length.
4. The cilia tend to beat in a co-ordinated rhythm.	4. The flagella usually beat independently.
5. The cilia move in a sweeping or pendular stroke.	5. The flagella exhibit undulatory motion.

Occurrence

The cilia are characteristic of group Ciliata in Protozoa and occur in the ciliated epithelial tissues in various organs of metazoa such as the lungs, excretory cells of flat worms etc. The flagella are found in many protozoans (Flagellates), spermatozoa of most metazoans, choanocytes of sponges, gastrodermal cells of coelentrates and certain cells of vertebrate nephron. In plant cells, flagella are confined to the zoospores and gametes of certain algae and aquatic fungi and to the spermatozoa of bryophytes and pteriodophytes, etc.

Structure of Cilia and Flagella

The structure of Eukaryotic Cilia and Flagella

An electron micrograph of a cross section of a cilium or flagellum is one of the most familiar images in cell biology. The entire ciliary or flagellar projection is covered by a membrane that is continuous with the plasma membrane of the cell. The core of the cilium is termed the *axoneme* and is generally connected to the surrounding membrane by filamentous projections. The axoneme of a cilium or flagellum contains a set of a continuous fibres that run longitudinally through the entire organelle. In nearly even case, there are nine peripheral fibres, each made of two subfibers joins together to form a doublet and two central fibers that occur as singlets, the fibers are composed of microtubules made of polymerized tubulin and all the microtubules of a cilium have the same polarity. The peripheral doublets are composed of one complete microtubule the A subfiber, and one incomplete microtubule, the B subfiber, that contains 10 or 11 subunits rather than the usual 13.

The early examination of whole mounts of flagella and cilia in both the light and electron microscopes revealed the fibrous nature of the organelle but it was not until study made by *Fawcett* and *Porter* (1954) that their cross-sectional organization was shown in detail. The central tubules were seen to be enclosed by projections that form the central sheath, and this was connected to the *A* subfibers of the peripheral doublets by a set of radial spokes. The doublets were seen to be connected to one another by a bridge which is composed of a highly elastic protein, nexin. Of particular importance was the observation that a pair of "arms" project from the *A* subfiber in a clockwise direction (when viewing the cilium from the base toward the tip). Recent studies on a wide variety of cilia and flagella using the quick freeze-deep etch technique have revealed that each of the arm consists of five discrete structural components with the shapes. If one disregards the arms of the *A* fiber, the cilium or flagellum can be bisected into two approximate minor halves by a plane passing through the length of the structure perpendicular to a line drawn between the two central fibers. This plane of symmery would cut one of the doublet in half (the number 1 doubtlet) and slice between two doublets (the 5 and 6 doublets) on the other side. As will be described below, there is a constant relationship between the locations of specific doublets and the nature of the organelle's beat. In some species, other minor components are also visible.

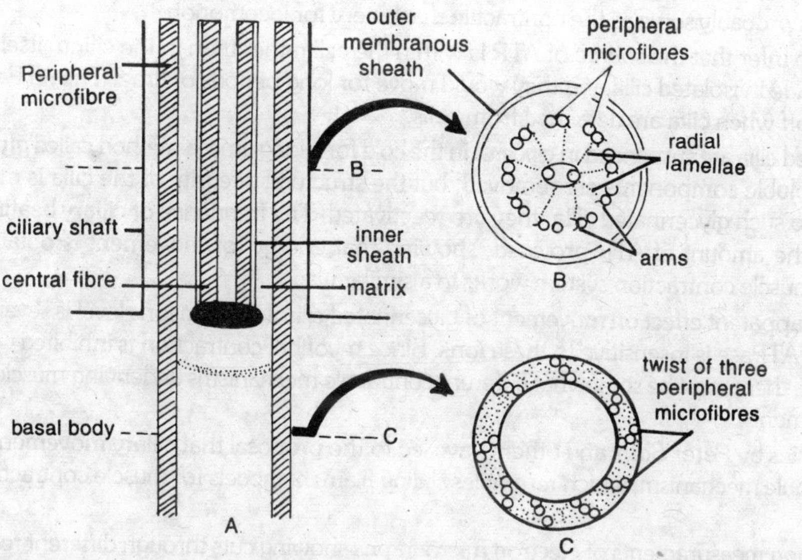

Fig. 19.1. Detailed ultrastructure of a cilium or flagellum.

While the 9 + 2 pattern has been highly conserved there are a number of notable evolutionary departures from this structural mode. For example a 9 + 1 pattern is found in flatworms, a 9 + 0 pattern is found in the eel and mayfly, and a 3 + 0 pattern in a sporozoan protozoan. Some flagella that lack central elements are able to generate a primitive type of motility, while others are non-motile. In some cases, flagella lacking the central tubule have a fibrous structure in its place.

Although the cross-sectional appearance of the axoneme reveals the major structures, it fails to indicate certain aspects of their organization that are best seen in a longitudinal section, i.e., one cut through the axoneme parallel to its long axis. Longitudinal sections indicate.

A and B microtubules of the triplets of the basal body. A terminal plate separates the basal body from the ciliary or flagellar axoneme and it is in the region of the terminal plate that the C microtubules of the basal body end and the central tubules of the cilia or flagella begin. Since the central tubules do not grow from microtubules of the basal body their polymerization must occur out of tubulin-containing amorphous material present in that region of the axoneme. The basal body usually has processes termed *rootlets,* that connect it to deeper layers of cytoplasm.

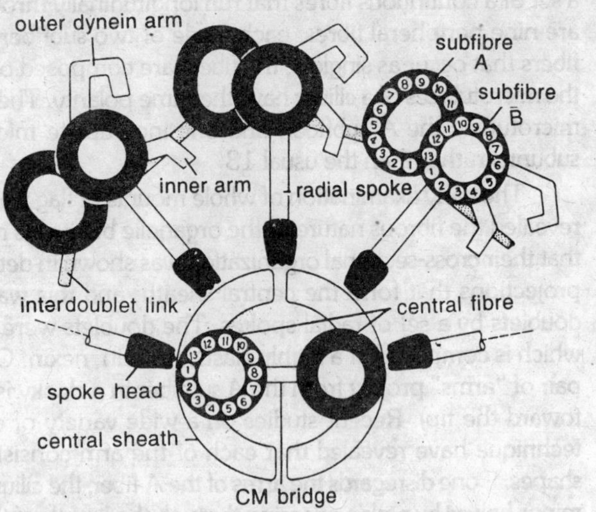

Fig. 19.2. Electron microscopic structure of T.S. of flagellum.

Ciliary Movement

Sliding Microtubule Mechanism

Cilia can move even after they are detached from cells; the cells themselves stop moving. Amputated cilia continue to swim until their ATP supply is exhausted. These and previous observations allow us to draw at least three conclusions:

1. Cells are moved by their cilia (or flagella) which are the primary organs of motility.
2. ATP is the energy source and its hydrolysis by ATPase makes free energy available for ciliary movement.
3. Ciliary fibers probably serve as the contractile machinery for locomotion.

We may also infer that the source of ATP is within the cell rather than in the cilium itself, otherwise ATP would be made in isolated cilia and they would move for long periods of time. The ATP supply must therefore be cut off when cilia are detached from cells.

When isolated cilia are suspended in glycerol in the cold for a long time (a method called *glycerination*), ATP and other soluble components are removed, but the structural integrity of the cilia is maintained. If ATP is supplied to such glycerinated cilia, they are reactivated. The frequency of ciliary beating is directly proportional to the amount of ATP provided, showing that energy for movement is obtained by ATP hydrolysis. The muscle contraction system works in a similar way.

There is no apparent effect on movement of glycerinated cilia if calcium ions level is severely reduced, showing that the ATPase is insensitive to these ions. Since myofibril contraction is inhibited when calcium ions are removed, there must be somewhat different contractile mechanisms underlying muscle contraction and ciliary movement.

Recent studies by *Peter Satir* and others have led to the proposal that ciliary movement is based on a sliding microtubule mechanism, which resembles sliding filament models for muscle contraction in certain ways.

From exacting measurements of electron micrographs showing cuts through different regions of bent and relaxed cilia, it has been proposed that ciliary subfiber doublets must slide past one another without contracting. As the microtubule doublets slide past one another, their movement causes a localized bend of the cilium. Doublet sliding results from cycles of attachment and detachment of the ATPase-containing dynein arms to the B-subfiber of the adjacent outer tubule doublet. This mechanism resembles sliding filaments in muscle where actin slides over myosin filaments by cyclic attaching and detaching of myosin heads to actin filaments. In cilia, interdoublet sliding is apparently converted into ciliary bending by cyclic interactions between the radial spokes and the central sheath. Individual tubules do not change in length or width, which shows that they themselves do not contract. Similar observations showed that muscle myofilaments remained constant in length and width, but sarcomeres contracted as sliding took place.

In addition to ATPase activity of the dynein arms of A-subfibers, it is important to know whether there actually is some ATPase activity as has been reported to occur at the swollen ends of the radial spokes where they make contact with the central sheath. It seems reasonable to expect ATPase activity there, since energy must be needed for intermittent interactions with the central sheath.

The sliding microtubule model satisfies conditions for movement in conventional 9 + 2 cilia and flagella and could provide a simple explanation for non-motility of modified cilia with a 9 + 0 pattern. Without the central pair of microtubules and the surrounding central sheath in 9 + 0 cilia, there is no place or radial spokes to undergo attachment-detachment cycles. There are other known modified fiber patterns, however, and it remains to be seen whether the sliding microtubules mechanism can explain ciliary movement in various types of modified cilia and flagella as neatly as it explains the 9 + 2 structural pattern.

The most likely source of ATP for ciliary movement are mitochondria, which are usually located very close to ciliary basal bodies (centrioles), ATP could easily be transferred from neighbouring mitochondria to the motility organelles. Sperm tails generally have a modified mitochondrion wrapped around the region between the head and tail of the sperm cell. Sperm tails stop moving when their ATP is exhausted or if ATP synthesis is inhibited when glycolysis and respiration are poisoned.

MODIFIED CILIA AND FLAGELLA

It was mentioned earlier that sperm tail ultrastructure varied from the conventional eukaryotic 9 + 2 pattern in some species. Many animal sperm have a 9 + 2 pattern of doublets and singlets but also contain another 9 components that are called *accessory fibers.*

These longitudinal components lie outside the ring of doublets and may be dense, massive structures. They are particularly prominent in mammalian sperm, but also occur in some other vertebrates and invertebrate groups. Little is known of the origin or function of accessory fibers, but they are usually found in species characterized by internal fertilization. Because of this correlation, studies were conducted to see if accessory fibers increased sperm motility in the viscous environment of the female reproductive tract. Evidence shows that accessory fibers contain an ATPase and a protein similar to one known in striated muscle has provided support for the suggestion that these structures make an essential contribution to sperm motility.

Specialized cells that respond to environmental stimuli such as light, sound, and odour are called sensory receptors. A modified cilium, subtended by a centriole, is found in many but not all of these cell types.

The central pair of singlet microtubules are usually missing or are extremely short in the ciliary shaft. Visual receptor cells of both rod and cone types have a ciliary structure that connects the outer and inner segments of the cell. A ring of 9 doublets encircles a hollow center where the singlet tubule pair would otherwise be located. In vertebrate visual receptors there may be upto 100 membrane limited discs arranged transversely in the outer segments. The discs are apparently formed from extensively folded ciliary membrane, which is derived from the cell plasma membrane just as in any cilium. The folded membrane system in visual receptor provides increased surface area for the conversion of light into electrical impulses. This represents an evolutionary advance in vertebrate animal eye organization.

Bacterial Flagella

Bacterial flagella are composed of two main parts, the elongated filament, which extends out into the medium and the basal part, which anchors the flagellum to the cell and generates the forces by which it moves. The extended part of the flagellar appratus is composed of a polymer of the protein flagellin. Purified preparations of flagellin can polymerize in vitro to form a flagellar filament, thereby providing an important system for the study of self-assembly. The base of the flagellum is much more complex being composed of several components and numerous proteins. The base of the flagellum is much more complex being composed of several components and numerous proteins. The inner ring of the base (M disk) is connected to the cell membrane, which in turn is responded for providing the energy utilized in the movement. The plasma membrane of the bacterium is in a high energy state as a result the maintenance of a proton gradient. This high energy state is directly responsible for driving the flagellum without the need for an ATP intermediate.

In a very real sense, a bacterium can be thought of as possessing an "electric motor" included within this motor is a rotating disk that may represent nature's first and only invention of a wheel as part of a biological machine.

The typical bacterial flagellum is approximately 150μ in diameter and 1 to 15 mμ in length and is capable of rotating in either a clockwise or a counter-clockwise direction. The nature of the rotation has been observed in experiments where the flagella are immobilized by attachment to a glass slide via an antibody linkage. Under these contains, the activity of the flagellar "motor" causes the bacterial cell to rotate, which, unlike the movement of very small flagella, can be followed in the light microscope. The direction of rotation of the flagella has been shown to have important consequences for movement of the cell. Consider the position of a bacterium, a small isolated cell without any type of advanced sensory apparatus to guide its way. If the ability to locomote is to be of any value to the organism, it need some type of guidance system to direct it toward more hospitable environments and away from delecterious ones. The behaviour of simple organisms is usually described in terms of taxes, i.e., responses either towards or away from a stimulus. On the basis of their movement toward positive stimuli and away from negative stimuli, these organisms will ultimately find themselves in the most appropriate region of their environment.

The basis for the locomotor behaviour of *E. coli*, which has a few flagella at one pole of the cell, appears to be determined by its ability to reverse the direction of flagellar rotation. Observations of bacteria indicate that they generally move in a straight line for a few seconds and then undergo a type of random tumbling motion which orients them in a new randomly chosen direction. The movement of a bacterium in a straight line occurs during counterclockwise rotation of its flagella, while the tumbling motion results from a clockwise rotation. When the flagella rotate in the counterclockwise direction, they remain associated with one another in a bundle and thereby act in a concerted manner as if they were a single multistranded organelles. However, when rotating in the clockwise direction, the flagella are separated from one another, and each acts independently, producing the tumbling behaviour. If a bacterium is exposed to a chemical that acts as an attractant (e.g., glucose), the rotation of the flagella will be maintained in a counterclockwise manner as long as the cell continues to move toward higher concentrations of the substance. In contrast, if a chemical known to repel bacteria is added, the clockwise rotation will be initiated, causing the bacterium to tumble until it moves off in a direction away from the source of the substance.

Formation of Cilia and Flagella from Centrioles

Eukaryote *cilia* (shorter appendages) are produced only from centrioles positioned at the cell periphery. These centrioles may have been formed near the cell surface, or they may have migrated there from some place of origin deeper within the cell.

In a very careful study by *I. Gibbons* and *A. Grimstone* reported in 1960, the ultrastructural relationship between centriole and flagellum was described as follows. The 9 subfiber triplet in the centriole extend to a transitional region just below the shaft of the flagellum. At some point in this zone near the flagellar shaft, the C-subfibers disappear. The remaining A-and-B-subfibers continue as doublets to a place just below the tip of the flagellum. At the same transition region where the C-subfibers end, a new pair of microtubules appears in the center of the ring of 9 subfiber doublets in the flagellum. This central pair of microtubules together with the encircling microtubule doublets make up the so-called "9 + 2" organization found in almost all eukaryotic cilia and flagella. The development sequence and continuity between centriole and cilium of flagellum has been described in other species and therefore appears to be a common plan of organization.

There is more than enough evidence to state that each cilium and flagellum forms directly from one existing centriole. For example, (1) cilia and flagella never form in the absence of a centriole; (2) if the centriole remains intact, a new cilium will be regenerated when a previous one has been removed; and (3) if the centriole has been destroyed or removed, the flagellum will degenerate and a new flagellum will not form unless a new centriole replaces the one that was removed.

Centrioles Display Polarity

New centrioles form only at the proximal end (farthest from the cell surface) of a mature centriole, while flagellar outgrowth occurs at its distal end (nearest to the cell surface). The organizing events are different in each of these two occurrences. While centriole and flagellum form a continuous structure, mature centriole and procentriole are always separated by a space of about 60-80 nm.

The Role of Calcium

Considering the prominent position of calcium in regulating other type of motile systems one would expect calcium ions to play some role in ciliary and flagellar activity and this expectation has been amply borne out. Although calcium ions are not required for the beating of the axoneme perse the Ca^{++} concentration generally has a controlling effect on the direction and or beat frequency of both cilia and flagella, which in turn, influences the behaviour of the entire cell. The concentration of calcium in the region of an axoneme is controlled by the surrounding plasma membrane. When a ciliated or flagellated cell is exposed to an appropriate stimulus there is a rapid increase in the permeability of the membrane to calcium and an influx of calcium ions from the extracellular medium. For a *paramecium*, the stimulation may be caused by the presence of an object in the path of the ciliate and the increasing concentration of calcium in the cilium leads to a reversal of the direction of the ciliary beat and the organism moves away from the obstacle. In the photosynthetic flagellate *Chlamydomonas*, changes in flagellar calcium concentration are responsible for moving the organism forward the light. In ferns, fertilization is mediated by the release of motile sperm cells. The presence of a chemical attractant in the medium affects the permeability of the sperm membrane to calcium.

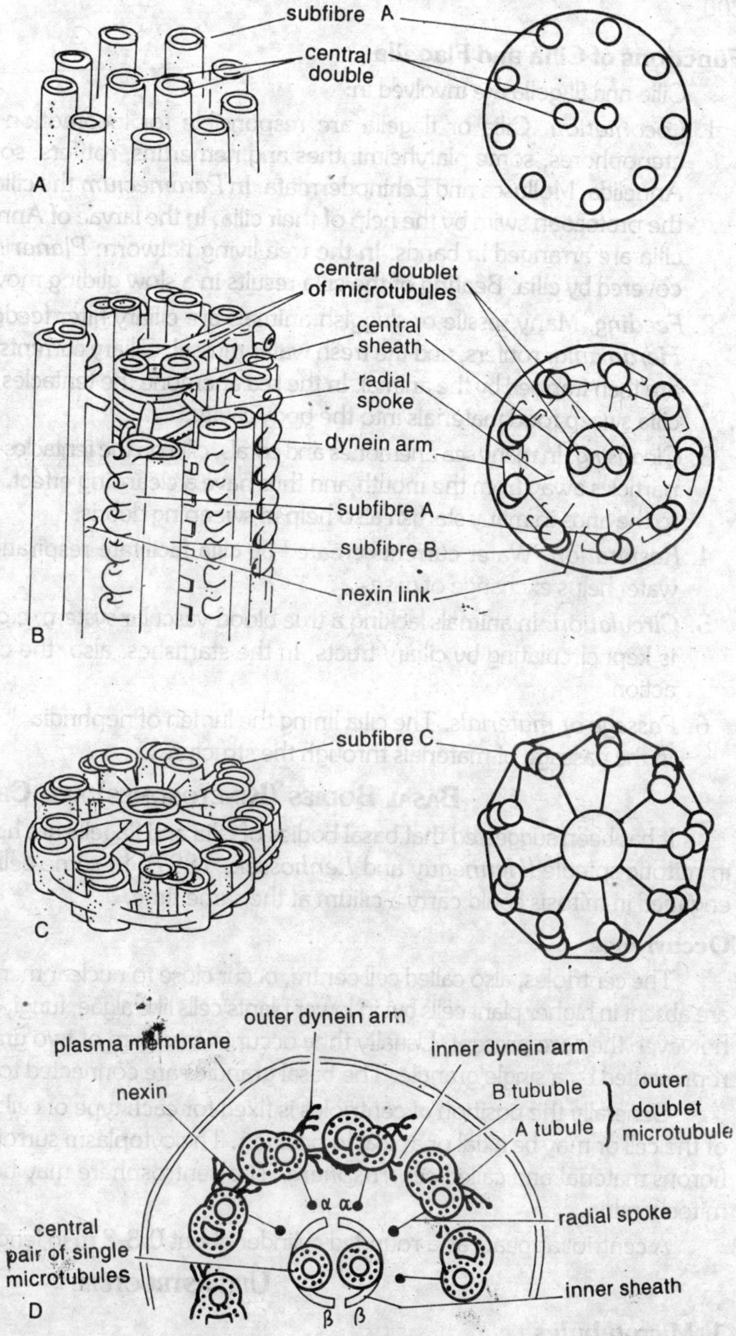

Fig. 19.3. Schematic diagram of a flagellar axoneme, showing relationships between its components at the flagellar tip (A), in the middle of the flagellum (B), and at the basal body (C). More detailed picture of flagellum in the cross section.

Functions of Cilia and Flagella

Cilia and flagella are involved in:

1. *Locomotion.* Cilia or flagella are responsible for locomotion in ciliate and flagellate Protozoa, ctenophores, some platyhelminthes and nemertins, rotifers, some snails and the ciliated larvae of Annelida, Mollusca and Echinodermata. In *Paramecium* the cilia cover the entire body surface, and the protozoan swim by the help of their cilia. In the larvae of Annelida, Mollusca and Echinodermata cilia are arranged in bands. In the free-living flatworm *Planaria* the ventral surface of the body is covered by cilia. Beating of the cilia results in a slow gliding movement over the substratum.
2. *Feeding.* Many sessile or sluggish animals are ciliary filter feeders, e.g., *Amphioxus,* the ascidian *Herdmania,* rotifers, and the fresh water mussel. Ciliary currents bring in fine particles of food which are then trapped by the animal. In the sea anemone the tentacles are bent towards the mouth and the cilia sweep food materials into the body cavity.
3. *Cleansing.* In many sea anemones and corals, cilia on the tentacles and the oral region sweep unrequired particles away from the mouth and thus have a cleansing effect. The beating of cilia from the mouth to the anus in many starfish also help in sweeping debris.
4. *Respiration.* Water currents created by cilia facilitate respiration. The continuous movements of water helps exchange of gases.
5. *Circulation.* In animals lacking a true blood-vascular system, e.g., some annelids, the coelomic fluid is kept circulating by ciliary tracts. In the starfishes, also, the coelomic fluid is circulated by ciliary action.
6. *Passage of materials.* The cilia lining the lumen of nephridia, kidney tubules, and genital ducts help in the passage of materials through the structures.

Basal Bodies (Kinetosomes) and Centrioles

It has been suggested that basal bodies of cilia and flagella are homologous with the centrioles found in mitotic spindle (*Henneguy* and *Lenhossek,* 1897). In some cells it was discovered that a centriole engaged in mitosis could carry a cilium at the same time.

Occurrence

The centrioles, also called cell centre, occur close to nuclear membrane in all the animals cells. These are absent in higher plant cells but in lower plants cells like algae, fungi, bryophytes, ferns and gymnosperms, however, they are present. Usually they occur in the form of two granules called *diplosome* but may be represented by a single granule. The basal granules are connected to the cilia or flagella.

Generally the position of centrioles is fixed for each type of cell. It may occupy the geometric centre of the cell or may be axial or apical in position. The cytoplasm surrounding the centrioles contains a fine fibrous material and called *centrosphere.* The centrosphere may be surrounded by Golgi complex and mitochondria.

A centriol appears as a rounded cylinder about 0.3-2 m in length and 0.15-0.25 m in diameter.

Ultrastructure

1. Microtubules

The wall of the centriole has nine groups of microtubules arranged in a circle. Each group is triplet formed of three tubules that are skewed toward the centre. The innermost tuble or subfibre designated as *A*, middle by *B* and the outer by *C*. Each subfibre is about 250Å in diameter and its wall is about 45Å in thickness. The *A* subfibre is completely rounded where as *B* subfibre and *C* subfibre are *C*-shaped. *B* subfibre shares the wall of *A* subfibre and *C*-subfibre share the wall of B subfibre.

The structure of *A* and *B,* and *C* subfibres, is similar to the structure of other microtubules (Stephens, 1970). The A-subfibre has 13 globular subunits around its perimeter. Each subunit is about 40-45Å in diameter. Three or four subunits of *A* subfibre are shared by subfibre-*B,* which in turn shares several of its subunit with the subfibre *C*.

Often the triplets are thought to run parallel to one another and to the long axis of the cylinder, but this is not always the case. In the basal bodies of some animals, the triplets get closer toward the proximal end, so that the diameter of the cylinder gets smaller. In centriole the three subtubules are arranged parallel to one another but turn in a long-pitched helix with respect to the cylinder axis.

The centeral microtubules are absent and there are no special arms.

2. Linkers

The *A* subfibre of each triplet is linked with *C*-subfibre of neighbouring triplet by a dense link. The *A-C* linker is made up of some dense material which is also made up of globular subunits. It has been suggested that the *A-C* linkers are responsible for the radial tilt of the triplet. In some species instead of *A-C* linkers, *A—A* linkers are formed, in other *C—C* linkers are present in addition to *A—C* linkers (e.g., *Pseudotrychonympha*).

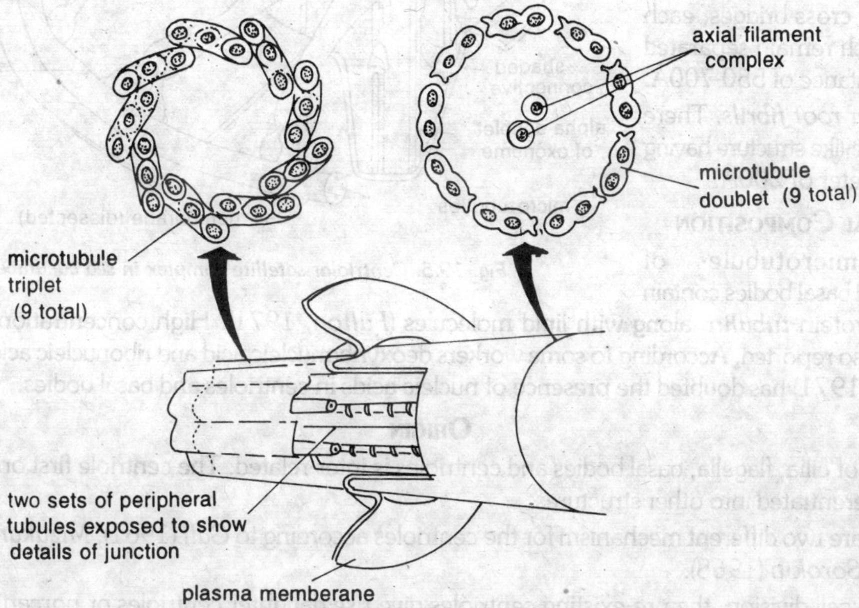

Fig. 19.4. Diagram illustrating the three-dimensional structure of a flagellum basal body.

3. Cartwheel Structure

At the proximal end of centrioles and basal bodies is found a cartwheel structure. The cartwheel structure consists of a *central rod* or *hub* in the centre and radiating from it are 9 *spokes*. Each spoke is connected to *A*-subfibre. The hub is about 25Å in diameter and may contain 1 or 2 longitiudinal flaments. Near outer end of each spoke there is a thickening X. An another thickness Y lies between two X-thickenings. These thickenings are connected to each other on inner side and on outer side with the triplet.

Associated Structures

1. Satellites

There occur nine amorphous plaques or spheres of electron dense material. These are attached to the centriole by short bridges made up of dense material. These satellites are found to change their position, shape etc., during mitoses and spermatogenesis.

2. Transitional Fibres

Two or three transitional fibres arise from B-subfibre of each triplet at the end of basal body and are found at the level of basal plate.

Basal bodies also have a variety of associated dense plaques and masses, and of rootlets, rhizoplasts and other filaments. Rootlets are fine structures arising from the basal bodies and terminate near the nucleus. The rootlets are of two types.

(i) *Striated root fibrils.* Striated root fibres contain regular cross bridges, each of which remain separated by a distance of 550-700Å.

(ii) *Tubular root fibrils.* There are tube-like structure having a diameter of 200Å.

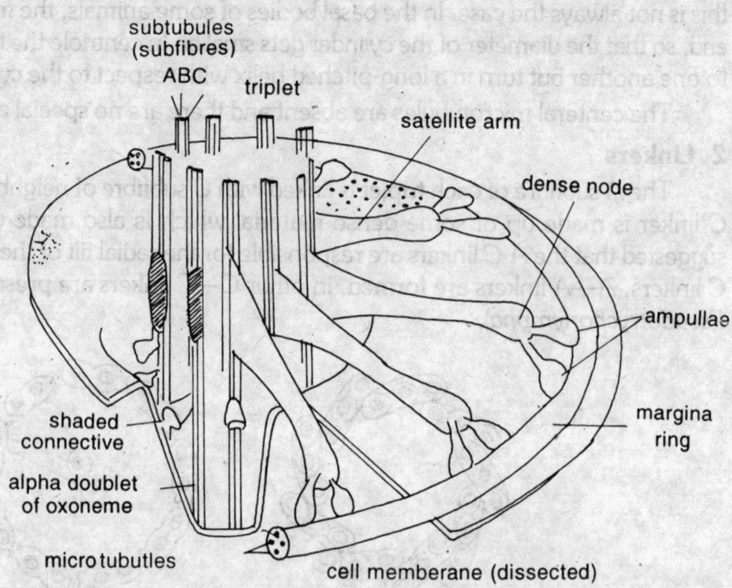

Fig. 19.5. Centriolar satellite complex in sea cucumber.

Chemical Composition

The microtubules of centriole and basal bodies contain a specific protein *tubulin.* along with lipid molecules (*Fulton,* 1971). High concentration of ATPase enzyme is also reported. According to some workers deoxyribonucleic acid and ribonucleic acid also found but *Fulton* (1971) has doubted the presence of nucleic acids in centrioles and basal bodies.

Origin

Origin of cilia, flagella, basal bodies and centrioles is inter-related. The centriole first originates and later on differentiated into other structures.

There are two different mechanism for the centrioles according to *Gall* (1961), *Mizukanni* and *Gall* (1966) and *Sorokin* (1968).

1. During cell division, the pre-existing centrioles give rise daughter centrioles or *porcentrioles.* The structures arise as annular structures which become lengthened into cylinders. From each cylinder of procentriole originate three subfibres and the maturation takes place in cytoplasm. This method of centriole formation occurs only during cell division.
2. In this method, the centriole orginates from the fibrogranular material which lies in it. The granules increase in size and accumulate to form spheroid bodies. These spheroid bodies give rise to procentrioles. These procentrioles after maturation form centrioles and these centrioles get arranged in rows just beneath the plasma membrane. After that each centriole produces another procentrioles from its lateral side and itself converted into the basal bodies. The basal bodies form rootlets from their basal ends and cilia or flagella from their apical ends.

The protein of cytoplasm where maturation of centriole takes place is called centrosome and lies near the nucleus.

20

NUCLEUS

The most prominent particulate feature of a cell when viewed under the microscope is the nucleus. Originally it was detected by *Leeuwenhock* in 1700 as refractile bodies in the centre of blood corpuscles of Salmon blood. These structures, which must have been nuclei, where seen with the simple lenses which were found by him in this spare time. *Fontana* in 1781 observed similar oval bodies inside the skin cells of an eel, and the full description was given by the Scottish botanist *Robert Brown* in 1835. It is an almost universal structure of cells at sometime during their life cycle, although such cells as the sieve tubes of vascular plants and the red blood cells of mammalian vertebrates may lose their nuclei by the time they are fully differentiated. The nucleus is not homogenous, but contains thread like structures whch were discovered some years later that these structures could be obtained with aniline dyes for which reason they become known as *chromosomes* (Gr. *Chroma*=colour and *Soma*=body). *Sutton* in 1903, first suggested that hereditary factors or genes are carried on chromosomes. And in 1871 for the first time it was isolated by *Miescher,* its morphological analysis was much less advanced than that of the cytoplasm. *Kossel* was able to demonstrate the presence of nucleic acid, and for this he was awarded the Nobel Prize in 1910.

Cell without nuclei have very limited features. The only common animal cell type without a nucleus, the mammalian red blood cell, lives only a few months; aside from its role in oxygen transport, it is extremely restricted in its metabolic activities. Egg cells from which nuclei have been experimentally removed may divide for a while, but the products of division never differentiate into specialized cell types, and eventually they die. Fragments without a nucleus, cut from such large acellular organisms as ameobae or the alga, *Acetabularia,* survive temporarily, but ultimately they die unless nuclei from other cells are transplanted into them.

The nucleus is essential to long-term continuation of metabolism and to the ability of cells to alter significantly their structure and function (as in differentiation). In large part, this reflects the primary role of the nucleus in producing the RNA required for protein synthesis. When cells change, their new functions and structures require new proteins. Even cells that are constant in metabolism and structure show continual replacement (turnover) of macromolecules and probably of organelles, including portions of the cytoplasmic proteinsynthesizing machinery.

MORPHOLOGY

Shape

In general the nucleus tends to be spherical, but may be fusiform, ellipsoidal, flattened, depending on cell shape and function. In young cells it is more often spherical and centrally located, but in differentiated ones it may be displaced and irregular in shape. The nucleus may be *elongated* as in columnar epithelium and muscle cells. In certain cases it becomes *irregular* and has often been observed in particular cases to undergo slow amoeboid changs of form in the living cells, e.g., cartilage cells, leucocytes or in animal ova. Nuclei (Nucleus-singular) of irregular or amoeboid form are frequent in cells characterized by very active metabolism, in which case the nuclei are often not only of large size, but show a marked further increase of surface by the formation of lobes, sacculations, or even in extreme cases of complex branches ramifying through the cell. An extreme example of this is offered by the spinning glands of certain insect larvae (Lepidoptera and Trichoptera), in which the nucleus, originally spheroidal, finally assumes labyrinthine appearance with convolutions occupying a large area in the cell.

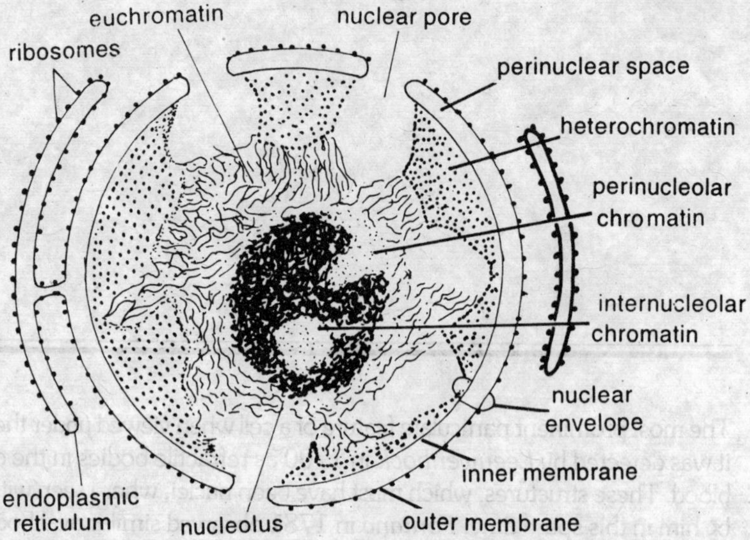

Fig. 20.1. The interphase nucleus.

In certain types of cells the surface of the nucleus may also be increased by its breaking up into more or less separate vesicles or *Karyomerites*, thus forming "Polymorphic" nuclei or nuclear nest. Vesicular or spherical shaped nuclei generally occur in the tissue cells of most multicellular animals and plants. Massive nuclei occur typically in the male germ cells of animals generally and of many lower plants (in coenocytic plants like *Vaucharia* etc.).

Position

The position of the nucleus is determined by many causes, such as surface tension, the position of the vacuole, the relative density of the cytoplasm in different portion of cell. In embryonic cells it almost always occupies the geometric centre. In non-vacuolated cells of plants even it occupies the center of the cytoplasmic mass. In adipose cells, in eggs rich in yolk the nucleus is forced toward the periphery by the accumulation of the *paraplasm*. In glandular cells it is located in the basal position, the granules occupying the apical cytoplasm. The position of the nucleus is also related to the function of the cell.

Size

In 1895 *Boveri* showed that the size of nuclei in merogic echinoderm larvae is dependent upon the number of chromosomes each contains. *Gates* 1909, however, *adduced* evidence to show that this rule is by no means universal. The size of the nucleus is variable, but in general it bears some relation to that of the cytoplasm. This may be expressed numerically in the so-called nucleoplasmic index (NP) by *O. Hertwig* (1906).

$$NP = \frac{Vn}{Vc - Vn}$$

where Vn being the nuclear volume and Vc being the volume of the cell

This NP index shows the relationship between the volume of cytoplasm and that of nucleus i.e., if former increases then the second also should increase. The ratio of the volume of the nucleus to that of the cytoplasm in a cell is called the 'nucleoplasmic ratio' or "Karyoplasmic ratio."

Number

The number of nuclei present in any mass of protoplasm depends largely upon the bulk of the mass since within limits a certain ratio of nuclear surface to cytoplasmic volume must be maintained for the proper action of the whole system. Generally all the cells are uninucleate or *mononucleate* i.e., one nucleus in one cell. But in certain cases *binucleate* condition occur as in *Paramecium caudatum,* here one nucleus is smaller which is micronucleus while the larger one is called the macro- or meganucleus. In few others the poly- or multinucleate condition exists as in *Opalina,* striated muscles.

STRUCTURE

The nucleus consists of four components: (1) Nuclear envelope (2) Nuclear sap, (3) nucleolus, and (4) Chromatin network.

1. Nuclear Envelope

The nuclear envelope has only recently enjoyed much investigation. As viewed with the standard phase contrast microscope, it appears as little more than a dark line, this line separating the nuclear contents from the cytoplasm. In most cells, it is observed to break down and reform as the cell enters and completes division. However, in some algae, protozoans, and fungi, it does not break down during division. Observation with the light microscope shows that the nuclear envelope serves as a barrier between the nucleus and the cytoplasm. The way in which the nuclear envelope regulates the flow of material is just beginning to be studied.

Ultra structural studies reveal that the nuclear envelope is actually composed of two membranes separated by a *perinuclear space* 110 to 400Å wide. The inner membrane appears to be in contact with the nuclear chromatin, which can be seen condensed along its surface, the outer membrane is continuous with the cytoplasmic endoplasmic reticulum and is often observed with ribosomes bound to its surface. The association of the outer membrane with the ER results in a continuity between the nucleus and the cytoplasm. However, this continuity is not complete because of the presence of the inner nuclear membrane, which still serves as a structural barrier.

Nuclear pores

Since ribonucleic acid synthesis occurs within the nucleus while protein synthesis takes place in the cytoplasm, it is evident that substances must be able to move outward. In fact, there is a flow of material in both directions through the nuclear pores, which, in spite of their size, are not wide open channels. The two nuclear membranes are fused together at intervals to form pores approximately 600Å in diameter. These were first seen by *Callan* and *Tomlin* (1950). It is estimated that pores occupy about 10% of the surface area of the nuclear membrane or the total number of pores per nucleus may vary from 100 to 5 x 10^7. These pores appear an roughly circular or polygonal areas.

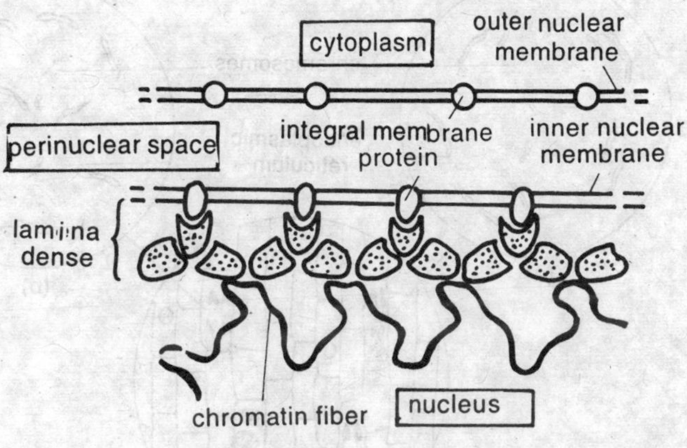

Fig. 20.2. Nuclear membranes and lamina densa.

However, the pores are not simply holes in the nuclear membrane. Much of the aperture is occupied by a cylindrical or ring-like arrangement (annulus) of moderately electron-dense granules or fibrillar material that appears to be continuous with the inner fibrous lamina. The annular material is organized as a *pore complex*. The annulus consists of eight annular granules on its nuclear and cytoplasmic side surface and a central granule in the centre. Fibres extend from the central granule and annular material. Some amorphous material forms a diaphragm over the pore. The significance of diaphragms and granules is uncertain since they do not appear to be universal constituents of nuclear pore. Perhaps they are transient structures which are present in certain physiological states and not in others.

Fibrous lamina

The nuclear membranes of many cells display an additional layer called *fibrous lamina*. It is apposed to the inner aspect of the inner nuclear membrane, facing the interior of the nucleus. It is composed of fine filaments made up of proteins. Probably they provide mechanical reinforcement to the nuclear membranes. According to some workers the fibrous lamina influences the exchange of materials between nucleus and cytoplasm. The degree of development of fibrous lamina varies greatly in different kinds of cells. In mammalian cells it is thin but highly developed in amoeba and certain other invertebrates. In amoeba, it has a honeycomb-like configuration and is 1000-1500Å in thickness.

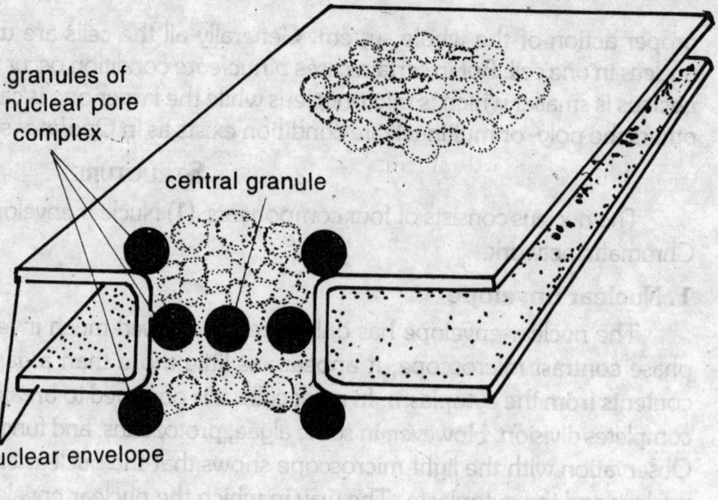

Fig. 20.3. Structure of nuclear pore complex.

Permeability of the Nuclear Envelope

Several experiments suggest that the pore complexes may be temporary or permanent openings in the nuclear envelope. By injecting colloidal gold particles, varying in size from 2.5 to 17nm, into the cytoplasm of amoebae, it was found that those with diameters upto 8.5 nm rapidly entered the nucleus. Particles with diameters of 8.9 to 10.6 nm penetrated more slowly, and the larger ones did not enter at all. These results indicate that the openings are smaller than the pore size would indicate.

Evidence has been obtained with tehse techniques, suggesting that the pores are pathways for the exchange of macromolecules. The annuli may regulate the exchange in relation to the size and possibly to the chemical nature of the penetrating substance. It is important to consider that the permeability of the nuclear envelope is not fixed, but vareis in different cell types and within a given cell—at least during the division cycle. Such differences are attributable to changes in the nature of annular material (*Feldher*, 1971).

The presence of pores in the nuclear envelope should be correlated with some of the electrochemical

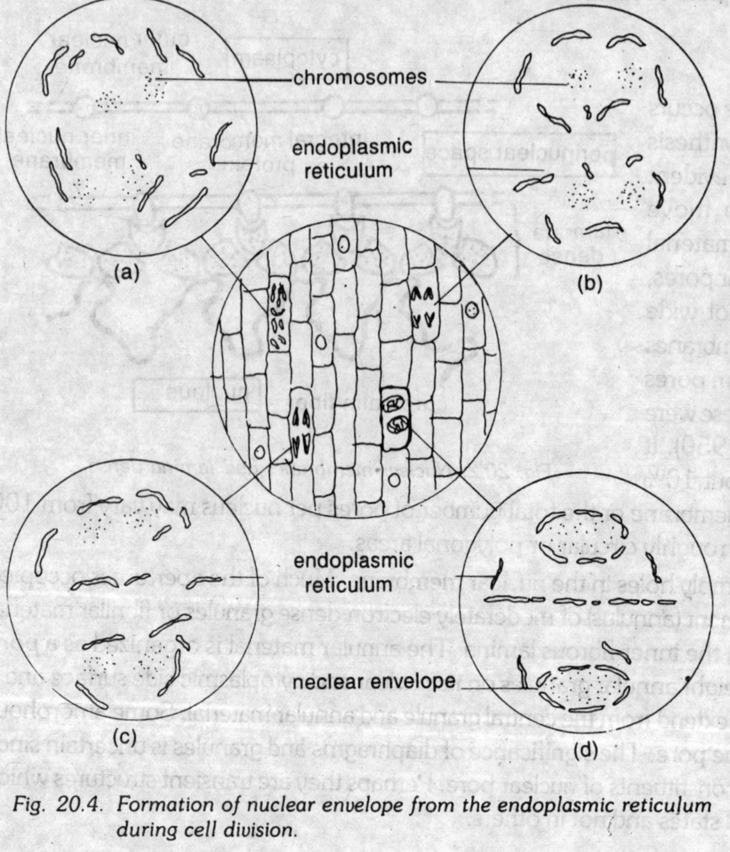

Fig. 20.4. Formation of nuclear envelope from the endoplasmic reticulum during cell division.

properties of this strtucture, which can be investigated with fine microelectrodes. Two types of nuclear envelopes have been recognized with this technique. When giant cells from the salivary gland of *Drosophila* are penetrated with a microlectrode, there is an abrpt change in potential at the plasma membrane (–12 mV); then, as the microelectrode enters the nucleus, there is another drop in negative potential at the nuclear membrane (– 13 mV). These results suggest that the nuclear envelope may be a diffusion barrier for ions as small as K^+, Na^+, or Cl^-. In the nuclear envelope present in oocytes, however, there is no detectable potential, thus indicating a free interchange of ions between the nucleus and the cytoplasm.

There are several morphological observation that suggest there is passage of ribonucleoproteins and other macromolecules across the nuclear pores. The relationship of the pores with channels in the nucleoplasm was mentioned earlier. Dense material extending through the nuclear pores has been observed in oocytes of Amphibia. Some of the material may correspond to ribosomal subunits, other to messanger RNA. The mechanism by which the passage of these substances is achieved is unknown. Cytochemical studies, however, have revealed the presence of an ATpase in the pores which may provide the necessary energy for the transfer of macromolecules.

Origin of nuclear membrane

The nuclear membrane is considered as a specialized cytoplasmic structure originating from the cytoplasmic membrane system. During telophase, it is formed by the accumulation of vesicles of endoplasmic reticulum around the chromosome group, which later on fuse to form a complete membrane.

There are basically two theories as to how the nuclear envelope reconstitutes after telophase, either the pair of lamellae is synthesize *de novo* at the sites where chromosomes and cytoplasm are in contact with each other (*Jones,* 1960) or parts of the already present endoplasmic reticulum adjoin the telophase nucleus and merge to form the nuclear envelope. In locust spermatocytes, for instance, the 0.5μ vesicles which are formed at the surface of the chromosomes, are indistinguishable from the vesicles of the endoplasmic reticulum (*Bohr,* 1959, 61). They gradually arrange themselves so as to lie parallel to the surface of the nucleus and then merge to form a continuous nuclear envelope.

2. Nuclear Sap

An unstained or light acidiophilic mass, the "Karolymph" or "Nuclear Sap" fills completely the nuclear space where the other components are found. It was *Claude* who in 1943 precipitated out the nuclear sap by histological fixative and stained with acid dyes. In eggs and in larger cells (e.g., *Acetabularia*), the nuclear sap is clearly visible. According to *Stich* (1951, 58) it gives positive cytochemical test for RNA, protein including-SH groups and glycoproteins.

Many metabolic pathways have also been demonstrated in nuclear sap, similar to cytoplasm. These include glycoysis, hexose monophosphate shunt, citric acid cycle, etc. The hexose monophosphate shunt is important since it supplies the nucleus with pentose. Further, in the nucleus NAD is synthesized which is the coenzyme of numerous dehydrogenases. Characteristic enzymes of the nucleus are DNA and RNA-polymerases.

Chemical Composition of Nucleus

The following chemical components have been analysed out from the isolated nuclear material:

(i) Nucleoproteins
  (a) Nucleo-protamine
  (b) Nucleo-histones
  (c) Non-histones or Acidic proteins
(ii) Nucleic acids
(iii) Enzymes
(iv) Lipids

(i) Nucleoproteins

Apart from nucleic acids the other important constituent of the cell nucleus is the protein components. The chromosomes of plants and animal cells, unlike the genetic material of bacteria, consists of DNA associated with proteins. In giant polytene chromosomes protein has been found in the banded and interbanded regions of gene activity. Proteins is associated with the extended loop of lampbrush chromosomes. Nucleoli contain high proportions of protein and there are of course protein present in the nuclear sap or nucleoplasm, and, in association with lipids, in the nuclear membranes.

The protein part of the nucleus is certainly very complex and has several components. Of these the best known are two strongly basic and simple proteins: the protamines and the histones. In addition to these there are acidic proteins, the so-called non-histone porteins and enzymes present.

(a) *Protamine or Nucleo-protamine.* It was *Miescher* who discovered the protamine in Salmon sperm. These are simple basic proteins having a very low molecular weight. These are very rich in basic amino acid *arginine* and are found in spermatozoa of some fishs and are tightly bounded to DNA by salt linkage. As the protamines are smaller than histones, which means that the chromosomes and nuclei can be packed into a smaller space, allowing greater mobility of the sperm. In general they consist of about 28 arginine polypeptides, which a total length of 100Å, and contain 19 arginines and 8 or 9 non-basic amino acids. During development there is a progressive replacement of the histones by protamine. This may be due to the higher affinity of protamine of DNA.

(b) *Nucleo-histones.* Histones were discovered in the goose erythrocytes by *Kossel* and soon thereafter were found in the thymus gland by *Lilientfeld.* Histones have also been found in certain plant cell nuclei, notably wheat and cedar nut embryo by *Mirsky.*

The nucleoproteins are also basic proteins having a molecular weight about 12000. In addition to about 13% *arginine, histones* contain other basic residues including *lysine* and *histidine.* Several histones of different composition have been isolated, and three main types have been characterized:

(i) Very lysine rich

(ii) Arginine rich

(iii) And slightly lysine rich.

These are heterogeneous and consists of several components. Histones are found in all nuclei of higher organisms, although they have been little studied in plants. Those found in plants, however, resemble the histones found in vertebrates.

The main function of histone ascribed is that of acting as a chromosomal 'glue' binding togehter the genetic units of DNA. It is also known that DNA—histones complexes are more resistent of denaturation by heat that DNA alone, and that proteins (mainly histones) in the nucleo-proteins complex can partially protect DNA from radiation damage. Evidence now indicates, however, that the really important role played by histones lies in representing the genetic activity of cells. It was suggested by *Stedman* and *Stedman* in 1950 that histones interact with DNA in a specific way, preventing it from acting as a template for RNA synthesis and thus preventing the transfer of genetic information to the cytoplasm.

(c) *Non-histones or Acidic Proteins.* Chemical analysis of isolated nuclei and chromatin threads led to the demonstration of another type of protein which is generally called non-histone protein. This protein contains *tryptophane* and has acidic properites. Another non-histone fraction has been identified from isolated chromatin threads. The insoluble residue appears under the microscope as spiral threads which maintains the characteristics of the interphase chromosomes. From a chemical standpoint it is very interesting that these residual chromosomes contain more ribonucleic acid (RNA) than DNA.

The considerable amount of non-histone proteins present in a metabolically active cell contrast markedly with the composition of the spermatozoan, which is much less active.

(ii) Nucleic acids

Nucleic acids, which are of two types: Ribonucleic acid (RNA and Deoxyribonucleic acid (DNA) have been observed in the nuclear region of the cell. (For the detailed account please see the relevant chapter.

(iii) Nuclear enzymes

A number of enzymes have been found in the isolated germinal vesicles of frog oocytes by *Brachet* and *Dupsiva*. Their actual concentration is difficult to ascertain because of high permeability of the nuclear membrane and enzymes are also liable to leak out of the isolated germinal vesicles during their isolation. According to recent studies the nuclear enzymes fall into two classes. Some have a general distribution where as others are found preferentially in certain tissues. In the group, only some enzymes related to the nucleoside metabolism, such as adenosine diaminase, nucleoside phosphorylase and guanase are found in high concentration. Other enzymes as esterases are present in varying concentration. Still others, such as alkaline phosphatase, nucleotide phosphatases and β-glucoronidase are either present in low concentration or lacking entirely. Of the special enzymes, catalase and arginase appear to be concentrated in some nuclei but are lacking in others.

(iv) Nuclear lipids

The lipid content of the nucleus has been investigated in isolated nuclei. Recently it has been reported that the lipo-protein complex in nuclei from ox spleen and chicken erythrocytes is approximately 10% lipid, these give positive tests for phospholipids and cholesterol.

3. Nucleolus

The nucleus contains a large, spherical and acidophilic dense granule known as the *nucleolus*. The nucleoli were discovered by *Fontona* in 1781. It was well established in 19th century that the size of the nucleolus is related with the synthetic activity of the cell. Therefore, the cell with little or no synthetic activities, e.g., sperm cells, blastomeres, muscle cells, etc., are found to contain smaller or no nucleoli, while the oocytes, neurons and secretory cells which synthesize the proteins or other substances contain comparatively large sized nucleoli. The number of the nucleoli in the nucleus depends on the species and the number of the chromosomes. The number of the nucleoli in the cells may be one, two or four. The position of the nucleolus in the nucleus is eccentric.

Fine Structure of Nucleolus

Regarding the fine structure of the nucleoli, it has been reported by *Estable* and *Sotelo* (1955) that they are formed of two parts—*a filamentous nucleolonema* and *a pars amorpha*. The nucleolonema is said to divide and to be equally distributed into the daughter nuclei during mitosis and is regarded as a permanent structure which persists throughout mitosis in association with the chromosome. It is believed that the filamentous part of the nucleoli contains DNA while the pars amorpha is partially made up of RNA. Pars amorpha undergoes the characteristic cycle of formation at telophase and disappearance at prophase.

The introduction of new embedding media and techniques allowed a better analysis of the nucleolar organization. Thus four principle components can be recognized:

1. *A fibrillar portion* approximately 50Å in diameter and upto 300-400Å long (*Marrinozzi*, 1963). In some cases these fibrils can be seen to have a double stranded structure (*Terzakis* 1965). These structures were also described as tubular elements of 15Å diameter.

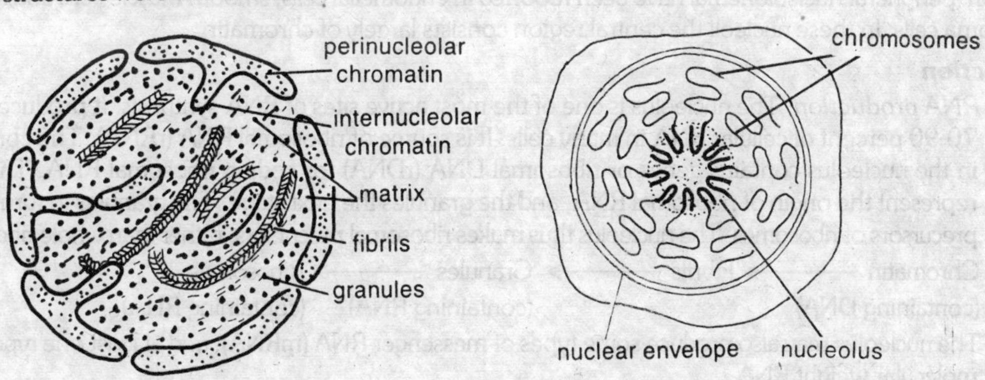

Fig. 20.5. A—Structure of nucleolus; B—Nucleolus containing loops of 10 chromosomes.

2. *A granular portion* composed of dense granules with a diameter averaging 150-200Å, more or less numerous along and between the strands of the fibrillar network (*Marrinozzi*, 1963).
3. An *amorphous region* of lwo electrondensity, found in some nucleoi (*Terzakis*, 1965) made of protein.
4. The *nucleolar associated chromatin* situated around the nucleolus and frequenlty having intra-nucleolar components seem to be constant in vertebrate cells, but their respective amounts or quantities may vary.

Chemistry

Cytochemical studies indicate that 5 to 10 percent of the nucleolus is RNA; the rest being the protein. The main protein components are phosphoproteins. No histones have been found in isolated nucleoli. There are evidences suggesting the presence of orthophosphate, which may serve as a precursor of the RNA phosphorus. Little is known about the enzyme contents of the nucleolus. However, the presence of acid phosphatase, nucleoside phosphorylase and NAD^+ synthesizing enzymes have been shown. RNA methylase has also been localized in the nucleolus of certain cells. The DNA is absent. Nucleolus may be surrounded by a ring of Feulgen-positive chromatin which actually represents the heterochromatic regions of the chromosomes.

Biogenesis of nucleolus or nucleolar cycle

The nucleolus as an organized body is lacking in continuity. It disappears at the beginning of cell division (prophase) and reappears at the end of cell division in telophase stage. A nucleolus is formed at a definite region by one or more chromosomes of a haploid set of chromosomes. Such chromosomes are known as nucleolar chromosomes. Majority of diploid species possess two nucleolar chromosomes in each diploid or somatic cell. But in man chromosomes numbering 13, 14, 15, 21 and 22 participate in the formation of nucleolus. The specific region of these chromosomes active in nucleolar formation is known as *nucleolar organizer* or *nucleolar zone.* Very often but not always, it is marked by secondary constriction. The nucleolar organizer carries genes for 18S and 28S ribosomal RNA.

The secondary constriction with nucleolar organizer is morphologically different from other secondary constriction. In pachytene stage of *Zea mays,* a darkly stained nucleolar organizing body is associated with the nucleolar zone.

Types of nucleoli

On the basis of distribution of ribonucleoprotein granules and fibrils three types of nucleoli can be distinguished (*Wilson*).

(1) Nucleoli with nucleolonemas, which are found in most cells called plasmosomes. (2) Compact nucleoli without nucleolonemas, which have been described in the salivary gland of sciarids and in the protozoan *Tetrahymena pyriformis.* In these nucleoli the ribonucleoprotein granules and fibrils are uniformly distributed. (3) Ring-shaped nucleoli with ribouncleoprotein granules and fibrils present only in the peripheral region (peripheral nucleolonema) have been reported in endothelial cells, smooth muscle cells and lymphosarcoma cells. In these nucleoli the central region consists largely of chromatin.

Function

(i) *RNA production.* The nucleolus is one of the most active sites of RNA synthesis. It produces about 70-90 percent of cellular RNA in many cells. It is source of ribosomal RNA (rRNA). The chromatin in the nucleolus contains genes or ribosomal DNA (rDNA) for coding ribosomal RNA. The fibrils represent the origin of ribosomal RNA, and the granules the next stage. The granules in turn are the precursors of ribosomes. The nucleolus thus makes ribosomal precursors, rather than whole ribosomes.

Chromatin ⟶ Fibrils ⟶ Granules ⟶ Ribosomes
(containing DNA) (containing RNA) (containing RNA).

The nucleolus may also produce some types of messenger RNA (mRNA), and at least one type of low molecular weight RNA.

(ii) *Protein synthesis. Maggio* (1960) and others have suggested that protein synthesis takes place in the nucleolus. If this is true the ribosomal proteins are produced in the nucleolus. Other studies, however, suggest that ribosomal proteins are synthesized in the cytoplasm.

(iii) *Ribosome Formation*. In eukaryotes the gene coding for RNA contains a chain of at least 100 to 1,000 repeating copies of DNA. This DNA is given off in the form of loops from the chromosomal fibre. The DNA loops are associated with proteins to form nucleoli.

The DNA serves as a template for 45S rRNA. Half the 45S rRNA is broken down to form 28S and 18S rRNA. The other half is broken down further to nucleotide level. Within the nucleolus the 28S rRNA combines with proteins made in cytoplasm to form the 60S ribosomal subunit. The 18S rRNA also associates with proteins to form the 40S subunit of the ribosome.

4. Chromatin Network

The chromatin appears as thread-like, coiled and elongated structure. These are stained with basic dyes like Feulgen's stain, acetocarmine etc. that is why these are termed as *chromatin fibres* or *chromatin substance* (Gr. *Chrome*=colour, *soma*=body). These are visible during interphase stage. During cell-division they become thick ribbon-like structures known as *chromosomes*. A detailed account on the chromosomes is given in chapter No 20.

Significance of Nucleus

Hammerling's experiment. J. Hammerling, a German biologist, experimentally proved the significance of nucleus. If a fragment containing the nucelus is cut from an *Acetabularia* of one species, characterized by a given morphology, the fragment will regenerate a whole cell that species. This regenerative ability permits experiments of the type illustrated in Figure in which nuclei of one species are combined with cytoplasma from different species. *Hammerling* conducted certain experiments using two species of a green alga, *Acetabularia*. The two species, namely *A. crenulata* and *A. mediterranea* used in this experiment differ in the shape of their caps. While in *A.crenulata* the cap has loose rays, in *A. mediterranea* an umbrella-like cap is found. The nucleus in both the species is situated in rhizoid at the bottom of stalk. If cap is cut off, it will develop again and its shape will be that of the original type. However, it after removing the caps, stalk of one species is grafted on rhizoid (containing the nucleus) of the other species, shape of cap will be determined by nucleus and not by stalk. If the nucleus belongs to *A. crenulata*, shape of cap will be of the *crenulata* type. If the nucleus comes from *A. mediterranea*, cap will be of *mediterranea* type. When both nuclei are present, shape of cap will be intermediate. The conclusion drawn from such experiments is that the nucleus produces material that enters the cytoplasm and participates in the control of cell growth and cell morphology. The crucial finding is tht the morphology of the regenerated cells ventrally becomes like that of the species from which the nucleus is taken. In the hybrid fragments with the nucleus from one species and most of the cytoplasm from the other species, old cytoplasmic material persists for a while and may influence cell form. Eventually, however, this is depleted and replaced by newly produced material from the nucleus.

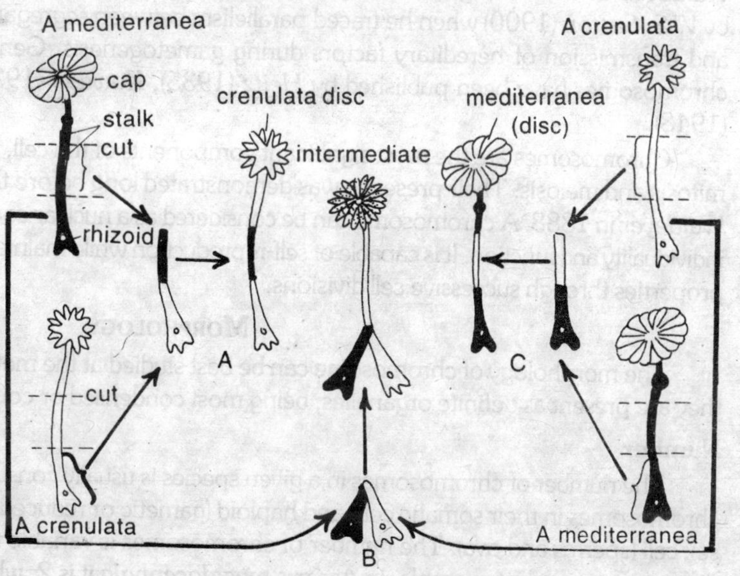

Fig. 20.6. Hamerling's experiment to show the role of nucleus in determination of characters.

21

CHROMOSOMES

Chromosomes were first seen by *Hofmeister* (1848) in the pollen mother cells of *Tradescantia* in the form of darkly stained bodies. The term chromosome (Gr. *chrom*=colour; *some*=body) was used by *Waldeyer* (1988) to designate their great affinity to basic dyes. Their functional significance was described by *W.S. Sutton* (1900) when he traced parallelism between segregation of chromosomes during meiosis and transmission of hereditary factors during gametogenesis. General reviews on the morphology of chromosomes have been published by *Heitz* (1935), *Kuwada* (1939), *Geitler* (1940) and *Kaufmann* (1948).

Chromosomes are the most significant components of the cell, particularly they are apparent during mitosis and meiosis. Their presence was demonstrated long before they were named "chromosomes" by *Waldeyer* in 1888. A chromosome can be considered as a nuclear component having special organisation, individuality and function. It is capable of self-reproduction while maintaining its morphologic and physiologic properties through successive cell divisions.

MORPHOLOGY

The morphology of chromosome can be best studied at the metaphase or anaphase of mitosis when they are present as definite organelles, being most condensed or coiled.

Number

The number of chromosomes in a given species is usually constant containing diploid number (2n) of chromosomes in their somatic cells and haploid (gametic or reduced) number (n) of chromosomes in their sex cell (sperms and ova). The number of chromosomes is variable from one to several hundreds among different species. For example, in *Ascaris megalocephala* it is 2, while in certain protozoans (*Aggregta*), there are more than 300 chromosomes; in *Paramecium* 30 to 40, in radiolarians as many as 1600, in *Hydra vulgaris* 32, *Musca domestica* 12, *Rana esculenta* 26, *Columba livia* 80, *Oryctolagus cuniculus* 44, *Gorilla gorilla* 48 and *Homo sapiens* (man) 46.

The chromosome number are also helpful for taxonomy. In the angiosperms the most frequent haploid number is 12 and members of this group have a range from 3 to 16. Similarly, in fungi, halploid number ranges from 3 to 8. In primates this haploid number is from 16 to 30. This haploid set of chromosomes present in the nucleus of gametes in generally called *genome* while in a diploid cell there will be two genomes. The diploid cells are the somatic cells in body. The diploid cells get the diploid set of chromosoems by the union of the haploid male and female gametes in the sexual reproduction.

Size

Chromosomes range, on an average from 0.5 to about 30μ in length and from 0.2 to 3μ in diameter. The relative number of chromosomes generally differ in the nucleus but at a time all chromosomes of a cell may be of the same size. Plant cell normally possess larger chromosomes than animal cells. *Trillium* has chromosomes which may reach upto the length of 32μ at metaphase. Monocotyledon plants usually have larger chromosomes, than the dicotyledon which contain greater number of chromosomes. Among the animals, grasshopper, crickets, mantids, newts and salamanders have large chromosomes.

Table 21.1. Showing the Diploid (2n) Number of Chromosomes in Some Important animals.

Common name	*Zoological name*	*Chromosomes*
Man	*Homo sapiens*	48 or 46 (?)
Rhesus monkey	*Maccaca mulatta*	48
Horse	*Equus caballus*	66 (64 (?)
Wild horse	*Equus przewalshii*	66
Pig	*Sus scrofa*	40
Ox	*Boa taurus*	60
Sheep	*Ovis aries*	54
Cat	*Felis maniculata*	38
Dog	*Canis familiaris*	78
Goat	*Capra hircus*	60
Ass	*Equus asinus*	66
Rat	*Rattus rattus*	42
Domestic mouse	*Mus musculatus*	40
Guinea pig	*Cavia cobaya*	64
Rabbit	*Oryctolagus cuniculus*	44
Opossum	*Didelyphys virginiana*	22
Frog (Indian)	*Rana tigrina*	24
Frog (British)	*Rana esculanta*	26
Hyla (Tree frog)	*Hyla sp.*	24
Toad	*Bufo vulgaris*	22
Platyfish	*Platypoecilus maculatus*	48
Gold fish	*Carassius auratus*	94
Pigeon	*Columba livia*	79, 80
Turkey	*Meleagris gallopavo*	81, 82
Chicken	*Gallus domesticus*	77, 78
Duck	*Anas indicus*	80
Aulocantha (Protozoan)	*Aulocantha*	1600
Hydra	*Hydra vulgaris*	32
Ascaris	*Ascaris megalocephala*	2
House fly	*Musca domestica*	12
Honey bee	*Apis mellifica*	32, 16
Cricket	*Gryllus*	23, 24
Mosquito	*Culex sp.*	6
Lac insect	*Bombyx mori*	32
Hermit crab	*Eupagurus*	254

Variation in size of the chromosomes can be induced by a number of environmental agents:

1. Cells dividing at low temperature have shorter, more compact chromosomes than those dividing at high temperature.

2. Colchicine is an alkaloid which interferes with spindle formation and cell division. In tends to shorten the chromosomes.
3. Rapid and repeated division tend to result in smaller chromosomes. It appears that rate of cell division proceeds more rapidly than the formation of chromatin material as usual.
4. In plants, the amount of phosphate in the nutritory medium has marked effect on the size of the chromosomes; high concentration gives larger chromosomes than those plants deficient of phosphates. Since the phosphate is an integral part of nucleic acid molecule, it would seem that the amount of nucleic acid in the chromosome can be varied to give alterations in size.

Table 21.2. Showing the Diploid Number of Chromosomes in Some Plants

Common name	*Botanical name*	*Chromosomes*
Yellow pine	*Pinus ponderosa*	24
Cabbage	*Brassica oleracea*	18
Radish	*Raphanus sativus*	18
Cucumber	*Cucumin sativus*	14
Cotton	*Gossipium hirsutum*	52
Plum	*Prunus domestica*	48
Cherry	*Prunus carasus*	32
Apple	*Malus sylvestris*	34, 51, 68
Pear	*Pyrus comming*	34, 51, 68
Pea (Garden)	*Pisum sativum*	14
Sweet pea	*Lathyrus odoratus*	14
Bean	*Phascolus vulgaris*	22
Scarlet	*Quercus coccinea*	24
Orange	*Citrus sinensis*	18, 27, 36
Sun flower	*Helanthus annuus*	34
Tobacco	*Nicotiana tobaccum*	48
Potato	*Solanum tuberosum*	18
Banana	*Musca paradisiace*	22, 44, 55, 77, 78
Corn	*Zea mays*	20
Onion	*Allium cepa*	16
Barley	*Hordeum vulgare*	14
Wheat (Summer)	*Triticum dicoccum*	28
Rye	*Secale cereale*	14
Rice	*Oryza sativa*	24

Shape

The shape of the chromosomes is changeable from phase to phase in the continuous process of the cell growth and cell division. In the resting phase or interphase stage of the cell, the chromosomes occur in the form of thin, coiled, elastic and contractile, thread-like stainable structures, the *chromatin threads*. In the metaphase and the anaphase, the chromosomes become thick and filamentous. Each chromosome contains a clear zone, known as *centromere* or *kinetocore*, along their length. The centromere divides the chromosomes into two parts, each part is called *chromosome arm*. The position of centromere vary from chromosome to chromosome and it provides different shapes to the later which are following:

1. *Telocentric*. The rod-like chromosomes which have the centromere on the proximal end are known as the telocentric chromosomes.

2. *Acrocentric.* The acrocentric chromosomes are also J-like in shape but these have the centromere at one end and thus giving a very short arm and an exceptionally long arm. The locusts (Acrididae) have the acrocentric chromosomes.

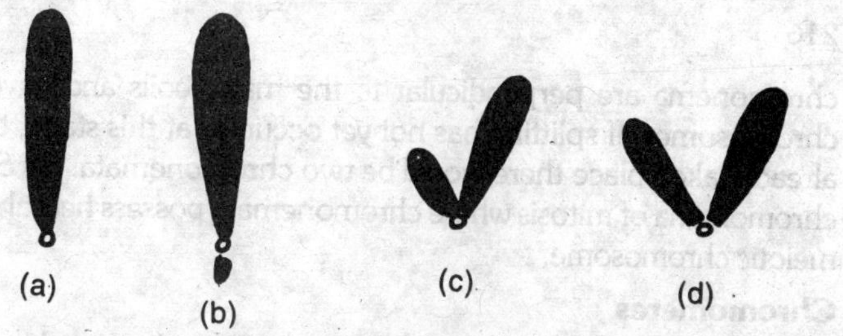

Fig. 21.1. Four morphologic types of chromosomes (a) Telocentric (b) Acrocentric (c) Submetacentric (d) Metacemtric.

3. *Submetacentric.* The submetacentric chromosomes are L-shaped. In these, the centromere occurs near the centre or at medium portion of the chromosome and thus forming two unequal arms.
4. *Metacentric.* The metacentric chromosomes are V-shaped and in these chromosomes the centromere occurs in the centre and forming two equal arms. The amphibians have metacentric chromosomes.

STRUCTURE OF THE CHROMOSOME

In earlier light microscopic descriptions the chromosome, or chromatid, was though to consist of a coiled thread called the chromonema lying in a matrix. The chromosome was supposed to be convered by a membranous cellicle. Electron microscopic studies later showed that there is no definite membranous pellicle surrounding the chromosome. Other structures present in the chromosome include the chromatids, centromere, secondary constrictions, nucleolar organizers, telomeres and satellites as given under the following heads:

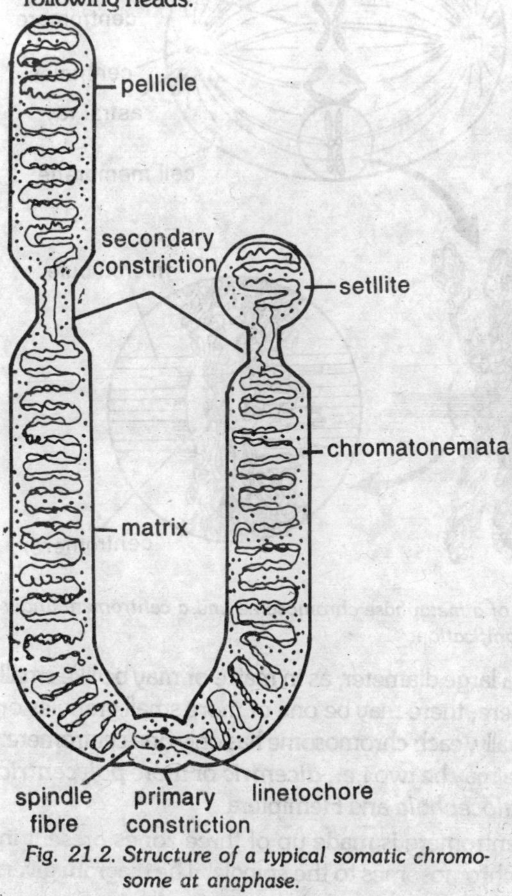

Fig. 21.2. Structure of a typical somatic chromosome at anaphase.

Chromatids

During metaphase a chromosome appears to possess two threads called *chromatids*, which become interwined in the matrix of chromosome. These two chromatids are held together at a point along their length in the region of constriction of the chromosome. These chromatids are really spirally coiled *chromonemata* (sing., chromonema) at metaphase. The coiled filament was first of all observed by *Baranetzky,* in 1880, in the pollen mother cells of *tradescantia* and was called chromonema by *Vejdovsky* in 1912. The chromonema may be composed of 2, 4 or more fibrils depending upon the species. This number of fibrils in the chromonema may depend on the different phases since at one phase it may contain one firbil and at other phase it may contain two or four fibrils. These fibrils of the chromonema are coiled with each other. The coils are of two types:

1. *Paranemic coils.* When the chromonemal fibrils are easily separable from each other, such coils are called paranemic coils.
2. *Plectonemic coils.* Here the chromonemal fibrils are closely interwined and they cannot be separated easily. Such coils are called plectonemic coils.

The degree of coiling of the chromonemal fibrils during cell divisions depend on the length of the chromosome. There are three types of coils: (i) *Major coils* of the chromonema posses 10-30 gyres. (ii) *Minor coils* of the chromonema are perpendicular to the major coils and have numerous gyres as observed in meiotic chromosomes. If splitting has not yet occurred at this stage, there will be a single chromonema, if it has already taken place there would be two chromonemata. (iii) *Standard* or *somatic coils* are found in the chromonema of mitosis where chromonemata possess helical structures, resembling the major coils of the meiotic chromosome.

Chromomeres

The chromonema of thin chromosomes of mitotic and meiotic prophases have been found to contain alternating thick and thin regions and thus giving the appearance of a necklace in which several beads occur on a string. The thick or bead-like structures of the chromonema are known as the *chromomeres* and the thin region in between the chromomeres is termed as the *inter-chromomeres*. The position of the chromomeres in the chromonema is found to be constant for a given chromosome.

The cytologists have given various interpretations about the chromomeres. Some consider chromomeres as condensed nucleo-protein material, while other postulated that the chromomeres are regions of the super-imposed coils. The later view has been confirmed by the electron microscopic observations. For long time most geneticists considered these chromomeres as *genes*, i.e., the units of heredity.

Centromere

It is indispensable part of the chromosome and forms the *primary constriction* of metaphase. Without centromeres chromosomes are unable to orient themselves properly on the metaphase plate; come, later on, at anaphase and ultimately, are eleminated. As centromeres occupy a constant position, therefore, the centromeres are responsible for the shape of the chromosomes. Thus, the shape of the chromosomes is determined by the primary constriction, situated at the meeting point of the arms of the chromosomes. Within the primary constriction, there is a clear zone, having a small granule or spherule. This clear region is known as centromere (Gr. *meros,* part) or *kinetochore* or *kinetomere.* Its function is in terms of movement. They are responsible for the formation of chromosomal fibers in the spindle. The structure of the centromere is ovoid, non-stainable, with a large diameter, as in maize or may be like small granules or spherules, as in *Tadescantia.* In the centromere, there may be one or more small granules or spherules, called *chromomeres* and the spindle fibres. Usually, each chromosome has only one centromere. In such cases, chromosome is called *monocentric.* There may be two i.e., *dicentric* or more *polycentric* or with a *diffused centromere,* as found in *Ascaris megalocephala* and Hemiptera.

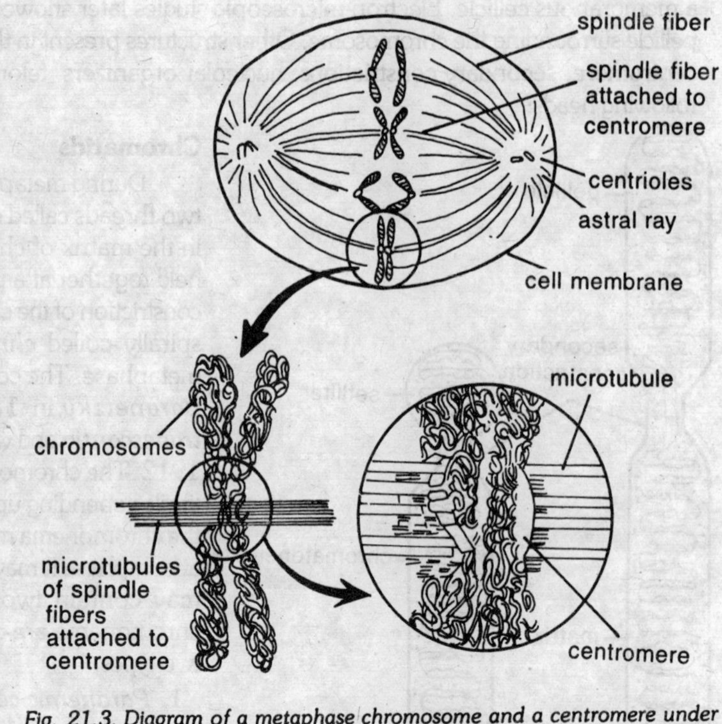

Fig. 21.3. Diagram of a metaphase chromosome and a centromere under high magnification.

After recent studies, it has been known that the centromere is made up of three zones present in duplicate. The middle zone matintain the relation of the chromosomes to the spindle. The diagram given below shows two sister-chromatids forming each metaphase-chromosome held by a region having a special division cycle.

The centromere is regarded to become functionally divided to the longitudinal axis of chromosome at the beginning of the anaphase. Its movements towards poles are governed by its attachment to the spindle. Sometime, divisions occur also at right angles to the longitudinal axis forming two segment-centromeres, to which are attached the two chromatids of each arm. This structure, made up of two arms, is known as *isochromosome;* the name was suggested by *Darlington* in 1939. *Mc Clintock* (1932) has reported that such breakage is also possible by X-rays. In such case, each fragment-part of the centromere is functional. It is also known that the centromere is a compound structure, whose parts are co-ordinated in division and movement. *Sachrader* (1936) and *Darlington* (1939) have suggested that the centromere can be considered to be homologous with the centrioles on structural as well as on theoretical basis.

Secondary Constriction

In addition to the primary constriction or centromere the arms of the chromosome may show one or more secondary constrictions (called *secondary constriction II*). These are different from nucleolar organizers (called secondary constriction I), although some cytologists also refer to the nucleolar organiser as the secondary constriction. The location of the secondary constriction II is constant for a particular chromosome, and is, therefore, useful for identification of chromosomes. It has been suggested that the secondary constrictions represent sites of breakage and subsequent fusion. In man secondary constrictions II are found on the long arms of chromosomes 1, 10, 13, 16 and Y.

Nucleolar Organiser (Secondary Constriction I)

Normally in each diploid set of chromosomes, two homologous chromosomes have additional 'constrictions' called nucleolar organisers. These are so-called because they are necessary for the formation of the nucleolus. The nucleolus is formed in the post-mitotic reconstruction phase. Under the light microscope the nucleolar organiser appears as a 'constriction' near one end of chromosome. The part of the chromosome beyond the nucleolar organiser is very short and appears like a sphere (*satellite*). In man chromosomes 13, 14, 15, 21, 22 and Y have nucleolar organisers and satellites. Chromosomes bearing satellites are called *SAT-chromosomes.* The prefix Sat-stands for 'Sine Acid thymonucleionico' (without thymonucleic acid or DNA), since the chromosome on staining shows relative deficiency of DNA in the nucleolar organiser region. There are at least two SAT-chromosomes in each diploid nucleus.

Telomeres

The ends of a chromosome act differently from the interstitial portions. If an end or telomere is broken off either spontaneously or by induction, it is usually lost from the nucleus in subsequent cell division because it lacks centromere. The broken end of the remaining free chromosome is unstable and may unite with another broken end of the chromosome in the vicinity. However, the broken end will not unite with normal end. In meiotic prophase the telomeres are sometimes attracted to the centriole and seen to migrate to the nuclear membrane near the centriole. This behaviour results in what has been described as *Bouquet stage.*

Marix of Chromosome

As supposed by some cytologists that chromonemata are embedded in the chromatic matrix which is bounded by a pellicle. However, recent observation from electron microscopic studies revealed the absence of pellicle. Matrix is not defined as the main mass of the chromosome, which is characteristically Feulgen-positive. It can be removed by enzymatic means leaving behind a Feulgen-negative residual chromosome.

Heteropycnosis

It has generally been observed during various stages of mitosis that certain chromosomes or parts of the chromosomes are not as such, but are more condensed that rest of the karyotype. This refers to the *heteropycnosis.* This phenomenon results in clumping of the chromosomes during cell division. In animals pycnotic sets, on elimination, culminate in a male whichwas cytologically haploid. Heteropycnosis may be positive, followed by over condensation or negative, showing under or no condensation. It has been also observed that a particular part of chromosome or entire chromosome may not exhibit the condensation or heteropycnosis in all the phases. Since heteropycnosis is peculiarity of the heterochromatin, it helps in demarcating it from the euchromatin. It is much more prevalent in sex chromosome, though others too exhibit it.

Euchromatin and Heterochromatin

Although, during interphase, the chromatin of the chromosomes spreads out in the form of the fine threads of linin but at certain regions, the chromatin remains condensed in the form of darkly stained chromatin mass. These regions are known as the heterochromatin regions or heterochromatin.

Heterochromatin is of two types:

1. Facultative heterochromatin, and
2. Constitutive heterochromatin.

1. Facultative heterochromatin

This represents a temporary state of inactivation of chromatin in which one chromosome of the pair becomes partially or totally heterochromatic. For example, in mammals one of the two X-chromosomes in female somatic cells becomes heterochromatic and forms the sex-chromatin of *Barr body* (*Bar* and *Bertram,* 1944). In male somatic cell, there is only one X chromosome and remains euchromatic (no barr body).

2. Constitutive heterochromatin

This type of heterochromatin present a more permanent feature and is found in both the chromosomes of a pair. It is very often found in the centromeric regions, telomeres, in the regions of nucleolar organizers or as bands in other regions of the chromosomes. It is closely associated with nucleoli in both plants and animals.

Fig. 21.4. Model of constitutive heterochromatin in a mammalian metaphase chromosome. A—Constitutive heterochromatin; B—Secondary constriction; I or nucleolar organizer; C—Primary constriction or centromere; D—Euchromatin; E—Secondary constriction II—possible site of 5S rRNA cistrons; F—Telomere.

Chromosome Banding

T.C. Hsu and other (1969) introduced new methods for staining chromosomes by which distinct patterns of stained bands and lightly stained interbands become evident. These staining methods were very important since they permitted each chromosome to be identified uniquely, even if the over all morphology was identical. Distinctions can now be made among the relatively similar A-group chromosomes. For example, we may now say 1 or 2 or 3 chromosome instead of mearly a A-group chromosome of *Denver system.*

Following are the methods for chromosome banding.

1. G-Banding

The most useful chromosome banding mathod is G-banding. This technique was developed by *Hsu* and *Arrighi*. It is observed that when the chromosomes are incubated in salive are stained with *Giemsa stain* or treated with urea or detergents. G-bands appear in the areas which are S-rich proteins. Giemsa stained preparations are more permanent and require ordinary microscope optics and illumination.

2. Q-Banding

This technique was developed by *Casperson*. It is observed, when the chromosomes are stained with *quinacrine mustrard* and observed through fluorescence-microscope, the regions of chromosomes rich in adenine and thymine get stained intensely. The guanine-cytosine regions remain unstained. These regions are called Q-bands. The defect of this staining is that, the stains fade after a short-time, moreover, special microscopic optics plus ultraviolet illumination are needed to see these bands.

3. C-Banding

This technique was developed by *Pardue* and *Gall*. The chromosomes are treated with strong sodium hydroxide followed by warm saline and than stained with Giemsa stain. C-bands are especially evident around the centromere and in other chromosomes that contain substential amounts of highly repetitive constitutive hetero-chromatin.

4. R-Banding

These bands appear when chromosomes are incubated in a buffer at high temperature and stained with Giemsa strain. R-bands correspond to the regions on chromosomes having proteins lacking sulphur. These are receprocal of G-bands.

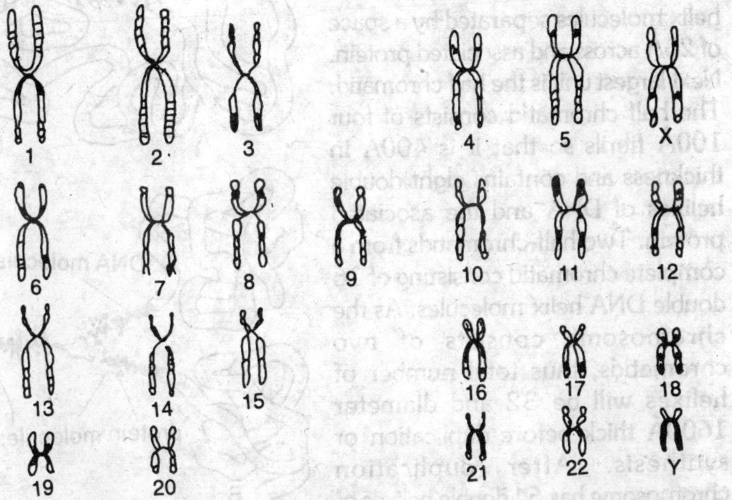

Fig. 21.5. Diagram showing the banding pattern of human chromosomes revealed with Giemsa stain. Since the banding is specific for each chromosome pair, this method, as well as, that involving fluorescent stains, is of great help in preparing a more refined analysis of normal and abnormal chromosomes.

Bending technique of chromosme staining is highly useful in working various types of chromosomal aberrations such as deletion, duplication, inversion or translocation. The greater certainty of identifying whole chromosomes or parts of chromosomes by G-bands often allows the investigator to know exactly which chromosomes are present and which chromosome parts have undergone structural rearrangement.

Banding also provides a means to compare karyotypes of related species and to describe differences that apparently have an evolutionary basis.

Ultrastructure of Chromosome

Two views have been proposed for ultra-structure of chromosomes:

(a) Multistranded View

This was proposed by *Ris* (1906). By electron microscope the smallest visible unit of the chromosome

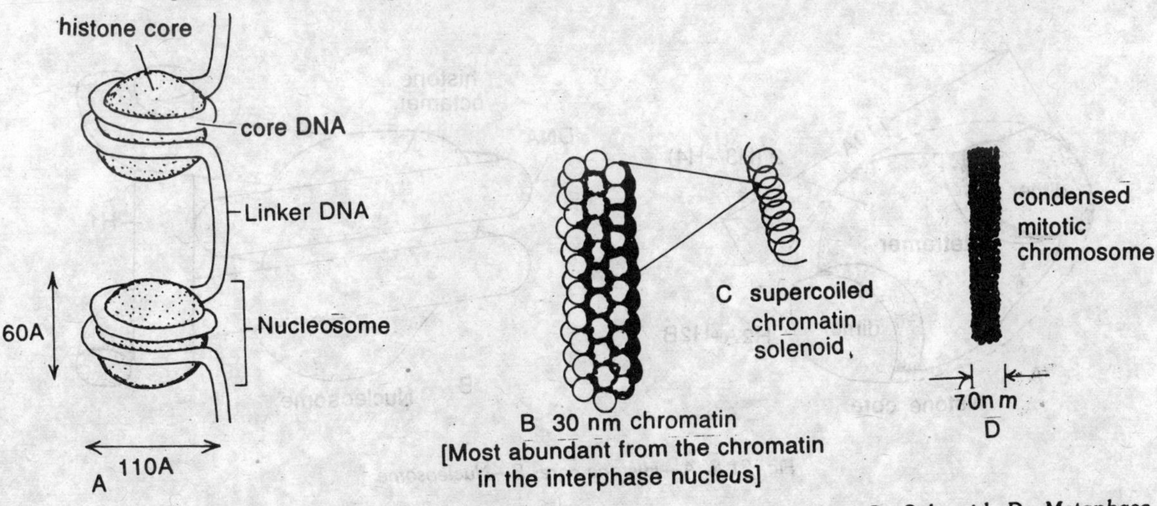

Fig. 21.6. A—Nucleosome model of chromosome (10-nm fiber); B—30-nm fiber; C—Solenoid; D—Metaphase chromosome.

is the fibril which is 100Å in thickness. This fibril contains two DNA double helix molecules separated by a space of 25Å across and associated protein. Next largest unit is the half chromatid. The half chromatid consists of four 100Å fibrils so that it is 400Å in thickness and contains eight double helixes of DNA and the asociated protein. Two half-chromatids from a complete chromatid consisting of 16 double DNA helix molecules. As the chromosome consists of two chromatids, thus total number of helixes will be 32 and diameter 1600Å thick before duplication or synthesis. After duplication chromosome has 64 double helixes of DNA with corresponding diameter of 3200Å. The number of DNA helixes in each unit above fibril level varies according to species. In summary, the chromosome is composed of numerous microfibrils, the smallest of which is a single nucleoprotein molecule.

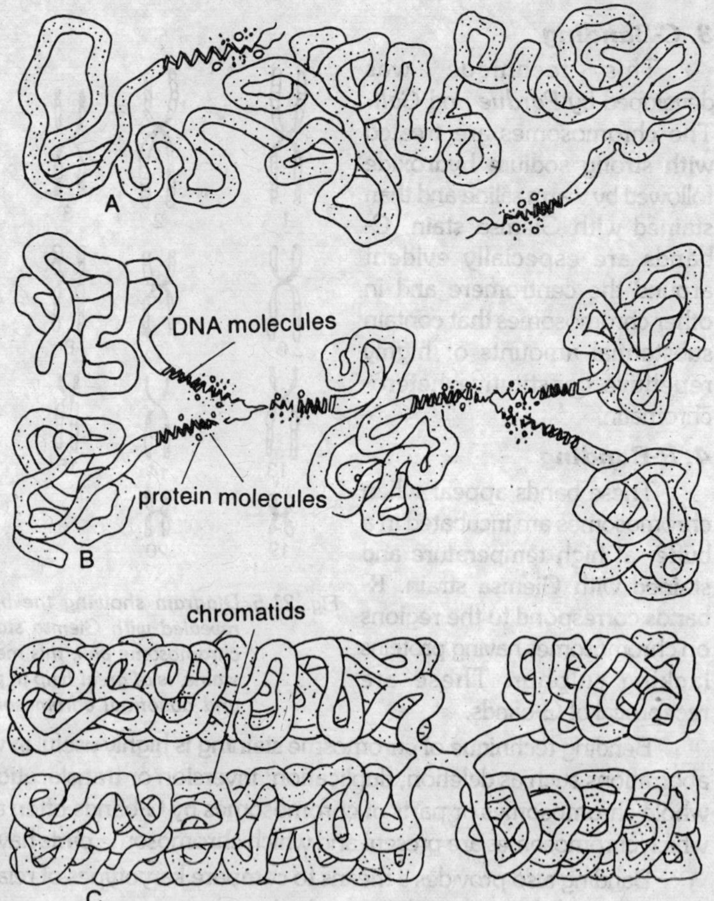

Fig. 21.7. Dupraw's folded fibre model of chromatin in interphase (A and B) and in metaphase.

(b) Folded-Fibril Model

DuPraw (1965) presented this model for the fine structure of chromosome. According to this model, a chromosome consists of single long chain of DNA and protein forming what is called as fibril. The fibril is folded many times and irregularly entwined to form the chromatid. This measure 250-300Å in thickness.

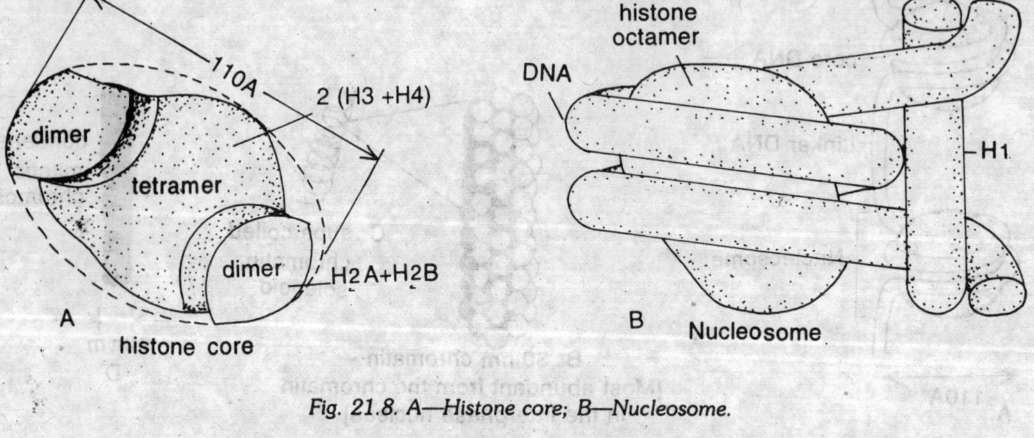

Fig. 21.8. A—Histone core; B—Nucleosome.

The Nucleosome sub-unit of chromatin

The chromatin is formed of repeating units, called *nucleosomes*. The term was given by *Oudet et.al.*, (1975). The nucleosome is made up of DNA and histone proteins. The proteins forms a *core particle* which is an octomer two molecules of each of the four histone proteins viz. H2a, H2b, H3 and H4. The surface of core particle is surounded by 1.75 turns of DNA (200 base pairs) The DNA linking the core particle is called linker DNA. An another histone protein, H1, is bounded to the *linker DNA*. (*Kornberg* and *Thomas*, 1974). The core particle measures 40Å height and 80Å in diameter. The entire nucleosome measures 55Å in height and 110Å in diameter.

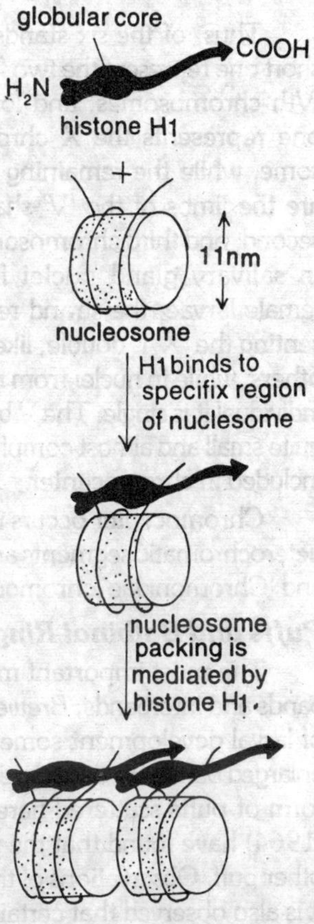

Fig. 21.9. Nucleosome packing to form solenoid.

Polytene Chromosomes

Balbiani in 1881 was the first to observe salivary gland chromosomes in the salivary glands of the *Chironomous larva*. This kind of giant chromosomes are strictly confined to certain type of somatic tissues in the insects belonging to the order Diptera. Usually they attain their largest size in the spherical nuclei of the larval salivary gland, but similar nuclei frequently exist in other tissues such as the lining cells of the gut and its derivatives, the malpighian tubules as well as in muscles and fat cells, etc. Thus the term "polytene" chromosome of *Koller* is more preferable than the general term of salivary gland chromosomes.

Structure

The structure of the salivary gland chromosome is of great cytogenetic interest. Along the entire length of the chromosome there is a series of *dark* bands alternating into other clear zones called *interbands*. The dark bands stain intensly and are Feulgen positive. Further more, they absorb ultraviolet light at 600Å. These bands may be considered as disks, since they occupy the whole diameter of the chromosome. They are of varying size. The longer bands have more complicated structure. They often form doublets, two bands located next to each other and of identical thickness and shape. the interbands are of fibrillar aspect, do not stain with basic dyes, are Feulgen negative and absorb very little ultraviolet light. Furthermore, they present a greater elasticity than the regions of the bands. The constancy in situation and distribution of the disks or bands in two homologous (paired) chromosomes is notable.

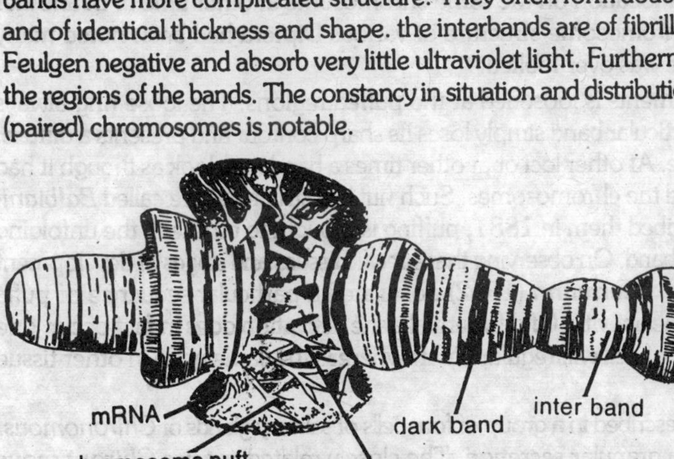

Fig. 21.10. Polytene chromosome of an insect.

In the case of *Drosophila melanogaster*, the chromosomes of each polytene nucleus, when flattened, appear as fine long strands and one quite short, attached to a central mass known as the *chromocenter*, to which single large *nucleolus* is also attached. The relationship between these strands are the light chromosomes of the ordinary mitotic set of this species is not at first obvious. The explanation depends on two facts: (1) The two members of each pair of chromosomes are closely fused throughout their length; (2) The centromeres of all the chromosomes together with the heterochromatic segments adjacent to them are all joined together to form the chromosomes.

Thus, of the six stands the short one represent the two fused IVth chromosomes, and longer one represents the X chromosome, while the remaining foui are the limbs of the 'V'-shaped second, and third chromosomes. In salivary gland nuclei from female larvae, the strand representing the 'X' is double, like the others, while in nuclei from male individual it is single. The V being quite small and almost completely included in chromocenter.

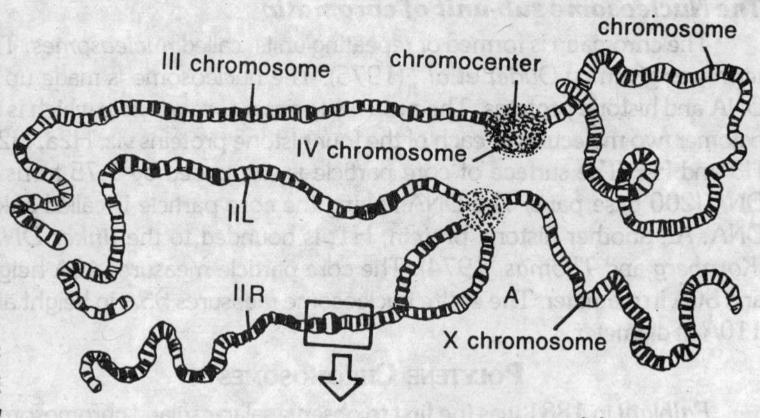

Fig. 21.11. Polytene chromosomes of Drosophila.

Chromocenter occurs in all species of *Drosophila*, and its size depending on whether the proximal heterochromatic segments are extensive or not. In some other groups of Diptera, of the families 'Sciaridae' and 'Chromonidae' chromocenter is absent.

Puffs and Balbinai Ring and Gene activity

The most important morphological peculiarity of polytene gland chromosome is the presence of bands and interbands. *Brewer, Pavan, Beermann, McChelke* and others have found that at certain stages of larval development some specific bands of the polytene chromosome show an enlargement. These enlarged bands are considered as ultimate units of heredity—the *genes* at work. These active genes take the form of puffs scattered here and there along the salivary gland chromosomes. *Beermann* and *Clever* (1964) have found that the puffs produce RNA and the RNA made in one puff differ from the RNA of other puff. Observations of the puffs have shown the patterns of gene activity in several developing insects. It is also observed that certain hormones and other substances can start, stop and prevent some of these activities.

The fine structure of individual band can differ with respect to puffs that are in one location on a chromosome in one tissue and in another location on the same chromosome at other time or in another tissue. These localized modification in chromosome structure of various Diptera had been noted many years earlier but their possible significance was over-looked.

The coherence of chromosome filaments is loosened at the puffed regions. The loose ring always starts at a single band. In small puffs, a particular band simply loses its sharp contour and presents a diffuse, out of focus appearance in the microscope. At other loci or at other times a band may look as though it had "exposed" into a large ring or loop around the chromosomes. Such nut like structures are called *Balbiani-rings,* after *E.G. Balbiani*, who first described them in 1881; puffing is thought to be due to the unfolding or uncoiling of individual chromosome in a band. On observing that specific tissues and stages of development are characterized by definite puff pattern, *Beermann* (1952) postulated a particular sequence of puffs represents a corresponding pattern of gene activity. Of differential gene activation does in fact occur, one could predict that genes in a specific type of cell will regularly puff whereas the same gene in other tissue will not puff.

A gene of the same nature has been described in a group of four cells of salivary glands of *Chironomous. Chironomous pallidivittatus* produces a granular secretion. The closely related species *Chironomous tentatus* gives off a clear, non-granular secretion from the same cells. In hybrids of these two species this nature follows simple Mendelian laws of heredity. *Beermann* and *Clever* (1964) were able to localized the difference in a group of fewer that 10 bands in one of the chromosomes of *Chironomous* and that chromosome is designated as IV chromosome. The granule producing cells of *C. pallidivittatus* have a puff associated with this group of bands, a puff that is entirely absent at the corresponding loci of chromosome IV in *Chironomous tentatus.* In hybrids the puff appear only on the chromosome coming from the *C.*

pallidiuittaus parent; the hybrid produces a far smaller number of granules than the parent. Moreover, the size of the puff is positively co-related with the number of granules. This reveals quite clearly the association between the puff and a cellular (specific) product. This study demonstrates a specific relationship between puff-gene and the specific function of a cell.

Theories Regarding the Structure of Polytene Chromosome

There are three theories for explaining the structure of polytene chromosome. Out of them, the third explanation is actually the combination of first theories:

1. Polytene chromosomes are the result of several cycle of intracellular chromosomal reproduction and consists of bundles of the folded ordinary chromosomes. This is the polytene theory sponsored by *Hertwig* (1935), *Cooper* (1938), *Painter* (1939) and *Beermann* (1952).
2. Polytene chromosomes are paired chromosomes which have enormous length and width by addition or incorporation of extra material not present in ordinary chromosomes. This is the earlier alveolar concept of *Metz* (1935) and theory proposed by *Kodani* (1942) and *Darlington* (1949).
3. Polytene chromosomes consists of bundles of chromonema, their size is due at least in part to the accumulation of extra material in the center of chromosomes, and or to an actual growth in length of the chromonema (*Koltzof*, 1934; *Painter*, 1934; *Calvin et. al.*, 1940; *Ris* and *Course*, 1954; *White*, 1945).

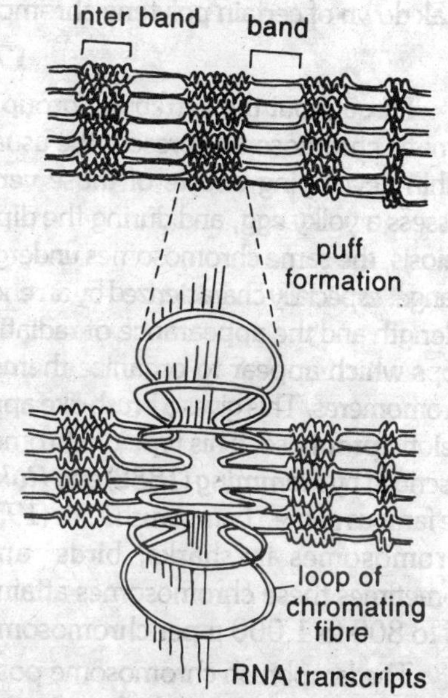

Fig. 21.12. A part of a giant polytene chromosome of Drosophila showing A-Bands and interbands; B-Puff formation.

Polytene theory

Painter (1941) believed that the increase in diameter is due to an enlargement and probably by a continuing duplication of the individual chromomeres but without a variable separation of individual chromonemata. Thus in the course of development each original chromomere becomes resolved by separation through stretching into a number of smaller chromomeres. The enlargement, duplication and agregation of homologous chromomeres produce the appearance of transverse chromatic bands. The chromosome, therefore, become multistranded as polytene but the individual chromonemata, which according to *Painter* may be as many as 1024, while *Beermann* (1952) estimated the degree of polytene to be as high as 16000 times.

Support to Polytene theory from the volume of DNA

One important point of evidence may be added in favour of the polytene theory. This concerns the DNA contents of salivary gland nuclei and is complementary to Hertwig's (1935) demonstration of nuclear size classes. *Kurnick* and *Herskowitz* (1952) have estimated from photometric measurements of individual nuclei stained with methyl green that the large salivary gland nuclei in *Drosophila melanogaster* have undergone about 420 folds increase in DNA content over that the diploid nuclei. Correcting few possible errors in the determination of the lowest values, *Kurnick* and *Herskowitz* assume that 8 cycles of internal duplication have resulted in chromosomes consisting of about 1000 strands. *Swift* and *Rasch* (1954) have obtained much mroe numerous data by photometric Feulgen dye determination as nuclei of *Drosophila* salivary gland cells and closely associated cell type. They found a geometrical series of 10 DNA classes, ranging from presumably diploid cell to 1024 ploid salivary gland cells. The DNA values fall into generally non-over lapping groups, indicating that they resulted from a series of successive synchronized duplication.

Although a good number of workers believe in the polytene theory of salivary gland chromosome but several interesting problems remain unresolved. One of these is concerned with the question to what extent different chromosomes in one nucleus or even different regions within a chromosome, can become polytene to different degrees. The "sex ratio" phenomenon analysed by *Sturtevart* and *Dobzhansky* (1936) represented a well substantiated case of this type. Other class which suggest this possibility have been reported by *Schultz* (1941) and *White* (1948) who discovered that certain chromosomes appear to be out of steps with the rest of the set when judged by their sizes and stainability. It will be eventually determined by means of photometric DNA measurements whether these cases are due to unequal polytene or merely to abnormal states of contraction of certain chromosomes. This later possibility rises during their development. The mechanism of different regions of the same chromosome is still completely obscure. The cause of the break down of certain polytene chromosomes into their constituent fibrils are equally unknown.

Lampbrush Chromosomes

Throughout the vertebrate group of animals the somatic chromosomes present the usual structure, but within developing oocyte of those vertebrates which possess a yolky egg, and during the diplotene stage of meiosis, the same chromosomes undergo a remarkable change, especially characterized by an enormous increase in length and the appearance of radiating hairs or side loops which appear to organize themselves from the chromomeres. This gives a brush-like appearance during meiotic prophase. This type of chromosome was first described by *Flemming* (1882) and *Rukert* (1892) gave the famous name, 'Lampbrush.' *Ris* (1951) found similar chromosomes in sharks, birds, amphibians etc. Sometimes these chromosomes attain maximum size up to 800 to 1,000 μ per chromosome.

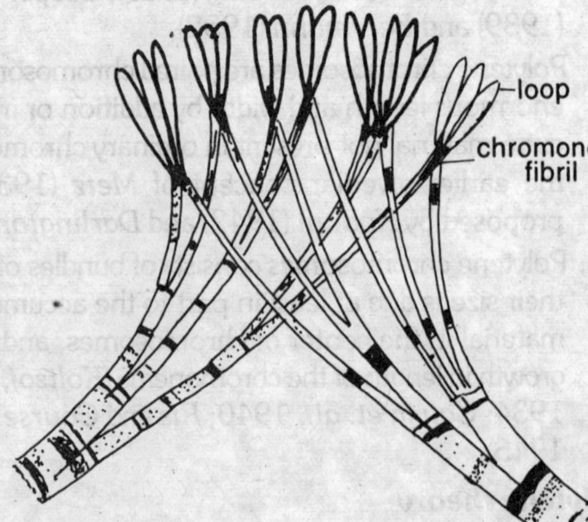

Fig. 21.13. A chromosomal loop of chromosome puff.

The lampbrush chromosome possesses a central chromosomal axis from which project a series of lateral loops. The loops appear to project out from the dense region. According to *Duryee*, each chromosome resembles a single plastic cylinder in which, at specific loci, are embedded chromatin granules. In a bivalent chromosome about 150-200 paired granules are present. These granules are of two sizes, viz., *smaller chromioles* and *larger chromioles*, later ones are ellipsoidal in appearance as though squeezed inside the matrix.

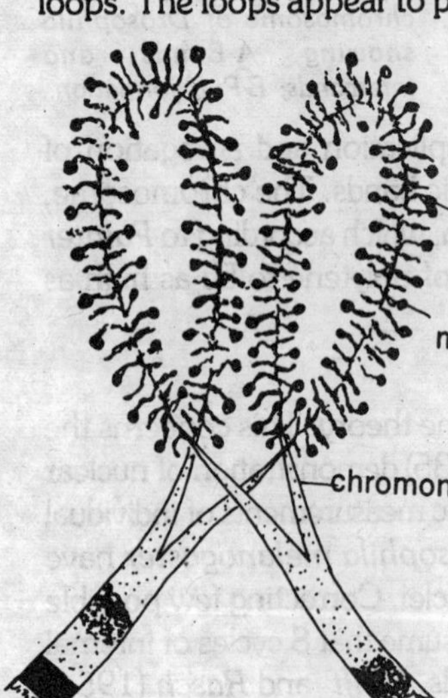

Fig. 21.14. Suggested form of the individual loops in chromosome puffs. Note the tiny fibrils with dense RNA granules attached to the loop.

The lateral loops give rise to brush-like appearance. The loops are considered to be chromatin material synthesized for external utilization, and not an integral portion of the chromonemata extended in the form of a major coil as proposed by *Ris* (1945). *Duryee* (1941) hypothesis of lateral synthesis is supported by the fact that stretching of the chromosome by micro-manipulation or contractions by calcium ions does not cause the loops to disappear or to be displaced and that dissolution of the loops by a variety of substances into granules does not effect the integrity of chromonemata. *Gall* (1956) showed quite conclusively through electron microscopy that loops are parts of the chromonemata and their apparent disappearance is due to the fact that prior to contraction they shed their coating of nucleic acid. The chromosome possesses a remarkable elasticity. The lateral loops extending from chromomeres are more fragile. *Gall* (1958) has interpreted that loops formation is a reversible physiological change which is probably non-genetic. However, he points out that loops exhibit variation in their morphology which also suggests, on the other hand, that perhaps each loop pair, represents a different genetic locus responsible for the formation of some particular cell product.

The lampbrush chromosome contains a continuous central main axis formed..... and protein. This varied from 30-50Å in diameter and is highly flexible. The loops axis is surrounded by the protein combined with RNA. Perhaps the loops are mainly concerned with synthesis of RNA, protein and yolk material.

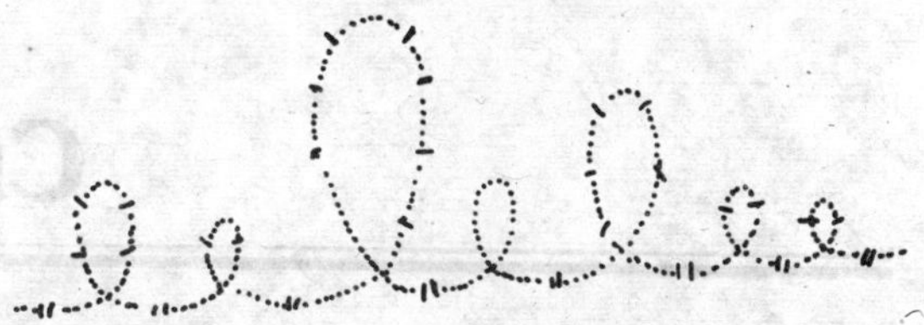

Fig. 21.15. Model of lampbrush chromosome.

To explain the large size of lampbrush loops, *Callan* and *Lloyd* (1960) postulated that each consists not of gene, but of a number of duplicate copies, linearly arranged, of one gene. The chromomeres, basic units of the organization in lampbrush chromosomes exist in two forms. There is a 'master' copy of particular gene in a chromomere which resembles to its identical 'slave' copy of gene. In this way loop contains number of duplicate copies only. The genes are isolated by a double line and single transverse lines indicating the ends of the duplicate copies. *Gall* and *Callan* observed that the lateral loops always have one thin and one much thicker end at the point of insertion in the chromomere. It is also believed that loop spun out from the chromomere reunites to it at the thick end, which shows heavy accumulation or RNA. *Callan* further suggested that only in 'slave' copies take part in RNA synthesis.

This ensures possibilities to synthesize large amount of ribose nucleic acid.

Nucleoli formation in lampbrush chromosome shows an unusual pattern. There may be several hundreds of nucleoil floating free in the nucleoplasm. The significance is not well-understood, but it is suspected that they must be synthesizing material for growth.

Accessory or Supernumerary Chromosomes

Nuclei of some animals and plants possess, in addition to the normal chromosomes, one or more accessory or super-numerary chromosomes. *Wilson* (1905) was the first cytologist to observe them in hemipteran insect, *Metapodius*. Since then, they have been reported in several insects and in many higher plants too.

In some cases, their nature and origin are certainly known. However, their ancestry is yet entirely unknown. The supernumerary chromosomes are usually of sizes smaller than those of there types. It is regarded that they perform some, as yet undermined, function, which is too slight to detect genetically.

22

CELL DIVISION (MITOSIS)

There are two types of organisms-acellular and multicellular. The growth and development of an individual depends exclusively on the growth and multiplication of the cells. It was *Virchow* who first of all adequately stated the cell division. In animal cell the cell division was studied in the form of segmentation division or cleavage by *Prevost* and *Dumas* in 1824. The mechanism of cell division was not precisely investigated until long afterward but *Remak* and *Kolliker* showed that the process involves a division of both the nucleus and the cytoplasm. The term *karyokinesis* was introduced by *Schleicher* (1878) to designate the changes of nucleus during division, and the term *cytokinesis* was introduced by *Whiterman* (1887) to designate the associated changes taking place in the cytoplasm.

Cell division is necessarily the avoidance of aging, and secondly for the segregation of an individual into semi-independent units which leads to efficiency. Thus, we see that cell division is a widespread phenomenon that is essential not only for the maintenance of life but also for the development of the organism itself.

Cell division can be conveniently described as:

(i) *Direct division*. Where the nucleus and cell body undergo a simple mass division into two parts. It is also called *amitosis*.

(ii) *Indirect division*. Here the nucleus undergoes complicated changes before it is divided into two daughter nuclei.

AMITOSIS

The amitosis or direct cell division is the means of asexual reproduction in acellular organisms like bacteria and protozoans and also a method of multiplication or growth in foetal membranes of some vertebrates. In amitosis type of cell division the splitting of nucleus is followed by cytoplasmic constriction.

During amitosis the nucleus elongates first and then assumes dumbell-shaped appearance. The depression or construction increases in size and ultimately divides the nucleus into two nuclei. The division of nucleus is followed by the constriction of cytoplasm which divides the cell into two equal or approximately similar halves. Therefore, without the occurrence of any nuclear event two daughter cells are formed.

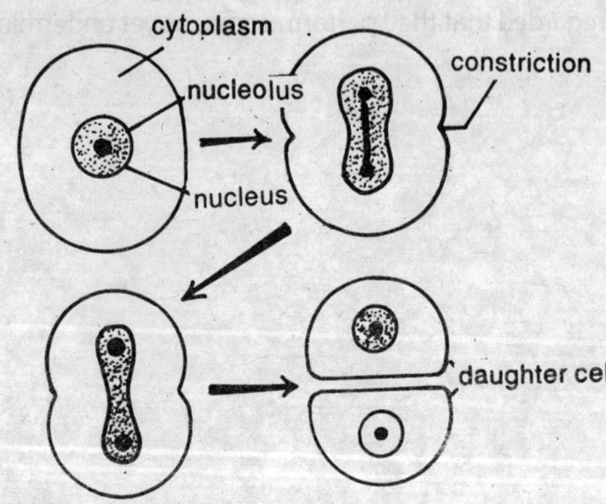

Fig. 22.1. Diagram to illustrate process of amitosis.

Mitosis

In mitosis, one cell divides into two, both of which are genetically identical to each other and to the parental cell. In other words, both, the chromosomes and the genes, are the same in all of the cells. This type of cell division is necessary if the organism and/or cell is to persist and survive. There are many facts to the need for cell division, and they vary depending on the particular biological function. For example, in liver tissue when some cells die or are damaged, others divide and provide new cell to replenish those that are lost. Other cells in the organism actually grow (increase in size), and it is possible that when they reach a point where there is too much cytoplasm far a given amount of nuclear material, they divide, and the whole process begins again. The phenomena of growth also involves an increase in numbers of cells. The increase in size of a tissue or organ is often due to numerical increase in cells rather than an increase in cell size. When subjected to the appropriate environmental and biochemical signals, these cells may be stimulated to differentiate upto a specific cell type. The sum total is that, as a result of all division, a certain degree of plasticity and immortality is provided to the organism. When the plasticity is lost, the organism undergoes the process of aging yet when the process of division is out of control, the organism literally "grows" to death!

The process of mitosis is a continuous one but on the basis of certain changes the same can be divided into several phases or stages.

Characteristic Features of the Mitotic Phases or Stages

In cell division, the first visible changes occur inthe nucleus, when the chromonemata condense into *chromosomes* (Gr. coloured bodies). This stage is called the *Prophase* (Gr. early figure).

Then the nuclear boundary disappears and the chromosomes line up in or near one plane in the cell. This is *Metaphase* (Gr. middle figure).

Following this each chromosome separates into two parts, and these two parts migrate away from each other to the ends of the cells. This is *Anaphase* (Gr. up and down figure.).

Next the chromosomes at each end of the cell reconstitute a nucleus. This is *Telophase* (Gr. final figure).

Then the cytoplasm divides into two daughter cells, bringing about the *Interphase* again.

Interphase

The period of metabolic activity during which cell division is not in a process, has been called the 'interphase.' This is frequently referred to as the 'resting phase' but this term is not appropriate because the cell is metabolically most active at this stage that is why *Berril* and *Huskins* (1936) referred it as the *energy phases*. The interphase is the period between the telophase of one division and the prophase of the new cell division. During this phase the cell does everything except division. It is during interphase that the genes self-duplicate and carry on their function of supervising synthesis.

The chromatin granules in the nucleus are not readily distinguishable in the living cell, but may be brought out by treatment with chemicals which kill, fix and stain them. They appear at first glance to be scattered throughout the nucleus, but careful study has produced evidence that they are arranged in a definite pattern as long coiled strands and appear as a thread like net work in ordinary stained preparation. "Nuclear sap" or, "Karyoplasm" fills the interstics between the chromosomes. One of more rounded bodies, the *nucleoli* are usually present. Between the nucleus and the surrounding cytoplasm, is the *nuclear membrane.* In the cytoplasm adjacent to the nucleus, there is a body, the *central body,* which consists of two granules or, after each granule has replicated, of two pairs of granules, the *centrioles*.

A typical cell cycle, including interphase lasts from 20-24 hrs. Interphase in the longest period in the cell cycle, and may last for several days in cells.

Interphase can be further divided into form sub-phases:

1. G_1-phase
2. S-phase

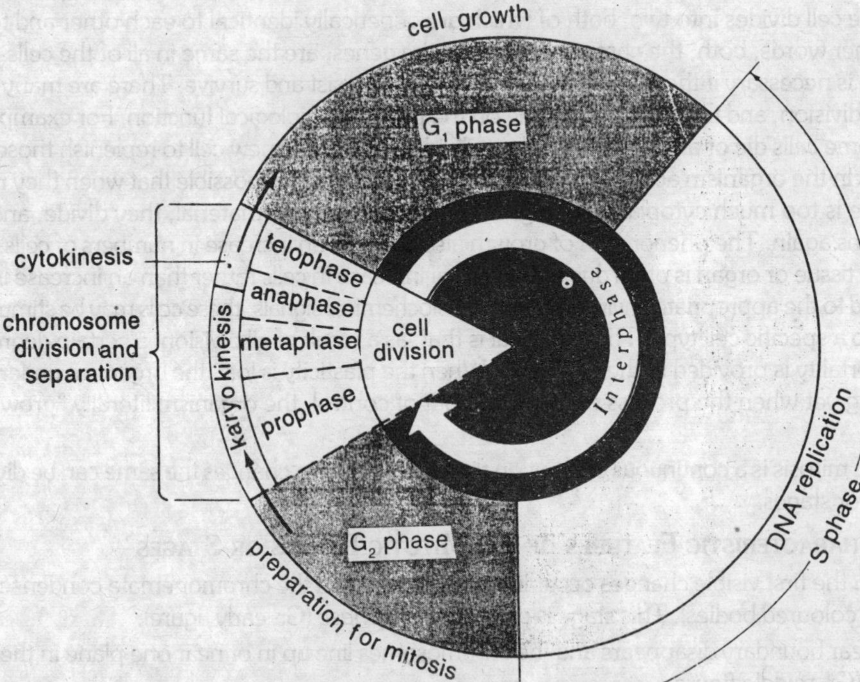

Fig. 22.2. Stages of the cell cycle.

3. G_2-phase
4. M-phase.

G_1-phase includes the synthesis and organization of the substrate and enzyme necessary for DNA synthesis. Therefore, G_1, is marked by the synthesis of RNA and protein.

G_1-phase is followed by the S-phase where the synthesis of DNA occurs.

During G_2-phase, all the metabolic activities are performed. M-phase is the period of chromosomal division.

The relative lengths of these phases differ in different organisms. A human cell in culture at 37°C, completes the mitotic cycle in about 20 hours and the M-phase lasts for only one hour. Temperature and cell environment plays an important role in determining the rate of cell division. Even the non-meristematic cells can sometimes be made to divide by changing the environmental conditions. Those cells which are not going to divide any more, have the mitotic cycle at the G_1 phase and start differentiating.

The cells shows following changes:

1. The cell, as a whole, attains the maximum growth and possesses synthesized proteins for energy for various divisions and processes.
2. The nuclear membrane is intact and the chromosomes are found in the form of more or less loosely coiled threads, somewhat closely appressed to the membrane. In this condition of chromosomes, most of the cytologists regard them to be duplicated, while some workers are of the opinion that they are multipartite.
3. The two centrioles, which are found at right angles to each other replicate into two each. *Mazia* (1961) has described that, if the replication is checked, then division will not take place.
4. For the future spindle condensation of protoplasm into a coherent area of jelly-like consistency also takes place. The spindle also starts growing and pushes the centrioles apart.

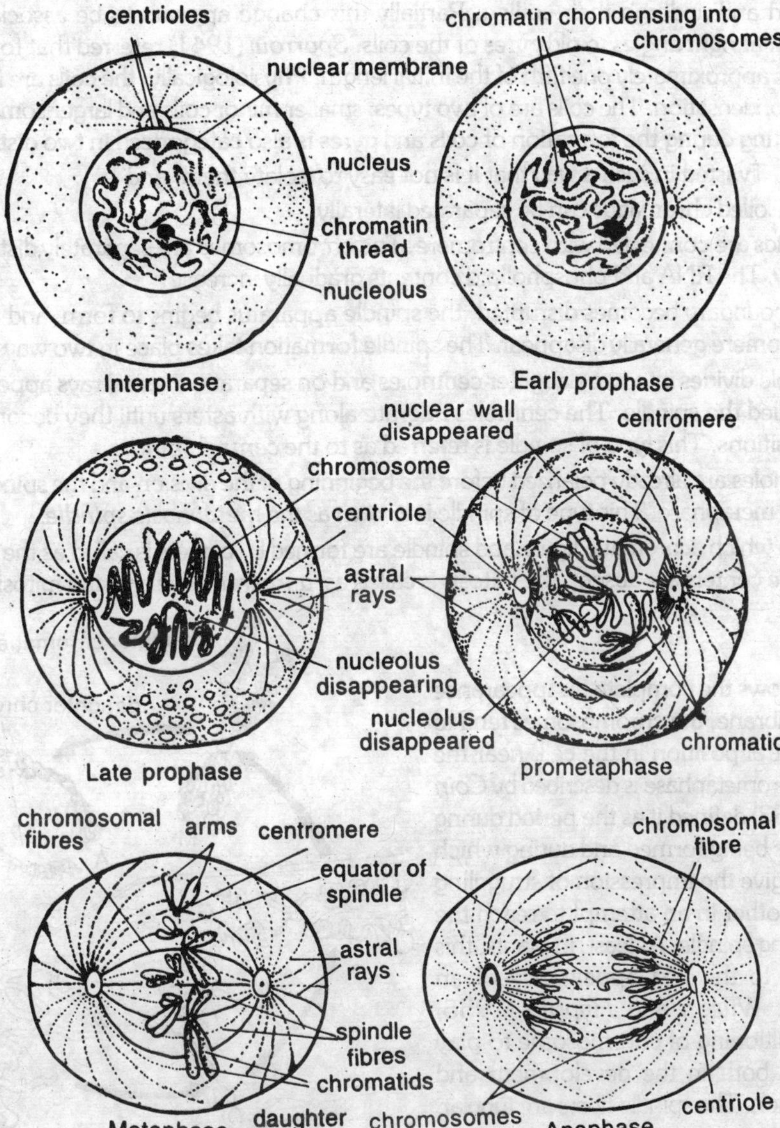

Fig. 22.3. Mitotic cell division in animal cells.

5. DNA synthesis occurs during autosynthetic interphase, when the chromosomes and disperesed.
6. Chromocentre are also conspicuous during the interphase.

Prophase

The prophase is the longest in M-phase and may take about one to several hours. In the neuroblast cells of grasshopper it takes about 102 minutes. Various important changes in this stage are as follows:

1. The cell tends to become spherical and increases its refractivity and turgidity by increasing its surface tension by removal of water.
2. The nucleus imbibes water from cytoplasm, and the chromosomes get themselves organized differently. Each chromosome shows its peculiar structure with marked isolation of chromatids, each of which undergoes a regular cycle of coi'ing. At the beginning of division the chromosomes start to contract,

get thickened and undergo their coiling. Partially this change appears to be associated with the development, at right angles to old gyres of the coils. *Sparrow* (1941) referred that total end to end contraction is approximately one fifth of the initial length. Physiologically, the coils are formed due to continuous condensation. The coils are of two types: smaller minor coils and larger somatic coils. The mode of twisting during the formation of coils and gyres is also categorized in two distinct varieties:

(i) *Plectonemic.* Twisted in such a way that it is not easy to isolate them, and

(ii) *Paranemic.* Coiled chromatids can be separated laterally.

3. The chromatids are connected with centromere. The chromosomes are separately distributed in the nuclear cavity. The RNA and phospholipid contents gradually increase.
4. The nuclear boundary becomes disrupted, the spindle apparatus begins to form, and the nucleolus and the centromere generally disappear. The spindle formation takes place in two ways viz.,

(i) Single centriole divides into two daughter centrioles and on separation, astral rays appear as delicate filaments, called the spindle. The centrioles migrate along with asters until they become located at antipodal positions. This type of spindle is referred as to the *central spindle.*

(ii) The two centrioles are already polarized before the beginning of the division and the spindle formation takes place at metaphase. This type of spindle is known as the *metaphasic spindle.*

The mitosis in which achromatic figure and spindle are formed by centers is called as the *amphiastral mitosis* and whence centers are absent, the mitosis is called as '*anastral*'. The anastral mitosis takes place in plants.

Prometaphase

This stage follows the complete disappearance of the nuclear membrane, the chromosomes tend to aggregate in a central position in the cell, near the equator. The term prometaphase is described by *Coin* (1964). *White* (1963) defined it as the period during which the spindle is being formed and during which the chromosomes give the impression of struggling and jostling one another in an attempt to reach the equator of developing spindle. At least, in plants, this stage corresponds to the first appearance of an organised spindle. *Wilson* and *Hyppio* (1955) considered this positioning of chromosome to play an important role, both in the development and functional organization. The spindle fibers are tubular, elastic, fibrous and proteinous in nature.

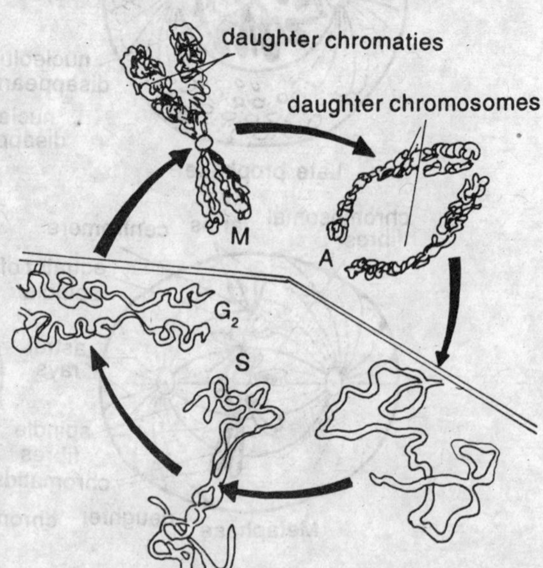

Fig. 22.4. Changes in chromosomes during cell cycle.

Metaphase

The metaphase chromosomes are sharply defined and descrete bodies and are tightly coiled. In this stage the chromosome number can be easily counted and it is possible to recognised the different chromosomes from their sizes, shape and gross structure. In early metaphase characteristic spindle-shaped figure appears in the clear nuclear region. This consists of fibrous radiations which extend from the central broad-portion called the *equator* and converge at two points on the opposite ends of the nuclear region, called the *poles.* The chromosomes which were till now scattered in a random manner in the central nuclear portion of the cell begin to show peculiar movements and arrange themselves in a single plane midway between the two poles of the spindle, forming an equatorial plate. To be more precise it is the centromeres of the chromosomes that are lined up at the equatorial plate. Usually, the smaller chromosome occur near the center of the *equatorial plate* and the larger ones near the outer ends. It is, however, not necessary that the two homologous should be near each other on the equatorial plate. The position of each chromosome is independent of other on the equatorial plate. The metaphase chromosome is a visibly double structure and is clearly seen to be split longitidinally into two exactly identical chromatids.

The centromere plays an essential role in the development of the spindle and the separation of the daughter chromosomes. The chromosomes are so arranged at the equator that one chromatid of each chromosome faces one pole and the other faces the opposite pole. Two spindle fibres are attached to the centromere of each chromosome, one on either side of it. These connect the chromosome to the two opposite poles of the spindle and are called *chromosomal* or *tactile fibres*. Other fibres of the spindle extend from pole to pole and are not attached to the chromosomes. These are called the *continuous spindle fibres*. The spindle fibres are composed mainly of protein, some ribonucleic acid (RNA) and lipids.

Form and arrangement of the chromosomes

The arrangement of chromosomes at the equator of spindle is not of one type in all organisms, as the arrangement depends upon the shape, size and number of chromosomes that differ in different organisms, because in some organisms they (chromosomes) are thread like (Urodela and in insects (Orthoptera and Diptera)). In Odonata, Coleoptera and Hemiptera they are shorter or rod shaped while in Orthropod they are rounded.

On the basis of their morphological study, one can divide them into three categories:

(i) *Straight rods or threads.* These chromosomes arise directly by shortening of the spireme thread.

(ii) *Loops, V's or hook forms.* These chromosomes are formed by a flexture at the middle point or near one end.

(iii) *Ovoidal or Spheriodal forms.* These forms arise by the extreme shortening of threads.

All the three forms mentioned above are double at metaphase due to their longitudinal splitting. The longitudinal splitting among rod shaped and thread shaped chromosomes, can be traced clearly but in spheroidal forms, it often appears as an apparently transverse constriction due to the extreme shortening of chromosomes.

The spindle attachment of the chromosomes

The mode of the attachment of the chromosomes to the spindle fibres based on the structure of chromosomes. But even though, the arrangement and the way of attachment is constant for every particular chromosome and that it is inherited from generation to generation. Chromosomes are attached to spindle fibres by centromeres. The chromosomes lacking centromere fail to attach with spindle fibres. The way of their attachment may be of two types:

(i) Terminal or telocentric

(ii) Non-terminal or atelocentric.

In terminal attachment the chromosomes may have their connection with spindle fibres at the free end. Non-terminal attachment may be at the middle point (*median*) or at an intermediate point *sub-median* or *sub-terminal*.

Anaphase

In anaphase the pairs of centromeres move apart along the spindle and carry one daughter chromosome of each pair to opposite poles. The spindle eventually grows longer. The movement of chro-

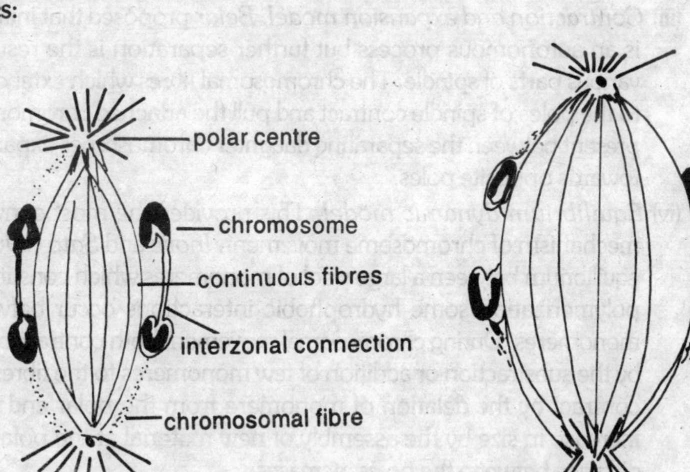

Fig. 22.5. Two types of Anaphasic spindles. A-Direct types, B-indirect type.

matids is complex. First separation of chromatid from centromere takes place and then a flow of current along the spindle governs their poleward movement. This stage lasts for very short duration, varying between 6 to 12 minutes. In the last anaphase the zone between the two sets of.chromosomes or the equatorial region gradually increases. In seems that the fibres stretch and are named *interzonal fibres*. The expansion of middle part has been referred as the stemmkoper or pushing body by *Bear.* The pushing body looks like a gel which pushes the chromosomal sets towards respective poles. During the poleward movements the chromosomes assume peculiar 'J' or 'V' shapes, depending upon the position of centromere. At this stage, 'J' are called *heterobranchials* and 'V' chromosomes as *isobranchalis*.

Movement of chromosomes

The movement of the chromosome are controlled by spindle fibres. Actually two processes may be at work; a continuous enlargement and elongation of the spindle and the shortening of the chromosomal fibres. Since the fibres shorten without becoming thicker, the process probably involves the removal of water or other molecules from fibres. A band of 'interzonal fibres' is often seen for a time after the separation has been accomplished, connecting the chromosomes which have pulled apart, and often including a remnant of the spindle.

Forces involved in the movement of chromosomes

Spingle fibres are responsible for the movement of chromosomes during anaphase i.e., from equator to the pole. Several models have been porposed by diferent workers to explain about the forces involved in the movement. Some of the them are given below:

(i) *Simple contraction model. Van Benden* (1883) suggested that chromosomes in the dividing cell are pulled towards the poles by the contraction of spindle fibres. *Swann* (1962) described that centromere secrete some substance which causes the contraction of spindle fibres. The main objection to the theory comes from the direct observations on chromosomes being carried even beyond the centriole to which the spindle fibres are attached. Further during cell division the whole cell elongates which is just opposite to the contraction model.

(ii) *Expansion Model. Watase* (1981) suggested that the spindle fibres exert pressure on the nucleus and chromosome become flattened on the metaphase plate. The chromosomal fibres become attached to the chromosomes, which are now pushed towards the opposite poles. For this reason, this is also known as *pushing model*.

(iii) *Contraction and expansion model. Belar* proposed that initial separation of daughter chromosomes is an autonomous process but further separation is the result of both contraction or expansion of various parts of spindle. The chromosomal fibres which extend from the centromeres of chromosomes to the poles of spindle contract and pull the attached chromosomes to the poles. The interzonal fibres present between the separating daughter chromosomes expand and push the daughter chromosomes towards opposite poles.

(iv) *Equilibrium dynamic model*. This provides the most convincing explanation about the possible mechanism of chromosome movement. *Inoue* and *Sato* (1967) have described the occurrence of an equilibrium between a large pool of monomeres which constitute the proteins of microtubules. During polymerization some hydrophobic interactions occur between the non-polar groups of protein monomeres. During chromosomal movement both contraction and elongation of spindle fibres occurs by the substraction or addition of few monomeres to the fibres. During anaphase chromosomal fibres contract by the delation of monomere from the polar end of the fibres and the continuous fibres increase in size by the assembly of new material at the polar ends. Thus the spindle elongates and distance between the poles increases.

(v) *Gliding model. Ostergen* and other are of the opinion that the movement of chromosomes is an active process in which chromosomal fibres glide as sail boats between the continuous fibres. *Ambrose* has suggested that motile forces for these movements could be electrosmosis and electrohoresis. Break-down of ATP is supposed to provide the magnitude required for spindle movement.

(vi) *Ratchet model. McInuosh, Helper, Van Wie* and others have described the presence of mechanical bridges between the microtubules similar to the ratchet connections in the muslces fibres. The chromosome fibres glide between the continuous fibres and draw the chromosomes apart. The energy is supplied by the breakdown of ATP.

(vii) *Electrical model. Lillie* and *Coller* (1936) suggested that the change in the membrane potential is due to local changes in permeability occurs near the poles and around the nuclear membrane and produces electric field. The chromosomes are negatively charged in prophase and the charged chromosomes easily migrate in the electric field.

(viii) *Push Model.* According to this concept chromosomes are pushed apart by the colloidal component of cell cytoplasm. The colloid imbibes water, swells up and pushes the two chromatids of the chromosomes apart. The spindle fibres act as tracks that guide the movement of chromosomes to the poles, preventing their scattering throughout the cell.

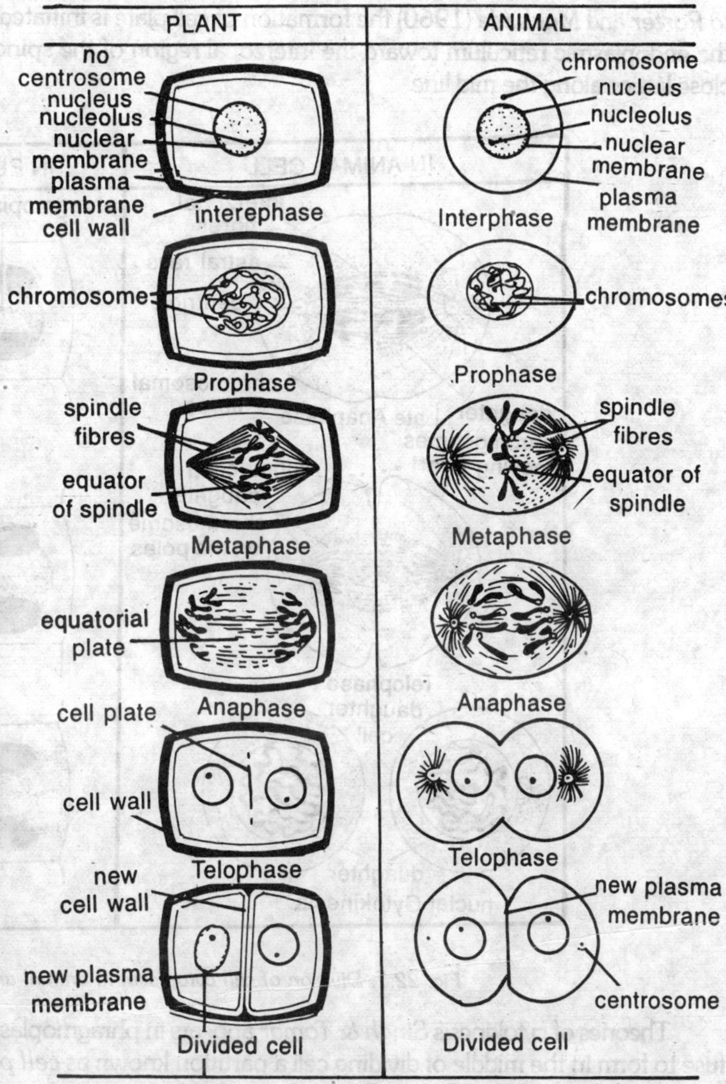

Fig. 22.6. Differences in mitosis in plants and animal cells.

Telophase

The two groups of chromosomes aggregate at the pole and on arrival they progressively lose their chromaticity. Decondensation takes place. Reformation of the nuclear membrane by certain unknown process take place. Perhaps new material is synthesized by RNA during telophase or it may be possible that the endoplasmic reticular system may be giving rise to a new membrane round the nucleus. All chromosomes become uncoiled. The spindle fibers are absorbed in the cytoplasm. The nucleolar organizer or SAT zone again forms the nucleolus. Finally, these change result into two daughter nuclei which are equivalent to the parent nucleus in all respects.

Cytokinesis in plants

Nuclear division or mitosis, as it is called it followed by the division of the cytoplasm. While the daughter nuclei are being organised at the poles, the mitotic spindle disappear except at the equator, where the continuous spindle fibres become more dense. This region is now called the *phragmoplast.* According to *Porter* and *Machade* (1960) the formation of cell plate is initiated by migration of tubular elements of the endoplasmic reticulum toward the interzonal region of the spindle where they spread out to form a close lattice along the mid line.

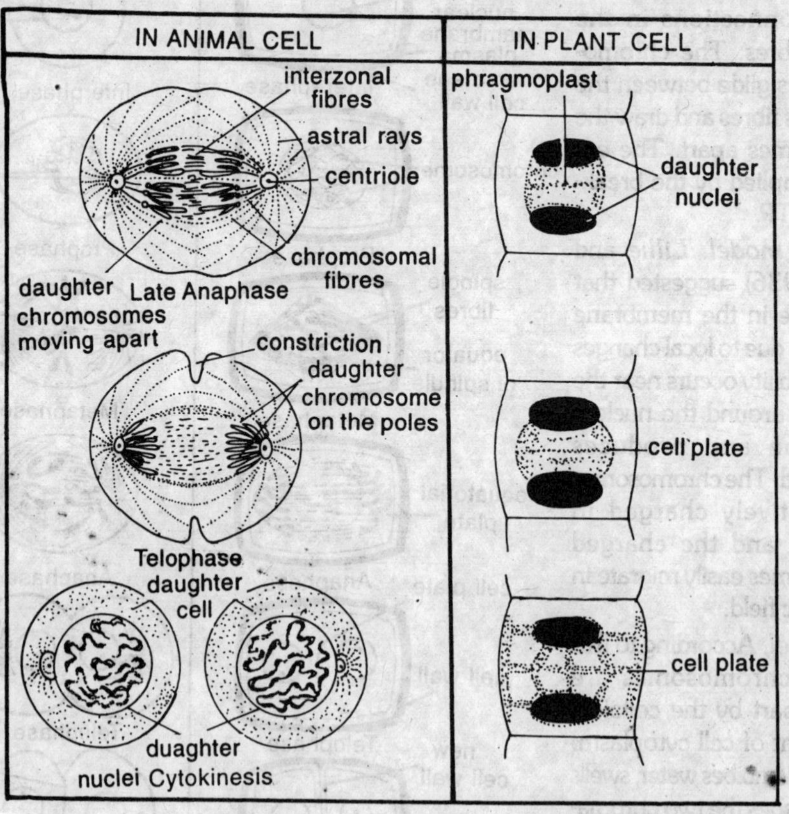

Fig. 22.7. Division of cell cytoplasm in animal and plant cell.

Theories of cytokinesis *Singh & Tomar* appears in phragmoplast and contain pectic substances which fuse to form in the middle of dividing cell a partition known as *cell plate.* More droplets are added to the cell plate to form the middle lamella which now begins to extend outwards until it reaches the outer wall of the original cell. Primary cell wall material is now deposited on either side of the middle lamella. This divides the cytoplasm into two roughly equal parts. Cell division is now complete and two cells are formed. The division of the cytoplasm is called cytokinesis.

CYTOKINESIS

Cytokinesis in Animal Cells

Cytokinesis starts by the appearance of a shallow furrow in the cytoplasm at the equator of the spindle. Slowly and slowly the furrow deepens and constricts the cytoplasm and the cell into two daughters. A number of theories have been proposed for the formation of furrow. These are as follows:

1. *Contractile Ring Theory.* According to *Swann* and *Mitchison* (1958) the cytoplasm around the equator of the spindle contains some contractile proteins. These proteins form a sort of ring at the equator. As the dividing cell elongates, the contractile ring contracts which results into the furrow formation.
2. *Expanding surface theory. Mitchison* (1922) proposed that a nuclear material is liberated by chromosomes which is responsible for the cellular expansion at the poles. As the polar areas expand, the

equator contracts which results in the appearance of furrow. The furrow divides the cell into two daughter cells. There are certain examples, showing that the furrow formation takes place in the absence of nuclei or chromosomes (*Nachtway*, 1965). This indicates that exciting material might have originated from other than nucleus.

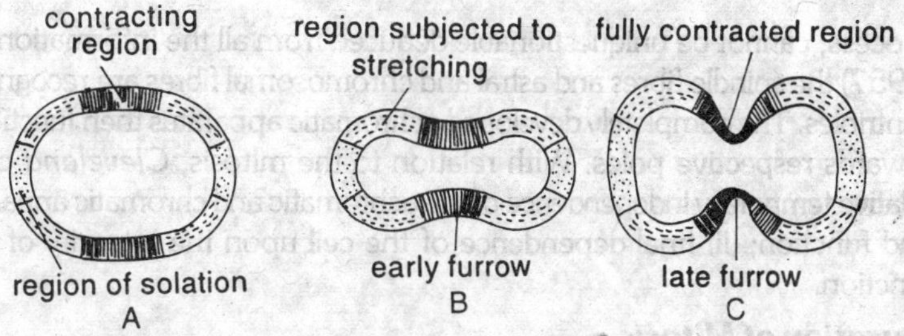

Fig. 22.8. Diagram to illustrate the contractile ring theory of Marsland for cytokinesis. A—Early telophase; B—Early furrow; C—Late furrow.

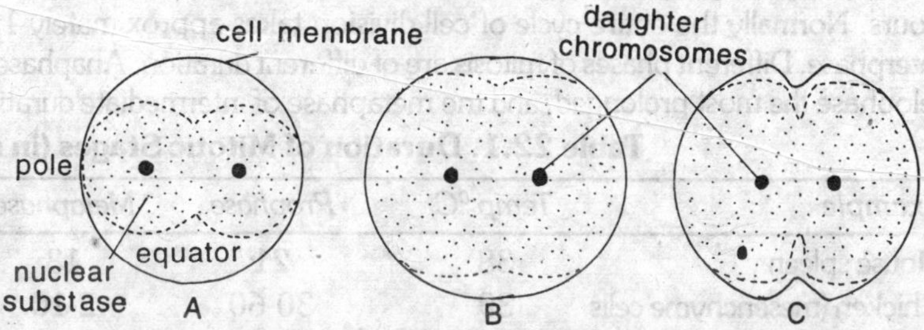

Fig. 22.9. Diagram to illustrate the expanding surface theory.

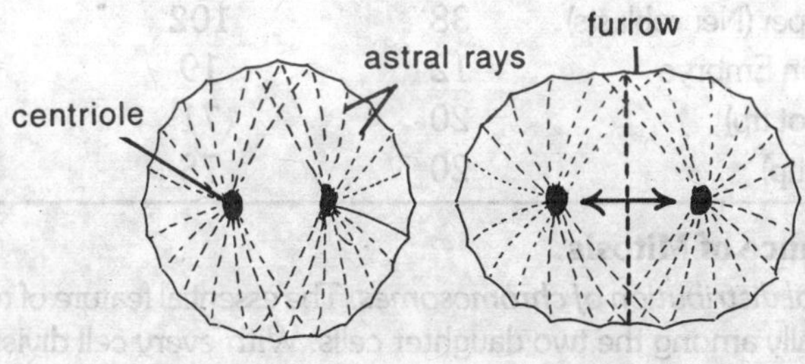

Fig. 22.10. Diagrammatic representation of Dan's spindle elongation theory.

3. *Spindle elongation theory*. This theory was proposed by *Dan* and *Dan* (1947) and *Dan* (1958). According to them the spindle and asters are responsible for the cytokinesis. During experiment, kaolin particles were attached with the egg membrane. It was observed that elongation of the cell at anaphase is accompanied by shrinkage of equatorial plane results two kaolin particles on either side of equator come close to each other. The driving force is believed to be the elongation of microtubules of the spindle which push the centres apart.
4. *Astral relaxation Theory*. *Wolpert* (1960, 1963) proposed that the surface of a cell is under uniform tension. During cell division, as the astral rays reach the poles they lower in the surface tension at the poles. The surface tension at the equator remains same. Because of low surface tension, the polar regions expand and cause the appearance of furrow at the equator.
5. *Vesicle Formation Theory*. According to *Threadgold* (1968), during anaphase, dividing cell lengthens. This occurs in inter-zonal region due to increased electron density. Plasma membrane also exhibits high electron density in the furrow simultaneously, the continuous spindle fibres, present in the inter zonal region, continue to increase its density and finally form an adielectronic fibrillary plate in the equator. As furrow progresses, a large, empty membrane-bound vesicle appears on each side of this plate. Later on small vesicles are added in the equatorial plane. At final stage, fusion occurs of all the large and small vesicles ot form a deep furrow, leaving the daughter cells to be connected by an intercellular bride. Further, a row of tiny vesicles arises which causes final separation of two daughter cells.

Role of Centriole during Mitosis

The centriole acts as an epicentre for the development of the central spindle. Whether the centriole produces the spindle fibres or serves to direct their formation or on the other hand is entirely passive in the process, cannot be uniquestionable deduced from all the informations at hand. According to *Cleveland* (1957) the spindle fibres and astral and chromosomal fibres are recognizable first in the close vicinity of the centrioles. The completely developed achromatic apparatus then functions to separate daughter chromatids towards respective poles. With relation to the mitosis, *Cleveland* concluded two important points: (i) relative temporary independence of the achromatic and chromatic apparatus as regards apparent duplication and function; (ii) final dependence of the cell upon the integrity of these two system and their united function.

Duration of Mitosis

The time required for mitosis differ with species and environment. Temperature and nutrition, in particular, are important factors. The entire sequence of phases may be completed in 6 minutes to many hours. Normally the entire cycle of cell division takes approximately 18 hours; about 17 hours for the interphase. Different phases of mitosis are of different duration. Anaphase is the shortest, the prophase and telophase the most prolonged, and the metaphase of intermediate duration.

Table 22.1. Duration of Mitotic Stages (in minutes)

Example	*Temp °C*	*Prophase*	*Metaphase*	*Anaphase*	*Telophase*
Mouse spleen	38	21	13	5	4
Chicken (mesenchyme cells in tissue culture)	39	30-60	2-10	3-7	2-10
Frog (Fibroblasts in culture)	20-24	32	20-29	6-11	
Grasshopper (Neuroblasts)	38	102	13	9	57
Sea Urchin Embryo	12	19	17	2	18
Onion (root tip)	20	71	6.5	2.4	3.8
Pea (root tip)	20	78	14.4	4.2	13.2

Significance of Mitosis

1. *Equal distribution of chromosomes.* The essential feature of mitosis is that chromosomes are distributed equally among the two daughter cells. With every cell division there is a division of chromosomes. The constant number of chromosomes in all the cells of the body is because of mitosis.
2. *Surface-volume ratio.* Mitosis restores the surface-volume ratio of the cell. A small cell has a greater amount of surface available in relation to volume than a large cell. As the cell increases in size the available surface area in relation to increased volume becomes less. By undergoing division the cell becomes smaller in size and the surface-volume ratio is restored.
3. *Nucleoplasmic index.* The growth of a multicellular organism is due to mitosis. A cell cannot grow in size to a large extent without disturbing the ratio between the nucleus and the cytoplasm. After a particular size has been reached the cell divides to restore the nucleoplasmic index. Thus growth takes place by an increase in the number of cells, rather than by increase in the size of the cells.
4. *Repair.* The repair of the body takes place because of addition of cells by mitosis. Dead cells of the upper layer of the epidermis, cells of the lining of the gut, and red blood corpuscles are constantly being replaced. It is estimated that in the human body about 500,000,000,000 cells are lost daily.

Abnormalities in Mitosis

Under unfavourable conditions the mitotic divisions become defective and thus various abnormalities are formed when exposed to physical and chemical agents like temperature, radiations, necrotics and enzyme inhibitors etc. Some abnormal mitotic divisions are given under the following heads:

1. C-Mitosis (abnormal spindle formation)

Brachet (1975) has described that colchicine inhibits mitosis by disorganizing spindle formation. The results are the formation of polyploid cells, and reduplication of chromosomes. Besides, during mitosis, sometimes cell nucleus is either deformed into a single spheriod compact mass (pycnosis) or becomes broken down (karyorrhexis).

2. Cytasteral mitosis

Wilson (1901) noticed the formation of numerous asters in the cytoplasm of unfertilized eggs called cytasters by placing them in hypertonic sea water. In the eggs of marine invertebrates, production of many small cytoplasmic asters is of frequent occurrence. The intermixing of nucleoplasm and hyaloplasm seems to be neccesary condition for the aster formation. *Costello* (1940) showed that asters can be produced only if germinal vesicles has broken down.

3. Multipolar and Catenar mitosis

Mitotic division showing several spindles and centrosomes are common in many protozoans and fish eggs. Multipolarity is usually caused by uneven division of centrosomes as well as irregular distribution of chromatids of the different spindles. It results in the formation of cells as aneuploidy (uneven chromosome number). Catenar mitosis have been described by *Dalcq* and *Simon* (1932) in amphibian eggs. During this, dividing cell forms a large number of asters are capable of dividing in the absence of spindle or nucleus in an autonomous way.

4. Achrosomal mitosis

Sometimes, mitosis in induced by various agents without use of nucleus or chromosomes. *Briggs et. al.* (1951), in fertilized eggs with heavily X-rayed sperms, removed the egg's maturation spindle by pricking and sucking. In this way, he obained very nice non-nucleate blastulae. Similarly, *Stauffer* (1945) obtained normal nonnucleated Axolotl blastula.

5. Anastra mitosis

In plant cells and many oocytes, generally asters around the centromeres are absent, thus producing anastral mitosis. *Bataillon and Tchou Su* (1933) described in details the anastral mitosis in amphibian interspecific hybrids.

Achromatic Figure

Achromatic figure consists of spindle fibers and astral rays. Astral rays are developed just before the distintegration of nuclear envelope during late prophase, while the formation of spindle rays can be traced only after the disappearance of nuclear membrane.

The spindle

In fixed preparations of dividing cells, fibers are seen forming a spindle shaped body lying between the centrioles. It is generally believed that spindle fibers are not visible in normal living cells. They

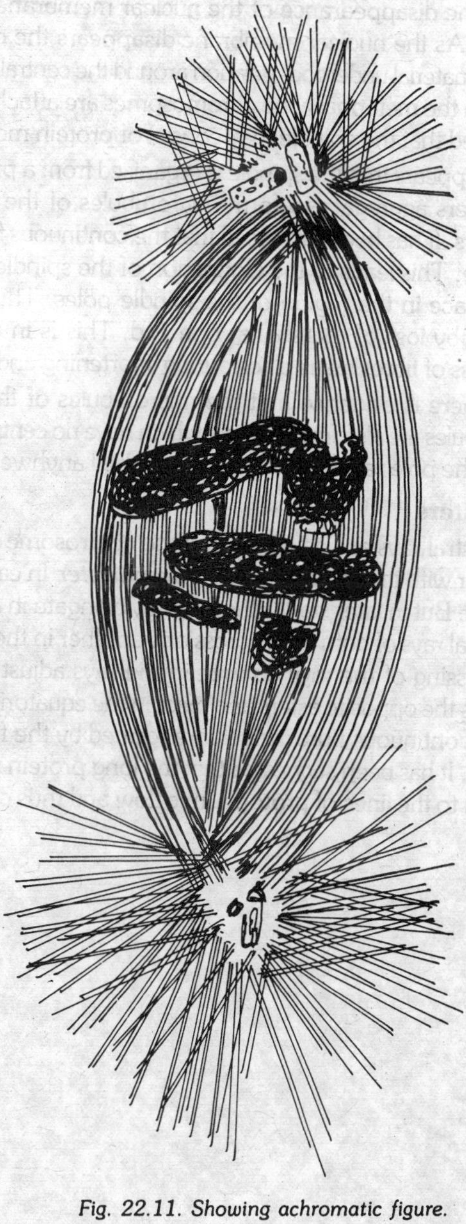

Fig. 22.11. Showing achromatic figure.

can be produced by adding acid to the medium, and, on removing the acid, the fibres disappear; as most fixatives contain acid, it is stated that the fibers are artificats. There has been considerable controversy as to the reality of fibers and various theories have been put forward to explain their presence in fixed material. *Schrader* in 1944 observed the spindle fibers in living cleavage cells of mite and certain flagellates.

The spindle is superficially similar to the spindle shaped area found between the poles in the magnetic field. The spindle consists of two kinds of fibers which have been supposed to differ widely in functional significance.

Those spindle fibers extending from the poles to the chromosomes are called half spindle, incomplete fibers or discontinuous fibers or chromosomal fibers. The remaining spindle fibers extend without interruption from pole to pole are called the continuous fibers or direct or complete fibers.

It is believed that the spindle is usually formed in two parts. A part arises between the centrosomes before the disappearance of the nuclear membrane; it is cytoplasmic in origin and is called the central spindle. As the nuclear membrane disappears the central spindle moves into nuclear region. Then the nuclear material undergoes gelation around the central spindle to from the 1/2 spindle fibers or chromosomal fibers. In the metaphase the chromosomes are attached to the 1/2 spindle fibers at the equator. *Schrader* suggested that the spindle may consist of protein molecules arranged in parallel lines.

It appears that the spindle is organized from a pre-existing pool of protein. It is postulated that protein monomers aggregate to form microtubules of the spindle fibres. The spindle fibres are thus oriented polymers. It has been proposed that the continuous fibres are elongated by the insertion of subunits within the fibre. This leads to the elongation of the spindle. Conversely, the shortening of chromosomal fibres takes place in the region of the spindle poles. Thus the fibres do not contract like a rubber band, but shorten by loss of material at one end. This is in agreement with observations that no change in the thickness of fibres takes place during shortening and elongation.

There is evidence that the microtubules of the chromosomal fibres originate at the kinetochore. Micotubules are also found in cells which have no centrioles e.g., those of higher plants. Here the microtubules end in the pole region and are not attached anyhwere.

The Asters

Astral rays are developed by the centrosome present in the viscinity of nuclear region. Astral rays together with the spindle form the *amphiaster.* In early stages the astral rays are straight, simple and few in number. But in later stages they rapidly elongate in all directions, and increase in number. In certain cases the astral rays of both asters cross one another in the equatorial region outside the spindle. Sooner or later, the crossing of the rays disappear and rays adjust themselves in such a way that their free ends move towards the opposite asters and meet in the equatorial plate thus giving the impression of continuous fibres. These continuous fibres which are formed by the fusion of astral rays are lying in the region outside the spindle. It has been suggested that the long protein molecules of the protoplasm become arranged in rows parallel to the lines of protoplasmic flow and thus constitute the astral rays.

23

CELL DIVISION (MEIOSIS)

Mitosis is a device of increasing the number of cells. During asexual reproduction it is the means of multiplying the individual. Therefore, in a sexually reproducing organisms, set of characters are transferred unchanged from parents to offspring, except of course accidents. There is uniparental inheritance due to mitosis. But in sexually reproducing organisms, gametes are produced as a result of meiosis. Two gametes unite to give rise to a zygote which divides mitotically to form the individual. More often than gametes from different parents unite to form a zygote, thus ensuring biparental inheritance. This result into a blending of characters from two parents into their offsprings and ensures variation which is important for evolution. Gametes contain half the number of chromosomes, as comapred to somatic cells. Union of two gametes restores the somatic number of chromosomes in the zygote. Gametic union results into the mixing of cytoplasms of two cells which is followed by the fusion of two nuclei. During meiosis or at the time of gamete formation, therefore, halving of the number of chromosomes or the amount of DNA is achieved by two successive divisions of cytoplasm and nuclei (meiosis I and meiosis II) accompanied by only one replication of chromosomes. For example, each somatic cells of man has 46 chromosomes. Somatic cells are derived from the zygote due to mitotic divisions. Zygote is the result of a fusion of two gametes, each of which contain only 23 chromosomes. Both the male and female gametes contain the same amount of DNA and equal number of morphologically similar chromosomes. The zygote and somatic cells derived therefrom, therefore, contain 23 pairs of homologous chromosomes, one of each pair derived from the male parent through the male gamete and the other from the female through the female gamete. The members of each pair are homologous to each other and non-homologous to the other chromosomes. In some organisms, the chromosome complements of male and female cells are slightly different and this forms the basis of sex differentiation. Those cells which possess two sets of chromosomes, are called *diploid* and they have 2n or *diploid* or *unreduced* or *zygotic* number of chromosomes. Gametes, on the other have n, or *haploid* or *reduced* or *gametic* number of chromosomes. In sexual reproducing organisms the diploid and haploid phases of the life cycle alternate and this phenomenon is called *alternation* of *generation.* In plants, there is an alternation of sporophytic (2n) and gametophytic (n) generation. Reduction division heralds the beginning of gametophytic generation and fusion of gametes that of the sporophytic phase. Thus, meiosis can be studied at the time of gamete formation which occurs in specialised organs i.e., the embryo sac and the pollen sac. A single cell undergoes meiosis in the embryo but in the pollen sac a large number of spore mother cells divide meiotically, almost synchronously, to produce haploid pollen grains. It is, therefore, easier to study meiosis is anthers, similarly testes in animals.

Meiosis consists of two divisions which take place one after the other, during which the number of chromosomes is halved. The two divisions are known as the first meiotic division or *heterotypic division* and the second meiotic division or *homotypic division.* The first meiotic division has the same stages as mitosis. These stages are known as prophase I, prometaphase I, metaphase I, anaphase I and telophase I.

Except for prophase, the other stages are similar to those of mitosis. Following telophase I is a period of interphase or rest. The second meiotic division consists of prophase II, metaphase II, anaphase II and telophase II.

I. Heterotypic or Ist Meiotic Division

This is the first half of meiosis and is true reductional in nature. Here the diploid cell divides into two haploid daughter cells. A already stated that the phases of this division one called prophase I, metaphase I, anaphase I and telophase I.

Prophase I

It is the longest stage and its duration makes from a few hours to a few days. The nucleus starts increasing in volume and this increase is much than the mitotic increase to some extent. This is due to an increase in hydration.

Nobel and *Ruttle* (1936) have called this stage as the *premeiotic spiral prophase.* Prophase I is further divided into six substages.

(i) Pre-leptotene
(ii) Leptotene
(iii) Zygotene
(iv) Pachytene
(v) Diplotene
(vi) Diakinesis

(i) Prepleptotene

As noted previously, replication of DNA has already taken place during interphase. The chromosomes at the beginning of meiosis are, therefore, note single but double strands. The beginning of meiotic prophase is marked by an increase in nuclear volume. During preleptotene the chromosomes are very thin and cannot be easily seen, except sometimes for the sex chromosomes.

(ii) Leptotene ("slender thread") or Leptonema

During this stage the chromosomes become more distinct, and their double nature is seen in many organisms. Leptotene extends from the time single unpaired chromosomes are first seen to the time when they begin to undergo homologous pairing. Under the light microscope the chromosomes appear as slender threads bearing a series of granule-like structures called *chromomeres.* Whole mount electron microscope preparations show that leptotene chromosomes have an axial filament to which chromatin fibres are attached as a series of lateral loops. In certain regions these loops appear in bunches, and from the chromomeres of light microscopy. The axial filaments are apparently composed to two independent protein fibrils, one for each chromatid. They later become transformed into the lateral elements of the synaptonemal complex. In the leptotene cells to the quail there is evidence of premeiotic pairing. Two homologous chromosomes become apposed to each other, chromomere for chromomere, even though no synaptonemal complex is formed. In leptotene cells there is a tendency for the ends of homologus chromosomes to be associated with each other at the nuclear membrane, where they are attached.

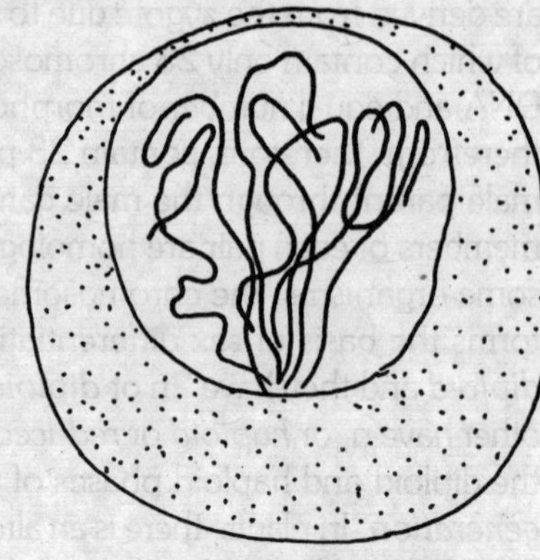

Fig. 23.1. Leptotene chromosomes arranged in a 'bouquet.'

Leptotene chromosomes may be irregularly arranged, or may be polarized toward the centrioles forming a 'bouquet.' Electron microscopic studies have shown that bouquet formation results when a group of chromosomes is attached close together on the nuclear membrane.

During leptotene the cytoplasm has many polyribosomes, but endoplasmic vesicles are few. In maize (*Zea mays*) the nucleus increases in size during leptotene, reaching its maximum size during pachytene. This increase in size is correlated with synthesis of RNA and protein.

(iii) Zygotene ("mating thread") or Zygoneme

The chromosomes becomes much shorter and thick and number of coils is also increased. Homologous chromosomes begin to pair by force of attraction. In synapsis or pairing, two homologous chromosomes come close to each other precisely chromomere to chromomere. The pairing starts from one point and proceeds onward in a zipper like fashion. Out of the two synaptic chromosomes, one is of paternal origin while the other is maternal. These are called *bivalents.* Since each chromosome is made of two chromatids, each pair of chromosomes consists of four chromatids thus representing *'tetrad'* of chromatids at this stage.

The pairing of zygonemal stage is completed in three different ways as follows.

(i) *Proterminal synapsis.* The two homologous chromosomes start pairing at the terminals, which gradually then progress towards the centromere.

(ii) *Procentric synapsis.* The pairing starts near the centromere and progresses onwards on either side.

(iii) *Random synapsis.* The pairing may be random at many points which then finally fuse together.

Two views have been advocated to explain the mode of initiation of synapsis. *Darlington* (1937) proposed precocity theory which explains that the chromosomes undergo synapsis due to singleness, but this theory does not explain the extra synthesis of DNA and duplication at leptotene stage. Another view is proposed by *Sax* (1932) which is referred as the retardation theory. It explains that pairing takes place due to the retardation or cessation of metabolic activities of the cell. During zygonema, the nucleus remains intact and the centriole moves a part initiating the aster formation.

(iv) Pachytene "thick thread" or Pachynema

It generally begins when the pairing of chromosomes ends. It represents one of the longer substages of prophase I and is characterised by the following important events:

When the pairing has completed, the apposition of homologues becomes so tight that it is difficult to identify two separate chromosomes. Within the major coils, and at right angles to them smaller minor coils appear. At the height of pachytene, the pairing chromosomes twist about one another as relational coils and each soon longitudinally splits if no splitting had occured. This coiling is further complicated by a coiling of two chromatids of each chromosome around each other. This coiling put the chromosomes under considerable strain. This means that at this stage, each pachytene element consists of four chromatids. The chromatids of each homologue are called *sister chromatids.* Formerly this element of four chromatids was called a *tetrad*-a name which has now been replaced by *bivalent.* When the chromosomes have longitudinally splitted, transverse breaks occur at the same level in two of the adjacent homologous chromatids. Soon both segments of the chromatids interchange and fuse together with those of homologues. In this process which is called *crossing over,* portions of the two chromatids with their genes are interchanged and two remain intact. The new chromatids thus interchanged will be mixed.

After this interchange, the parts begin to separate, repelling each other. However, this separation is not at all complete since the homologous chromosomes remain joined at their point of interchange. This point, where the two are joined, is called the *chiasma* (pl. chiasmata). This chiasma is an immediate consequence of crossing over and are found in all plants and animal with few exceptions (e.g., males and females silkworm). The number of chiasmata is variable but for one bivalent atleast one chiasma is found. Bivalents with 2, 3 and 4 chiasmata are rare. The highest number of chiasmata has been observed in the long chromosome of the broad bean, *Vicia faba,* where there are probably 12 chiasmata in a bivalent. The average number of chiasmata is known as *chiasma frequency.* This number seems to depend upon the length of the chromosome. Short acrocentric chromosomes have a greater number of chiasmata than metacentric chromosomes. In small chromosomes, chiasmata are localized at the distal ends. Nucleus remains as such, being somewhat larger than the proceeding substage.

Cytomixis

It is a botanical term which refers to the passage of chromatin from one microsporocyte (pollen mother cell) to another during early prophase. This phenomenon is so widespread in the normal species of hybrids that one wonders if it is a general phenomenon.

(v) Diplotene or Diplonema

The separations of homologous chromosomes initiates diplotene. The synaptic forces of attraction between them lapse due to breakage at one or more point, so that the homologous chromosomes uncoil and start separating. But the separation is incomplete since the homologues are in, contact at one or more where the crossing over has already taken place. These points of contact are points known as chiasmata (sing-chiasma; meaning cross) which present crosshaped apperance. The chiasma is the morphological equivalent of genetic crossing over. Their number and position varies in the same pair of chromosomes and in different cells of the same individual.

By the end of diplotene, the chiasmata begin to move along the length of chromosomes from the centromere towards the end. This displacement of chiasmata is termed as terminalization. When the terminalization of chiasmata is complete the homologous chromosomes are kept in contact by terminal chiasmata. The degree of terminalization is generally expressed as a coefficient to terminalization (T).

$$T = \frac{\text{Number of terminal chiasmata}}{\text{Total number of chiasmata}}$$

The average number of chiasmata in bivalents is known as frequency of chiasmata (fq).

$$\text{Frequency (fq)} = \frac{\text{Total number of chiasmata}}{\text{Total number of bivalents}}$$

According to *Darlington* two types of repelling forces operate on the chromosomes at diplotene. One of the forces is electroegatively charged and operates on the surface of the chromosome throughout its length and the other with electropositive charge is localized on the centromere. The former controls the repulsion of the chromosomes and the later causes distal movement of the chiasmata.

Diakinesis

At this stage the chromosomes are best seen as very short and thick bivalents. The contraction is nearly maximum and the chromosomal pairs are well spread throughout the cell as thought by mutual repulsion. The chromatids of each chromosome are very close to each other and their identity is not distinguishable. Terminalization may be completed in the substage. The bivalents migrate to the periphery of the nucleus. The nucleus generally disappears during this stage but may persist, in reduced size until anaphase. The nuclear membrane disrupts and eventually disappears thus all the chromosomes are set free in the cytoplasm. The spindle fibres are organized with an establishment of the pole.

Prometaphase I

Spiralization reaches its virtual maximum in this stage; the nuclear membrane disappears, and the chromosomes become arranged at the equator of the cell to begin the metaphase. Nuclear membrane disappears and spindle and astral rays and formed during this stage.

Metaphase I

The bivalent chromosomes arranged themselves in the equatorial plane so that the two members of each homologous pair are found with their centromeres directed toward opposite poles. Whereas in mitosis there is a single centromere which unites the chromosomes, in meiosis there are two centromeres. Each homologous chromosome has a centromere to which the daughter chromatids are united. The repulsion of the centromere is accentuated, and the chromosome is found on the verge of dividing.

Anaphase I

In this the chromosomes are pulled apart by the movement of their centromeres which remain undivided, with the result that each tetrad breaks up into two into two component pairs of chromatids or dyads. The dyads separate and move along the tactile spindle fibres to the opposite poles. This is the true reduction division.

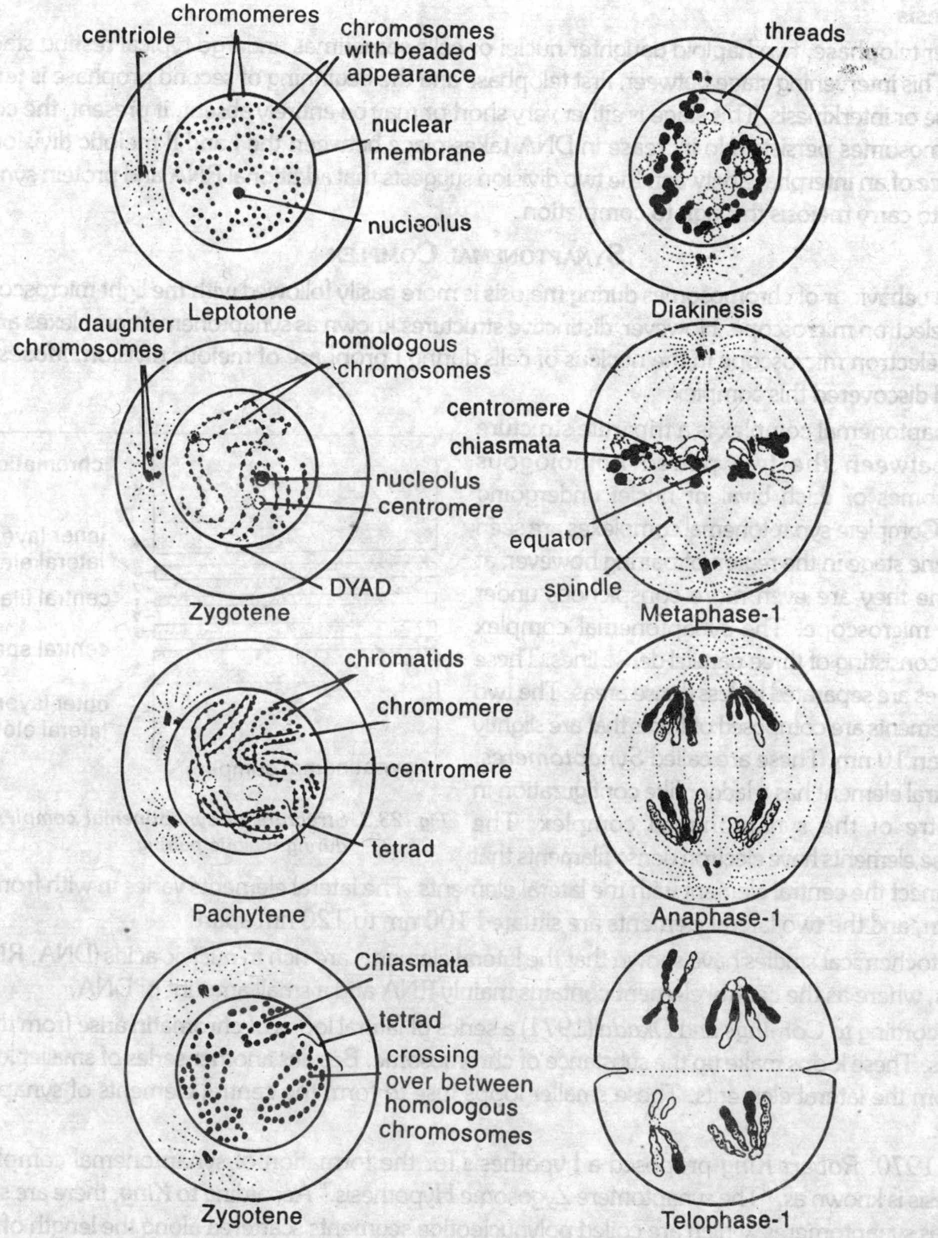

Fig. 23.2. Diagrammatic representation of different stages in the first meiotic division.

Telophase I

The group of chromosomes at each pole forms a nucleus. Each daughter nucleus has half the number of chromosomes as compared to the parent nucleus. Unlike the mitotic telophase each chromosome consists of two distinctly separated chromatids united only at their centromeres. This is followed by the formation of a cell plate between the separated sets of chromosomes, dividing the mother cell into two daughter cells. Sometimes, cytokinesis fails to occur after the first meiotic division. In such cases the second meiotic division takes place in the common cytoplasm of the mother cell.

Interkinesis

After telophase, two haploid daughter nuclei or cells sometimes undergo typical resting stage as in mitosis. This intervening stage between first telophase and the beginning of second prophase is termed as interphase or interkinesis. This stage is either very short or may be entirely absent, if present, the coiling of the chromosomes persists. No increase in DNA takes place between the I and II meiotic divisions. The occurrence of an interphase between the two division suggests that additional RNA and protein synthesis is required to carry meiosis through to completion.

SYNAPTONEMAL COMPLEX

The behaviour of chromosomes during meiosis is more easily followed with the light microscope than with the electron microscope. However, distinctive structures known as synaptonemal complexes are visible with the electron microscope in the nucleus of cells during I prophase of meiotic division. *Moses* (1950) first of all discovered this complex.

Synaptonemal complex is a tripartite structure found between the two paired homologous chromosomes of each bivalent nuclei undergoing meiosis. Complete synaptonemal complexes are seen at zygotene stage in the region of pairing however, at pachytene they are even more conspicuous under electron microscope. The synaptonemal complex appears consisting of three parallel dense lines. These dense lines are separated by less dense areas. The two lateral elements are composed of fibres that are slightly wider than 10 nm. These are called *Synaptomeres.* The central element has a ladder-like configuration in the centre of the synaptonemal complex. The transverse elements have electron dense filaments that interconnect the central element with the lateral elements. The lateral elements varies in with from 20 nm to 30 nm, and the two lateral elements are situated 100 nm to 120 nm apart.

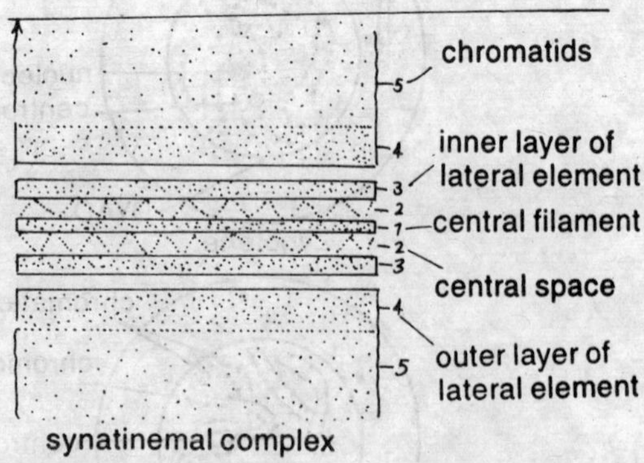

Fig. 23.3. Structure of synaptinemal complex formed during meiotic pairing.

Cytochemical studies have shown that the lateral elements are rich in nucleic acids (DNA, RNA,) and proteins, where as the central element contains mainly RNA and a small amount of DNA.

According to *Comings* and *Okada* (1971) a series of lateral loops of chromatin arise from the lateral elements. These loops make up the substance of chromosome. Besides another series of smaller loops also arise from the lateral elements. These smaller loops fuse to form the central elements of synaptonemal complex.

In 1970, *Robert King* proposed a hypothesis for the formation of synaptonemal complex. This hypothesis is known as, "The synaptomere-Zygosome Hypothesis." According to *King*, there are structures known as synaptomeres, which are coiled polynucleotide segments scattered along the length of a pair of synapsed chromosomes lying in close proximity to one another. Each synaptomere is made out of A, B and C segments. The A and C segments are called lateral segments and the B is central. The lateral segments of synaptomeres pair with the respective segments of the adjacent synaptomeres. The B segments are directed towards the central element of synaptonemal complex. These are the sites where zygosomes are attached. The zygosomes are rod-shaped subunits. These are formed in the nucleoplasm and visualized as protein molecules. The zygosomes have a folded head by which they can attached to the B segment of the synaptomere. The tail ends of zygosomes contain charged sites. These charges allow the zygosomes to bind laterally with adjacent zygosomes in a ladder-like manner.

Functions of synaptonemal Complex. A number of functions have been assigned to the synaptonemal complex. These are:

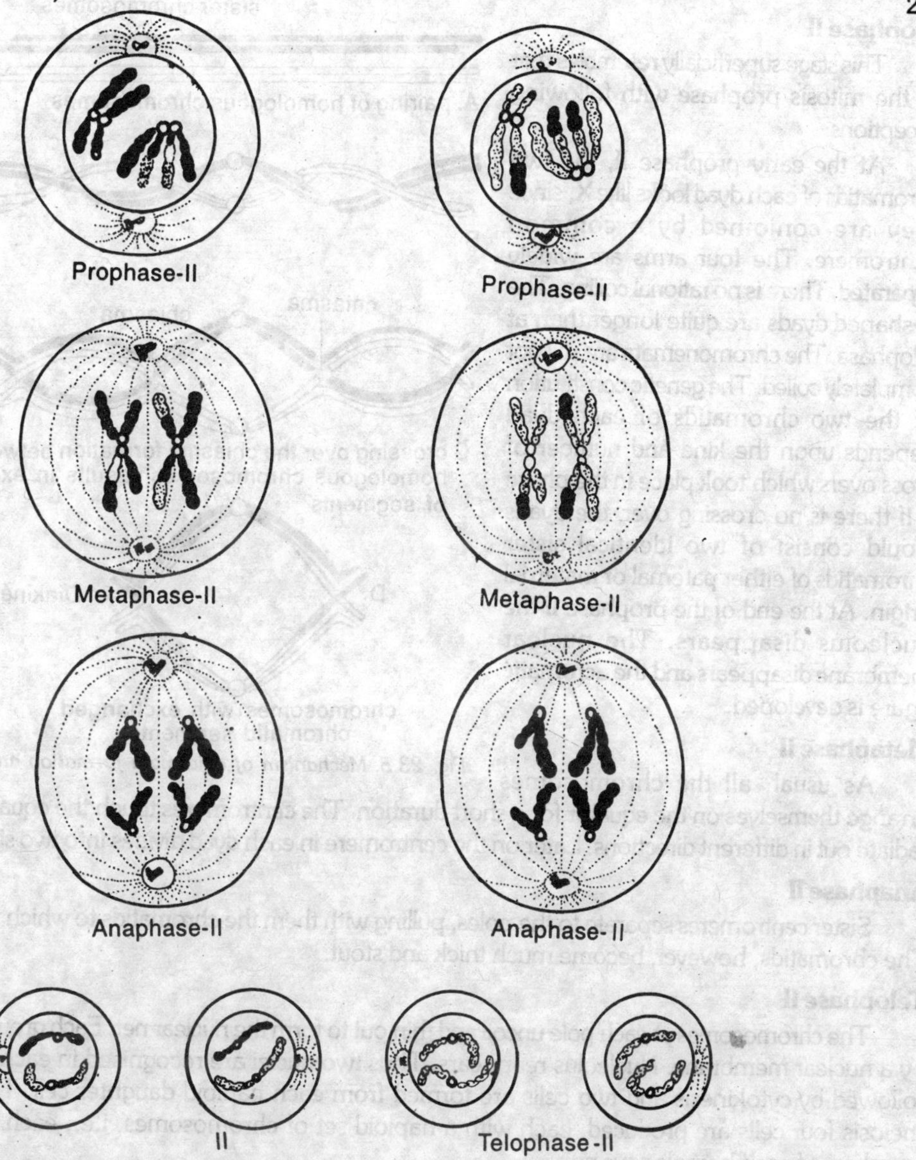

Fig. 23.4. Diagrammatic representation of homeotypic division of meiosis.

1. According to *Moses,* it is a prerequisite to chiasme formation and crossing over. But alone it is not sufficient. The presence of synaptonemal complex does not necessarily mean that chiasmata are present.
2. *King* (1970) suggested that synaptonemal complex may orient non-sister chromatids of homologous chromosomes in a manner that facilitates enzymatically induced exchanges between their DNA molecules.
3. Several evidences indicate that it is more directly related to the process of recombination. It has been observed that a small amount of DNA is synthesized at meiotic prophase. If this synthesis is inhibited, the function of synaptonemal complex is arrested.

II. Homotypic or 2nd Meiotic Division

Before the start of this division, most of the activities of heterotypic division cease for sometime. This resting stage is termed as the interkinesis. 2nd meiotic division involves following steps:

Prophase II

This stage superficially resembles that of the mitosis prophase with following exceptions:

At the early prophase II, the two chromatids of each dyad looks like X, since they are conjoined by a common centromere. The four arms are widely separated. There is no rational coiling. The X-shaped dyads are quite longer than at telophase. The chromonemata are still not completely coiled. The genetic constitution of the two chromatids of each dyad depends upon the kind and number of cross overs which took place in the phase I. If there is no crossing over, the dyads would consist of two identical sister chromatids of either paternal or maternal origin. At the end of the prophase II the nucleolus disappears. The nuclear membrane disappears and the acromatic figure is developed.

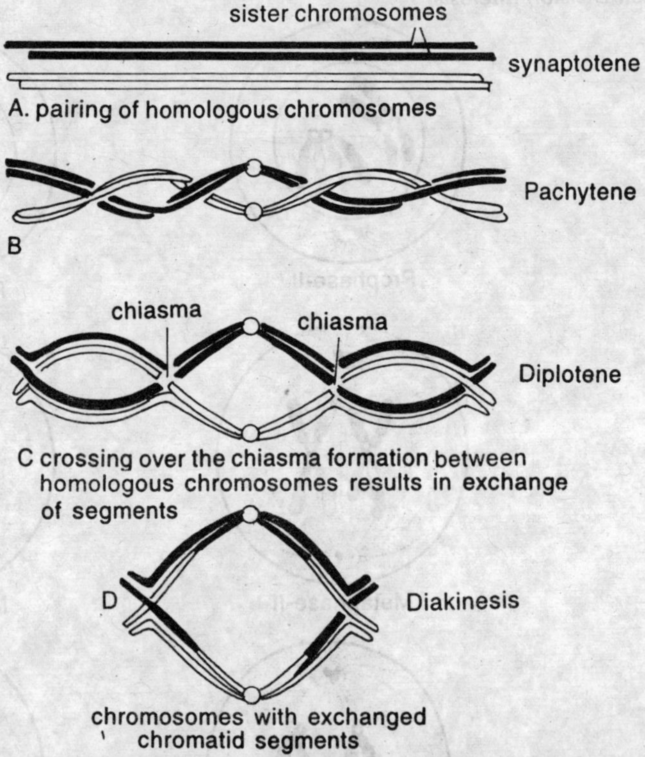

Fig. 23.5. Mechanism of chiasmata formation and crossing over.

Metaphase II

As usual, all the chromosomes arrange themselves on the equator for a short duration. The centromeres touch the equator but their arms radiate out in different directions. Later on the centromere in each dyad divides into two sister centromeres.

Anaphase II

Sister centromeres separate to the poles, pulling with them the chromatids to which they are attached. The chromatids, however, become much thick and stout.

Telophase II

The chromosomes at each pole uncoil and thin out to form the nuclear net. Each group gets surrounded by a nuclear membrane. Nucleolus reappears. Thus two nuclei are recognised in each cell. This is soon followed by cytokinesis and two cells are formed from each haploid daughter cell. Thus as a result of meiosis four cells are produced, each with a haploid set of chromosomes, i.e., each contains just one members of each homologous pair.

Cytokinesis

Sometimes meiosis I is followed by cytokinesis and sometimes it is deferred until the end of meiosis II. The details of the process are same as during mitosis. Cell plate formation results into a tetrad of cells with reduced chromosome numbers. The arrangement of cells in a tetrad is different in different organisms, but is characteristic of the species.

Like mitosis, meiosis too is a dynamic process and its different stages merge into subsequent ones. There is no sharp demarcation between different stages. DNA synthesis takes place during interphase, prior to prophase I. Duplication of chromosomes takes place during interphase itself and the chromosomes that appear during leptotene are double and not single. Number of chromosomes is reduced during meiosis I. If the reduction process is examined more closely, it becomes clear that there is qualitative reduction during meiosis I of those part of chromosomes which are not involved in exchange. The homologous chromosomes or the maternal and paternal chromosomes segregate or undergo qualitative reduction during meiosis I. Their quantitative reduction takes place during meiosis II, when their centromeres divide and the sister chromatids separate. But the reverse is true of those parts of chromatids which are exchange due to the formation of chiasmata. For these parts quantitative reduction takes place during meiosis I and qualitative reduction during meiosis II. Thus, complete of both qualitative and quantitative reduction takes place as a result of meiosis I and II and none of them alone can bring about complete reduction. It is, therefore, not proper to say that meiosis I is reductional and meiosis II equational. Duration of meiosis varies a great deal in different organisms. So does the duration of different stages in an organism. Prophase I lasts for the longer duration.

Type of Meiosis

There are three types of meiosis, *zygotic or initial meiosis, gametic or terminal meiosis* and *sporic or indeterminate meiosis.*

1. Zygotic or initial meiosis (haplotone pattern) occurs in lower plants. Fertilization is immediately followed by meiosis, giving rise to cells with the haploid chromosome number. The zygote is the only diploid stage in the life cycle.
2. Gametic or terminal meiosis (diplotonic pattern) occurs in animals. The cells resulting from the two meiotic divisions are transformed into sperms or eggs without further cell division. The process is as follows: Primordial germ cells divide repeatedly to give rise to many generations of daughter cells, spermatogonia (in males) or oogonia (in females). The spermatogonia and oogonia are transformed into primary spermatocytes and primary oocytes, respectively. Each primary spermatocyte undergoes the first maturation (meiotic) division and forms two secondary spermatocytes. Each secondary spermatocyte undergoes the second maturation (meiotic) division to give riseto two spermatids, which are then transformed into spermatozoa. In the female during the first meiotic divison the primary oocyte transformed into a large secondary oocyte and a smaller cell, the first polar body. The secondary oocyte undergoes the second meiotic division and divides into an ootid and a second polar body. The first polar body may or may not divide into two. The ootid is the ovum proper.
3. Sporic or indeterminate meiosis (diplohaplotonic pattern) is characteristic of higher plants and some thallophytes, but is not known in animals. In the diploid sporophyte, meiosis takes place at some inderminate point between the zygote and gametic formation. As a result of meiosis haploid megaspores and microspores are formed. The megaspores and microspores undergo repeated mitotic divisions and form the male and female gametophyte, respectively. Certain cells of the gametophyte diferentiate and form the gametes. In plants the cells that undergo meiosis are the microspore mother cell and the megaspore mother cell.

Significance of Meiosis

The process of meiosis keeps the chromosome number constant from one generation to other. This also involves reshuffling the hereditary characteristic which arises from reassortment of the parental chromosomes though crossing over. *Carothers* (1913) and others mentioned that in the metaphase I of meiosis it is usually a matter of chance whether the centromere of the paternal or the bivalent lies towards the particular pole. This results in random origin of new set on separation. There may be many new combinations. In fact, since each chromosome can go to either pole, the total possible number of arrangements will be 2n where n is the haploid number of chromosomes.

Darlington (1931) has suggested that the meiosis can be regarded as a precocious mitotic, and that the force which retains daughter chromatids together in mitotic prophase also remains the homologous chromosomes pairs in the earlier stages (*Ambrose*).

Role of Centriole During Meiosis

The centrioles perform precisely controlled synchronous function during meiosis. Where meiosis occurs in two steps, that is, two nuclear divisions take place in sequence, the centriole reproduces and functions to form two achromatic figures. In the first division of meiosis, the centromeres do not duplicate but rather homologous chromosomes pair together and move to opposite poles, while in the second meiotic division the centromeres of chromatids are separated. A scheme for relating various key events is presented briefly as under:

(i) Non-duplication of centromeres in meiosis-I

(ii) Non-duplication of chromosomes in meiosis-II, and

(iii) Duplication and function of centrioles in both divisions.

By a complete repression of the duplication simultaneous function of the chromatic apparatus brings about the reduction. The certriolar function is super-imposed upon the duplication process of the chromatic apparatus in a sequence which leads to the loss of a generation of chromatic apparatus. When meiosis is gametic and fertilization follows after, the immediate formation of zygote as a cell might contain the maternal as well as paternal centrioles, having endowed as it were, with the extra centriole generation. According to *Daniel Mazia* the kinetochore is an essential parts of the chromosomes so far as the mitosis is concerned. In gametogenesis there is a gain in total chromatic material, since these types of cell behaviour lead to duplication and subsequent reamassing of chromosomes. The extra set of centriole is rejected normally in fertilization or autogamy. Thus a cell results in a chromatic apparatus. This is accomplished by meiosis: one division meiosis gives the centriole a chance to 'catch up' in a single duplication.

Table 23.1. Showing Differences between Mitosis and Meiosis

Mitosis	*Meiosis*
1. Prolonged resting stage.	1. This stage is very short.
2. The prophase is short without any substages.	2. The prophase is subdivided into six stages, viz., Proleptonema Leptonema, Zygonema, Pachynema, Diplonema and diakinesis.
3. The chromosomes split longitudinally into sister chromatids during the preceding resting stage. From early prophase they appear as double tetrads.	3. The chromosomes do not split longitudinally.
4. The split homologues are not attracted towards each other.	4. At zygonema or succeeding stages, in prophase of first division, the homologous chromosomes are attracted towards each other and they form bivalents.
5. Synapsis and tetrad stages do not occur.	5. The chromosomes split after synapsis to form sister chromatids which are arranged in tetrads.
6. No chromosomal crossing over take place.	6. The crossing over is a prominent feature which results in exchange of genetic material.
7. The chromatids are coiled in plectonemic manner.	7. Coiling is paranemic type, and much evident.
8. Doubling of chromosomes takes place early in prophase.	8. Doubling of chromosomes occurs in pachytene stage of prophase.
9. Nucleolus and nuclear membrane disappear later on and the formation of nuclear spindle is completed with the commencement of metaphase.	9. The nuclear membrane and nucleolus disappear and the formation of acromatic figure is completed with diakinesis simultaneously.
10. The nucleus enlarges to some extent.	10. The nucleus is greatly enlarged.
11. At metaphase, the centrosome of each chromosome divides and the sister chromatids move towards the opposite poles. The somatic number of chromosomes is maintained without any reduction.	11. At metaphase I, the centrosome does not divide but the homologous chromosomes move towards opposite poles. The chromosome number is reduced to half by reduction division.

12. The chromosomes are seen in dyad consisting of two chromatids.	12. The chromosomes are in the form of tetrads having four chromatids.
13. Each chromosome of paternal and maternal origin is longitudinally divided and the separation to opposite side results in identical cells at the end of mitosis.	13. At metaphase I, the orientation of chromosome at the equatorial plate is in such pattern that either paternal or maternal chromosome goes to either pole. The orientation of one pair is independent of the other.
14. During orientation of dyads the centromeres face towards the equator and the arms directed towards the poles.	14. During the orientation of tetrads, the centromeres are placed equidistant and face the pole while arms are directed towards the equator.
15. The centromere of each bivalent undergoes division at metaphase.	15. No division of centromere occurs, but homologous chromosomes simply separate to form two dyads.
16. The initial movement of chromosomes at late metaphase is due to repulsion between the centromeric parts.	16. The main repulsive force lies between the centromeres of homologous chromosomes.
17. At anaphase the chromosomes are less thick and short.	17. The chromosomes are very thick and short.
18. At anaphase further separation of chromosomes is due to the formation of stem body.	18. Most of the changes are similar to mitosis.
19. At telophase two daughter nuclei, each with diploid number of chromosomes, are reconstituted.	19. At this phase two daughter nuclei are haploid.
20. There is no evidence of second division in mitosis.	20. There are two divisions for each meiotic division The second division follows after a short gap, interkinesis. The second division is homotypic by nature, and results into four cells, in the form of tetrads.
21. In the mitosis neither the chromosomal number nor the quality is changed and the daughter cells carry 2n chromosomes.	21. At the end of meiosis each of tertrads is haploid and quality is changed.
22. This division takes place in growing somatic cells.	22. This division takes place only in gonadial cells and result in the formation of gametes.

24

GAMETOGENESIS

The constancy in the number of chromosomes in the body cells of sexually reproducing organisms posed a problem. It should be obvious that in sexually reproducing organisms a reduction in the number would have to occur sometime in life cycle in order to maintain the constant number of chromosomes; if such a reduction did not occur, the number of chromosomes would double with each fertilization. The process in which the number of chromosomes reduces is known as *meiosis* and involves two rapid divisions. It occurs in animals in the formation of eggs and sperms and in plants in the formation of spores.

The process is essentially the same wherever it occurs in the formation of eggs in animals, in the formation of sperms in animals, and in the formation of spores in plants. In all of the preceeding discussion on the life cycles of both plants and animals, the process of meiosis has been continuously referred to without any discussion of the details of the process. As has been pointed out in above account, the cells in the life cycles of animals are all diploid (2n), with the exception of the eggs and sperms that are haploid (n).

The meiotic process in the formation of eggs in animals differs from that in the formation of sperms in one important respect. In the formation of sperm cells, each cell that undergoes meiosis, forms four functional sperm cells. Each sperm is composed of essentially of a *head* and a *tail,* which effects the movements of the sperm. In the formation of the eggs, on the other hand, only one functional ovum is formed from each cell that undergoes meiosis.

ORIGIN AND DIFFERENTIATION OF GERM CELLS

The origin of the germ cells during ontogeny, and their differentiation, have long occupied a central position of interest for obvious reasons. In most of higher animals the germ cells seem to arise in the "germinal epithelium" which occupies a localized region of the peritoneal or coelomic epithelium. The formation of germ cells can be traced during the early stages of cleavages. In such cases the primordial germ cells seem to lead an independent life within the soma, almost as if they were parasitic or symbiotic organisms. Zygote is supposed to be the stem cell which gives rise by divisions to both germ cells and somatic cells. From the stem cells arise the primordial germ cells which are then converted into gonia to form gonads. These cells show the same general characters in both sexes, but are sooner or later differentiated as primary oogonia or spermatogonia.

Formation of Sperm

Sperms are formed by the germ cells which are derived from the germinal epithelium of the embryo. The germ cels and somatic cells are always diploid in nature. In higher animals where the definite sex organs are present, the male sex organs are the testes which are composed of a number of germ cells. The formation of spermatids from germ cells is called *spermatogenesis* while their metamorphosis into sperm is called *spermiogenesis.* All the germ cells do not take part in spermatogenesis while few of them simply provide nourishment to the cell forming spermatids, these cells are called *nutritive cells* or *Sertoli cells*, while the remaining cells are called the *primary germ cells* or *primordial cells*.

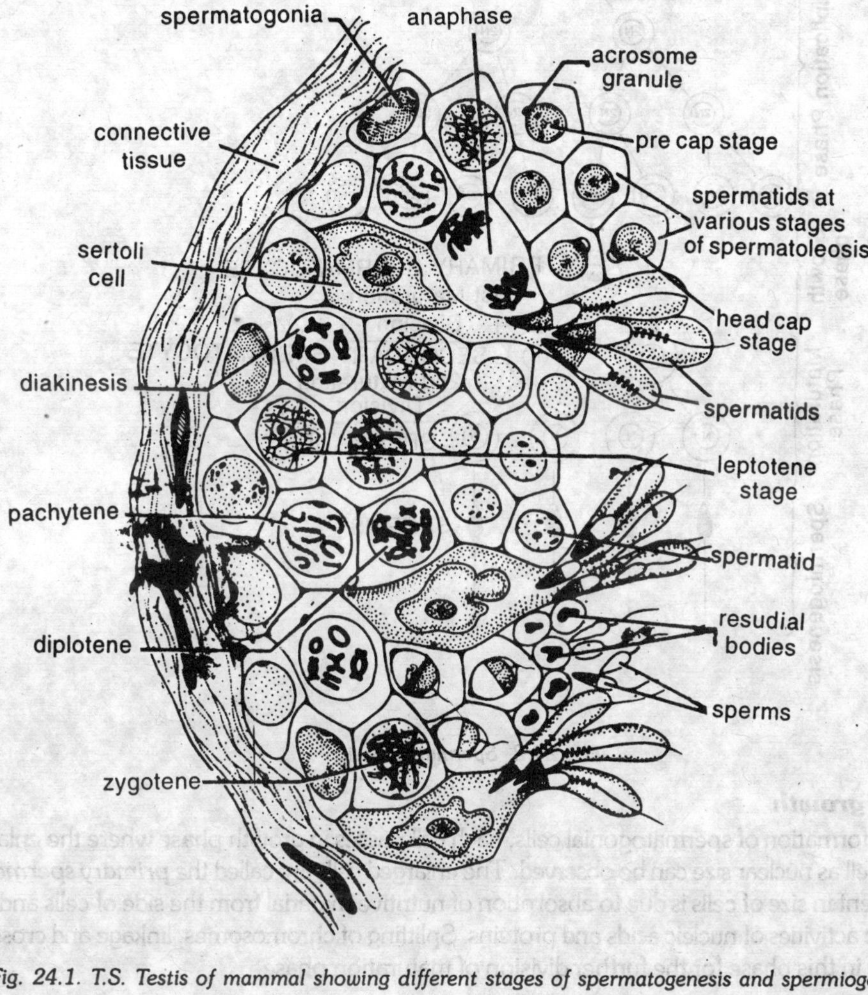

Fig. 24.1. T.S. Testis of mammal showing different stages of spermatogenesis and spermiogenesis.

Spermatogenesis

The process of spermatogenesis can be studied under the following heads:

(i) Multiplication phase or phase of multiplication.

(ii) Growth phase or phase of growth.

(iii) Maturation phase or phase of maturation.

All the germ cells are diploid having a nucleus in the middle of each cell, and in cytoplasm a number of cytoplasmic organelles of are embedded such as Golgi bodies, Mitochondria, Endoplasmic reticulum, Centrosome containing centrioles etc.

(i) Phase of multiplication

As explained above that all the cells of testes do not form spermatids. Those cells which help in nutrition are called the nutritive cells or Sertoli cells and the remaining cells are promordial or primary germ cells. All the primary germ cells divide by repeated mitotic divisions to produce a number of cells having the same number of chromosomes (diploid). Thus a large number of *spermatogonia* are formed by the mitotic divisions of primary germ cells.

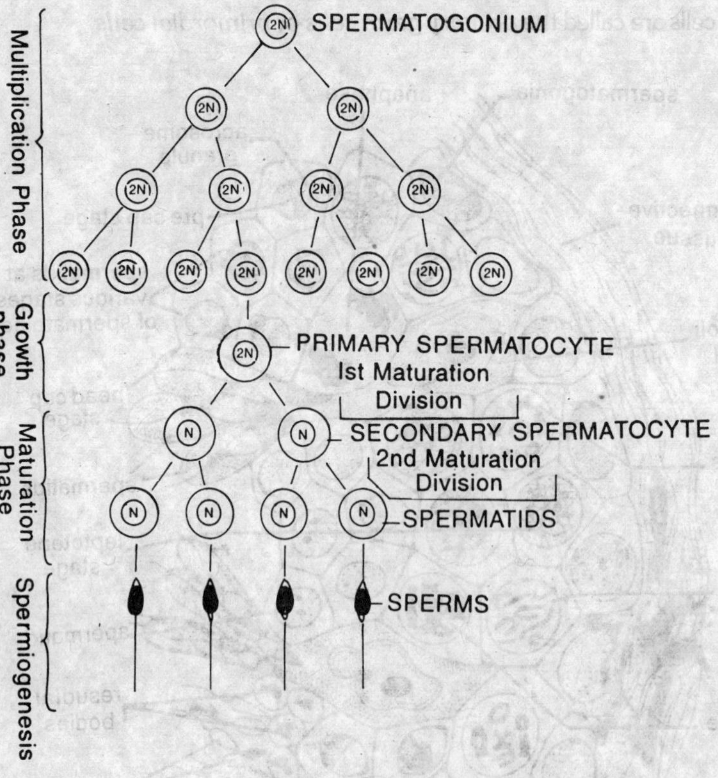

Fig. 24.2. Spermatogenesis.

(ii) Phase of growth

After the formation of spermatogonial cells, each cell passes in growth phase where the enlargement of cell size as well as nuclear size can be observed. The enlarged cells are called the *primary spermatocytes*. This enlargement in size of cells is due to absorption of nutritive material from the side of cells and also due to the synthetic activities of nucleic acids and proteins. Splitting of chromosomes, linkage and crossing over also take place in this phase for the further division of maturation phase.

(ii) Phase of maturation

After the enlargement of the cell during phase of growth, the primary spermatocytes jump to the phase of maturation. Here the maturation or reduction division occurs. Reduction division is double as it is composed of two divisions:

(a) First meiotic or first maturation division which istrue reductional in nature.

(b) Second meiotic or second maturation division which is true mitotic in nature.

Two haploid *secondary spermatocytes* are formed from one diploid primary spermatocyte by first meiotic division, and from each haploid secondary spermatocytes, two haploid *spermatids* are formed by the second meiotic division. Thus from one primary diploid spermatocyte, four haploid spermatids are formed during the phase of maturation. The spermatids do not divide further, so after their formation from germ cells, they undergo metamorphosis to form sperms.

Metamorphosis or Spermiogenesis

During spermiogenesis a number of haploid spherical spermatids are formed. Each spermatid contains a haploid nucleus, some what in the middle and is surrounded by the cytoplasm and plasma membrane. In the cytoplasm are embedded a number of cytoplasmic organelles such as Golgi bodies, Mitochondira, Centrosome etc. These various organelles undergo certain modification during the process when spermatids form the spermatozoa. These changes are well-studied in Hemipterans.

Role of Centriole during Spermiogenesis

Centrosome is generally present in the vicinity of nucleus towards the anterior side. Centrosome may have one or two centrioles. If there is one centriole, it divides during the process to form two centrioles. Then the membrane encircling the centrioles, breaks down. Thus both the centrioles become free lying in the cytoplasm. One of these centrioles remains against the nuclear membrane and is called the *proximal centriole* while the other one which is away from nuclear membrane is called the distal centriole. From the *distal centriole* a delicate filament which is *axial filament* emerges out piercing the plasma membrane. After the formation of *axial filament,* both the centrioles along with axial filament move towards the posterior side of the nucleus, where they take their final position. Here the *distal centriole* which may form a ring divides into two, one of which forming the *end knob* and the other passing back as a ring to the end of the connecting piece.

The axial filament, when viewed under the ordinary microscope, is seen to be composed of single fibre but when seen by electron microscope eleven fibres can be observed easily, arranged in two rings, out of which two fibrils are single and are present in the middle while remaining nine double firbisl are arranged against the periphery. But in mammalian sperm one additional ring of nine single fibres are present just out side the ring of nine tubular doublets.

Role of Mitochondria in Spermiogenesis

The helicular portion of the collar region of the sperm is formed by the mtochondria of spermatid. The helicular portion is called the *mitochondrial helix* or *spiral*. The majority of workers believe that the mitochondrial spiral is a continuous formed by the fusion of mitochondria end to end. But according to *Nath* (1965) all the mitochondria of spermatid do not fuse but retain their individuality.

The formation of mitochondrial helix is well studied in insects where all the mitochondria form a more or less compact body known as *nebenkern*. Soon after its formation, the inner contents of the nebenkern is differentiated into dark stained portion or areas forming a series of cencentric plates alternating with layers of the lighter substance. Then the nenbenkern elongates in length and at the same time a furrow is developed which moves inwards and eventually cuts the nebenkern into two portions. Both the portions constitute two elongated ribbon like structures

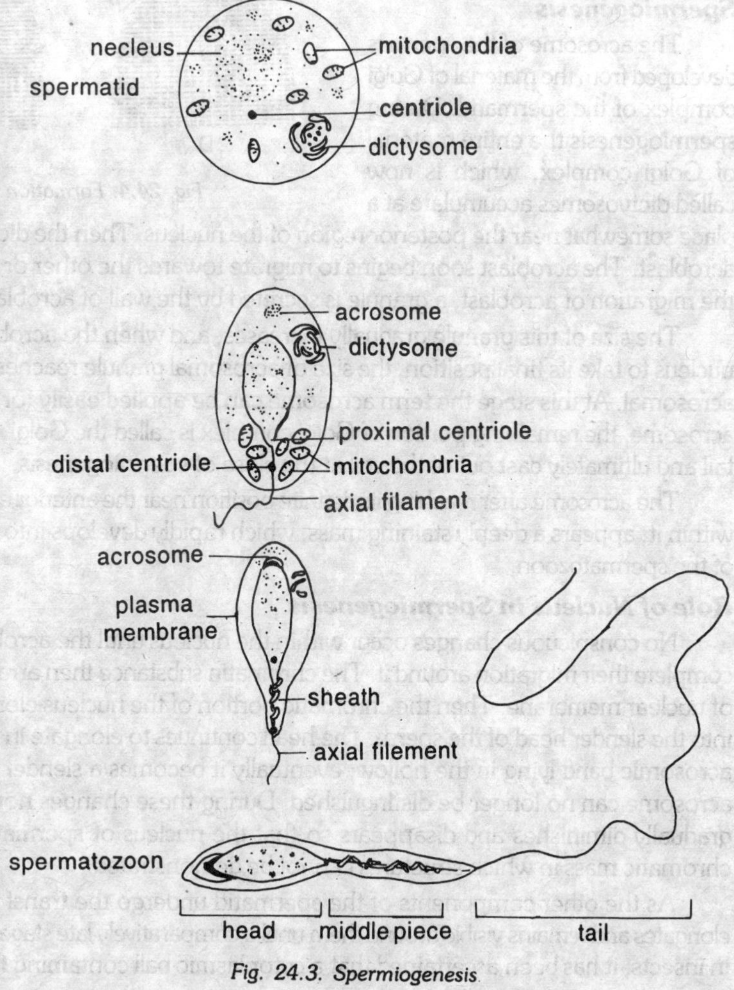

Fig. 24.3. Spermiogenesis

having lighter and denser portions respectively. Then both these filaments coil around each other but in such a precise manner that the axial filament is covered on either side by the coiling of filaments which now at this stage constitutes the mitochondrial helix or spiral.

According to *Burgos* and *Fawcett* (1955) during the spermiogenesis the internal structure of mitochondria undergoes a marked change, the cristae which are perpendicular to the outer membrane in normal mitochondria become irregular in outline and many fill the internal canal of spiral (*Fawcett*, 1955) as in opossum with *Yasnzumi* (1956) has reported that in mitochondrial spiral of albino rat, the cristae are entirely absent.

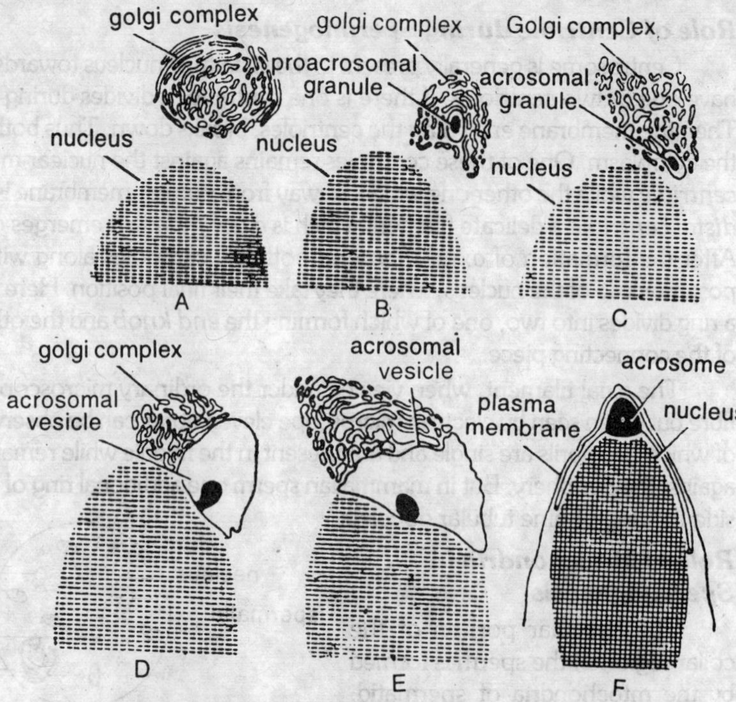

Fig. 24.4. Formation of acrosome of sperm.

Role of Golgi complex in Spermiogenesis

The acrosome of the sperm is developed from the material of Golgi complex of the spermatid. During spermiogenesis the entire material of Golgi complex, which is now called dictyosomes accumulate at a place somewhat near the posterior region of the nucleus. Then the dictyosomes fuse to form a single large acroblast. The acroblast soon begins to migrate towards the other or anterior side of the nucleus. During the migration of acroblast, a granule is secreted by the wall of acroblast which is acrosomal granule.

The size of this granule gradually increases, and when the acroblast reaches the anterior side of the nucleus to take its final position, the size of acrosomal granule reaches its peak and fills the entire space of acrosomal. At this stage the term acrosome can be applied easily for the acrosomal granule. Besides the acrosome, the remaining portion of Golgi complex is called the Golgi remnant, passes backward along the tail and ultimately cast out of the cell at the close of spermiogenesis.

The acrosome after reaching its definite position near the anterior end undergoes a further differentation, within it, appears a deeply staining mass, which rapidly develops into the pointed perferotorium at the tip of the spermatozoon.

Role of Nucleus in Spermiogenesis

No conspicuous changes occur within the nucleus until the acroblast and its differentiated acrosome complete their migration around it. The chromatin substance then arrange itself in a layer against the inside of nuclear membrane. Then the chromatic portion of the nucleus elongates and begins its transformation into the slender head of the sperm. The head continues to elongate in the form of a narrow spoon with the acrosomic band lying in the hollow; eventually it becomes a slender pointed body in which nucleus and acrosome can no longer be distinguished. During these changes non-chromatic constituents of nucleus gradually diminishes and disappears so that the nucleus of spermatozoon is an extreme concentrated chromatic mass in which structures cannot be demonstrated.

As the other components of the spermatid undergo the transformation, the cytoplasmic cell body elongates and remains visible around them until a comparatively late stage. Towards the close of spermogenesis, in insects, it has been ascertained that a cytoplasmic ball containing Golgi remnant and other granules of unknown nature is sloughed off and ingested by the nutritive epithelial cells lining the spermatocyte. In other animals a cytoplasmic mass is like-wise eleminated but the amount of active protoplasmic contained in it, is uncertain.

Structure of a Mature Sperm

The body of a mature sperm can be divided into following parts:

1. Head

The head is the uppermost portion of the sperm bearing acrosome developed by Golgi complex. It may be spherical as in teleosts and certain invertebrates, rod shaped or filliform as in insects; flattened or spoon shaped as in Guinea pig, man and many other mammals; hook shaped as in rat etc. Head of the sperm is further differentiated into two parts:

(a) *Acrosomal part.* The anterior most region of head which is smaller than the nuclear part is called the acrosomal part. It exhibits a great diversity of form; varying from a small granule or knob to a large conical or spine like process. The acrosome of acrosomal part is developed from the Golgi complex of the spermatid.

(b) *Nuclear part.* constitutes of nucleus and occupies the major portion of head. Nucleus of sperm is composed of mainly a solid and homogenous mass of basichromatin. The development of the sperm shows that both nucleus and acrosome are covered by a very thin cytoplasmic covering, perhaps a cell membrane.

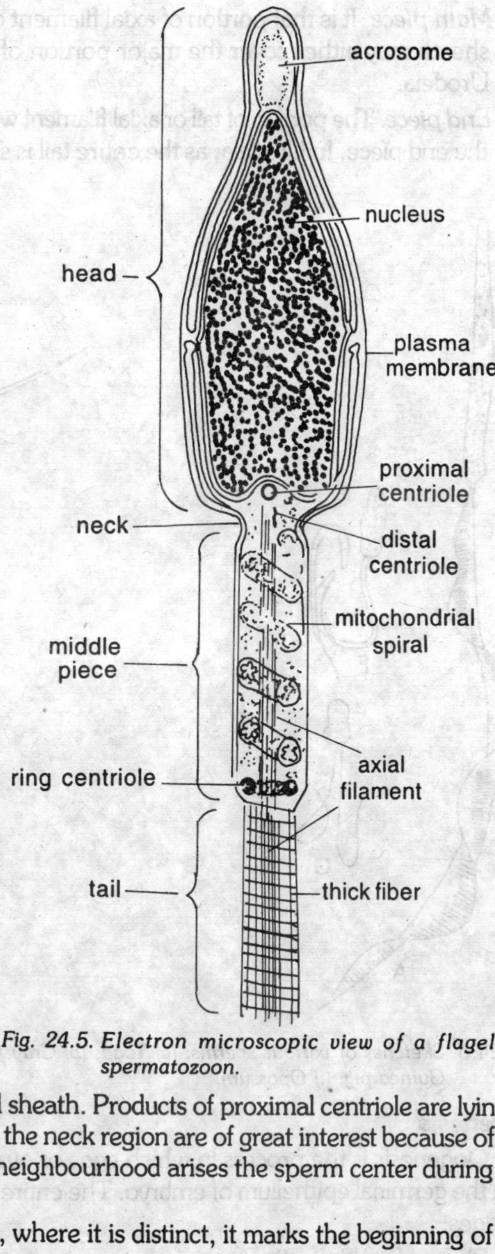

Fig. 24.5. Electron microscopic view of a flagellate spermatozoon.

2. Neck

It is the portion of sperm which is present between head and collar region. It forms a stalk by which the flagellum is attached to the head. This region often appears not to be surrounded by a mitochondrial sheath. Products of proximal centriole are lying in this region. The relations of the centriolar apparatus to the neck region are of great interest because of the evidence that from this region of the sperm, or in its neighbourhood arises the sperm center during the fertilization of the egg.

Neck region is not distinct in mammalian sperm, where it is distinct, it marks the beginning of the axial filament.

3. The Middle piece

It is the portion of sperm where the axial filament is covered by the helicular structure of mitochondria. Mitochondrial helix is connected by the products of distal centrioles on either ends. This mitochondrial helix provides energy for the movement of sperm during fertilization. The energy resides in mitochondrial helix in the form of ATP.

4. Tail

Generally tail is present in all the sperms which is in the form of an elongated filament (axial filament) and is extruded out from the distal centriole during spermiogenesis. But in certain cases as in nematodes and arthropods (myripods, arachnids, crustaceans etc.) the sperms are not provided with tail.

Tail is further differentiated into two parts of pieces:

(a) *Main piece.* It is that portion of axial filament or tail which is enveloped by cytoplasmic sheath. This sheath may either cover the major portion of the filament or the entire tail may be sheathed as in Urodela.

(b) *End piece.* The portion of tail or axial filament which is not enveloped by cytoplasmic sheath, constitutes the end piece. In Urodela, as the entire tail is sheathed, end piece is lacking.

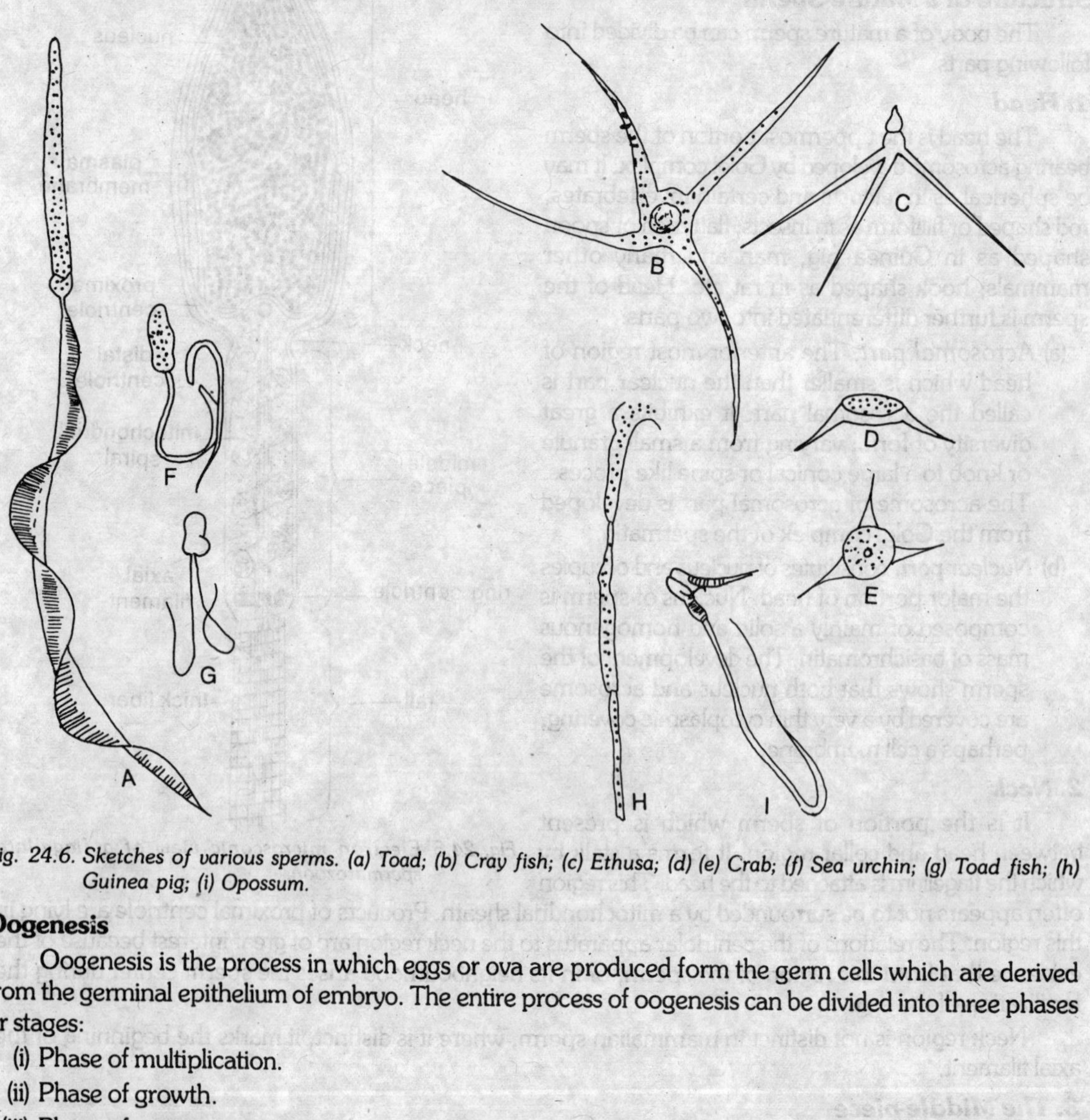

Fig. 24.6. Sketches of various sperms. (a) Toad; (b) Cray fish; (c) Ethusa; (d) (e) Crab; (f) Sea urchin; (g) Toad fish; (h) Guinea pig; (i) Opossum.

Oogenesis

Oogenesis is the process in which eggs or ova are produced form the germ cells which are derived from the germinal epithelium of embryo. The entire process of oogenesis can be divided into three phases or stages:

(i) Phase of multiplication.

(ii) Phase of growth.

(iii) Phase of maturation.

(i) Phase of multiplication

During this phase primary germ cells divide in such a way that chromosome number in daughter cells remains as such as they (germ cells) divide mitotically to produce a large number of daughter cells which are the *oogonial cells.* Oogonial cells contain the diploid number of chromosomes as they are produced from the diploid primary germ cells by mitotic divisions.

(ii) Phase of growth

Phase of multiplication is replaced by growth phase, where the oogonial cells simply enlarge in size due to the absorption of nutritive material from the surrounding medium and also due to the synthesis of nucleic acid and proteins. The phase of growth can also be labelled as Ist prophase of reduction division as the replication of chromosomes, linkage and crossing over occur during this phase. After these changes each oogonial cell is then called the *primary oocyte*. All the primary oocytes are once again diploid as none of the division has been traced during this phase, so there is no disturbance in the number of chromosomes.

Phase of Maturation

The phase of growth is replaced by phase of maturation. During this phase the reduction division occurs which is double in nature, as it is composed of first meiotic division which is true reduction division and second meiotic division which is true mitotic in nature.

But in oogenesis both the divisions are incomplete as in both the cases nucleus divides properly but the division of cytoplasm is not regular. In first meiotic division the diploid nucleus of primary oocyte divides to form two haploid nuclei, but the cytoplasm does not divide. Then one of the daughter nucleus moves towards the periphery and penetrate out of the vitelline membrane to lie beneath the second egg membrane, the chorion. This migrated body constitutes the *first polar body* while the remaining structure with haploid nucleus is called *secondary oocyte*. Then the haploid nucleus of secondary oocyte divides by second meiotic divison to form two daughter haploid nuclei. Again one of these nuclei migrates towards the periphery to constitute the *second polar body*. After the extrusion of second polar body the developing structure is called the *ootid*. The first polar body may or may not divide to form two polar bodies. The polar bodies disintegrate later on. Here in oogenesis only one haploid ootid is formed from one diploid oogonial cell.

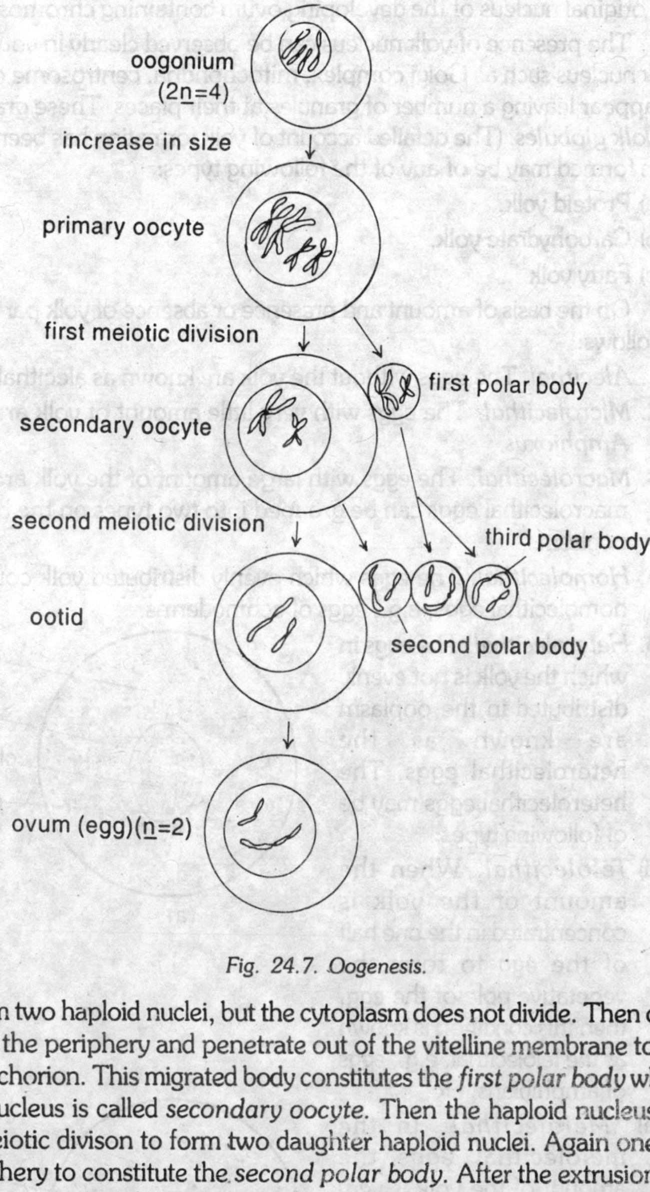

Fig. 24.7. Oogenesis.

Changes in the Cytoplasm

An early oocyte is spherical in outline having a nucleus in the middle, while the remaining portion of the oocyte is filled with cytoplasm along with a number of cytoplasmic organelles such as Mitochondria, Golgi complex, cell center etc. All these organelles take an important role in *vitellogenesis*. Cytocenrum or cell center or centrosome is situated in the vicinity of nucleus enclosing one or two centrioles. The word *archoplasm* or *idiosome* can be applied for the denser cytoplasm in which centrioles are embedded in centrosome. Then soon the nucleus which is haploid in nature is surrounded by rings or circles developed by the arrangement of Golgi complex, mitochondria and centrosome in circles. The entire circular rings thus formed can be labelled as *Balbianic vitelline rings* or *Yolk nucleus.* Yolk nucleus may either surrounded the original nucleus of the developing ovum containing chromosomes, or may lie adjacent to it.

The presence of yolk nucleus can be observed clearly in young oocyte. Then the entire organelles of yolk nucleus such as Golgi complex, mitochondria, centrosome etc., disperse through the cytoplasm and disappear leaving a number of granules at their places. These granules or globules are the *yolk granules* or *yolk globules.* (The detailed account of yolk formation has been given under separate heading). The yolk thus formed may be of any of the following types:

(a) Proteid yolk.

(b) Carbohydrate yolk.

(c) Fatty yolk.

On the basis of amount and presence or absence of yolk particles the eggs of animals can be classified as follows:

1. *Alecithal.* The eggs without the yolk are known as alecithal eggs. e.g., Eutherian mammals.
2. *Microlecithal.* The eggs with very little amount of yolk are known as the microlecithal eggs, e.g., *Amphioxus.*
3. *Macrolecithal.* The eggs with large amount of the yolk are known as the macrolecithal eggs. The macrolecithal eggs can be grouped into two types on the basis of the distribution of the yolk in the ooplasm:

A. *Homolecithal.* The eggs which evenly distributed yolk contents in the ooplasm are known as the homolecithal eggs, e.g., eggs of echinoderms.

B. *Heterolecithal.* The eggs in which the yolk is not evenly distributed in the ooplasm are known as the heterolecithal eggs. The heterolecithal eggs may be of following types:

(i) *Telolecithal.* When the amount of the yolk is concentrated in the one half of the egg to form the vegetative pole of the egg, then this condition is known as the telolecithal, e.g., eggs of amphibians, etc.

(ii) *Meiolecithal.* In the meiolecithal eggs the amount of the yolk is very large and it occupies the largest portion of the egg except a small disc shaped portion of the cytoplasm. The cytoplasm contains the zygote nucleus and is known as the germinal disc. e.g., eggs of fishes, reptiles and birds.

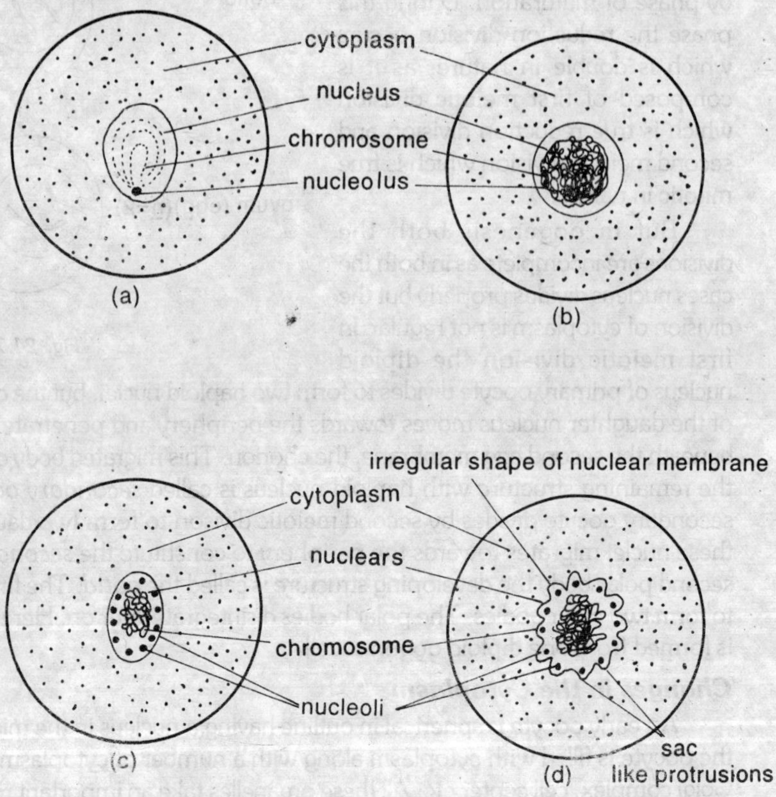

Fig. 24.8. Showing the bouquet arrangement of chromosomes (A) and the protrusion of nucleoli in the cytoplasm (BCD).

(iii) *Centrolecithal.* In the centrolecithal eggs the yolk accumulates in the centre of the ooplasm, e.g., eggs of insects.

Changes in the Nucleus

No conspicuous changes have been traced upto the formation of primary oocyte. After its formation a sequence of changes have been observed. When the oogonial cells enlarge in size to enter in the *phase of growth* or *vegetative phase* and form the primary oocytes, the chromatin material assumes definite chromosomes. The free ends of all the chromosomes than arrange themselves in such a precise manner that they face towards one pole. This stage is called the *bouquet stage.* The chromosomal threads which constitute the bouquet arrangement are also known as *leptotene threads.* The leptotene threads now conjugate to give rise haploid number of bivalents. At the same time numerous side branches or side loops are developed by the chromosome. Such chromosomes bearing side loops are called *lampbrush chromosome.*

Nucleolus also plays an imporant role in the synthetic activity taking place in the nucleus. The small nucleus may give rise to several nucleoli by fragmentation and budding. All these nucleoli come to lie near the nuclear membrane.

Due to these changes the outline of nucleus become irregular which shows sac like protrusions into which lie nucleolar fragments from where they pass out into the cytoplasm.

Egg Membranes

Three kinds of egg membranes have been observed so far:

(i) Primary egg membrane
(ii) Secondary egg membrane
(iii) Tertiary egg membrane.

(i) Primary egg membrane

It is innermost membrane of the egg which is secreted by the egg itself oogenesis. It is also called *vitelline membrane.* The vitelline membrane is formed by the hardening of the outer peripheral layer of cytoplasm of the oocyte. Some time it is the only egg membrane present as in Hydra, earthworm and in mammals (placentals).

(ii) Secondary egg membrane

More commonly outside the primary vitelline, are present secondary and tertiary membranes. The secondary membranes are formed by the accessory or follicular cells of the ovary closely associated with the egg. The membrane which is secreted by ovary is called the chorion as in insect eggs and certain cephalopod eggs. In certain cases where it is tough, is provided with one or more than one aperture or micropyle for the entry of sperms.

(iii) Tertiary egg membrane

In the eggs of frogs, birds, reptiles and oviparous mammals, secondary egg membrane is enveloped by another hard or gelatinous material which constitutes the tertiary egg membrane. In frog it is made up of 4-6 gelatinous rings. In birds and reptiles, the white hard shell constitutes the tertiary egg membrane and is secreted by the inner wall of oviduct.

Comparison of Spermatogenesis and Oogenesis

From the above account it is quite clear that both the processes are alike to each other. The following comparison can be made between the two processes:

Similarities

(i) Both the processes start from a primordial germ cell.
(ii) Both pass through three phases, namely multiplication, growth and maturation.

(iii) The process of synapsis or pairing of homologous chromosomes occurs in both the processes during the growth phase.

(iv) In the maturation phase, both have two divisions.

Differences

(i) The spermatogenesis occurs in the testes while the oogenesis in the ovary cells.

(ii) The process of spermatogenesis results in the formation of four similar sperm cells from each primary spermatocyte. All these four are functional, while oogenesis results in the formation of only one large cell, from each primary oocyte, the egg cell and three non-functional small cells or polar bodies (polocytes).

(iii) Mature sperm cells are very small and are also very active. On the other hand mature egg cell, has the large size and is very passive and receptive.

(iv) All sperm cells are functional and due to much greater number of multiplication, the total number of sperms produced is enormously greater than the numberof eggs.

(v) Complete process of spermatogenesis finishes in the testes while the oogenesis is completed outside the ovary.

25

FERTILIZATION

In most of living organism sexual reproduction takes place. In this the gametogenesis results male and female gametes. The union of the male and female gametes is known as the *fertilization* (L.,*fertilis*-to bear). The fertilization is the only method for the production of the diploid zygotes in vertebrates. In the process of fertilization the haploid male gamete (spermatozoa or pollen in plants), which carries the paternal genetic information, unites with the haploid female gamete (ovum or egg), which carries genetic informations of female parent, to form a diploid zygote. The zygote ultimately, produces a diploid multicellular organism by the several repeated and organised mitotic divisions and cellular differentiation.

External and Internal Fertilization

The fertilization always occurs in the aquatic media such as sea water, fresh water or intra-somatic (body) fluid of the maternal individual. If the fertilization occurs outside the body of the organism it is known as *external fertilization* and if it occurs inside the body of the organism then it is termed as *internal fertilization*. The external fertilization is common in various invertebrates and chordates, while the internal fertilization occurs only in those animals which possess specialized sex organs for receiving and transmitting the sperms, e.g., reptiles, birds, mammals and angiosperms, etc.

MECHANISM OF FERTILIZATION

The mechanism of fertiIization is complicated one and is completed in the following stages:

1. Approach of the Spermatozoon to the Egg

Whether fertilization is external or internal, the first step is the encounter of spermatozoon and the ovum, which is brought about by swimming movements of spermatozoa. Encounter between sperm and ovum is purely accidental, because there are a few cases in which the sperms are guided towards ovum by chemical substances.

2. Agglutination

In the presence of water (water in which eggs have been lying for sometime), the spermatozoa become "sticky" and adhere to the surface of the egg to its membrane and even to each other. This mutual adhesion of the spermatozoa results in their clumping or agglutination which is easily observed with the sperms of some animals, specially in sea urchin. The agglutination and sticking of sperms to eggs of the same species is found in most animals. The agglutination depends to a large extent on environmental conditions and on certain chemicals such as "*fertilizin*" which is generally present in the surface layer of the egg. The fertilizin is glyco-protein or mucopolysaccharide in chemical composition. The molecules of the fertilizin are quite large having the molecular weight about 300,000 and each molecule may have more than one "active group" by which it may become attached to the surface of the spermatozoa.

Another chemical substance present in the surface layer of the cytoplasm of spermatozoa is the *antifertilizin,* having a low molecular weight than fertilizin. Molecular weight of antifertilizin is about 10,000. The antifertilizin can be extracted out from the spermatozoa by heating, freezing or acidifying the water. Both fertilizin and antifertilizin are having certain specific peculiarity as they combine in a specific manner i.e., the egg fertilizin of any species reacts best with the sperm antifertilizin of the same species. These can react with other species even, but the rate of reaction may be much weaker, this is possible only in the case when two species are fairly nearly related to one another.

The reaction between the fertilizin and antifertilizin molecules accounts for both the agglutination of the spermatozoa by egg water and the adhesion of the spermatozoa to the egg membrane or egg surface. The fertilizin particles have more than one reactive spots on their surfaces so that one fertilizin particle may become attached to two or more spermatozoa, thus binding together.

This reaction or adhesion of a spermatozoon to the surface of the egg or to the egg membrane is the result of the establishment of a bond between the antifertilizin molecules embedded in the surface layer of the cytoplasm or the egg membrane.

3. Penetration of the Sperm

The penetration of the sperm is based on the chemical mechanism. A substance which is an enzymatic in nature is produced by the spermatozoon is known under the general name of sperm lysins.

The sperm lysins are produced by the acrosome of the spermatozoon. This is in accord with the origin of the acrosome; we have seen that the acrosome is developed from the Golgi complex of the spermatid. There is clear evidence that in normal animal cell at least some of the secretory activity is performed by the Golgi complex. The production of sperm lysins is thus one of the aspects of the secretory activity of these bodies. The sperm lysins certainly differ from one animal group to another.

The sperm lysins make a path clear for the entrance of sperm by dissolving the egg membrane. In mammals the egg

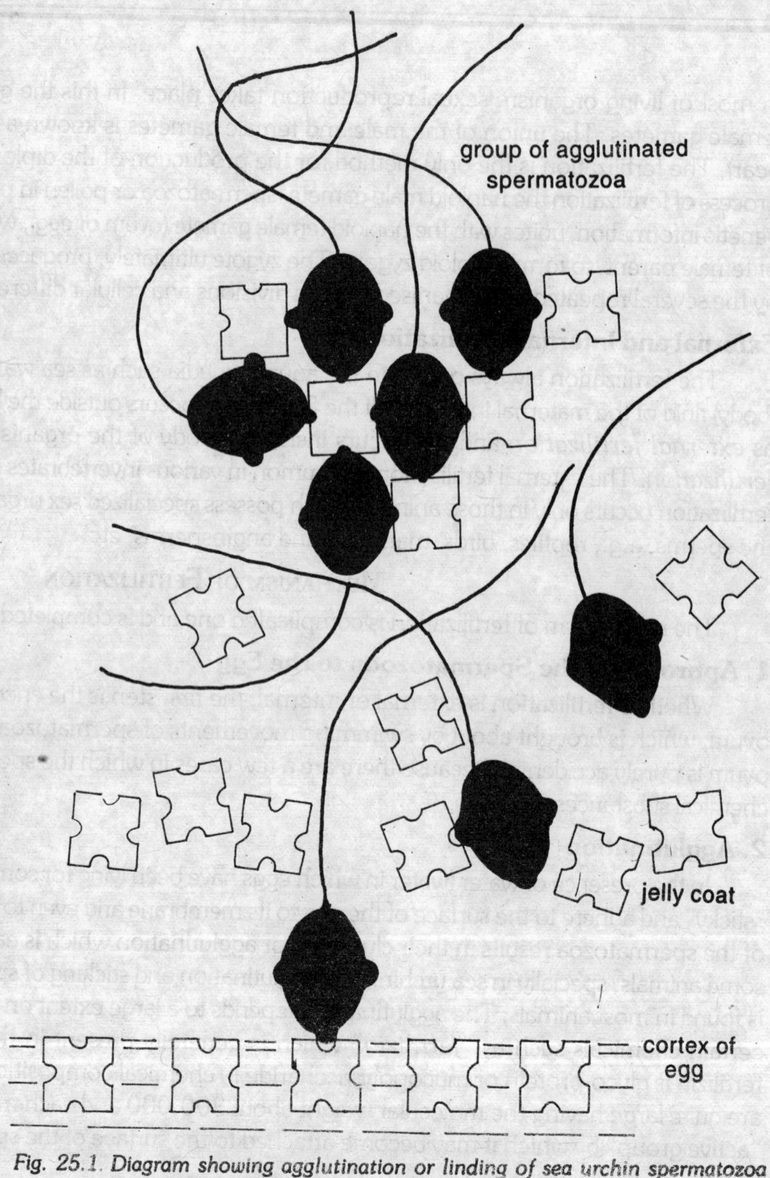

Fig. 25.1. Diagram showing agglutination or linding of sea urchin spermatozoa by fertilizin particles.

is surrounded by follicular cells which constitute the zona radiata. The penetration of the sperm through the zona radiata is also supported by another enzymatic substance called *hyaluronidase*, which is produced by it and dissolves the mucopolysaccharide, hyaluronic acid. The substance (mucopolysaccharide hyaluronic acid) cementing the follicle cells together, which is dissolved by hyaluronidase enzyme produced by the sperm. It also makes a narrow canal through the zona pellucida, which lies inside the corona radiata, by dissolving the substance of the zona, and so reaches the surface of the egg.

In certain cases the eggs are having very thick and resistance membranes, such as the chorion. Such eggs are not directly penetrated by the sperms. Here in such cases that sperm must penetrate through a special canal called micropyle or outgrowth left in the egg membrane, the receptor cone.

In certain cases, the egg membrane is said to be dissolved by the acidification of water. As in sea urchin's egg, seawater is acidified by CO_2 produced by the spermatozoa during their respiration.

Formation of acrosomal filament

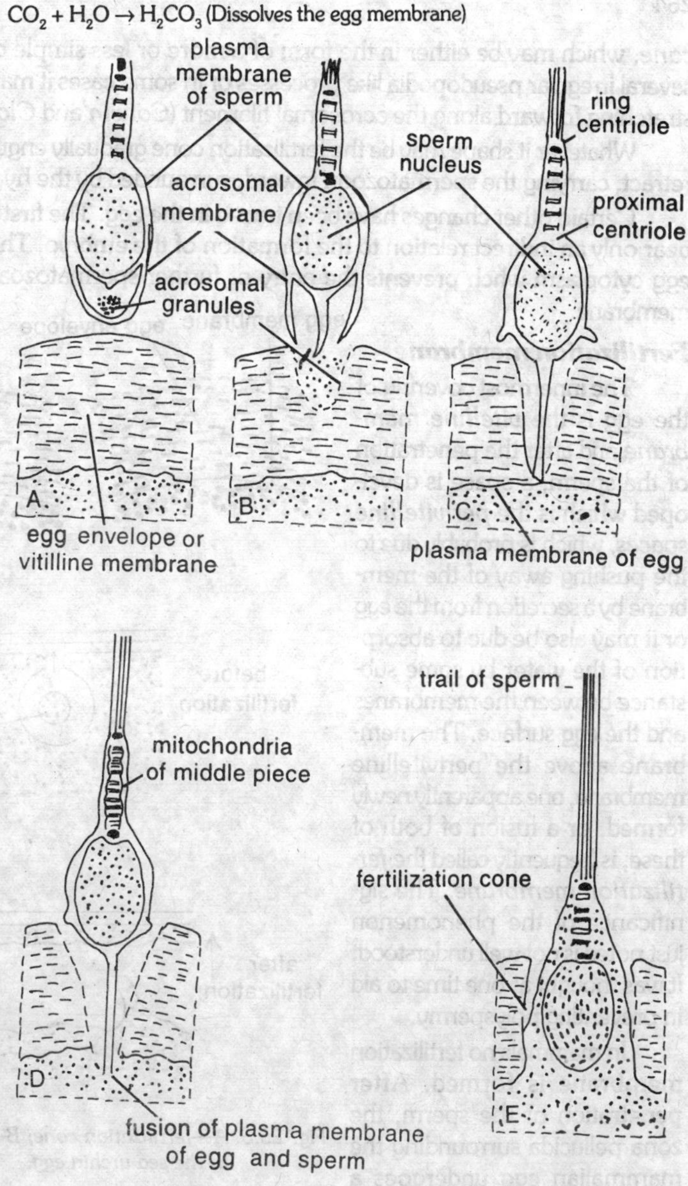

Fig. 25.2. Hypothetical sketch of the entrance of a sperm into an ovum.

Not only the agglutination of the spermatozoa is caused by the eggs, but even under the influence of fertilizin, we find certain other changes in the structure of spermatozoa particularly in the peripheral portion which collapses and becomes extruded and partially dissolved in water (possibly releasing the lysins). The central transparent part of the acrosome elongates and becomes transformed into a long (1 to 75 μ) thin filament or possibly a tube, which has considerable rigidity and produces forward from the head of the spermatozoon. This structure has been called by *Dan Wada* (1955) the *acrosomal filament*. The formation of such filaments activate the sperms to penetrate into the egg. These acrosomal filaments are pushed through the jelly (if present), as the spermatozoon approaches the egg, and through the vitelline membrane, the path way far it being cleared by the action of the lysine. Eventually the tip of the acrosomal filament touches the surface membrane of the egg cytoplasm and triggers off the next phase of the process of fertilization.

4. The Reaction of the Egg

As soon as the acrosomal filament of the spermatozoon touches the surface of the egg, the cytoplasm of the egg bulges forward at the point of contact, producing a process of hyaline cytoplasm, the *fertilization cone*, which may be either in the form of a more or less simple conical protrusion, or it may consists of several irregular pseudopodia like processes or in some cases it may take the form of a cytoplasmic cylinder stretching forward along the acrosomal filament (*Colwin* and *Clowin* 1957).

Whatever it shape may be the fertilization cone gradually engulfs the spermatozoon and then begins to retract, carrying the spermatozoon inward, surrounded by the hyaline cytoplasm of the fertilization cone.

Certain other changes have been traced in the egg. The first observable change in the egg, however, bear only an indirect relation to the formation of the embryo. They are: (i) a change in the surface of the egg cytoplasm whch prevents the entry of further spermatozoa, and (ii) the elevation of a fertilization membrane.

Fertilization membrane

The innermost covering of the egg is the *vitelline membrane* and after the penetration of the sperm, a space is developed which is the *perivitelline spaces,* which is probably due to the pushing away of the membrane by a secretion from the egg or it may also be due to absorption of the water by some substance between the membranes and the egg surface. The membrane above the perivitelline membrane, one apparently newly formed, or a fusion of both of these, is frequently called the *fertilization membrane.* The significance of the phenomenon just noted is not well understood. It was thought at one time to aid in preventing polyspermy.

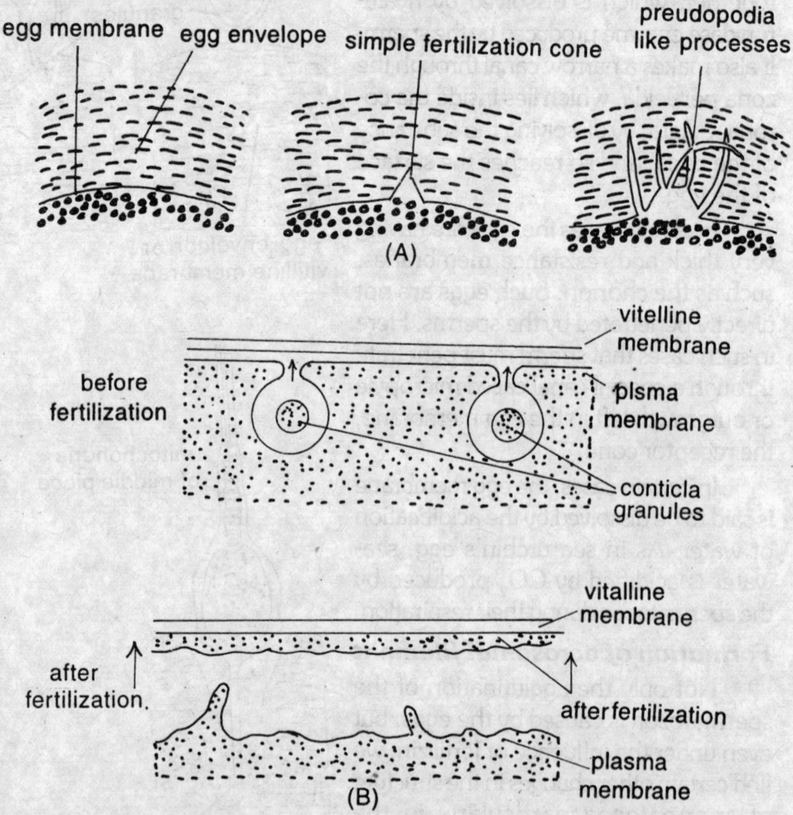

Fig. 25.3. A—fertilization cone; B—the formation of the fertilization membrane in the sea urchin egg.

In mammals no fertilization membrane is formed. After penetration of the sperm, the zona pellucida surrounding the mammalian egg undergoes a change called the zona reaction. This change, which prevents the entry of other sperms, starts from the point of sperm entry and proceeds to the opposite pole. It is possible for a second sperm to enter the opposite pole of the egg immediately after entry of the first sperm, resulting in polyspermy.

Paths of the Pronuclei

The first accurate study of the paths of the pronuclei during the fertilization was made by *Roux* in frog, where he marked the course of the sperm by two compartments:

(i) A penetration path

(ii) Copulation path.

Penetration path is the path travelled by the sperm (male pronucleus) towards the female pronucleus (in case when female pronucleus is present in the middle of the egg). It is nearly vertical to the surface while in the copulation path the male nucleus moves towards the point of union with the egg nucleus. Here in latter case the female pronucleus is generally accentric in position, that is why both these paths form a considerable angle.

A third path which is the *cleavage path* may also be added, along which the united nuclei move together to the definite position of the first cleavage nucleus. All these paths or facts can be well traced in the sea urchin egg where the egg nucleus at first occupies an accentric position near the point at which the polar bodies are formed, but before fertilization may wander to any position.

Generally the sperm penetrates and enters the egg at any point (but it is supposed to be predetermined), the sperm nucleus generally first moves towards the centre, that show no constant relation to the position of the egg nucleus. So this much of the path travelled by the male nucleus after penetration is called the penetration path (already stated). Then the sperm or male pronucleus moves towards the egg or female pronucleus which is excentrically placed, making copulation path. And from the meeting point the opposed pronuclei move slowly along the cleavage path to their final position, which is always in the egg axis and slightly eccentric towards the animal pole.

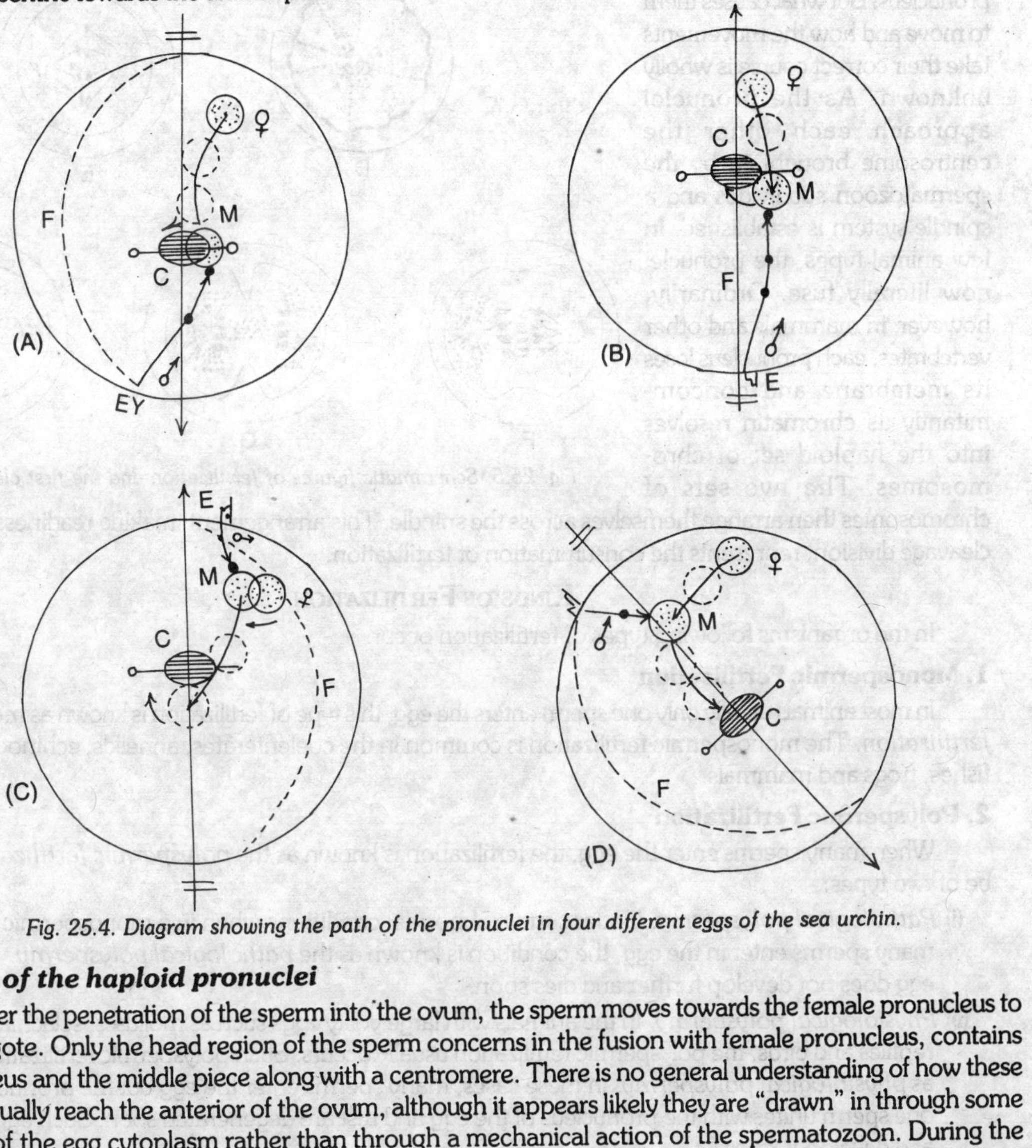

Fig. 25.4. Diagram showing the path of the pronuclei in four different eggs of the sea urchin.

Fusion of the haploid pronuclei

After the penetration of the sperm into the ovum, the sperm moves towards the female pronucleus to form zygote. Only the head region of the sperm concerns in the fusion with female pronucleus, contains the nucleus and the middle piece along with a centromere. There is no general understanding of how these parts actually reach the anterior of the ovum, although it appears likely they are "drawn" in through some activity of the egg cytoplasm rather than through a mechanical action of the spermatozoon. During the penetration or entrance of the sperm, the tail is ordinarily discarded out side the eggs. But generally in mammals (mouse) it remains attached with the sperm but it soon degenerates and plays no role in subsequent events.

Entry of spermatozoon in the egg vertebrate incites the second maturation division. And the egg nucleus becomes ready for union after the production of the second polar body. The male pronucleus always move towards the female pronucleus, while the later in turn often moves to meet the male pronucleus. But what causes them to move and how the movements take their correct course is wholly unknown. As the pronuclei approach each other the centrosome brought in by the spermatozoon subdivides and a spindle system is established. In few animal types, the pronuclei now literally fuse. Ordinarily, however, in mammals and other vertebrates, each pronucleus loses its membrane and concommitantly its chromatin resolves into the haploid set of chromosomes. The two sets of chromosomes then arrange themselves across the spindle. This arrangement, making readiness for the first cleavage division, represents the consummation of fertilization.

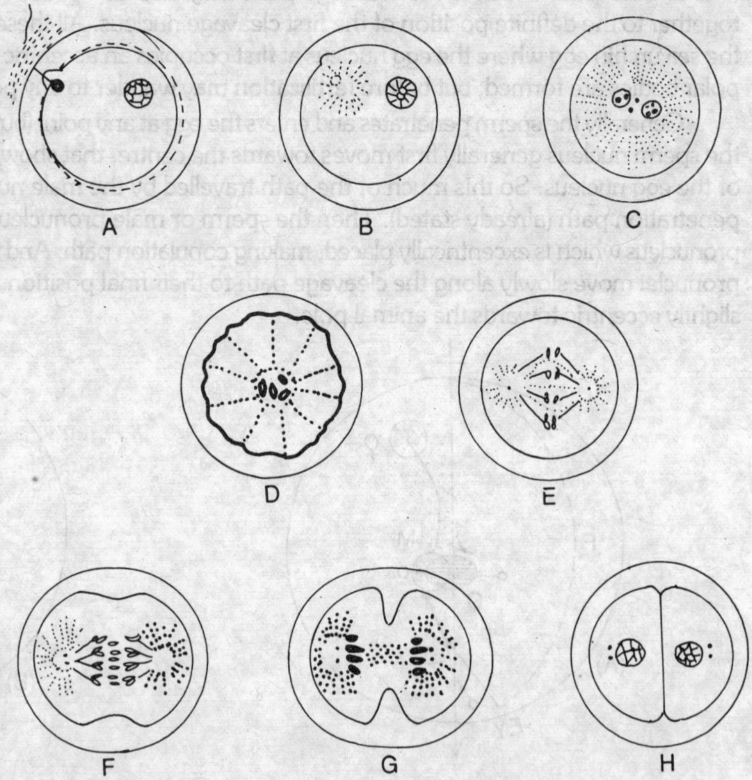

Fig. 25.5. Schemmatic figures of fertilization and the first cleavage.

Kinds of Fertilization

In the organisms following types of fertilization occur:

1. Monospermic Fertilization

In most animals usually only one sperm enters the egg, this type of fertilization is known as *monospermic fertilization*. The monospermic fertilization is common in the coelenterates, annelids, echinoderms, bony fishes, frogs and mammals.

2. Polyspermic Fertilization

When many sperms enter the egg, the fertilization is known as the *polyspermic fertilization*. It may be of two types:

(i) *Pathological polyspermy*. Under certain abnormal conditions when in a monospermic type of egg many sperms enter in the egg, the condition is known as the *pathological polyspermy*. This type of egg does not develop further and dies soon.

(ii) *Physiological polyspermy*. In the animals with large yolky eggs such as molluscs, selachians, urodels, reptiles and birds, the polyspermic fertilization usually occurs. Such polyspermic fertilization is known as *physiological polyspermy*. In these cases, many sperms enter the egg but the pronucleus of only one sperm unites with the pronucleus of the egg and rest are degenerated soon. Such eggs are viable and develop further.

3. Polyandry

When two male pronuclei unite with a female pronucleus, the union is known as polyandry, e.g., man and rat.

4. Polygamy

When two egg pronuclei unite with single male pronucleus the phenomenon is known as polygamy, e.g., sea urchins, polychaete worm, urodels and rabbits.

5. Gynogenesis

When only sperm activates the egg but its pronucleus does not unite with the egg pronucleus, e.g., planarians and nematodes.

Significance of Fertilization

The results of fertilization are:

1. The entrance of sperm activates the secondary oocyte to complete its second maturation division.
2. Induces the movement of cytoplasm and the pigment granules and marks the fertilization track.
3. Induces the 2nd centriole and intiates spindle formation.
4. Restores the diploid number of chromosomes and recombines the maternal and paternal genetic traits.
5. Induces change in the periphery of egg and this change precludes the entry of other sperms.
6. This causes separation of vitelline membrane which allows the rotation of egg inside.
7. This enhances the metabolic activities and thus renews the vigour and vitality.
8. The amount of cytoplasm lost due to the separation of polycytes is restores.
9. Establishes a new plane of egg axis.
10. Causes rearrangement and organization of egg cytoplasm into distinct 'organ forming areas.'

Polyspermy

Normally the eggs are fertilized each by one sperm, though more than one sperm may enter the egg. So when one egg is fertilized by one sperm the condition monospermy occurs as in echinoderm, nematodes, molluscs or mammals. The term polyspermy is applied in those cases where the eggs are penetrated by more than one sperm. This polyspermy mainly confined to eggs heavily laden with yolk (insects, elasmobranchs, amphibians, reptiles, birds etc.) Generally in polyspermy, only one sperm nucleus conjugates with the egg nucleus while the other sooner or later degenerates without conjugation either with the egg nucleus or with one another. In some cases, the degeneration takes place very early (urodeles, insects). While in still others the supernumerary sperm nuclei may enlarge, assume a vesicular form and divide repeatedly by mitosis producing a number of small nuclei having haploid number of chromosomes. Here in such cases the normal distribution of the chromosomes is affected at each mitosis, hence the resulting embryo contains nuclei of two kinds, diploid ones descended from the original fused nuclei, and haploid ones derived from supernumerary sperm nuclei. Larvae having such abnormalities are abnormal in various degrees and incapable of complete normal development.

26

PARTHENOGENESIS

Parthenogenesis is understood here as the production of an embryo from a female gamete without the concurrence of a male gamete, or with or without eventual development into an adult. The parthenogenetic organism is called a *parthenogenone.* Another type of reproduction, where there is a fusion of reproductive cells of opposite sex is called *Amphimixis. Apomixis* means that there is development without fusion of reproductive cells of opposite sex and in animals the embryo is derived from the female parent only. In animals apoximis and parthenogenesis are synonyms. *Automixis* in animals means a fusion of two nuclei of the same sex, as in fusion of an oocyte with a polar body and can sometimes accompany amphimixis or apomixis. According to the mode of sex determination in parthenogenesis, three terms are used; *arrhenotoky,* in which unfertilized eggs develop into males, fertilized eggs into females, *thelytoky,* in which unfertilized eggs develop into females, and *deuterotoky* or *amphitoky* in which unfertilized eggs develop into both sexes.

Occurrence

Parthenogenesis occurs sporadically in a large number of plant groups, but never seems to have become widely established in any of the major groups of animal kingdom. *Bonnet* (1972) first observed the occurrence of parthenogenesis in aphids. Since then hundreds of examples of parthenogenesis have been observed in liverflukes, rotifers, nematodes, annelids, crustanceans and certain orders of insect i.e., *Hymenoptera* (bees and wasps), *Coleoptera* (beetles). In some groups, the parthenogenesis occurs as a regular feature in the life history and alternates with the sexual generation, whereas certain individuals propagate exclusively by parthenogenesis.

Parthenogenesis may occur as a regular phenomenon in the life history of certain individuals or the eggs may be induced by artificial means. Hence, parthenogenesis may be of two different types—*natural parthenogenesis* and *artificial* or *induced parthenogenesis.*

NATURAL PARTHENOGENESIS

The occurrence of parthenogenesis in nature can be studied under the *natural parthenogenesis.* Study of natural parthenogenesis can be done in bees, wasps and ants where the type occurs regularly in the life cycle. Natural or spontaneous parthenogenesis is of rare occurrence among higher animals specially among mammals, but, however, *Dempsey* (1939) reports that normal and abnormal maturation figures are found in ova from atretic guinea pig follicles, and these ova may divide into complex structures resembling trophoblastic giant cells. *Thibault* (1949, 1952) has made a general attack on reports of spontaneous parthenogenesis development in ovarian or tubular mammalian eggs. Natural parthenogenesis is of two types.

(i) Haploid Parthenogenesis (Arrhenotoky)

(ii) Diploid Parthenogenesis (Thelytoky)

(i) Haploid Parthenogenesis (Arrhenotoky)

In rotifers, arachanids, honeybees and certain other insects, males are developed parthenogenetically i.e., in all these cases if the offspring is developed from the egg without fusing with sperm, the offspring will be male. As we know that eggs and sperms formed during gametogenesis, are always haploid in nature so all the male offsprings thus developed from unfertilized eggs are haploid. While the diploid eggs always develop into female offsprings.

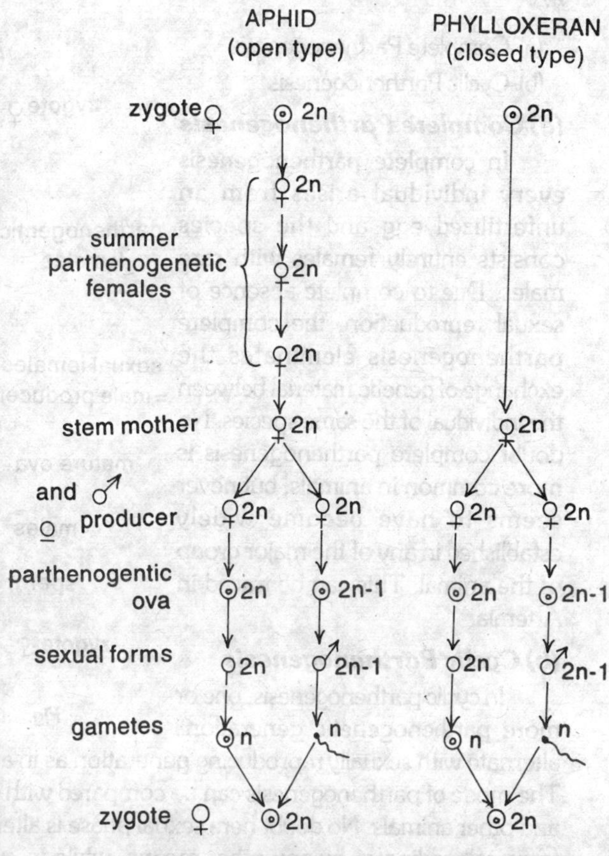

Fig. 26.1. Diagram of mixed and haploid parthenogenesis.

Arrhenotoky in Insects

Among insects, specially Hymenoptera, Homoptera, Coleoptera and Thyasanoptera groups show the haploid parthenogenesis. The classical case of haploid parthenogenesis is offered by the honeybee. Here the females are developed as a result of sexual reproduction by fusion of gametes but males are haploid and develop parthenogenetically from the unfertilized haploid eggs. Males containing 16 number of chromosomes, while diploid females contain 32 number of chromosomes. During oogenesis all the eggs produced contain haploid number of chromosomes i.e., 16, while all the sperms produced contain the same number of chromosomes as in adult males i.e., 16. This is due to the non-occurrence of reduction division. Thus females develop from fertilized eggs where both male and female haploid gametes fuse to form zygote of 32 chromosomes, while as males contain 16 chromosomes, so they directly develop from unfertilized haploid eggs.

Another beautiful illustration can be traced among parasitic wasps' *Habrobracon.* This group is of same nature as seen in honey bee. Here again males are haploid while females are diploid and also develop in the same fashion.

Arrhenotoky in Rotifers

As in gall-flies or aphids, in rotifers, all the fertilized eggs produce parthenogenetic females, the eggs of which form but a single polycyte and develop by diploid parthenogenesis (Thelytoky). There occurs certain complicated alternations of parthenogenetic and sexual generation. The dwarf males are supposed to develop from unfertilized and haploid eggs.

Arrhenotoky in Acarina

This group also shows arrhenotoky or haploid parthenogenesis, specially in ticks and mites. In "red-spider" male diploidy was demonstrated by *Schrader* (1923). While male diploidy was also reported in 1935 by *Opperman. Whiting* (1945) concluded that arrhenotoky probably arose only once in the evolution of the group.

(ii) Diploid Parthenogenesis (Thelytoky)

In diploid parthenogenesis, all the ova produced are diploid in nature as during oogenesis the reduction division is abortive and is checked up and all the diploid ova develop into offsprings without fertilization. Diploid parthenogenesis can be studied under the two separate headings:

(a) Complete Parthenogenesis

(b) Cyclic Parthenogenesis

(a) Complete Parthenogenesis

In complete parthenogenesis every individual arises from an unfertilized egg and the species consists entirely females with rare males. Due to complete absence of sexual reproduction, the complete parthenogenesis eleminates the exchange of genetic material between the individual of the same species. No doubt complete parthenogenesis is more common in animals, but never seems to have become widely established in any of the major group of the animal. This can be traced in Artemia.

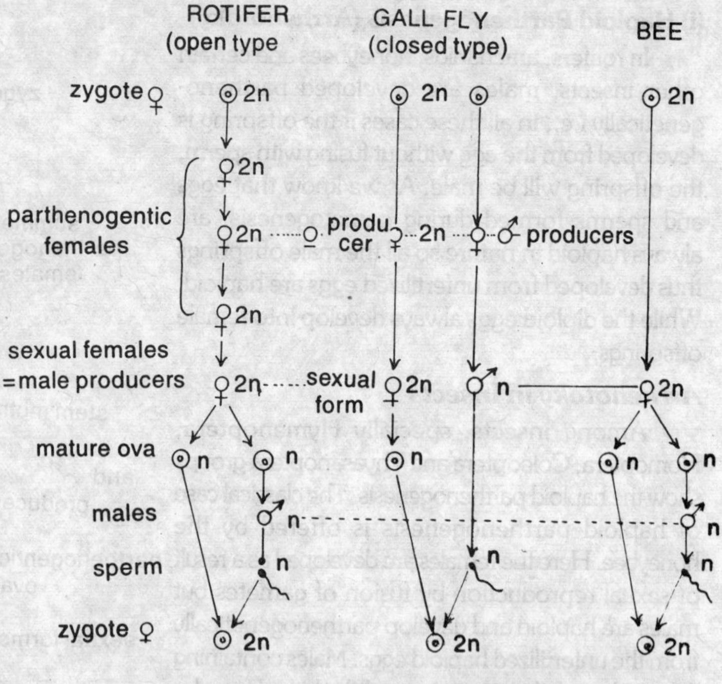

Fig. 26.2. Diagram of diploid cyclic parthenogenesis.

(b) Cyclic Parthenogenesis

In cyclic parthenogenesis, one or more parthenogenetic generations alternate with sexually reproducing generation as in aphids, wasps, gall flies, rotifers and certain parasites. The mode of parthenogenesis can be compared with the alternation of generation of certain coelenterates and other animals. No doubt here sexual phase is alternated with asexual phase which may be either in the form of budding or by any other means, while in cyclic parthenogenesis, sexual phase alternates with parthenogenetic phase.

Cytological Study of Diploid Parthenogenesis

On the basis of cytological studies, thelytoky can be explained under two separate headings:

(A) Meiotic thelytoky

(B) Ameiotic thelytoky

(A) Meiotic Thelytoky

Generally haploid eggs are formed by a process of oogenesis from diploid germ cells. The haploid nature of gamete (egg) is controlled by reduction division which occurs in the phase of maturation. This division is double in nature and is composed of first meiotic division and second mitotic division. So the mature ovum always possess haploid number of chromosomes. But in meiotic thelytoky the reduction of chromosomes is compensated by doubling of chromosome number (diplosis) at some stage of life cycle, so that diploid nature is maintained. These eggs develop parthenogenetically into young individuals without fertilization. Thus may also be called *automictic parthenogenesis*.

The diplosis or doubling of chromosomes occurs by two processes:

(i) By restitution

In this method the doubling of chromosomes occurs during oogenesis at phase of maturation where Ist and 2nd meiotic divisions occur. During Ist meiotic division, chromosomes arrange themselves at the equator, and at anaphase even their movement towards opposite poles have been traced but cytokinesis fails to occur. This process is called restitution. Or as a whole one can say that in this method the doubling of chromosomes occurs due to the complete lacking of Ist meiotic division. So Ist polar body is not formed during Ist maturation division but as the 2nd meiotic division is regular, the formation of 2nd polar body can be seen. As already explained tht the haploid nature of gametes is controlled by Ist meiotic division of maturation phase which is completely omitted in this process thus doubling the number of chromosomes. Diplosis is found in several lepidopterans.

(ii) By auto-fertilization

In this method the diplosis occurs again during oogenesis. Polar bodies are formed by Ist as well as by 2nd meiotic division. Polar bodies are haploid in nature and after their formation the diploid nature of developing ovum changes to haploid. In certain cases any of the haploid polar bodies fuses with the haploid voum, thus again diploid nature is produced. This mutual fusion of polar body nucleus and egg nucleus is called auto-fertilization.

Occurrence of Meiotic Parthenogenesis

The meiotic parthenogenesis is found in a large number of animals of various groups, few of them are given below:

(i) Hymenopteran type. Diploid parthenogenesis also occurs in this group and is supposed to have arisen from haploid parthenogenesis through a modification of the maturation division in the egg so as to produce diploid egg nucleus instead of a haploid one. In parasitic wasp '*Nemeritis cotesceus*' the first meiotic division is incomplete and separated chromosomes again come together by a process of restitution, thus meiotic thelytoky occurs.

(ii) *Lepidopteran type.* Meiotic thelytoky has been studied in some lepidopterans by *Seiler* (1948), *Narbel* (1950), and *Pardi* (1950), which occurs in different ways. In *Solenobia triquetrella* three 'races' are known : (a) a diploid bisexual race; (b) a diploid thelytokous race; (c) and a tetraploid parthenogenetic race. The eggs of parthenogenetic females (diploid and tetraploid undergo normal Ist and 2nd maturation divisions resulting the formation of haploid ovum and the two polar bodies. But further mechanism is not normal, as four blastomeric nuclei are formed by Ist and 2nd meiotic cleavage divisions. These four blastomeric nuclei are haploid in nature as they develop from haploid nucleus of ovum through incomplete mitotic division and here cytokinesis fails to occur. Then all these four nuclei fuse in pairs to restore the original somatic (diploid) number of chromosomes.

(iii) *Artemia salina.* In this case during oogenesis Ist maturation division occurs in normal way producing Ist polar body. No doubt during 2nd maturation division nucleus divides into two to produce second polar body but even though 2nd polar body fails to occur and the diploid nature is restored by the fusion of products of second maturation division.

(B) Ameiotic thelytoky

This is also called *apomictic parthenogenesis* and occurs in a number of crustacean and insects. Here in this type reduction or meiosis division is entirely suppressed and the maturation division becomes mitotic in nature.

Occurrence of Ameiotic Parthenogenesis

The ameiotic parthenogenesis is found in animals especially arthropods, few of them are given below:

(i) *Isopoda.* Certain isopods show polyploid parthenogenesis as in genus *Trichoniscus.* The diploid forms consists of about equal numbers of females and males, and reproduce sexually while in triploid females, the chromosomes do not pair during oogenesis, so only one maturation division which is mitotic in nature occurs. Thus the egg which develops without fertilization possesses diploid number of chromosomes.

(ii) *Daphnia. Mortimer* in 1936 has described the ameiotic parthenogenesis in *Daphnia pulex.*

(iii) *Long horned grass-hopper.* In this case, no males occur and thus the reproduction is purely thelytokous type. *Mattehy* (1941), observed the formation of only one polar body, while the formation of 2nd polar body is omitted and thus diploid eggs are produced.

OBLIGATORY PARTHENOGENESIS

In a few species of animals, parthenogenesis is the only method of reproduction. This type of parthenogenesis is called *obligatory* or *complete* or *total parthenogenesis.* In this form of parthenogenesis meiosis is either totally lacking or is followed by a compensatory process. Populations having total parthenogenesis consist almost entirely of females. Males when present are very rare and are genetically non-functional. In cases where there is no meiosis, segregation and recombination of genes is absent. Thus, the populations show no variations, and are in effect clones.

In obligatory thelytoky meiosis may be totally eliminated (*apomictic* or *ameiotic thelytoky*) or may be present (*automictic* or *meiotic thelytoky*) with a later compensating process. In *apomictic thelytoky* there is no meiosis. It may or may not be replaced by one or two mitotic divisions. In most cases there is a single mitotic maturation division, although sometimes there may be two (e.g., in the roach, *Pycnoscelus surinamensis*). In automictic thelytoky meiosis is present and results in halving of the chromosome number. However, the somatic number is restored by fusion of ovum with one of the polar bodies, or by fusion of the cleavage nuclei. Since there is no meiosis in apomictic thelytoky, there is no synopsis and crossing over. Hence there is no genetic combination and variation. In automictic thelytoky, however, occurrence of meiosis provides a greater possibility of genetic variation.

Obligatory thelytoky has not proved very successful in evolution because of its limitations in producing variations. Populations in which there is little variation tend to be eliminated when environmental conditions become adverse. Obligatory parthenogenesis has arisen sporadically in animals, and there are about a thousand species known. It is found in some species of earthworms, Bedlloid rotifers, grasshoppers, roaches, phasmids, moths, gallflies, fishes salamanders etc. As mentioned previously populations having obligatory parthenogenesis consists almost entirely of females. In the rose gall fly *Rhodites rosae* and in the phasmid *Bacillus rossili* males are extremely rare. In the gall fly *Cynips kollari*, the praying mantis *Brunneria borelis* and some rotifers males have never been found. These species reproduce exclusively by unfertilized eggs.

Transitional Intermediate Cases

Certain cases in the life cycle of many organisms are found to be intermediate between parthenogenesis and fertilization such as:

1. *Partial fertilization.* In partial fertilization, an egg is fertilized by a part of the sperm. For example, in sea-urcihin egg, *Boveri* has described the fertilization of ovum by sperm aster alone. The sperm nucleus fuses with the egg nucleus in two cell stage, or later.
2. *Gynogenesis.* Here the sperm penetrates the egg but its nucleus does not fuse with that of ovum and as such it does not participate in the formation of the embryo. The entry of sperm in the ovum initiates division in the egg, but the sperm nucleus remains secluded and degenerates later on. Example is *Rahabdites aberrans.*
3. *Androgenesis.* In androgenesis, the nucleated egg is activated by the sperm and the development takes place without the formation of zygote under the instructions of sperm nucleus. For example, if eggs of frog or toad are irradiated with radium and then fertilized with normal sperm the egg nucleus does not participate in the development. Therefore, the zygote nucleus is haploid and of paternal origin.
4. *Merogony.* In this case the egg fragments devoid of complete nucleus develop when fertilized with normal sperm. If sea urchin eggs are broken into small pieces without nuclei, and then are fertilized with the normal sperm, these develop into dwarf larvae. Such type of development is called *merogony.*
5. *Transitional cases.* Certain transitional cases are found between the diploid and haploid parthenogenesis. In soem hemipterans and hymenopterans the development begins with haploid eggs, but chromosome number, later on, becomes diploid. In them the eggs are produced by complete meiosis and are haploid. These remain unfertilized, but start developing. During development these restore diploid number, possibly be a secondary splitting of chromosomes after the maturation division.

ARTIFICIAL PARTHENOGENESIS

Parthenogenesis development can be seen by using the artificial means. A number of experiments have been performed on animals like amphibians, birds, and mammals and even on higher invertebrates like annelids, molluscs and echinoderms. Here in this type the eggs are treated by artificial means.

Frog's egg can be stimulated to develop parthenogenetically by pricking them slightly with a needle that has been dipped in frog's blood.

For parthenogenetic development, eggs can also be treated by hypertonic sea water, fatty acids, foreign blood, serum and other chemicals.

Physical agents like increase of osmotic pressure, thermal changes, electrical stimulate, shaking and puncturing can stimulate the eggs to develop without fertilization.

Eggs can develop parthenogenetically even after receiving the sperms but of different species, as is seen, when the eggs of sea urchin were treated with the sperms of annelids.

Stimulus for Artificial Parthenogenesis

Normally, the egg is activated and induced to develop by the sperm. Under experimental conditions, however, a variety of agents can be used to bring about activation of the egg and initiate development.

1. *Physical stimulus.* It is provided by pricking the eggs slightly with a fine needle.
2. *Temperature.* Stimulus may be provided by exposing the eggs to high or low temperature.
3. *Chemical stimulus.* Several chemicals can be used to induce the unfertilized egg to divide. These are neutral salts like potassium chloride, weak acids and weak bases; carbon dioxide; certain organic acids like butyric acid, lactic acid, oleic acid and a number of other fatty acids; a variety of fat solvents such as toluene, ether and acetone; hypo-and hypertonic sea water and foreign blood serum.
4. *Radiations.* Ultraviolet radiations and emissions from radium also induce parthenogenic development of unfertilized eggs.
5. *Stimulus provided by the entry of sperm.* Even the sperm from a totally different group, which does not contribute its nucleus to the egg, can stimulate the egg of some different group. For example, when sperm from annelids are made to enter the eggs of echinoderms or sea urchin eggs are fertilized with the sperm of starfish, the nuclei of sperms and eggs do not fuse, but the eggs start dividing.

Significance of Parthenogenesis

1. *As a method of reproduction.* In certain cases where fertilization fails to occur and for the continuation of race, the unfertilized eggs develop parthenogenetically which is not doubt simple and easier method of reproduction that sexual reproduction.
2. *A device for high multiplication.* As explained above that the process of parthenogenesis is easier than sexual reproduction and due to simplicity of the process, a large number of individuals can be produced. So in one sense can say that the parthenogenetic generation exceeds the sexual generation in the number of individuals.
3. *Elimination of non-adaptive characters.* The genetic consequences of parthenogenesis are not suitable for the organisms. Natural selection in parthenogenetic forms eliminates the non-adaptive combination of genes.
4. *Escape from sterility.* It was considered by *Darlington* in 1939, the parthenogenesis as an "escape from sterility." The sterility in the generation may be due to triploidy, pentaploidy or to incompatibility of gametes arising from genetic cause.
5. *Sex determination.* Parthenogenesis provides a method of sex determination. As in case of bees, if the egg develop parthenogenetically, all will be the males while fertilized egg develop in females. Other examples are ants, wasps and aphids etc.

Disadvantage

The individuals of parthenogenetic generations fail to adopt themselves in the environment and they may be better out in the struggle for existence, as parthenogenesis does not allow the recombination and exchange of genes. If some non-adaptive characters are present, the whole generation possesses the same non-adaptive characters and therefore they have less chances for survival.

27

NUCLEOCYTOPLASMIC INTERACTION

The nucleus and cytoplasm are two main constituents of an eukaryotic cell and form an integrated physiological system. All the activities of the cell are governed by the interaction between nucleus and cytoplasm. Nucleus without cytoplasm or cytoplasm without nucleus cannot function and this fact has been proved in many nuclear transplant experiments. Thus the cellular function is a product of the co-operation between the nucleus and cytoplasm.

METHODS OF NUCLEOCYTOPLASMIC INTERACTION

For understanding the nucleo-cytoplasmic relationship a number of methods have been devised to attempt the study of nuclear activity free of interferring cytoplasmic activity and a study of cytoplasmic activity in the absence of nucleus.

There are three groups of methods for the study of nucleocytoplasmic interaction. These are:

1. Cell Enucleation

This method is employed to know those activities which are under nuclear control. In the absence of nucleus such activities are not seen in the experimental cell.

Eunucleation can be done by any one of the following three ways:

(i) In this method the nucleus is drawn out with the help of microtool (*Tarter*, 1961) by cutting through the cell surface.

(ii) The second method is simply to push the nucleus out of the cell with a probe (*Defornbrune*, 1949).

(iii) A third method used really is to suck out nucleus with a micropipette (*Kopec*, 1959).

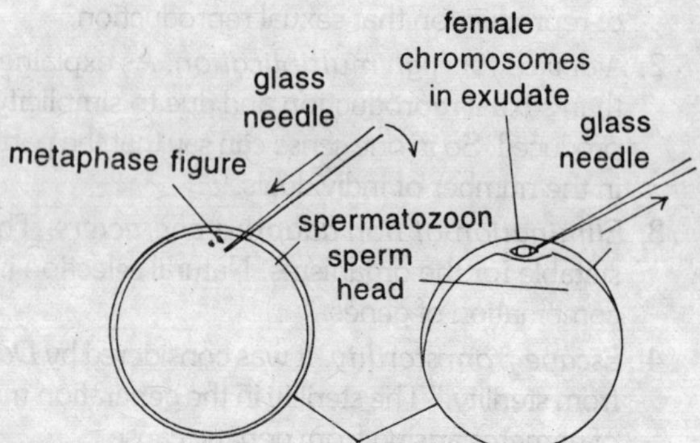

Fig. 27.1. Enucleation of the frog's egg with capillary piptte.

2. Isolation of Nucleus

The essential step of most isolation procedures are, to break up cells by the homogenization process, that will not seriously damage the nuclei and than to separate the nuclei from the remaining cellular material by centrifugation.

3. Transplantation of Nuclei

There are two methods of nuclear transfer in living cells (a) Natural, and (b) Microsurgical. In the natural transfer the passage of nuclear material from one organism to the cytoplasm of another organism occurs by normal sexual process as in *Paramecium* and certain other organisms. In the second method the nuclei may also be transplanted by microsurgical means *Kings* and *Bridges,* (1953) transplanted nuclei from embryonic amphibian cells into eunucleate oocytes.

The nucleocytoplasmic interaction can be studied under following heads:

1. Action of cytoplasm on nucleus.
2. Protein synthesis.

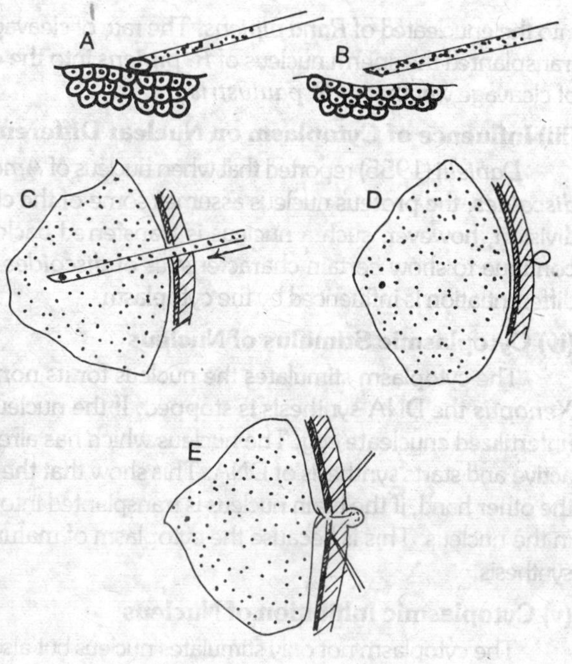

Fig. 27.2. Diagram illustrating the step involved in the transplantation of blastile of frog nucleus into enucleated frog's egg.

Action of Cytoplasm on Nucleus

The cytoplasm influences the nucleus in a mysterious way. Embryologists have demonstrated particular interest in the problem of how cytoplasm influences the nucleus. The cytoplasm can perform a diversity of activities independent of nucleus. This is perhaps the most dramatically observed in *Acetabularia,* which can synthesize a variety of macromolecules, grow and differentiate (*Brachet,* 1961) for a week in the absence of nucleus. If one accepts that cytoplasm is dependent on the nucleus for genetic informations, it would follow that enucleate cells perhaps dies because the informations, previously acquired from nucleus are unstable and are no longer available to direct the synthesis of essential components of the cell. On the other hand, the nucleus depends on (cytoplasm for protection from the severity of the environment) as well as for precursors and energy.

The influence of cytoplasm on nucleus can be illustrated in several ways, a few of them are given as below:

(i) Influence of Cytoplasm on Maturation of Nucleus

The nuclear size is related to the age of *Acetabularia.* In young *Acetabularia* the nucleus is smaller in size where as in mature *Acetabularia* it is larger in size. In an experiment if the nucleus from a young *Acetabularia* is transplanted to an eunucleated mature *Acetabularia,* the nucleus undergoes rapid transformations and assumes the role of mature form. This result clearly demonstrates that the maturation of nucleus is dependent on the cytoplasm.

In *Amoeba,* when the cytoplasmic mass increases to a maximum size, it induces to mature and divide the nucleus. The nucleus can be kept in immature state and devoid of nuclear division for a long time by a periodic removal of cytoplasm from the cell (*Hartmann,* 1924; *Prescolt,* 1956). This proves that the maturation of nucleus is entirely influenced by the cytoplasm.

(ii) Influence of Cytoplasm on the Mitotic Events

Hammerling (1953) observed that in the *Acetabularia,* the nuclear division takes place only after the formation of cap. He removed the cap just before the cap was completely formed. A new cap begin to form. He again repeated the same act. In this way *Hammerling* was able to keep the nucleus in an immature stage for a long time. He demonstrates by his experiment that division of nucleus depends on cytoplasm.

Moore (1941) performed a number of transplantation experiments to demonstrate that the early rate of cleavage and the timing of cleavage are influenced by the cytoplasm or in other words the rate of mitotic division is influenced by cytoplasm. *Moore* in his experiments transplanted sperm nucleus of *Rana paulustris* into the enucleated of *Rana pipiens*. The rate of cleavage was that of *R. pipiens*. In another experiment he transplanted the sperm nucleus of *R. pipiens* into the enucleated egg of *R. paulustris*. This time the rate of cleavage wet that of *R. paulustris*.

(iii) Influence of Cytoplasm on Nuclear Differentation

Danielli (1955) reported that when nucleus of *Amoeba proteus* is transplanted into enucleate *Amoeba discoides*, the proteus nucleus assumes some of the characters of *discoides* nucleus. If, after several cell division, however, such a nucleus is transferred back into an enucleate *Amoeba proteus*, the nucleus continue to show certain characteristics of *discoides* nucleus. This experiment proves that the nuclear differentiation is influenced by the cytoplasm.

(iv) Cytoplasmic Stimulus of Nucleus

The cytoplasm stimulates the nucleus for its normal metabolic activity. In the mature brain cells of *Xenopus* the DNA synthesis is stopped. If the nucleus from such brain cell is transplanted into mature unfertilized enucleate egg. The nucleus which has already stopped DNA synthesis immediately becomes active and starts synthesis of DNA. This show that the dormant nucleus is activated by the cytoplasm. On the other hand, if the brain nucleus is transplanted into immature enucleate egg, similar activity is not seen in the nucleus. This is because the cytoplasm of mature egg alone contains factors that can initiate DNA synthesis.

(v) Cytoplasmic Inhibition of Nucleus

The cytoplasm not only stimulates nucleus but also inhibits nucleus. The activity of nucleus is inhibited by cytoplasmic factors atleast in some stages of cell. In *Xenopus*, during oogenesis the synthesis of rRNA is very high but it is retarded during maturation. This activity is stopped until the initiation of gastrulation. In young tadpole larvae rRNA synthesis takes place vigorously and the nucleolus is more prominant. If the nucleus taken from tadpole is transplanted to a mature egg. In about 40 minutes the nucleolus disappears and rRNA synthesis is inhibited until the onset of gastrulation. This experiment clearly demonstrates that the genes of nucleus controlling rRNA synthesis are inhibited by the repressors released from egg cytoplasm.

(iv) Influence of the Cytoplasm on Localized Chromosomal Activity

The localized chromosomal swelling is known as *puffs* or *Balbiani rings* in the polytene chromosomes of Diptera larvae. The puff represents the active stage of a gene. The puffs location in the same chromosome varies during different developmental stages of the animal. *Kroeger* (1960) transplanted the nucleus of salivary gland of larval *drosophila* into the egg cytoplasm. The chromosomes of transplanted nucleus exhibited a unique and predictable puffing pattern. The change in puffing pattern was due to the action of egg cytoplasm on the transplanted nucleus.

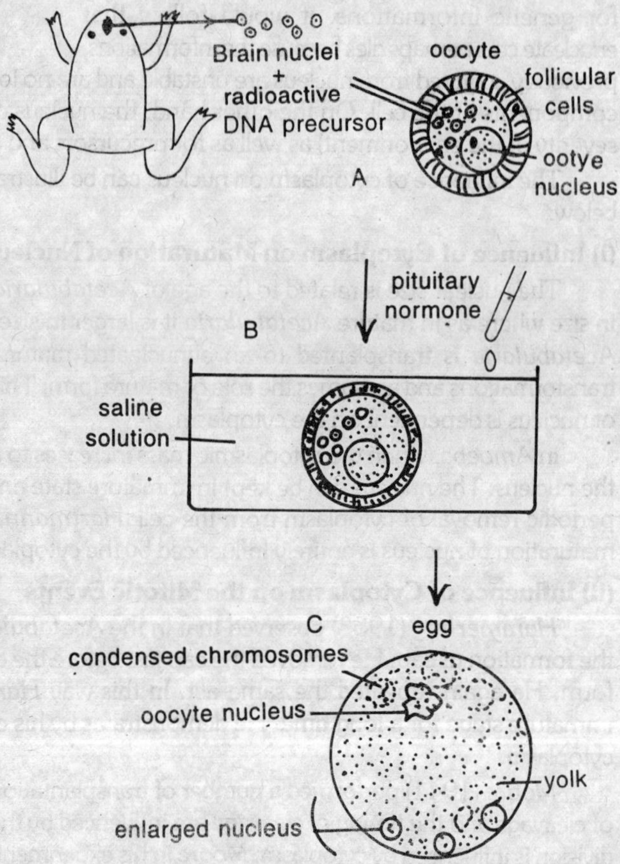

Fig. 27.3. Diagram showing the inducing factors in the opplasm.

Clever and *Karison* (1969) injected ecdysone into the larva of *Chironomous*. The ecdysone hormone stimulates the metamorphosis in insects larvae. After about 30 minutes of injection the chromosomes of larval *Chironomous* developed puffs which were identical to one occuring during the pupation. This experiment clearly shows that the hormone acts first on the cytoplasm and through it on the genes.

(vii) Influence of Cytoplasm on Chromosomal Elimination

In some animals the number of chromosomes is determined by the cytoplasm. The gall midget (*Mayetiola destructor*) is unique in possessing different number of chromosomes in somatic and germ cells. There are only eight chromosomes in the somatic cells where as germ cell contain about 40 chromosomes. It is due to difference in the cytoplasmic environments.

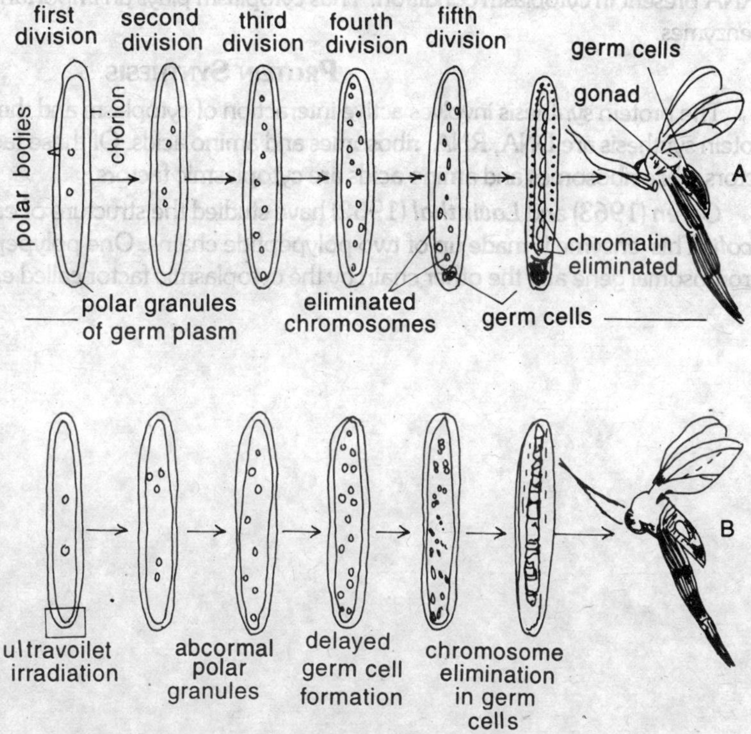

Fig. 27.4. A—Germ cell specialization in Moyetiola. The germ cells are separated at third and fourth division. The somatic nuclei after fourth division show chromosome elimination. Each nucleus loses 32 chromosomes and retain only 8 chromosomes. The germ cells, on the contrary, retain all the 40 chromosomes. B—The posterior end of the egg containing germplasm is tradiated with ultraviolet rays. The nuclei entering this region also show chromosomal elimination.

The cleavage is a peculiar type. During cleavage the zygote nucleus (40 chromosomes) divides but there is no division of cytoplasm. At the 8 nucleus stage, one of the nucleus moves to the posterior part of egg. In this part the cytoplasm is called *germplasm* having special properties. In the following division 16 nuclei are formed. At the same time, the germplasm divides in such a manner that two cells are formed. Each cell bears a nucleus. These two cells lead to the formation of germ cells (sperms or ova) while remaining 14 cells become the somatic cells. During the fifth cleavage the somatic nuclei alone divide. This mitotic division is unusual type. During this division only 8 chromosomes undergo duplication in each nucleus. Remaining 32 chromosomes do not duplicate and are eliminated from the nuclei into cytoplasm. These chromosomes than distintigrate and disappear in cytoplasm. Thus each somatic cell bears only 8 chromosomes whereas each germ cell bear 40 chromosomes.

To determine the cause of such chromosomal elemination in somatic cells *Bantock* conducted experiments. He irradiated the germplasm with ultraviolet rays just before the nucleus migrated into it. On migation the behaviour of the nucleus was different and it also suffered the elimination of chromosomes. These cells still developed into primary germ cells but they failed to develop into eggs or sperms in the adults.

This experiment clearly shows that cytoplasm has effect on the elimination of chromosomes.

(viii) Cytoplasmic Role in Inactivating 'X' Chromosome

Barr (1949) showed that in female mammals there are two x-chromosomes. They differ in appearance. One of them is active while other is *heteropycnotic* i.e., condensed and deeply stained inactive chromosome.

This heterpycnotic chromosome is called *Barr body* or *sex chromatin*. Autoradiographic studies of this chromosome has shown that it does nto participate in nucleic acid synthesis. Both the X-chromosomes are active in early stage but one is inactivated by cytoplasm. This was explained by *Lyon* (1961) it is due to cytoplasmic changes in the early development. Hence this view has been called *Lyon hypothesis*.

(ix) Cytoplasmic Regulation of Enzyme Synthesis

Acetabularia has a unique property of synthesizing the enzyme *phosphatase* at a *pH* of 12 even in enucleated condition. This was studied by *Spencer* and *Harris* (1964). *Acetabularia* synthesis three kinds ofphosphatases at the *pH* of 5, 8.5 and 12 respectively. At *pH* 12 the synthesis of phosphatase occur only when the cap is being formed even in enucleated condition. It is sure that this enzyme is synthesized by mRNA present in cytoplasm condition. Thus cytoplasm plays an important role in controlling the synthesis of enzymes.

Protein Synthesis

The protein synthesis involves active interaction of cytoplasm and the nucleus. The factors involved in protein synthesis are DNA, RNA, ribosomes and amino acids. Of these factors DNA and RNA are nuclear factors while ribosomes and amino acids are cytoplasmic factors.

Garen (1963) and *Lavinthal* (1962) have studied the structure of enzyme *alkaline phosphatase* in *E.coli*. This enzyme is made up of two polypeptide chains. One polypeptide chain is determined by the chromosomal gene and the other chain by the cytoplasmic factor called *episomes*.

28

NUCLEIC ACID (DEOXYRIBO-NUCLEIC ACID)

The fundamental investigations which led to the discovery of the nucleic acids were made by *Fredrich Micscher* (1844-95) who may be regarded as the founder of our knowledge of the chemistry of the cell nucleus. In early work carried out in 1868, in the laboratory of Hoppe Seyler in Tubingen, he isolated the nuclei from pus cells obtained from discarded surgical bandages and showed the presence in them of an unusual phosphorus containing compound that he called '*nuclein*' and which we now know to have been nucleoprotein. It was subsequently shown that nucleic acids were normal constitutents of all cells and tissues and Miescher's investigations of the nucleic acid were continued by *Altmann,* who is 1889 described as method for the preparation of protein free nucleic acids from animal tissues and from yeast. The term *nucleic acid* was introduced by *Altmann. Albrecht Kossel, P.A. Levine* and *Walter Jones* described the chemical nature of nucleic acid. According to them the nucleic acid si composed of phosphoric acid, a sugar and nitrogenous bases. *Franklin W. Stahl* presented first evidence that nucleic acid forms the genetic material. *Chargaff* (1961) described the occurrence of nitrogenous bases in equal proportion. *Dotty* (1961) has emphasized much on the physical properties of DNA. In 1962 *Willkins, Wlatson* and *Crick* proposed a model to explain the double helicular structure of DNA. They also shared the Nobel prize for the same.

Deoxyribonucleic Acid or DNA

DNA is found in the cells of all living organisms except plant viruses, where RNA forms the genetic material and DNA is absent. In bacteriophages and viruses there is a single molecule of DNA, which remains coiled and is enclosed in the protein coat. In bacteria, mitochondria and plastids of eukaryotic cells DNA is circular and lies naked in the cytoplasm. In the nuclei of eukaryotic cells DNA occurs in the form of long spirally coiled and unbranched threads. The number of DNA molecules is equivalent to the number of chromosomes per cell. In them DNA is found in combination with proteins forming nucleoproteins—the *chromatin* material and is enclosed in the nucleus.

DNA is a polymer, the monomeres of which are the deoxyribonucleoside monophosphates. Most DNA have high molecular weight, varying from 10^6 to 10^9 or more. Chemical analysis of DNA molecules from different kinds of orgnaisms suggest that in one kind of DNA the number of adenine (A) nucleotides was always the same as the number of thymine (T) nucleotides. Similarly the number of guanine (G) nucleotide was always the same as the number of cytosine (C) nucleotides. This evidence suggest that nucleotides occur in pairs in the DNA molecule. Further, quantity of A present in one organisms's DNA (and it varies from species to species) is equal precisely to the amount of T. Similarly the amount of G present is always equal to the amount of C.

Physical Structure of DNA

According to the model proposed by *Watson* and *Crick,* the DNA contains two polynucleotide chains running antiparallel and twisted around each other in the form of a regular double helix. The diameter of the helix is about 20Å and each chain makes a complete turn every 34Å. Each chain is polynucleotide in which the sugar of each nucleotide is linked by a phosphate group to the sugar of the adjacent nucleotide. There are 10 nucleotides on each chain in every turn of the helix. The two helices are coiled in such a way that they cannot be separated without unwinding. The two chains of helix are of opposite polarity. If one chain runs 3'-5' direction, then the other will run as 5'-5' direction.

The molecular weight for purified DNA is found in the range 1000,000 to 120 x 10^6. The amount of DNA in a mammalian cell corresponds to about 25,000 molecules of molecular weight 120 x 10^6, a total of about 10,000 million nucleotides.

Monomeric Components of DNA

Before any account is given of the structure of DNA proper it is necessary to describe the structure of its component parts. The DNA molecule is a polymer consisting of several thousand pairs of nucleotide monomeres. Each nucleotide consists of the pentose sugar, *deoxyribose, a phosphate* group and a *nitrogenous base,* which may be either a *purine* or a *pyrimidine.*

1. Deoxyribose sugar

Levine (1909) identified ribose sugar as a main constituent of nucleic acid structure. *Mori* (1929) isolated sugar from the guanine nucleoside and showed that it is deoxyribose sugar. It is pentose type which in DNA molecule shows the absence of one oxygen molecule from carbon-2 position of ribose sugar. Both deoxyribose and ribose (pentose sugars of nucleic acids) have a pentagonal ring with five carbons, among which two (i.e., 3' and 5') are attached to phosphoric acid and three (1') to the base. This sugar is called *deoxyribose* and it simply acts as a support column to which bases are attached. When the pentose sugars occur in nucleic acids nucleotides the carbon atoms an numbered as 1', 2', 3' etc. to avoid confusion with the numbering of the ring atoms of the bases.

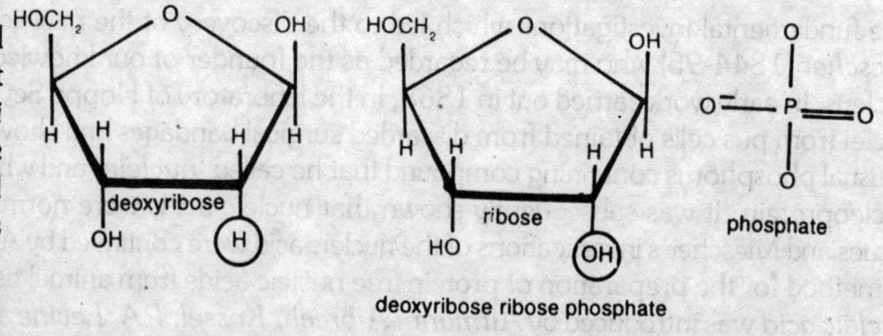

Fig. 28.1. Pentose sugars and phosphate.

2. Phosphate

In the DNA strand the phosphate groups alternate with deoxyribose. Each phosphate group is joined to carbon atom 3' of one deoxyribose and to carbon atom 5' of another. Thus each strand has a 3' end and a 5' end. The two strands are oriented in opposite directions. The 3' end of one strand corresponds to the 5' end of the other. Consequently the oxygen atoms of deoxyribose point in opposite directions in the two strands. *Phosphodiester bonds* are formed between the sugars of two different nucleotides and a phosphate group.

3. Nitrogenous bases

The nitrogenous bases of nucleic acid are of two types : *Purine* and *Pyrimidine.* Purine bases comprise mainly adenine (A) and guanine (G) which are common to both DNA and RNA while pyrimidine bases comprise cytosine (C) and thymine (T) in DNA and in the case of RNA, thymine (T) is replaced by uracil (U).

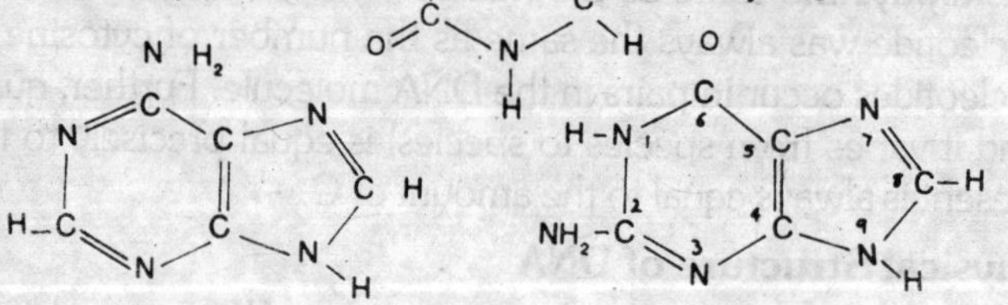

Fig. 28.2. Common nitrogenous bases of nucleic acids.

These bases are arranged linearly as the back bones of chains which (back bones) are composed of alternating sugar and phosphate. The bases have specific way of combining, if C is found in one chain G will be opposite to it and in other if A is on one T must be on the other. Or in other words adenine (A) is always paired with thymine (T) and guanine (G) with cytosine (C).

Among bases A-T and G-C, two hydrogen bonds are formed between A and T and three hydrogen bonds are formed between C and G. This hydrogen bond formation precludes A-C or G-T pairs. Only these arrangements A-T and G-C are possible, for two purines would occupy too much space to allow a regular helix and correspondingly, two pyrimidines would occupy too little. The stickness of these pairing rules result in a complementary relation between the sequences of bases on the two interwinded chains. For example, if we have a sequence ATGTC on one chain, the opposite chain must have a sequence TACAG. The hydrogen atom with its positive charge is shared between an oxygen atom and a nitrogen atom, both with slight negative charges. Although hydrogen bonds are weak, the fact that there are so many gives stability to the DNA molecule. The weak hydrogen bonding enables the two strands of the DNA to separate during replication.

The bases are joined to the pentoses by N-C *glycosidic bonds.* For the purine, the glycosidic bond is between the C_1 position of the pentose sugar and the N_9 position of the bases. For the pyrimidines the linkage joints the C_1 and N_3 positions.

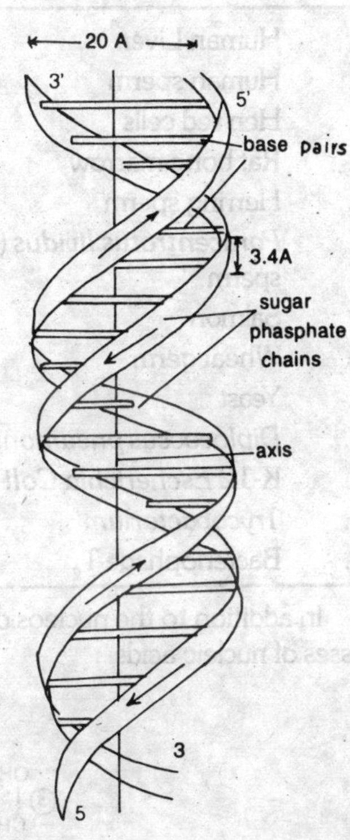

Fig. 28.3. Watson and Crick's model of DNA showing double helical structure.

Nucleosides

When a purine or a pyrimdine base is linked to ribose (in case of RNA) or deoxyribose (in case of DNA) the resulting compound is known as a *nucleoside.* Thus adenine condenses with ribose to form the nucleoside adenosine, guanine forms guanosine, cytosine forms cytidine and uracil form uridine. These ribonucleosides can be formed on partial hydrolysis of RNA. The ribonucleoside from hypoxathnine is named inosine. The nucleosides derived from 2-deoxyribose are known as deoxyribonucleosides-deoxyadenosine, deoxyguanosine, deoxycytidine, deoxythymidine and so on.

Table. 28.1. Four Nitrogen Bases, Nucleosides and Nucleotides of DNA Molecule

	Nitrogen base	*Base + deoxyribose =deoxyribonucleoside*	*Deoxyribonucleoside + Phosphoric acid = Deoxyribonucleotide*	*Abbreviation for nucleotide*
1.	Adenine (A)	Deoxyadenosine	Deoxyadenylic acid (Deoxyadenosine monophosphate)	3'-dAMP
2.	Guanine (G)	Deoxyguanosine	Deoxyguanylic acid (Deoxyguanosine monophosphate)	3'dGMP
3.	Cytosine (C)	Deoxycytidine	Deoxycytidylic acid (Deoxycytidine monophosphate)	5'-dCMP
4.	Thymine (T)	Deoxyhymidine	Deoxy Thymidylic acid (Deoxythymidine monophosphate)	5'-dTMP

Table 28.2. Showing Amount of Various bases in Different Tissues

S.No.	Source	Adenine	Guanine	Cytosine	Thymine	$\frac{A+T}{G+C}$
1	2	3	4	5	6	7
1.	Human Liver	30.3	19.5	19.9	30.3	1.53
2.	Human sperm	30.7	19.3	18.8	31.2	1.62
3.	Hen red cells	28.8	20.5	21.5	29.2	1.38
4.	Rat bone marrow	28.6	21.4	21.5	28.4	1.33
5.	Herring sperm	27.8	22.2	22.6	27.5	1.23
6.	*Paracentrotus lividus* (sea urchin) sperm	32.8	17.7	18.4	32.1	1.85
7.	Salmon	29.7	20.8	20.4	29.1	1.43
8.	Wheat germ	26.5	23.5	23.0	27.0	1.19
9.	Yeast	31.3	18.7	17.1	32.9	1.79
10.	*Diplococcus pneumonial*	29.8	20.5	18.0	31.6	1.59
11.	K-12 *Escherichia Coli*	26.0	24.9	25.2	23.9	1.00
12.	*Trycobactarium*	15.1	34.9	35.4	14.6	0.42
13.	Bacteriophage T_2	32.5	18.2	16.7	32.6	1.86

In addition to the nucleosides listed above several others are found in very small amounts in certain classes of nucleic acids.

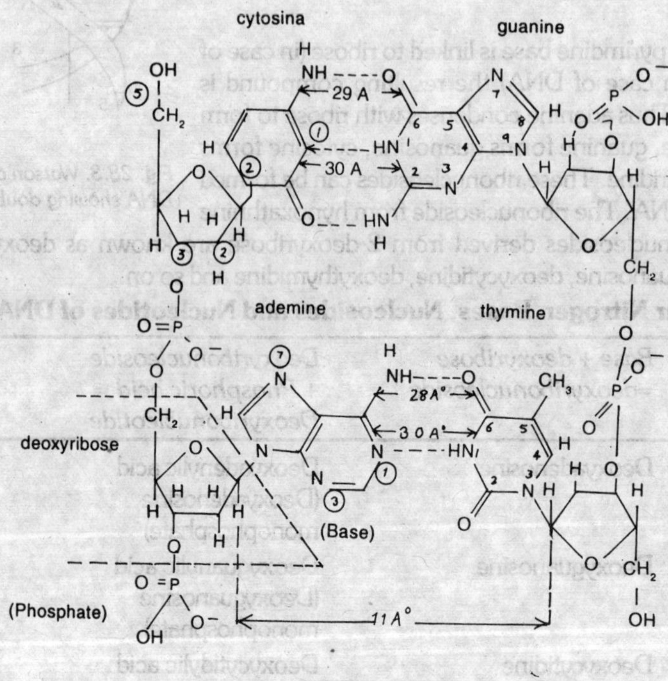

Fig. 28.4. Segment of a DNA molecule.

Nucleotides

When a nucleoside is linked to phosphoric acid the resulting compound is known as a *nucleotide*. They are all phosphoric acid esters of the nucleosides. Those derived from ribonucleosides are usually referred to as *ribonucleotides* and those from deoxyribonucleosides as *deoxyribonucleotides*. These terms are sometimes abbreviated to ribose, ribotide, deoxyriboside and deoxyribotide

Table 28.3. The Relative (G +C) Content of DNA's from Various Sources

Source of DNA	*Per cent (G + C)*
Dictyostelium (slime mould)	22
M. pyogenes	34
Bacillus cereus	37
B. megaterium	38
Haemophilus influenzae	39
Saccharomyces cerevisiae	39
Calf thymus	40
Rat liver	40
Bull sperm	41
Diplococcus pneumoniae	42
Chicken liver	43
Mouse spleen	44
Salmon sperm	44
B. subtilis	44
T_1 *bacteriophage*	46
Esch. coli	51
T_3 *bacteriophage*	53
Neurospora crassa	54
Pseudomonas aeruginosa	68
Micrococcus lysodeikticus	72
Herpes simplex virus	72
Mycobacterium phlei	73

The Primary Structure of DNA

As stated already that nucleic acids are polymers made up of hundreds, thousands of millions of nucleotides coupled together by *phosphodiester linkages*. In the case of DNA where C-4' in the sugar is occupied in ring formation and C-2' carries no hydroxyl group, only the hydroxyl groups at positions 3' and 5' are available for internucleotide linkages. The absence of a hydroxyl group at C-2' makes cyclic phosphate formation impossible and DNA is therefore not hydrolysed by alkali in the same way as in RNA. The primary structure of the polynucleotide chain in DNA is shown in the following fig.

The representation of polynucleotide chains by complete formulae is clumsy and it has become customary to use the schematic system illustrated in the above figure, where the chain is abbreviated as on the right. The vertical line denotes the carbon chain of the sugar with the base attached at C-1'. The diagonal line from the middle of the vertical line indicates the phosphate link of C-3' while that at the end of the vertical line remote from the base denotes the phosphate link at C-5'. This system may be used for either RNA or DNA.

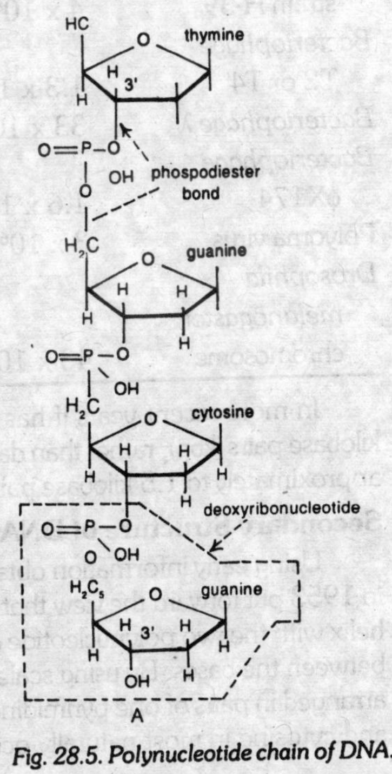

Fig. 28.5. Polynucleotide chain of DNA.

Molecular Weight of DNA

The molecular weight of DNA molecules are very difficult to determine accurately by the methods of classic chemistry since they range from 10^6 to more than 10^{10}. Conventional analytical equilibrium ultracentrifugation is unsatisfactory since the available instruments are not stable at speeds low enough to balance centrifugal forces against diffusion forces. The best absolute method of DNA molecular weight determination are light scattering on low angle instruments equilibrium analytical ultracentrifugation in Caesium chloride gradient and viscoelastic relaxation. Most other methods have an empirical bases and rely on the few light scattering experiments which have been performed for their calibration. These more empirical methods include the measurement of intrinsic viscosity, sedimentation rate, electron microscopy, electrophoresis and radio autography. Generally a combination of two or more of these methods can be used. The advent of the laser has led to the development of a new technique for determining diffusion coefficients of large molecules from the Doppler shift in the wavelength of the scattered laser night. This technique together with conventional analytic velocity centrifugation to determine sedimentation coefficients should lead to reliable absolute molecular weight determinations. The determination of the molecular weight of DNA has also been complicated by the difficulties experienced in the preparation of whole DNA molecules since high molecular weight DNA is very susceptible to hydrodynamic shearing forces and to the action of contaminating nucleases. Thus all molecular weight determinations on an unknown DNA must be carried out in nuclease free conditions, without pipetting or any other manipulation which puts shear stress on the DNA, any methods which might themselves shear the DNA, such as the viscosity determinations, must be used with caution. The size of large DNA molecules from individual chromosomes of eukaryotic cells has recently been determined by viscoelastic relaxation, and if appear that in *Drosophila* at least the DNA is contained in one complete molecule per chromosomes.

Table 28.4. DNA Molecular Weights

Source	*Mol. wt*	*Length*	*Number nucleotide pairs*	*Conformation*
E.coli chromosome	2×10^9	1 mm	3×10^6	Cyclic, duplex
H. influenzae chromosome	8×10^{-3}	300 μm	12×10^5	—
Mycoplasma PPLO strain H-39	4×10^8	150 μm	6×10^5	—
Bacteriophage T2 or T4	1.3×10^3	50 μm	2×10^5	Linear, duplex
Bacteriophage λ	33×10^6	13 μm	0.5×10^5	Linear, duplex
Bacteriophage ϕX174	1.6×10^6	0.6* μ	—	Cyclic, single-stranded
Polyoma virus	3×10^6	1.1 μm	4.6×10^3	Cyclic, duplex
Drosophila melanogaster chromosome	43×10^9	20 mm	65×10^6	—

In more recent years if has become customary to describe the size of DNA molecules in terms of kilobase pairs (kbp), rather than daltons or molecular weight. A molecular weight of one million corresponds approximately to 1.5 kilobase pairs.

Secondary Structure of DNA

Using early information obtained by X-ray diffraction and chemical observations, *Watson and Crick* in 1953 put forward the view that the DNA molecule is *double stranded* and in the form of a right handed helix with the two polynucleotide chains would round the same axis and held together by hydrogen bonds between the bases. By using scale models they were able to show that the bases couuld fit in if they were arranged in pairs of one pyrimidine and the purine. The equivalence of adenine and thymine, and guanine and cytosine in most naturally occuring DNA molecules. First abserved by *Chargaff* suggested that the most likely hydrogen bonding configuration of base pairs of these two types. These base pairing arrangement have been confirmed as the only ones possible in the B-form of DNA. Although other base pairing arrangements have since been suggested they have only been shown to occur in RNA where they are involved in secondary and tertiary structure stabilization.

The most important consequence of the base pairing configuration formed in DNA is that the order in which the bases occur in one chain automatically determines the order in which they occur in the other, *complementary chain.* Apart from this essential condition, there are no restrictions on the sequence of the bases along the chains.

The shorthand method of describing two DNA strands has been to have the strand of 5'→3' polarity on the top line of the sequence with the complementary strand of opposite polarity lying below. For example:

5' - AGGTC -3'

3' - TCCAG -5'

The pairs of bases are flat and may be stacked on above the other like a pile of plates so that the molecule is readily represented as a spiral staircase with the base pairs forming the treads. The two polynucleotide chains are of opposite polarity in the sense that the internucleotide linkage in one strand is 3'→5' while in the other it is 5'→3'. The two helices are right handed (i.e., (a) turn in a clockwise direction as viewed from the near end or (b) turn in the direction of the fingers of the right hand when the thumb indicates the line of sight) and cannot be separated without unwinding. The pitch of the duplex is 3.4 nm and since there are 10 base pairs on each turn of the helix, there is a distance of 0.34 nm between each base pair. The diameter of the helix is 2 nm.

The base sugar linkage is normally *anti* with the C8 of the purine or the C6 of the pyrimidine ring over the sugar. The sugar puckev is variable but a base pair usually has the sugar attached to the pyrimidine in the C3' *endo* configuration. There is lot of water associated with DNA. Upto 72 molecules of water per 12 base pairs are found largely in the *minor groove* where they form a spine.

The stacked bases form a hydrophobic core but the amino and keto groups point upto the grooves and allow interaction with the solvent. The sugar backbone points out to give hydrophilic with one negative charge per phosphate.

While the basic model put forward of *Watson* and *Crick* remains close to the accepted structure of the DNA molecule in solution, the more refined X-ray diffraction studies of *Wilkins* and his colleagues demonstrated that DNA fibres can have three possible structures with the B structures corresponding closely to the original Watson-Crick model. The A and C structures are also right handed double helices but differ in the pitch and in the number of bases per turn. In both A and C conformations the bases are not flat but are tilted. Their biological significance is not clear but the A conformation is believed to be very close to the structure adopted by double stranded RNA and by DNA-RNA hybrids. Because of the presence of extra 2' hydroxyl group, RNA appear to be unstable to adopt the B conformation. Thus when it is engaged as a template for making RNA, the DNA molecule must adopt the A conformation. Various organic solvents and proteins have been suggested to force the DNA from B to the A form, and the transition from the B conformation to the C conformation appear to occur in concentrated salt solution and ethylene glycol.

Various forms of DNA

A

B

C

D

Z

The Tertiary Structure of DNA

The DNA from many sources, particularly cellular chromosomal DNA, cannot be isolated in the form of intact molecules because of its large size and consequent sensitivity to shear and enzymatic degradation.

However, the DNA from many viruses from mitochondria and chloroplast and from a few other sources (e.g. bacteral plasmid DNA) can be isolated in an ungraded form. The examination of such ungraded DNA has shown that the double helical secondary structure of DNA can in some instances be further contained giving rise to tertiary structure such as super coiled and open cyclic molecules. Linear DNA occurs naturally and is also a product of the degradation or cyclic DNA. Single stranded DNA, both in linear and cyclic forms, is found in some viruses, and can be formed artifically by denaturing double stranded DNA.

TYPES OF DNA'S

Eukaryotic DNA

It is the filamentous types of DNA found in all eukaryotic cells of animals and plants. In these cells, DNA synthesis takes place in a restricted time period during the cell cycle. In mitosis the cells divide and form two daughter cells. This is then followed by a gap period designated G1 during which no DNA synthesis occurs. The length of the various gap periods and indeed the length of the entire cell cyclic depends very much upon the cell type. The synthesis of DNA occurs in S phase and then follows a second shorter gap period known as G_2. The cells then go into mitosis again.

Prokaryotic DNA

Also called the *circular* a *superhelicular DNA*.The DNA molecules of prokaryotes and most viruses are circular. A circular molecule may be a covelently closed circle, which consists of two unbroken complementary, single strands, or it may be a *nicked circle,* which has one or more interruption (nicks) in one or both strands. With few exceptions, convalently closed circles are twisted. Such a circle is said to be a *superhelix* or a *supercoil*.

Extranuclear DNA

Also known as non-chromosomal or cytoplasmic DNA. In cytoplasm this DNA is associated with some of cell organelles as given belows:

(a) Mitochondria DNA

Mitochondrial DNA is usually found in cyclic double stranded, supercoiled molecules, the exceptions being the linear mitochondrial DNA1 molecules from *Tetrahymena* and *Parameciui*. Marmialian mitochondrial DNA molecules are not packaged into nucleosomes. They are about 15 kbp long and can therefore code for 15 to 20 proteins. However, some yeast ones are considerably larger and *Saccharomyces* mitochondrial DNA is 75 kbp long. Plant mitochondrial DNA is much longer still. Yeast mitochondrial DNA molecules have been widely studied since they are readily amenable to genetic analysis. Mitochondrial DNA can only codes for a small proportion of mitochondrial proteins.

Mitochondrial DNA from several mammals, *Xenopus* and *Drosophila,* has been sequenced and the sequence analysed. There are genes from two ribosomal RNAs, 22 tRNAs and for 13 protein most or all of which are involved in electron transport. The only region of mammalian mitochondrial DNA which is non-coding is the *D-loop region* involved in the initiation of DNA replication. This is also the region at which transcription of both strands is initiated. Transcription continues uninterrupted around the cyclic molecule and the transcripts are the processed to give individual messenger, ribosomal and tRNAs. The tRNA forms the punctuation between the various protein coding regions and provides the signals for the processing enzymes.

Plant mitochondrial DNA is apparently much more complex than animal mitochondrial DNA. It consists of per mutation of basic structure related to each other by recombination.

(b) Chloroplast DNA

Chloroplast DNA is in general much larger than mitochondrial DNA, being in the molecular weight range 100 million (150 kbp). In contrast to mitochondria which have from one to ten molecules of DNA per organelle, chloroplasts tend to have a very large number of copies of the DNA molecules in each organelle, in some cases greater than one hundred. Like mitochondrial DNA, the DNA in chloroplast carries the coding information for essential membrane components, tRNA and rRNA. All known chloroplast DNA molecules are cyclic and supercoiled.

(c) Kinetoplast DNA

The kinetoplast is part of highly specialized mitochondrion found in certain groups of flagellated protozoa such as trypanosome. Its DNA (kDNA) consists of an interlinked series of many thousands of cyclic DNA molecules which vary in size from 0.6 kbp to 2.4 kbp depending from the type of trypanosomes from which they are obtained. These components, whch are known as *minicircles,* are further interlinked with a much smaller number of larger circular DNA molecules of above 30 kbp in size known as *maxicircles* in the majority of system analyzed. While maxicircles appear to perform the connectional functions of mitochondrial DNA in trypanosomes, minicircles are microheterogeneous in sequence and size and there is no evidence to suggest that they are ever transcribed so that it is possible that they may fulfil some structural rather than coding role.

Unique DNA

In human cells it is generally classified as that DNA which has a $Cot_{1/2}$ value of about 1000 moles of nucleotide seconds $litre^{-1}$. It comprise about half of the total haploid DNA content and is thought to consist of the sequences coding for most enzyme functions for which there is only one or a small number of genes per haploid genome. The genes coding for the various chains of globin or the enzyme glucose 6-phosphates fall into this category.

Repetitive DNA

In this type of DNA certain bases or nucleotides are repeated many times. Among the repetitive DNA is a fraction which reanneals with a Cot value of between 100 and 1000 moles of nucleotides seconds $litre^{-1}$. The sequences represented in this group are generally thought to be those coding for proteins which form major structural components of the cell such as the histones. The genes for rRNA and tRNA also fall into this category.

Satellite DNA

At the other extreme DNA with a $Cot_{1/2}$ value as low as 10^{-3} moles of nucleotide seconds $litre^{-1}$ consists largely of *satellite DNA.* This represents highly repeated sequences of which there may be a million or more copies per haploid genome, which are usually quite short and are arranged in tandem arrays. The origin of the name satellite relates to the method of its isolation on caesium chloride buoyant density gradients of sheared DNA where it will sometimes form a satellite band separate from main DNA band, due to its differing content of adenine and thymine residues. The simplest known DNA is poly [d (A-T)], which occurs in certain crabs. Other satellites can have any number upto several hundred base pair which are repeated in tandem fashion alsong the genome.

The distribution of satellite DNA among chromosomes varies. Some chromosomes have virtually no satellite sequences while others are largely composed of satellite sequences. In general, satellite DNA appears to be concentrated near the centromere of the chromosomes in the heterochromatin fraction. DNA sequence analysis has shown that the basic repeat unit of satellite DNA is itself made up of sub repeats. For example, the major momse satellite has a repeating structure of 234 base pairs made up of four related 58 and 60 bp segments each in turn made up of 28 and 30 bp sequences. The satellite between and within the related species are themselves related in an evolutionary sense by cyclic rounds of multiplication and divergence of an initial short sequence.

Fold-back DNA

Also known as *palindromic DNA.* It is a special class of DNA sequences comprising 3-6% eukaryotic DNA. The size range is from 300 to 1200 base pairs and the molecules have a $Cot_{1/2}$ value of less than 10^{-5} moles of nucleotide second $litre^{-1}$. This palindromic DNA is represented in all frequency classes and is widely distributed throughout the metaphase chromosomes. Some palindromic DNA arises when two copies of the *Alu* sequences are present close to one another in opposite orientations and it was a result of the ability of such DNA to renative instantaneously.

Z-DNA or Left Handed DNA

On the basis of the arrangement of pentose sugar forming chains of DNA two forms of DNA have been reported. Usually the sugar chain has the pentose sugar with ascending sequence of carbon atoms. This is the right handed DNA or B-DNA. Recently in 1979, another form of DNA have pentose sugar with ascending arrangement of carbon atom has been reported and named Z-DNA or left handed DNA.

Tautomeric form of DNA

An important chemical feature of DNA is the position of the hydrogen atoms in the purine and pyrimidine bases. Before 1953, many chemists through that some of these hydrogen atoms randomly moved from one ring nitrogen or oxygen atoms to another and os could not be assigned a fixed location. Now we realize that althoguh such movements, called *tautomeric shifts* do occur, these hydrogens have preferred atomic locations. The nitrogen atoms attached to the purine and pyrimidine rings are usually in the *amino* (NH_2) form and only rarely assume the *imino* (NH) configuration. Likewise, the oxygen atoms attached to the C6 atoms of guanine and thymine normally have the *keto* (C-O) form and only rarely take up the *enol* (COH) configuration.

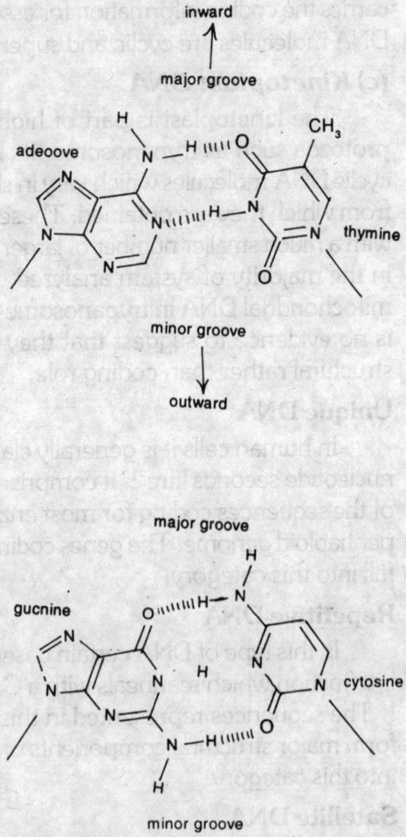

Fig. 28.6. The relative directions in which the base-pair nitrogens and oxygens project in each groove, the nitrogens and oxygens of the floor of the major groove project toward the center of the DNA molecule, while those of the minor groove project toward the outer edge of the DNA molecule.

It is essential to the biological functioning of DNA that the hydrogen atoms have relatively stable locations. If the hydrogen atoms had no fixed positions, adenine could often pair with cytosine and guanine with thymine, and sequence of the bases on the two interwined chains would not necessarily be complementary. If that were the case, the DNA could not function as a genetic molecule, for it is the complementary relationship between the opposing chain sequences that gives DNA its capacity for self-replication. Replication of the double helix involves strand separation followed by formation of complementary DNA chains using the now free single strands as templates to attract the appropriate partner bases dictated by AT and GC base-pairing rules. The imino forms adenine and cytosine and the enol forms of guanine and thymine must indeed occur only very rarely compared to their amino and keto alternatives. Otherwise, the many errors (mutations) incurred during DNA replication would be incompatible with orderly cell growth and division.

Denaturation and renaturation of DNA

Single-stranded DNA

Single-stranded DNA molecules rarely occur naturally. The best known exmaple is the DNA molecule from the small spherical bacteriophage ϕX174 first isolated by *Sinsheimer* in 1959 but it is also found in the filamentous bacteriophages such as fd. Among the animal viruses single-stranded DNA is found in the parvo viruses. Some of these naturally occurring single-stranded DNA's are circular and other linear. The former can be distinguished by their insensitivity to exonucleases. In general they are comparatively small molecules with less than 10,000 nucleotides. The molar proportions of the bases do not show the usual equivalence of A and T, and G and C, required for double helix formation and the bases react readily with formaldehyde since the amino groups are not protected by hydrogen bondings as they are in double stranded DNA. Single-stranded DNA molecules are very much more flexible than double-stranded molecules of the same size since they lack the rigid double helical structure. In general they behave in solution in manner similar to RNA molecules.

The helix-coil Transition

When double stranded DNA molecules are subjected to extremes of temperature or pH, the hydrogen bonds in the double helix are ruptured and the DNA collapses into two single-stranded molecules. If heat is used as the denaturant, the temperature at which this collapse occurs is known as the T_m or transition temperature. The absorption at 260 nm of any polynucleotide is due to that of its component bases. However, this absorption tends to be suppressed in the double-stranded DNA molecule where the stacked above one another and inhibited from swinging out freely in solution in hydrogen bonds, and consequently it is very much lower than that of equimolar amounts of the component bases or nucleotides free in solution. This inhibition of absorption is to some extent relieved when the DNA molecule goes through the helix-coil transition and the bases are no longer so rigidly stacked although they still interact to some extent. As a consequence of this absorption of DNA solutions rises by about 20-30 per cent as the DNA undergoes the melting process, and the transition process is usually measured by ultraviolet spectroscopy. The increase in absorbance on melting is known as the *hyperchromic effect.*

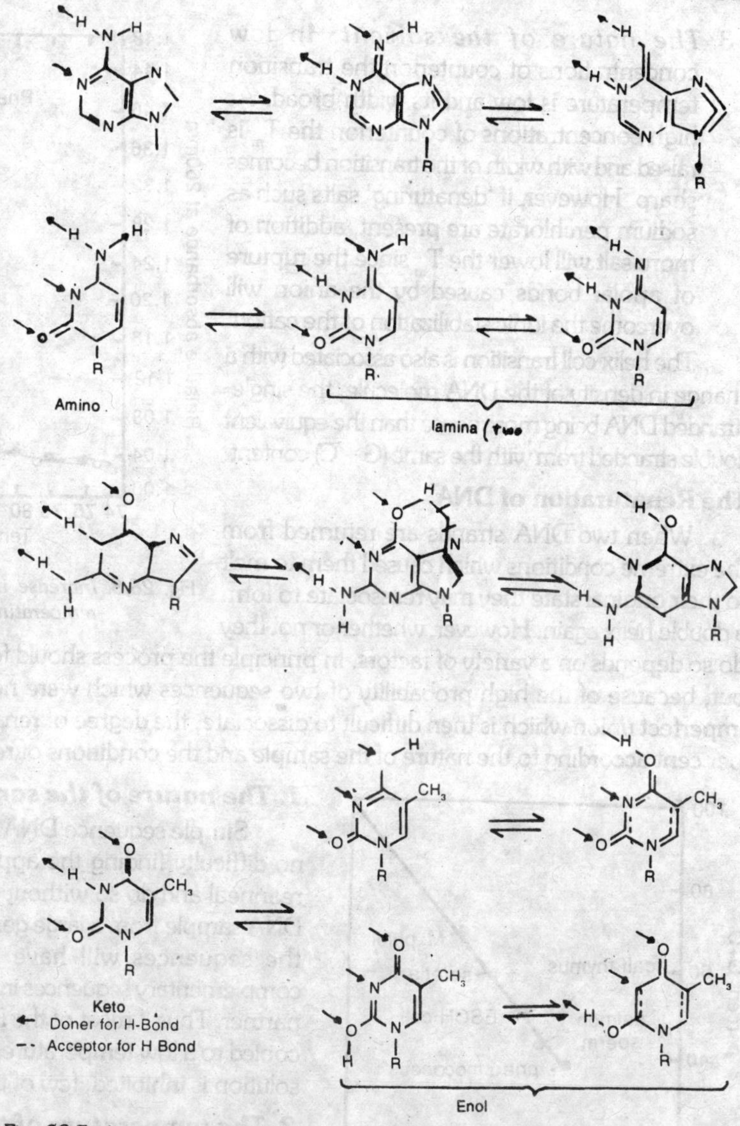

Fig. 28.7. Amino=imino and keto=enol tautomerism. (a) The A and C nucleotide bases are usually in the amino form but rarely form the imino configuration. (b) The bases G and T are usually in the keto form but are rarely found in the enol configuration. Atoms shown in colour may participate in hydrogen-bond formation.

Such a transition can be characterized by several factors:

1. *The nature of the DNA.* Homogenous DNA, such a viral DNA, melts over a short temperature range but heterogenous DNA melts over a longer range.
2. *The (G + C) content of the DNA.* The melting temperature of any DNA can be related to its (G + C) content since this base pair confer extra stability on the molecule. In DNA molecules such as those from bacteriophage l, in which some regions are richer in (G + C) than others, the transitions of both regions can be observed.

3. *The nature of the solvent.* In low concentrations of counterion the transition temperature is low and its width broad. At high concentrations of counterion the T_m is raised and with width of the transition becomes sharp. However, if 'denaturing' salts such as sodium perchlorate are present, addition of more salt will lower the T_m since the rupture of apolar bonds caused by the anion will overcome the ionic stabilization of the cation.

The helix-coil transition is also associated with a change in density of the DNA molecule, the single-stranded DNA being more dense than the equivalent double stranded from with the same (G + C) content.

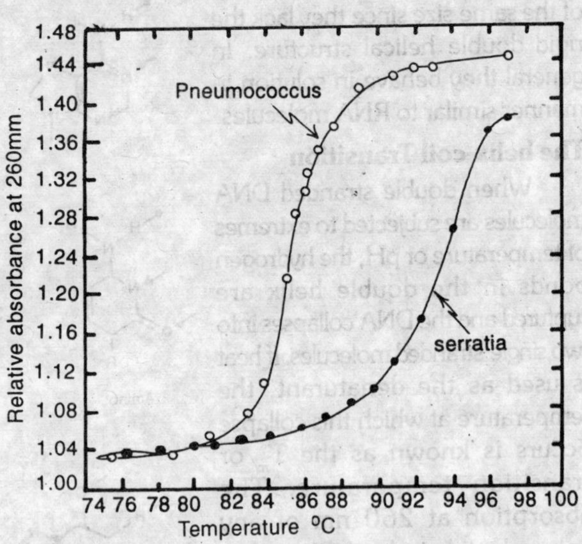

Fig. 28.8. Increase in absorbance at 260 nm with rising temperature for DNA from two different sources.

The Renaturation of DNA

When two DNA strands are returned from the extreme conditions which caused them to melt to their original state they may reassociate to form a double helix again. However, whether or not they do so depends on a variety of factors. In principle the process should follow simple second-order kinetics but, because of the high probability of two sequences which were not previously matched forming an imperfect union which is then difficult to dissociate, the degree of renaturation can vary from less than 1 per cent according to the nature of the sample and the conditions of renaturation.

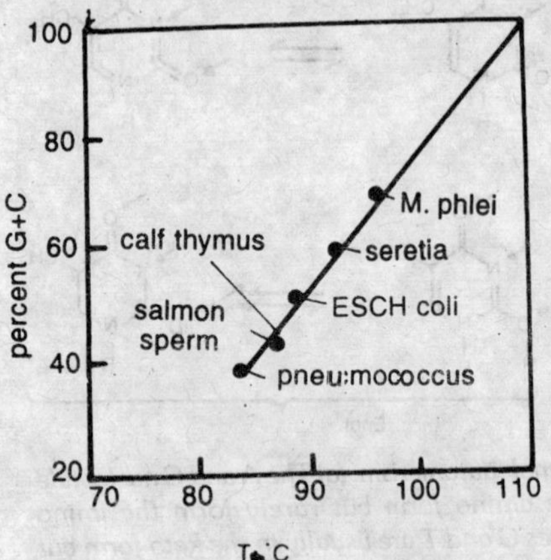

Fig. 28.9. Relationship between the content of guanine plus cytosine and melting temperature (T_m) in DNA's from various sources.

1. The nature of the sample

Simple sequence DNA molecules such as $d(G)_a\ d(C)_n$ have no difficulty finding the appropriate sequence with which the reanneal and do so without difficulty. However, if a eukaryotic DNA sample from a large genome is reannealed, clearly some of the sequences will have to encounter many other non-complementary sequences in solution before they find the correct partner. Thus if most of the DNA from eukaryotic cells is quickly cooled to a low temperature, so that the diffusion of the DNA in solution is inhibited, few of the single strands will reassociate.

2. The temperature of the reassociation process

At very low temperature (4°C) not only a diffusion limited but, for a DNA molecule which has become mismatched with a strand having only a few complementary bases, the opportunities to break away and continue its search for th correct complementary strands are reduced. Consequently DNA which is heated beyond the T_m and then quickly cooled to a low temperature will be denatured whereas solutions maintained at high temperatures below the T_m may renature.

3. The size of the DNA fragments

Large string-like fragments of single-stranded DNA encounter diffusion problems and frequently cannot reanneal correctly since this effectively reduces their opportunities for finding complementary strands. For this reason DNA is often sheared for reannealing experiments.

4. The ionic strength of the solution

Two highly charged DNA molecules are likely to repel one another, and the presence of salt is necessary to mask this repulsion in renaturation.

5. The concentration of the DNA

Clearly, at higher concentrations the probability of two complementary strands encountering each other is raised.

6. The time allowed for reannealing experiments

If renaturation is allowed under ideal conditions, two DNA samples of identical concentration should take different times to reanneal according to the genome size. For this reason the term *Cot value* has been defined for the study of reannealing of DNA and also for the study of the formations of DNA-RNA hybrids. *Co* represents the DNA concentration and *t* represents time in seconds. This is again monitored by ultraviolet spectroscopy or, more frequently, by hydroxyapatite chromatography since this can be used to separate double- and single-stranded DNA molecules and can be used at higher concentrations and hence for shorter times.

Single-Stranded DNA or Viral Chromosomes

At first it was thought that all DNA molecules are double stranded except during replication, when a small region around the replicating fork is temporarily in a non-hydrogen-bonded, single-stranded form. It therefore came as quite a surprise when experiments revealed that the DNA of several groups of small bacterial viruses exists as single-stranded molecules in which the amount of A is not equal to the amount of T and the amount of G does not equal the amount of C. Among these single-stranded phages are the spherically shaped ϕX174 and S13 virsuses and the rod-shaped f1 and M13 viruses. The chromosomes of the parvoviruses, minute virsuses that infect many vertebrate organisms, are also single stranded DNA molecules.

When these single-stranded viral chromosomes enter their host cells, they serve as templates for the formation of complementary strands. The resulting double helices in turn serve as templates for new single strands that then become incorporated into new virus particles. Thus, the fundamental mechanism for ordering nucleotides during the synthesis of single-stranded DNA is basically the same as that used for double-hellical DNA. Nucleotide selection always occurs by attraction of the complementary base. Single-stranded DNA replication usually differs from double-helical replication in that it uses only one of the two complementary strands (the—strand) as a template for the progeny (+) strand.

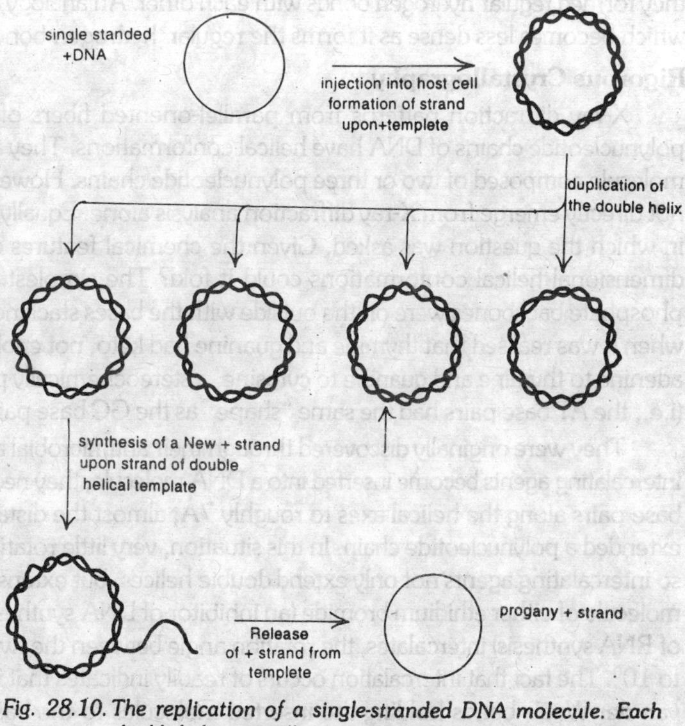

Fig. 28.10. The replication of a single-stranded DNA molecule. Each double helix generally servers as the template for the formation of a large number of new + strands.

Although single-stranded DNA can serve as the chromosome of a small virus, it would not be very effective in storing genetic information within the chromosomes of cells. In bacteria and in many eukaryotic cells (e.g., the haploid phase of yeast cells), there usually is only one copy of a given gene. If interphase chromosomes contained single-stranded DNA, with the double helix existing only briefly during mitosis, then the daughter cells would contain two completely different sets of genetic informations. This follows from the fact that the two complementary chains do not have identical base sequences and hence would code for entirely different amino acid sequences. The double-strand ensures that each daughter cell will contain the same genetic information. The double-stranded form also permits effective DNA repair mechanisms to exist. For example, when one strand is damaged by exposure to X-rays, the remaining strand can provide the genetic information to rebuild the damaged section. It is probably no accident that only very small viral chromosomes are single-stranded. If they were to become, say, the size of T2, they would present too large a target for chain-damaging events.

Structure of Single-stranded DNA

Single-stranded DNA molecules have a strong tendency of fold back on themselves to from irregular double-helical hairpin loops whenever their sequences permit signficant numbers of nucleotides to base-pair. Imperfect as these loop are, they nonetheless are energetically favoured under physiological salt conditions, since they allow much more effective stacking of the flat hydrophobic surfaces of the bases then is possible in any fully extended single-stranded structure. Often, as many as half the bases of single-stranded DNA are so hydrogen-bonded, and their molecules are highly compacted. If, however, there are only minimal numbers of, neutralizing cations (e.g., less than or equal to 0.01 M Na^+) then the electrostatic repulsion between the partially unneutralized phosphates will kept the single-stranded chains highly extended, and there will be fewer hairpin loops.

Single-stranded DNA is always denser than double-helical DNA. Much more complete hydrogen bonding to water is possible for the atoms of the highly regular double helix than for the atoms of the inherently irregularly shaped single-stranded DNA. In the absence of base-pairing hydrogen bonds, Van der Waal's interactions take over and bring the atoms of DNA closer together than would be the case if they formed regular hydrogen bonds with each other. An analogy can be made with the behaviour of water, which becomes less dense as it forms the regular, hydrogen-bonded lattice of crystalline ice.

Rigorous Crystallography

X-ray diffraction patterns from parallel-oriented fibers of DNA provided the first clues that the polynucleotide chains of DNA have helical conformations. They also told us that DNA was a multichained molecule composed of two or three polynucleotide chains. However, the double-helical nature of DNA did not directly emerge from X-ray diffraction analysis alone. Equally important was the use of model building in which the question was asked. Given the chemical features of a single DNA chain, into what three-dimensional helical conformations could it fold? The simplest models were those in which the sugar-phosphate backbones were on the outside with the bases stacking in the center. The double helix emerged when it was realised that thymine and guanine had keto, not enol, configurations and that by base-pairing adenine to thymine and guanine to cytosine, a stereochemically pleasing, regular helical molecule resulted (i.e., the AT base pairs had the same "shape" as the GC base pairs).

They were originally discovered through their antimicrobial and anticancer attributes. Each time these intercalating agents become inserted into a DNA molecule, they necessarily *increase* the spacing of successive base pairs along the helical axes to roughly 7Å, almost the distance between phosphate atoms in a fully extended a polynucleotide chain. In this situation, very little rotation is possible around the helical axis, and so intercalating agents not only extend double helices, but extensively unwind them. For example, when a molecule of either ethidium bromide (an inhibitor of DNA synthesis) or actinomycin D (a powerful inhibitor of RNA synthesis) intercalates, the rotation angle between the two adjacent base pairs is reduced from 36^0 to 10^0. The fact that intercalation occurs of readily indicates that it must be energetically favoured, with the van der Waals bonds holding the inserted molecules to the base pairs beings ronger than those found between conventionally stacked base pairs. Intercalation is additional evidence for the *metastability* of the double-helical structure—its ability to temporarily assume many inherently unstable configurations that normally quickly revert back to the standard B conformation. One such metastable variant must be short stretches of extended chains with gaps separating adjacent base pairs, into which an intercalating agent may bind.

The Chromosomes of Viruses, E.coli and Yeast are Single DNA Molecules

The first estimates of the average molecular weights of DNA centered at about a million, the size needed to encompass in average-size bacterial gene coding for some 300 to 400 amino acid. Therefore, it seemed natural to equate single DNA molecules with single genes. However, these early reports were inaccurate owing to DNA breakage during its isolation and study. Now it is clear that virtually all undergraded DNA molecules contain the information of a least several genes. The most certain molecular weight values come from DNA-containing viruses. Regardless of whether the DNA content is relatively small or large, each virus particle contains a single DNA molecule. For example, all the DNA of the small monkey (simian) virus SV40 is present within a single molecule of molecular weight ~ 3×10^6 daltons (5×10^3 base pairs, or 5 kbp), while the DNA molecule of the large bacterial virus T2 has a molecular weight of $1.2. \times 10^8$ dalton (2×10^3 kbp). In these cases, the entire viral chromosome, rather than separate genes, corresponds to a single DNA molecules. Likewise, the chromosome of an *E. coli* cells is a single DNA molecule whose molecular weight is about 2.5×10^9 daltons (about 4×10^3 kbp) and whose extended length is, roughly 1 mm.

DNA molecules of the yeast *Saccharomyces cerevistiae* range is size between the T2 and *E.coli* DNAs, with their average size being that expected if each of the yeast's 17 chromosomes contain one DNA molecule. Thus, the centromeres or yeast chromosomes (if not those of all eukaryotic chromosomes) do not represent discontinuities between separate DNA molecules, but instead are specialized regions of DNA evolved to interact with the spindle bodies upon which chromosomal segregation occurs during cell division. As of yet, there is no direct proof that the very much larger chromosomes of higher plants and animals (often 50 times larger than the *E. coli* chromosomes) also contain only one DNA molecule. However, we now expect that the one chromosome—one DNA molecule rule will hold for the chromosomes of all organisms. In any case, some DNA molecules are larger than any other biological molecules by several powers of ten.

Circular Versus Linear DNA Molecules

Our evidence, which came largely from autoradiography and electron microscopy, initially suggested that all DNA molecules are linear and have two free ends. But when it became possible to take a better look at undergraded DNA, many DNA molecules were found to be circular. For instance, the small monkey DNA virus SV40 has a 5000-base-pair circular double-helical chromosome; the similarly short chromosomes of single-stranded phases are likewise circular, as are almost all autonomously replicating plasmid DNAs. Most, if not all, bacterial chromosomes are also circular. We suspect that the DNA found in the rare circular chromosomes of higher cells are likewise circular molecules. Circular shapes were initially puzzling, since they seemed to present obstacles to the untwisting of double helices during DNA replication. Only when specific enzymes that first snip and then join DNA chains were discovered did the untwisting dilemma disappear.

Equal important is the realization that linear DNA molecules do not have the uncomplicated structures first envisioned for them. As we see inthe next chapter, replicating the ends of DNA molecules is not a straight forward process. Thus, all linear DNA molecules have evolved special tricks to replicate their ends, as well as to prevent their free ends from either being nibbled back by cellular enzymes or being enzymatically ligated together to form even larger chromosomes. For example, the ends of the linear adenovirus chromosomes have specialized proteins sovalently attached to the 5' ends of strands; and poxvirus chromosomes have closed hairpin loops at their ends, with their basic structures being that of largely self-complementary circular single strands. Still other linear viral chromosomes (e.g., T2 and T7) have the same sequence at both ends, allowing the ends of the chromosome to recombine together during DNA replication to form very long, end-to-end aggregates (concatamers).

Moreover, some DNA molecules that are linear when isolated from a virus particle (e.g., phase λ) are found as circles inside the host cell. This tells us that the linear and circular forms of such DNA molecules are interconvertible. Starting with a closed circle, a specific enzyme introduces two breaks, one in the +

strand and one in the – strand. As thse breaks are very close to each other, the intervening hydrogen bonds occasionally are broken by thermal agitation, and then the circle unfolds. Most importantly, the resulting linear form contains single-stranded ends that have complementary nucleotide sequences (:*stickly ends*"). Thus, at a later time the linear form can base-pair and resume a circular configuration. If the mission phosphodiester bonds are then reformed, the covalent circle is regenerated.

There is solid evidence that phase λ DNA interconverts between its circular and linear forms using precisely this mechanism. Later, we shall see that the duplication of such "sticky" DNAs involves circular replicative intermediates. Thus, the linear form is most likely an adaptation for injection of the viral chromosome through the narrow phase tail.

Supercoiling of Circular DNA

As DNA is a very flexible structure, its exact molecular parameters are a function of both the surrouding ionic environment and the nature of the DNA-binding proteins with which it is complexed. Because their ends are free, linear DNA molecules can freely rotate to accommodate changes in the number of times the two chains of the double helix twist about each other. But once the two ends become covalently linked to form circular DNA molecules, the absolute number of times the chains twist about each other (the *linkage number*) cannot change. For the most part, changes in the average number of base pairs per turn of the double helix will necessarily be accomodated by the formation of the appropriate number of supercoils in the opposite direction. *Supercoiling* refers to the further twisting of double-helical DNA molecules. Untwisting of the double helix usually leads to supercoiling in the negative (left-handed) direction while ovetwisting leads to positive (right-handed) supercoiling.

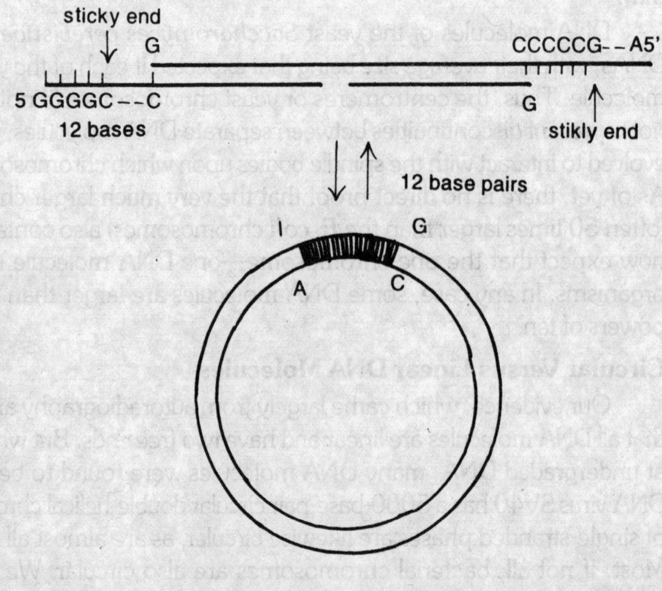

Fig. 28.11. Interconversion of the linear and circular forms of λ DNA.

The supercoiled state is inherently less stable than uncoiled DNA, and the cutting, or *nicking*, of a single strand instantly converts a supercoiled molecules into its simple (*relaxed*) circular state. Because circular DNA molecules become increasingly compacted as they become more supercoiled, relaxed DNA is easily distinguished from supercoiled DNA by its slower sedimentation in a centrifugal field and by the longer time it takes to move through an agarose gel. At the present time gel electrophoresis affords the most direct way to measure supercoiling, since it can separate molecules that differ by only single turns.

Protein Containing Supercoils

Just becuase isolated DNA molecules would energetically prefer to be in the relaxed state does not mean that the "natural state" of DNA is relaxed. In fact, virtually all DNA within both prokaryotic and eukaryotic cells exists in the negative supercoiled state. As such it seldom, if ever, occurs as free DNA, but is complexed with specific DNA-binding proteins to form compacted molecules called *chromatin*. How this compaction occurs in best understood for eukaryotic chromatin, whose most prominent DNA-binding components are the histones. *Histones* are relatively small, positively charged, arginine- and lysine rich proteins that aggregate together to form discrete ellipsoid-shaped packets (histone cores) around which the DNA supercoils. The resulting *nucleosomes* give to chromatin its beaded appearance.

Since the same collection of histones binds to nearly all sections of DNA, histones are thought to play essentially structural roles, as opposed to enzymatic or regulatory roles. Only about half the mass of chromatin proteins is histone. The remaining proteins are not nearly as well characterized with their exact functions yet to be determined.

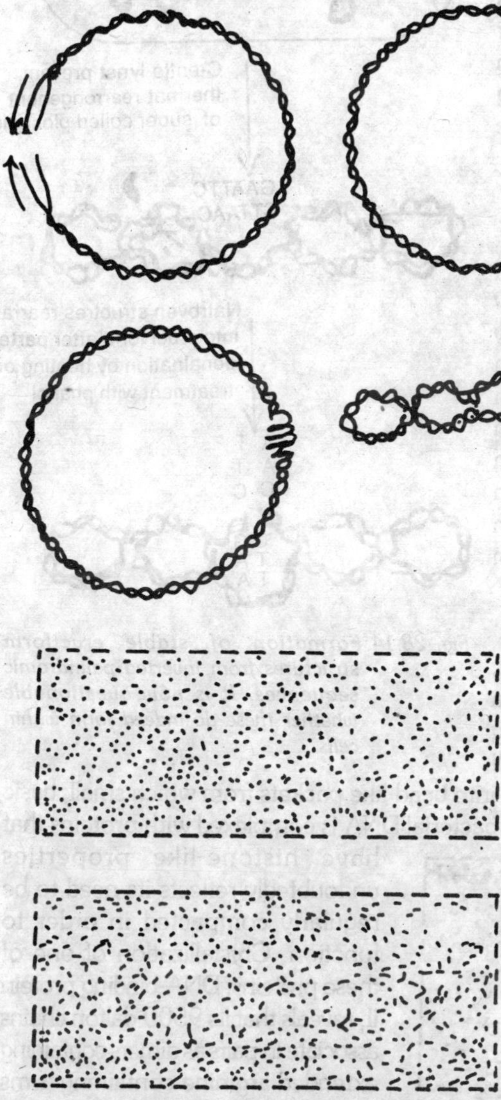

Fig. 28.12. Generation of supercoiled DNA by cutting and ligating. In a circular DNA molecule, the number of times the DNA strands twist about each other is fixed. If both strands are cut (1) and rotated in the opposite direction from that of the twist of the helix (2), and if the cut ends are then rejoined (3) then the DNA molecule negatively supercoiled. (4) The upper electrons micrograph is relaxed (non-supercoiled) DNA molecule of bacteriophage PM2. The lower electron micrograph shows the phages in its supertwisted form.

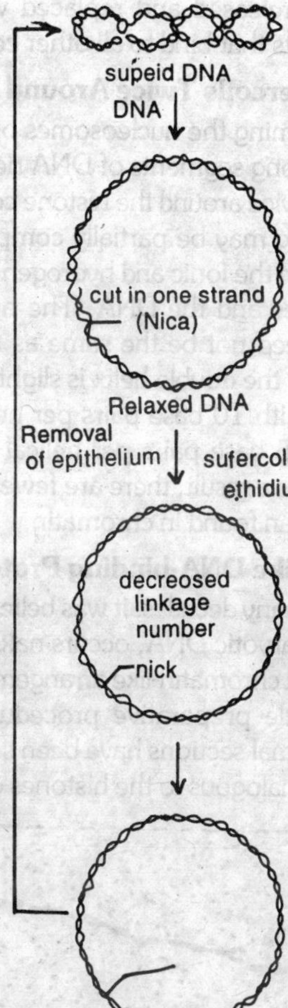

Fig. 28.13. Use of nicking and intercalation to interconvert supercoiled and relaxed DNA. When one strand is nicked, a supercoiled circular DNA molecule relaxes. Ethidium intercalated into the DNA molecule incerases the spacing of base pairs, unwind the double helix by 26^0, and thus decrease the linkage number. DNA ligase can rejoin the free ends. When ethidium is removed, the DNA molecule supercoils negatively.

Viral DNA chromosomes are also complexed with proteins, both when multiplying within cells and when packaged into virus particles. The SV 40 viral chromosome is at all time compacted into 24 histone-containing nucleosomes identical it structure to those existing in the chromosomes of their host cells. When freed from its protein components, SV 40 DNA is found as a negative supercoil. In contract, no histones are complexed with the DNA found in adenovirus particles. These vertebrate viruses encode a unique DNA binding protein to help package their DNA within their protein capsules. When, however, an adenovirus DNA functions within a cell, its own DNA-binding protein is released and replaced with the same histone components that bind to all other cellular DNA.

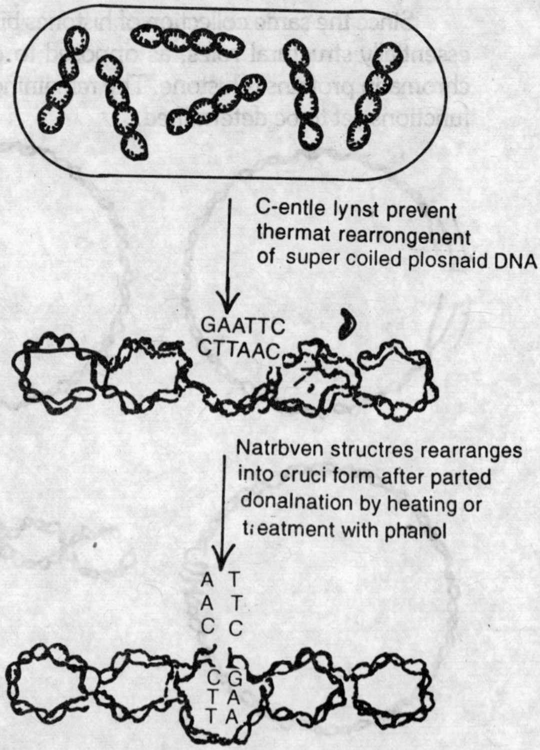

Fig. 28.14. Formation of stable cruciform structures from inverted palindromic sequences. It is now questionable whether these do indeed form within cells.

DNA Supercoils Twice Around Each Nucleosome

In forming the nucleosomes of eukaryotic cells, 200-base-pair-long segments of DNA negatively (left-handedly) supercoil twice around the histone cores. The energy lost by supercoiling may be partially compensated by the energy gained from the ionic and hydrogen bonds formed between the histones and the DNA. The number of supercoils in solution need not be the same as found in chromatin. In chromatin, the double helix is slightly more twisted than in solution, with 10 base pairs per helical turn in chromatin versus 10.5 base pairs per helical turn in naked DNA in solution. As a result, there are fewer super helices found in solution than found in chromatin.

Histone-like DNA-binding Proteins in Prokaryotes

For many decades, it was believed that bacterial DNA, unlike eukaryotic DNA, occurs naked and does not have a compacted chromatin-like arrangement. But recently, using more gentle preparative procedures, condensed *E.coli* chromosomal sections have been seen that are organized into bead-like packets from which small, basic proteins analogous to the histones can be extracted. That bacterial DNA is complexed with proteins that have histone-like properties undoubtedly reflects its need to be regularly compacted in order to function. Crystallization of one of these proteins, DNA-binding protein II, reveals that its 9500-dalton chains associate in pairs as dimers containing extended arginine containing arms able to interact with the phosphates of one turn of the DNA backbone. So it is likely that dimers sited adjacently along prokaryotic DNA are oriented into a helical arrangement with the DNA bound on the outside. Interestingly, the average *degree of supercoiling* (i.e., the number of super-helical twists per ten base pairs) is about 0.05 for all naturally occurring DNA supercoils, regardless of their source. This may in part be because the underlying chromatin structure is so similar within both prokaryotic and eukaryotic cells.

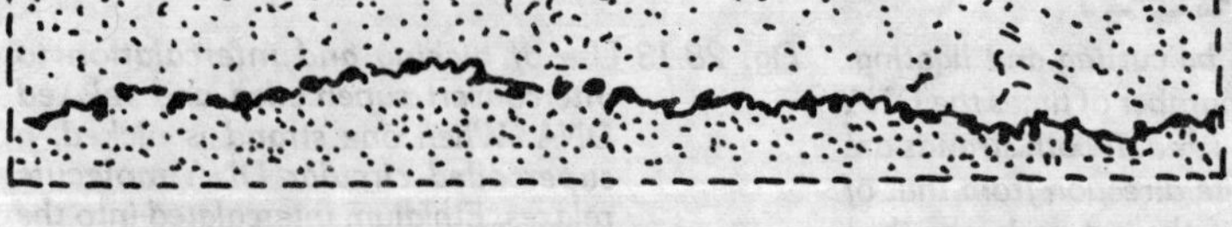

Fig. 28.15. Nucleosomes give chromatin strands a beadlike appearance. Magnification 100,000.

Topo-isomerases Change the Linkage Numbers of Supercoiled DNAs

That supercoiling not only occurs but is biologically very important is affirmed by the existence of the enzyme topoisomerase II, which specifically generates the negative supercoils that characterize so much of chromosomal DNA. *Topoisomerases* are a group of enzymes that convert (isomerize) one topological version of DNA into another. They do so by changing the linkage number, which is the number of times two DNA chains twist around each other. Prokaryotic topoisomerase II (sometimes called *gyrase*) uses the energy of ATP to generate negative supercoils by untwisting DNA in the left handed direction. Its eukaryotic counterpart, however, may be capable of using ATP to generate negative supercoils only in the presence of other, yet to be described proteins. The supercoiling action of topoisomerase II is counterbalanced by a second enzyme, topoisomerase I, which converts supercoiled DNA to the unstrained, energetically more favourable relaxed state. The relative amounts of topoisomerase I and topoisomerase II in cells are finely tuned, tending to create just the right amount of negative supercoiling. Thus, mutations that lower the number of topoisomerase I molecules are viable only if the number of topoisomerase II molecules also decreases.

Both topoisomerase I and II work by catalyzing the breakage and rejoining of DNA phosphodiester bonds. Unlike virtually all other enzymes, their action does not lead to new patterns of covalent bonds. Rather, their role is to create temporary gaps in polynucleotide chains. When so acting topoisomerases do not create free ends but instead become themselves covalently attached to one of the two broken ends (depending on the specific enzymes, either the 3' and 5' end). Through such catalysis, a tyrosine group on the topoisomerase becomes linked to the terminal pohosphate group of the cut polynucleotide chain. The free energy present in the original phosphodiester bond is thus preserved so that it can be used to rejoin the broken chain.

Table 28.5. DNA Topoisomerases

	Prokaryotic (E.coli)		Eukaryotic (Hela)	
	Type I	*Type II*	*Type I*	*Type II*
Approximate MW	100,000	400,000	100,000	309,000
Subunits	Monomer	Tetramer (A_2B_2)	?	Dimer (subunit MW=172,000)
Gene	top A	*gyr A* Polypeptide MW 105,000; activity inhibited by nalidixic acid and oxolinic acid *gyr B* Polypeptide MW 95,000; activity inhibited by cumer-mycin and novobiocin	?	?
Covalent intermediate with DNA	At 5' P	Subunit A Tryosine covalent bond to 5' P	At 3' P	At 5' P
Relaxation	Mg^{2+} required; ATP-independent	ATP-independent	No Mg^{2+} required; ATP-independent	ATP-dependent
Negative supercoiling	None	ATP-dependent	None	None

When a DNA chain is broken by a topoisomerase, its broken ends do not fall apart but are held together by the topoisomerase. At no time do the broken ends have the capacity to rotate freely. If that happened, topoisomerase would be limited to only completely relaxing double helices. Instead, individual topoisomerases function to make discreate changes of double helices. Instead, individual topoisomerases function to make discrete changes of either one positive turn (topoisomerease I) or two negative turns (topoisomerase II) in the linkage number. Such discrete steps result from DNA chains passing through either transient single-stranded breaks (topoisomerase I) or transient double-stranded breaks (topoisomerase II).

Topoisomerases are constructed in such a way that they can undergo reversible conformational changes that create cavities through which DNA chains can pass. The ATP requirement for bacterial topoisomerase II (gyrase)-induced supercoiling most likely reflects the use of ATP in mediating the necessary conformational changes that lead to chain passage. Topoisomerase I, however, relaxes supercoiled DNA without requiring a cyclical input of energy. Complete understanding of how topoisomerase I and II work can occur only when their precise structures and determined through X-ray crystallographic procedures. Even now, however, the prediction can be made that double helices wrap themselves around topoisomerase II in positive supercoils prior to the start of the cutting-rejoining cycle. Without this restraint on the orientation of the DNA, topoisomerase II would break as well as make negative supercoils, and the net result of its action would be eventual relaxation of the double helical substrate to the uncoiled state.

Looped and Supercoiled Structure

The fact that linear DNA molecules have free ends would seem to preclude their possession of supercoiled segments. Yet, examination of gently isolated linear eukaryotic chromosomes shows that their chromatin is organized into a large number of successive looped domains organized along a scaffold containing two major DNA-binding proteins. Somehow, attachment ot the scaffold prevents the rotation of one domain from being transmitted to adjacent sections, since the DNA within each of these loops independently supercoils and may be under different torsional strain. Now there is much evidence both from *Drosophila's* giant salivary chromosomes and from the looped lamph rush chromosomes of the salamander that the individual loops represent functional units of chromatin, with given loops usually being transcribed into one or more very long RNA molecules. How the protein scaffold might maintain independently supercoiled loops was a complete mystery until recently. Now it is less so, following the discovery that a major component of the scaffold is none other than topoisomerase II.

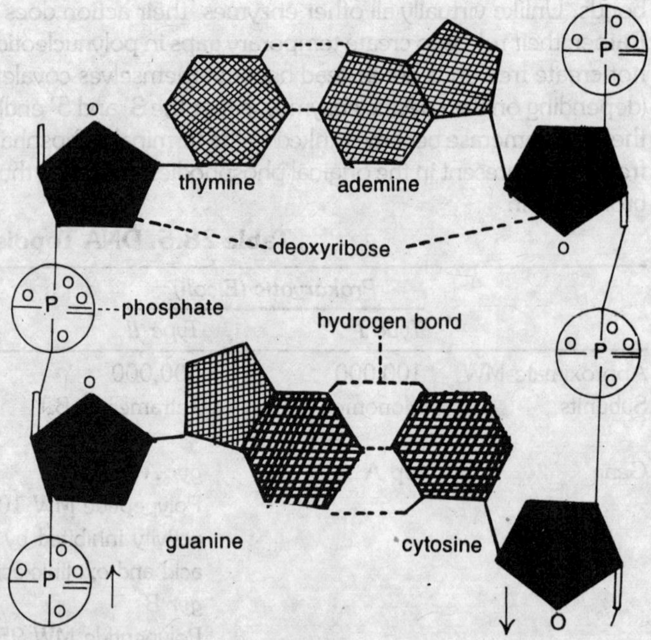

Fig. 28.16. Showing antiparallel nature of backbones of DNA.

Not only are the long, linear chromosomes of higher organisms divided into looped, functional domains, but the circular *E. coli* DNA chromosome also is somehow divided into some 50 looped domains (each about 20 μm long) whose supercoiling is independently controlled. No scaffold has yet been isolated from any bacterial chromosome. Indeed, the best way now to search for it may be to look for topoisomerase II activity in very gently ruptured cells.

Biological Replication of DNA

Study of replication of DNA is one of the interesting aspect of molecular control of heredity. Various study of the duplication of genetic material was presented from time to time in different organisms. But most of them are discarded because no concrete ideas have been proposed till *Watson* and *Crick* proposed a detailed scheme for the molecular sructure of DNA. Later on *Korenberg, Stahl* and *Taylor* have confirmed the Watson-Crick hypothesis. They also made a suggestion for its mode of replication which follows in a natural and attractive fashion from the proposed structure. During the replication, the DNA duplicates exactly into two chains. In the beginning both the strands of double stranded DNA, apart from each other by the disolution of hydrogen bonds and thus uncoiling of the chains occur. And then new chains, "the complementary chains" are built on them so that the newly formed daughter chains are the mirror images of each other.

Mechanism of DNA Replication

The entire process of DNA replication involves following steps in *E.coli:*

1. *Recognition of the initiation point.* Replication of DNA begins at a specific point, called initiation point. Specific initiator proteins are required to recognise the initiation point on DNA.

   The initiator protein along with DNA directed RNA polymerase initiates the synthesis of RNA primer for the formation of new DNA chain.

2. *Unwinding of DNA.* The unwinding proteins bind to the nicked strand of the duplex and open up a loop or buble, separating the two strands of DNA duplex.

3. *RNA priming.* The DNA directed RNA polymerase now synthesizes the primer strands of RNA (*RNA primer*). The priming RNA strands are complementary to the two strands of DNA and are formed to 50 to 100 nucleotides.

4. *Formation of DNA on RNA primers.* The new strands of DNA are formed in the 5'-3' direction from the 3'-5' template DNA by the addition of deoxyribonucleotides to the 3' end of RNA primer. The addition is affected by DNA polymerase III in the presence of ATP. The unwinding proteins separate the duplex strands ahead of the replication fork.

   The leading strand of DNA synthesized in 3'-5' direction. The following strand of DNA is formed in its opposite direction. The DNA strands are synthesized in short segments, consisting of 1,000 to 2,000 nucleotides. These segments are called *Okazaki fragments.*

5. *Excision of RNA primers.* Once a small segment of an Okazaki fragment has been formed, the nucleotides of RNA primer are removed from the 5' end one by one by the action of 5'-3' exonuclease activity of DNA polymerase-I.

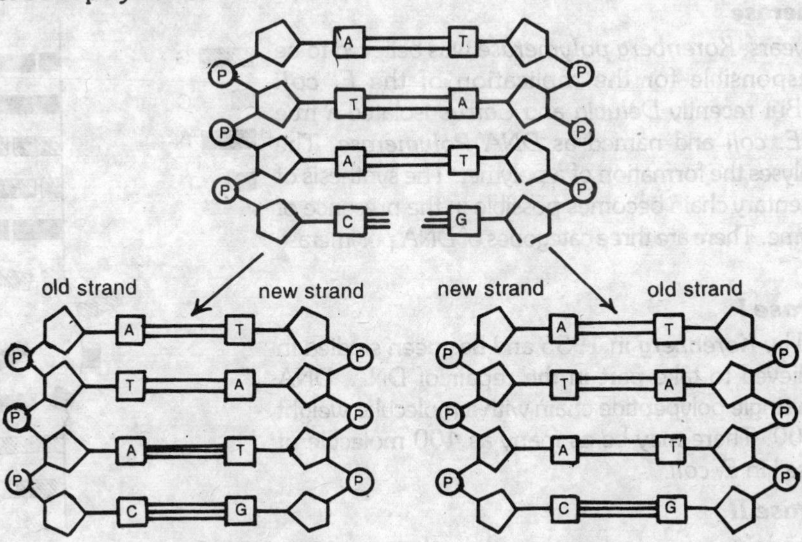

Fig. 28.17. Process of DNA duplication.

6. *Joining of Okazaki fragments.* The gaps left between Okazaki fragments are filled with complementary deoxyribonucleotide residues by DNA polymerase-I. Finally the adjacent 5' and 3' ends are joined by *DNA-ligase.*

In eukaryotic cells, the mechanism of DNA replication is expected to be more complex than in prokaryotes. The replication of eukaryotic DNA begins at multiple points of origin. Their RNA primer is formed of about 10 nucleotides and the Okazaki fragments are much shorter (about 100 to 150 nucleotides).

Enzymes for DNA Synthesis

Different prokaryotic and eukaryotic cells have been formed to contain three kinds of enzymes that act on DNA namely nucleases, polymerases and ligases.

1. Nuclease Enzymes

These enzymes attack on the DNA molecules directly. There are two types of nuclease enzyme:

(a) Exonuclease enzyme

The exonuclease begins its activity either at a free 3'-OH end of a polynucleotide and progressively cleave the bands as the 3'-OH side of the phosphodiester back bone or its begins at a free 5'-P end and digests the polynucleotides in a 5'-3', direction. In both the cases the enzyme travels along the chain in a stepwise manner.

(b) Endonuclease enzyme

Endonuclease enzyme also attacks one of the two sides of phosphodiester linkages, but they react only with those bonds that occur within the interior of a polynucleotide chain. If the polynucleotide chain is single stranded (e.g., viral DNA) such an attack will obviously cut the chain into two pieces. If however, the polynucleotide strand is a member of a DNA helix (e.g., prokaryotic and eukaryotic DNA), a single endonucleolytic cut will create a nick in the helix; the helix remain in one piece but it now possesses a gap that contains two free ends, which can serve as substrates for exonucleases. A nicked double helix suffers a localized disrupted of its secondary structure. The molecules becomes free to bend or rotate around its intact strand and the two broken ends are free to dangle. The increased molecular motion in the region of the nick will very likely interrupt hydrogen bonding between bases in the vicinity of the nick, thus effecting a limited "unraveling" of the helix.

2. DNA Polymerase

For many years, *Korenberg polymerase* was believed to be the enzyme responsible for the replication of the *E. coli* chromosomes. But recently *Delucia* and *Cairus* isolated a true replicase from *E. coli* and named as *DNA Polymerase*. The polymerase catalyses the formation of a polymer. The synthesis of a new complementary chain becomes possible in the presence of polymerase enzyme. There are three categories of DNA polymerase enzymes.

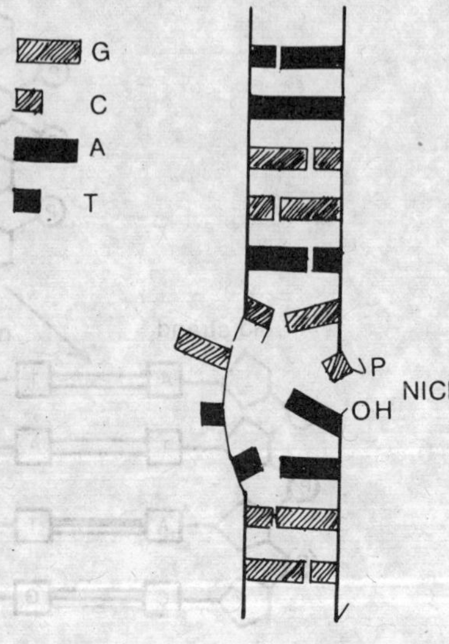

Fig. 28.18. A nicked duplex DNA molecule such as is formed by an endonuclease enzyme.

DNA polymerase I

Discovered by *Korenberg* in 1955 and has been studied in *E. coli*. It is believed to take part in the repair of DNA. DNA polymerase I is a single polypeptide chain with a molecular weight of about 109,000. There may be as many as 400 molecule of DNA polymerase I in *E. coli*.

DNA polymerase II

It is also made up of a single polypeptide chain having a molecular weight of about 90,000. The number of DNA polymerase II molecule is about 40 in *E.coli*. The enzyme can elongate Okazaki segments even in the absence of polymerase I.

DNA polymerase III

It is considered to be a true replicase of *E.coli*. It consists of two polypeptide chains with molecular weight of 140,000 and 40,000. It is most effective and can polymerize about 15000 nucleotides per minute in *E. coli*.

3. DNA Ligases

DNA ligase enzymes are capable of catalyzing phosphodiester bond formation between free 3'-OH and free 5'-P groups of nick of DNA which is created by endonuclease enzyme, thereby restoring an intact DNA duplex. Many DNA ligases have already been discovered. The ligase enzyme from *E.coli* requires the presence of oxidized nicotinamide adenine dinucleotide (NAD^+) a cofactor, whereas, the ligase enzyme specified by T_4 bacteriophage requires ATP to bring about the joining reaction.

RNA primer in DNA replication

Short segments of RNA are first synthesized before actual DNA replication starts. These RNA segments are later excised and filled with DNA with the held of enzyme DNA polymerase I. It was found that if RNA synthesis is stopped by a drug rifampicin, DNA replication stops thus providing the role of RNA primers in DNA replication. There are several other evidences which established this point.

TYPES OF REPLICATION

Three different types of replication of DNA have been discovered:

1. Semiconservative
2. Conservative
3. Dispersive

1. Semiconservative

Here the replication of DNA involves the unwinding of the double helix, but no rupture of the separated polynuclotide strands. Thus the primary DNA is causevned during the replication process. According to this mechanism, the unwinding of the original strands would be accompanied or followed by the guided synthesis of the complementary strands of each of the two separated polyucleotide strand. The linearly assembly, catalyzed by enzymes, of the nucleotides of the complementary sequence would be guided by their paining with the Watson-Crick partners on the original strands. In this manner, the proper nucleotide sequence would be automatically assumed. The DNA helix would thereby regenerated each of the strand of the original helical duplex uniting with its freshly synthesized complement to form a total of two new double stranded structures. The semi-conservative model thus predicts that entire polynucleotide strands of the parental DNA will be passed on to the progeny in intact form. Each molecule of progeny DNA will receive either one half or none of its nucleotides from the parental DNA.

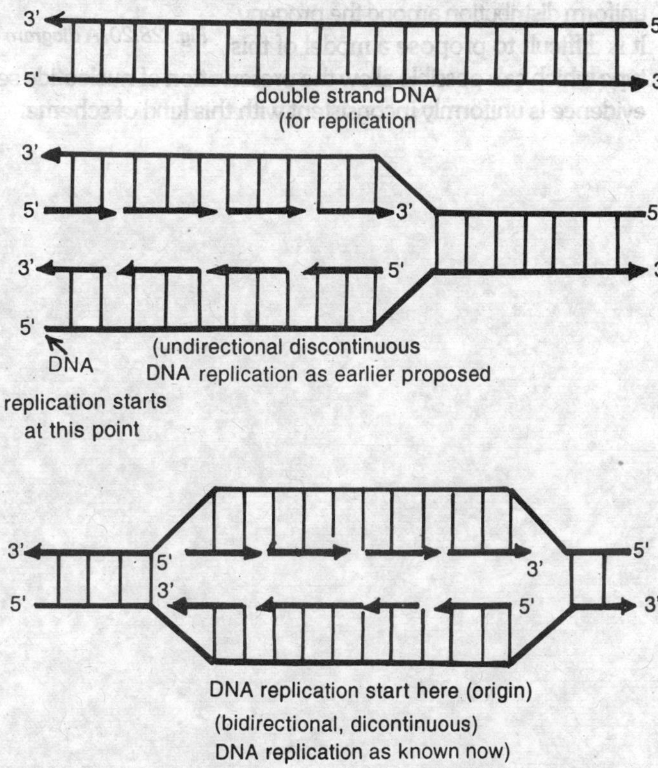

Fig. 28.19. Discontinuous DNA replication.

2. Conservative

According to this mechanism both the primary and secondary structure of the parental DNA are conserved. Thus the replication process would involve neither the rupture of the individual strands nor the unwinding of the helical duplex. The details of these propose that the sequences of hydrogen bonded base pairs, rathter than individual bases, serves the genetic determinant. According to this picture, particular base pairs might preferentially bind specific nucleotides and determine their sequential position in the daughter polynucleotide.

The conservative model allows for no redistribution of parental DNA among the progeny. The original DNA is transferred as an intact unit or not at all.

3. Dispersive

Into this class fall all mechanisms which postulate a general break up of the parental DNA and its more or less uniform distribution among the progeny. It is difficult to propose a model of this type which can possible allow the preservation of nucleotide sequence. Moreover, the existing experimental evidence is uniformly inconsistent with this kind of scheme.

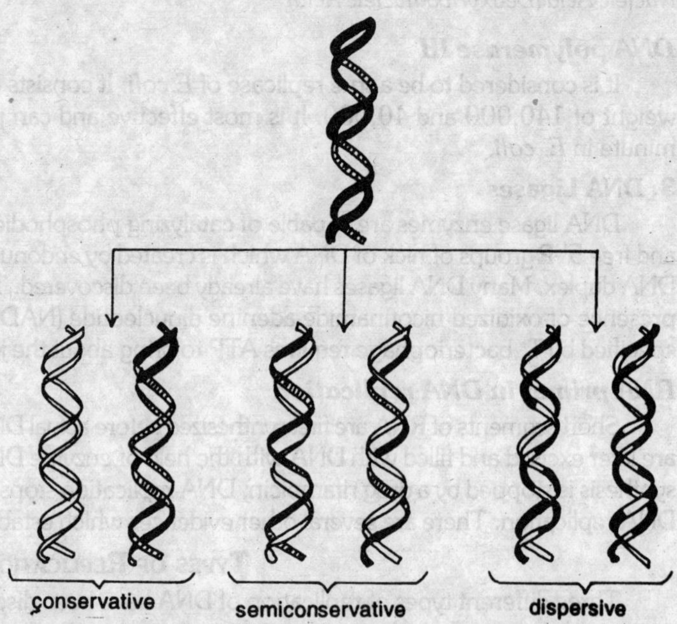

Fig. 28.20. A diagram illustrating three possible modes of DNA replication.

29

BIOLOGICAL REPLICATION OF DNA

The Replication of DNA

When the double helix was first discovered, the feature that most excited the world of genetics was the complementary relationship between the sequences of bases on its interwined polynucleotide chains. It seemed unimaginable that such a complementary structure would not be utlized as the basis for DNA replication. In fact, it was the self-complementary nature of DNA that finally led most geneticists and biolgoists to accept Avry's conclusion that DNA, not some form of protein, was the carrier of genetic information.

Earlier in our discussion of how templates act, we emphasized that two identical surfaces will not attract each other and that it is much easier to visualize the attraction of groups with opposite shape or charge. Thus, without any detailed structural knowledge, we might guess that a molecule as complicated as the gene could not be directly copied. Instead, replication would involve the formation of a molecule complementary in shape, and this, in turn, would serve as a template to make a replica of the original molecule. So some geneticists, in the days before detailed knowledge of protein or nucleic acid structure, wondered whether DNA served as a template for a specific protein that in turn served as a template for a corresponding DNA molecule.

But as soon as the self-complementary nature of DNA became known, the idea that protein templates might play a role in DNA replication was discarded, for it was immensely simpler to postulate that each of the two strands of very parental DNA molecule served as a template for the formation of a daughter strand complement. Although from the start this hypothesis seemed too good not to be true, experimental support nonetheless had to be generated. Happily within five years of the discovery of the double helix, there emerged both descisive evidence for the separation of the complementary strands during DNA replication and firm enzymological proof that DNA alone can function as the template for the synthesis of new DNA strands.

With these results, the problem of how genes replicate was in one sense solved. But in another and equally valid sense, the study of DNA replication had only begun. As this chapter will show, the replication of even the simplest DNA molecule is a complicated, multistep process, involving many more enzymes than we initially anticipated following the discovery of the first DNA polymerizing enzyme. Fortunately, it now seem likely that all the key enzymatic steps involved in the replication of both linear and circular DNA have been described and that the general picture of DNA replication presented here will not be seriously modified.

Strand Separation Requires Untwisting of the Double Helix

In even the smallest DNA molecules, the two chains of the double helix are twisted around each other many times and must be untwisted an equal number of times for DNA replication to occur. This feature initially troubled many scientists who were otherwise enamored of the implications of DNA's double-helical structure. But untwisting linear DNA molecules need not lead to insuperable problems if the unreplicated sections of DNA can rotate around their axis, much like the rotating cables of speedometers. Since the DNA molecule is very thin, if should encounter virtually no frictional resistance, and thus the energy input needed for rotation should be negligible.

No conceptual difficulty arises from the need to break the hydrogen bonds joining the base pairs in the parental templates. Although these bonds are highly specific, they are also relatively weak, and no enzymes are required to make or break them. Hydrogen bonds thus provide highly suitable means of specifically holding together a template and its complementary replica.

Base Pairing should Permit very Accurate Replication

It has already been stated that the chemistry of the various amino acid side groups argued against their employment as templates. Just the opposite, however, is true of the purine and pyrimidine bases, because each of them can form several hydrogen bonds. These bonds are ideal tempate forces, for they are highly specific, in contrast to Van der Waals interactions, whose attractive forces are both weaker and virtually independent of particular chemical groupings.

The average energy of a hydrogen bond is absent 3 kcal/mol, or eight times the average energy of thermal motion of molecules at room temperature. This value gives us an estimate of how often a given group (e.g., an amino group) will be hydrogen-bonded. Under ordinary cellular conditions, the ratio of bonding to non-bonding is about 10^4:1. This means that two molecules held together by several hydrogen bonds are almost never found in the non-bonded form. Because of the great specificity of the hydrogen bonds, the frequency with which adenine will attach to thymine (10^{-4} X 10^{-4} since two hydrogen bonds are involved). The replication of the GC pair should be even more accurate, since it is held together by three hydrogen bonds rather than two. However, the actual selection process never reaches this high level of accuracy, because occasionally (perhaps about once in 10^4 times) the base selected will be the wrong (imino or enol) tautomeric form, which must lead to mispairing. Now we realize that only through proofreading of the initially selected bases does the DNA replication process achieve its measured level of accuracy (error levels between 10^{-8} and 10^{-12}).

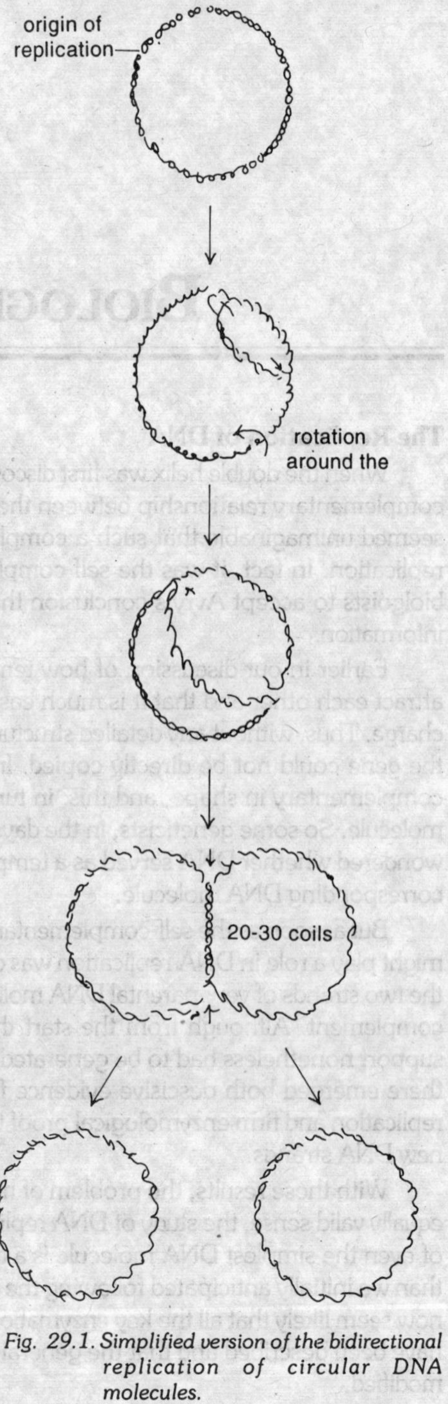

Fig. 29.1. Simplified version of the bidirectional replication of circular DNA molecules.

Θ-Shaped Intermediates form during the Replication of Circular DNA

Duplication of circular DNA does not involve the temporary creation of linear DNA. Visualization of the replicating intermediates of many ring chromosomes reveals that the parental strands maintain a circular form throughout replication. They always possess a 0-like shape that comes into existence by the initiation of a replicating bubble at the origin of replication. Circular λ DNA, for example, has its replication origin within gene O, and synthesis proceeds outwardly in both the clockwise and counterclockwise directions.

From the moment of its discovery, circular DNA posed the question of how two covalently closed interwined strands could unravel during replication. As long as each chain remains intact, even minor untwisting of one section of a circular double helix forces the creation of positive supercoils in the opposing direction. The semiconservative replication of such DNA therefore demands temporary cuts in one or both of its polynucleotide backbones. The problem thus becomes how and when the breaks are made. About this there could only be speculation until the discovery of the topoisomerases, DNA nicking/closing enzymes that can either further supercoil or relax DNA. The transient single-stranded cut made by topoisomerase I, for example, can create a molecular swivel around which supercoiled parental DNA strands can twist.

Now it seems clear that both parental strands remain intact throughout most of the replication cycle. The unreplicated portion of the 0-shaped intermediate becomes periodically supercoiled in either of two possible ways: Negative superhelices (untwists) maybe made by topoisomerase II (gyrase) in anticipation of subsequent chain elongation, and positive superhelices may be made by elongation of replicating bubbles in the absence of chain cuts. The later, positive supertwisting is clearly limited in scope and should halt replication bubble growth owing to the steric impossibility of creating further positive supercoils. Such halted bubbles could resume growth if a topoisomerase type I molecule made the single-stranded cuts that convert the highly supercoiled unreplicated section into the relaxed configuration. Alternatively, the essential cutting agent needed for further movement may be topoisomerase II (gyrase), acting to untwist the positively supertwisted DNA.

Overall Chain Growth Occurs in both 5'→3' and3'→5' Directions

The opposing chain directions (5'-3' and 3'-5') of the two strands of a double helix mean that the two daughter strands being synthesized at each replicating form must also run in opposite directions. Therefore, the overall direction of chain growth must be 5'→3' for one daughter strand and 3'→5' for the other daughter strand. However, all known forms of DNA polymerase, the only type of enzyme that can add nucleotide precursors of DNA, extend chains only in the 5'-3' direction, since the chemical reaction catalyzed by these enzymes allows a nucleotide triphosphate to react only with the free 3'-OH end of a growing polynucleotide strand. The question thus arose whether other replication enzymes had been missed, specifically, those that would add nucleotides onto free 5' ends. Despite a very extensive search, no enzyme with such specificity has ever been found.

Small Fragments are Precursors of many Long Chains: The Leading Versus the Lagging Strand

Resolution of the apparently paradoxical 3'→5' direction of synthesis came with the discovery that deoxyribonucleotide substrates are not always added directly to the very long daughter polynucleotide. Instead, they are sometimes found as parts of much shorter DNA chains 9100 to 1000 bases long) that later link up to the main daughter strands through the action of the polydeoxyribonucleotide-joining enzyme, DNA ligase. This finding opened up the possibility that a chain whose final direction of growth is 3'→5' might, in fact, be formed by the joining of smaller chains (often called *Okazaki fragments,* after their discoverer), each of which grows in the coventional 5'→3' manner. In this way, the same set of enzymes could be responsible for making all progeny strands. Proof that this indeed in the case came from direct studies of the ends of growing Okazaki fragments to which precursor nucleotides had recently been added. All such experiments show growth exclusively at the 3' end.

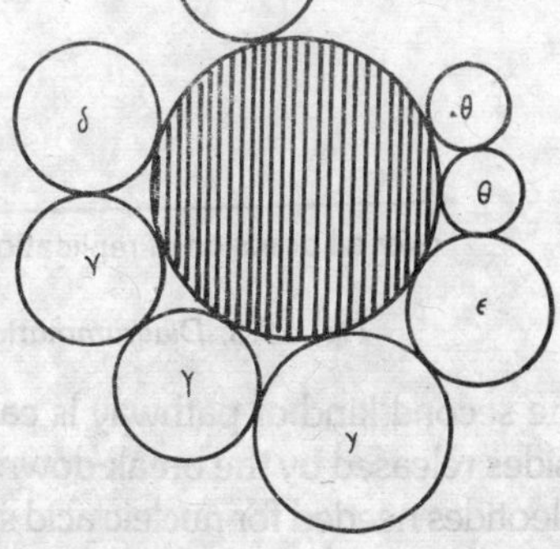

Fig. 29.2. DNA polymerase-III holoenzyme. It consists of subunits α, β, γ, ε, θ, δ and ϵ.

Now it is very clear that the Okazaki fragments are precursors only to half of the daughter chains—those that grow overall in the 3'→5' direction. Synthesis of the strand that grows overall in the 5'→3' direction for the most part precedes that of the strand appearing to grow 3'-5'. This difference in rates of growth between the leading (5'→3') and lagging (3'→5') strands produces a small section of single-stranded parental DNA at the replication fork. It becomes converted to double-stranded DNA upon the initiation of an Okazaki fragment whose synthesis continues until it reaches the terminus of the previously synthesize Okazaki fragment, to which it then becomes joined. Okazaki fragments only commence synthesis at precise points along single stranded DNA templates and it is the distance between these signals that determine the average length of Okazaki fragments formed during replication of any particular DNA.

DNA Replication in Prokaryotes

Nucleotide Synthesis

To be efficient, virtually all cells use two different kinds of pathway for the synthesis of nucleotides. One is a *de novo pathway,* in which a sugar (ribose 5'-phosphate), certain amino acids, carbon dioxide, and NH_3 are combined in a series of reactions to form the nucleotides directly. In this case neither the free purines or pyrimidines nor the nucleosides are intermediates in the pathway.

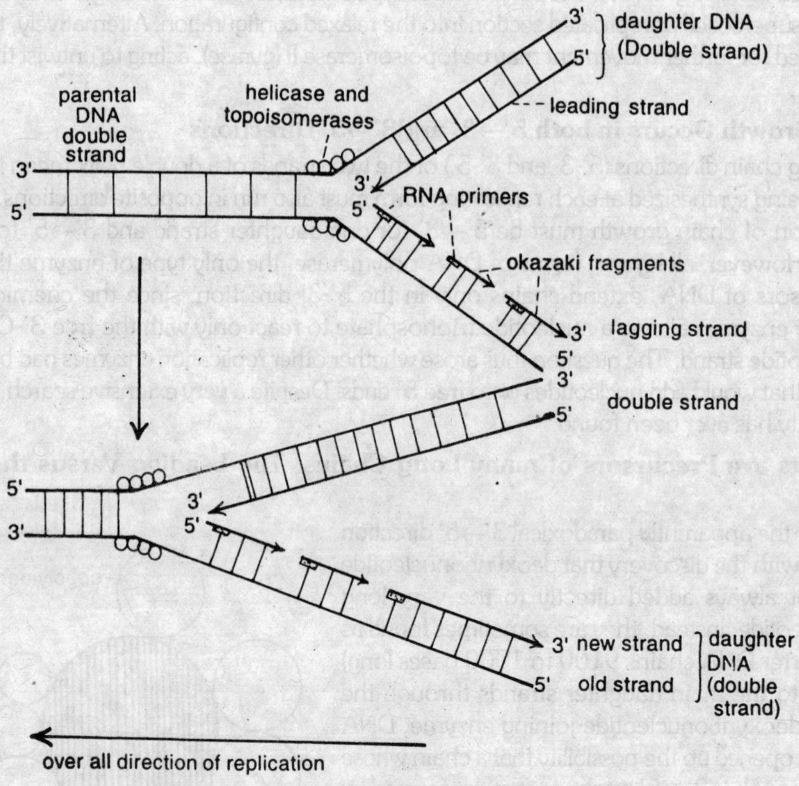

Fig. 29.3. Diagrammatic representation of discontinuous synthesis of DNA.

The second kind of pathway is called a *salvage pathway,* in which the purines, pyrimidines, and nucleosides released by the break-down of nucleic acids can, by a variety of routes, be converted back to the nucleotides needed for nucleic acid synthesis. Both pathways are important to the cell, and how active each pathway is depends upon the resources available to the cell.

Purine Biosynthesis

The purine biosynthetic pathway follows essentially the same sequence of reactions in a wide range of prokaryotic and eukaryotic organisms, including *E. coli*, yeast, plants and man. The purine ring structure is assembled on the sugar, ribose 5'-phospahte. A number of components are used to produce the purine ring, including some amino acids. The primary product of the purine biosynthesis pathway is the ribonuclotide inosine 5'-monophosphate (IMP) and it is from this compound that the adenine and guanine ribonucleotides are derived. As will be described later, these later two compounds are phosphorylated further to convert them to the immediate RNA and DNA precursors.

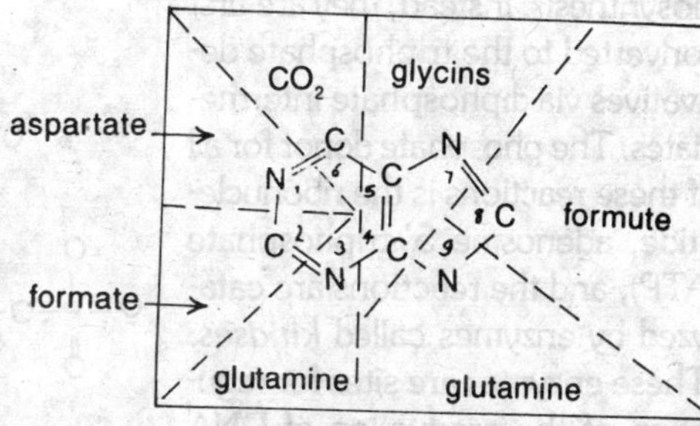

Fig. 29.4. Origin of the carbon and nitrogen atoms of the purine ring.

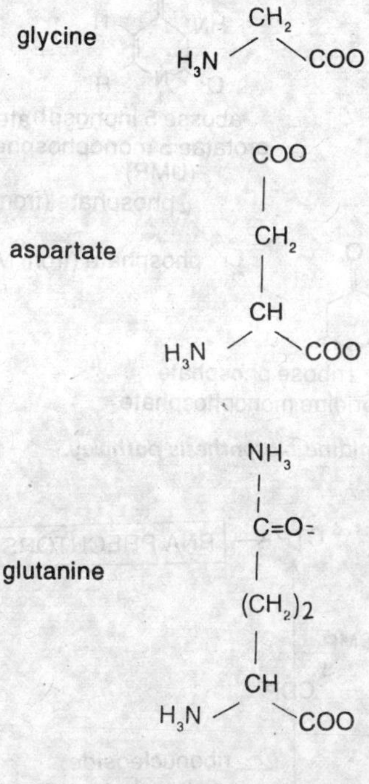

Fig. 29.5. Structure of amino acids used in the formation of purine (and pyrimidine) rings.

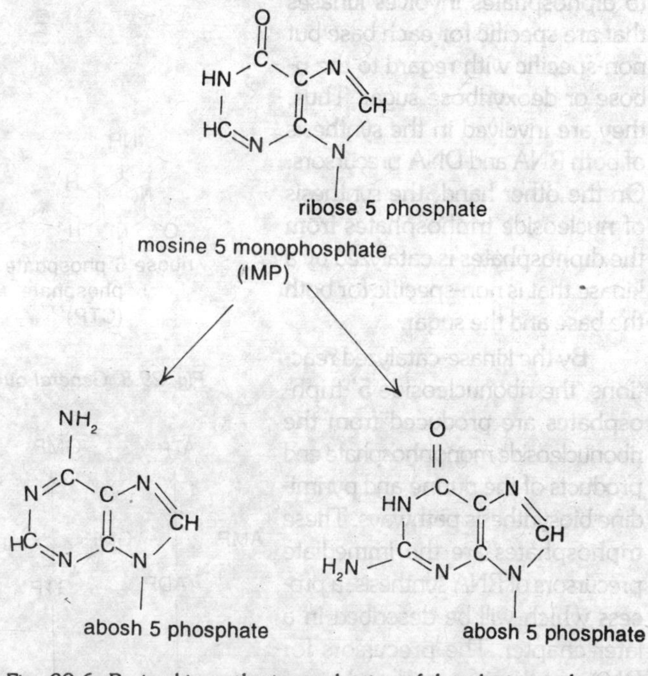

Fig. 29.6. Purine biosynthesis: production of the adenine and guanine nucleotides from inosine 5'-monophosphate.

Pyrimidine Biosynthesis

The biosynthesis of pyrimidine nucleotides differs in general from that of the purine nucleotides in that the pyrimidine ring is assembled first and then is attached to the ribose-5'-phosphate. As was the case with the purine ring, a number of compounds are used in the construction of the pyrimidine ring. The pyrimidine biosynthesis pathway is also shown diagrammatically. The key pyrimidine intermediate in this pathway is orotate, which becomes attached to an activated form of ribose 5'-phosphate, phosphoribosyl pyrophosphate (PRPP), to form a nucleotide, orotidine 5'-monophosphate (OMP). Removal of a carbon dioxide molecule generates uridine 5'-monophosphate (UMP). The cytosine nucleotide then is produced from UMP by the addition of an amino (NH_2) group.

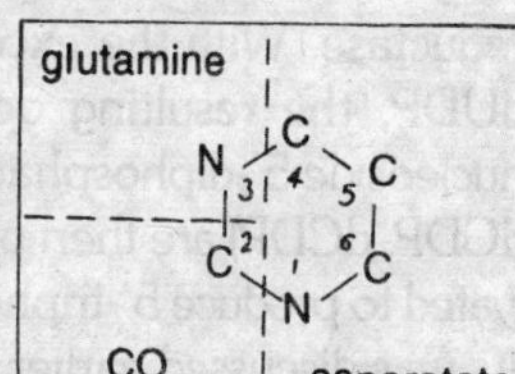

Fig. 29.7. Origin of the carbon and nitrogen atoms of the pyrimidine ring.

The nucleoside monophosphates generated by the de novo purine and pyrimidine biosynthesis pathways do not participate directly in nucleic acid biosynthesis. Instead, they are first converted to the triphosphate derivatives via diphosphate intermediates. The phosphate donor for all of these reactions is the ribonucleotide, adenosine 5'-triphosphate (ATP), and the reactions are catalyzed by enzymes called *kinases.* (These enzymes are sites for regulation of the production of DNA and RNA precursors). The conversion of nucleoside monophosphates to diphosphates involves kinases that are specific for each base but non-specific with regard to the ribose or deoxyribose sugar. Thus, they are involved in the synthesis of both RNA and DNA precursors. On the other hand, the synthesis of nucleoside triphosphates from the diphosphates is catalyzed by a kinase that is non-specific for both the base and the sugar.

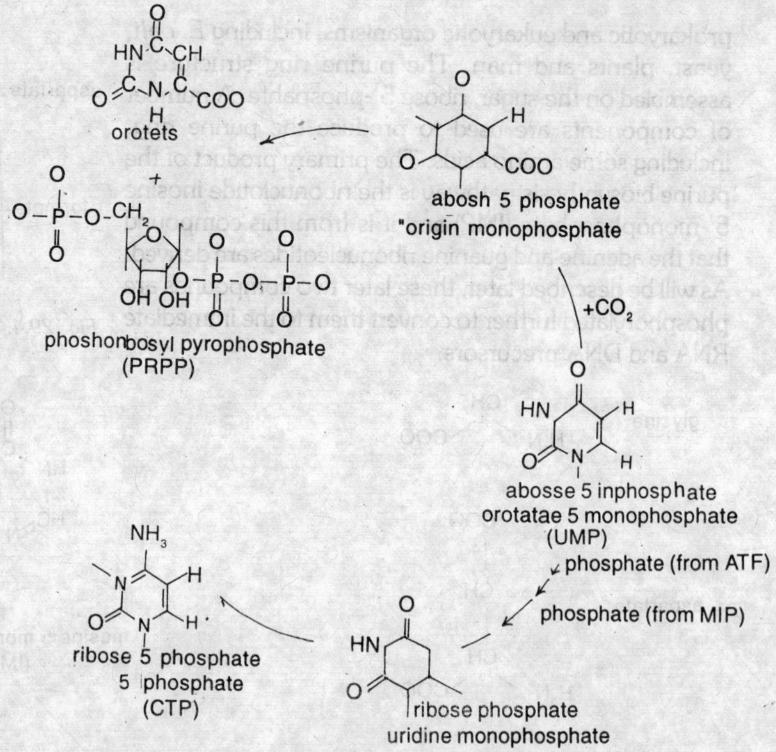

Fig. 29.8. General outline the pyrimidine biosynthesis pathway.

By the kinase-catalyzed reactions, the ribonucleoside 5'-triphosphates are produced from the ribonucleoside monophosphate end products of the purine and pyrimidine biosynthesis pathways. These triphosphates are the immediate precursors of RNA synthesis, a process which will be described in a later chapter. The precursors for DNA synthesis are also derived from the ribonucleoside 5'-monophosphates. First, the diphosphates are produced and then the ribose sugar is reduced to the deoxyribose sugar in a reaction catalyzed by the enzyme ribonucleoside phosphate reductase. With the exception of dUDP, the resulting deoxyribonucleoside 5'-diphosphate (dADP, dGDP, dCDP) are then phosphorlyated to produce 5'-triphosphates. As was discussed earlier, the base uracil is specific for RNA and this pyrimidine is replaced in DNA by thymine, which is actually 5'-methyluracil. To produce the thymine deoxyribonucleotide, dUDP is dephosphorylated to dUMP and the 5-position carbon of the pyrimidine ring is then methylated in a reaction catalyzed by thymidylate synthetase. The

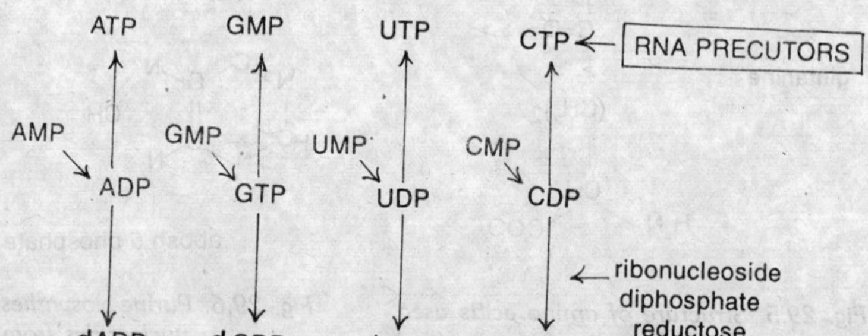

Fig. 29.9. Formation of ribonucleotide precursors for RNA and deoxyribonucleotide precursors for DNA.

resulting deoxythymidine 5'-monophosphate is phosphorylated to produce the triphosphate. In this way, the for DNA precursors, dATP, dGTP, dTTP and dCTP, are produced in reactions catalyzed by the kinases discussed previously.

Since the synthesis of RNA and DNA precursors are intimately related, one might expect that the regulatory elements of the system would be complex and indeed that is the case. For example, since cells must distribute their resources appropriately between RNA and DNA synthesis, one of the pivotal enzymes is ribonucleoside diphosphate reductase. This enzyme receives complex inhibitory and stimulatory signals by a variety of deoxyribonucleotides so as to provide a balanced supply of precursors for DNA synthesis. The synthesis of the enzyme can also be repressed by the deoxyribonucleotides if their poor sizes become large.

DNA Synthesis in Vitro

It was shown that the basic building blocks for DNA synthesis are the deoxyribonucleoside 5'-triphosphate, and the precursors are polymerized into DNA. In DNA synthesis, the polynucleotide chain grows from one end by the stepwise addition of deoxyribonucleoside triphosphates. In this, the direction of synthesis is determined by the fact that the 5'-triphosphate precursors must bond to the 3'-OH of the preceding deoxyribose. Before discussing recent information about DNA synthesis, we shall review the historical work of A. Kornberg and his colleagues that resulted in the in vitro synthesis of DNA under defined conditions.

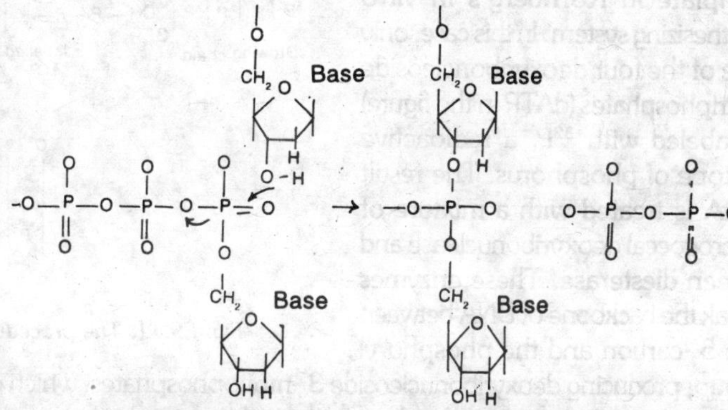

Fig. 29.10. Mechanism of DNA Polymerization.

In his experiments of the 1950s, Kornberg determined that the following ingredients were required for the in vitro synthesis of DNA.

1. A mixture of all four deoxyribonucleoside 5'-triphosphates.
2. Magnesium ions (Mg^{++})
3. A purified enzyme, DNA polymerase, obtained from cell-free extracts of *E. coli.*
4. High-molecular-weight DNA.

In his initial experiments with the above ingredients, *Kornberg* got an approximately 20-fold increase in the amount of DNA over that added. The reaction itself continued until one of the 5'-triphosphate precursors ran out.

That all four ingredients are required for DNA synthesis in vitro was shown by experiments in which one ingredient at a time was omitted from the reaction mixture. In each case no DNA synthesis occurred. Similar omission experiments showed that all four deoxyribonucleoside 5'-triphosphates must be present in order for DNA synthesis to proceed. The key to the reaction, though, is the DNA polymerase, which promotes synthesis by effecting the 3'-5'-phosphodiester linkages.

Role of DNA in the Reaction

Two possibilities were proposed originally for the role of high-molecular weight DNA in the *in vitro* synthesis of DNA:

1. The DNA acts as a primer, a growing point for the terminal random addition of nucleotides.
2. The DNA acts as a template; that is, the base sequence of the DNA chain specifies the complementary base sequence of the new DNA chain with which it is bound by hydrogen bonds in the double helix structure. The requirement for all four DNA precursors to be present for any DNA synthesis to occur suggests that DNA has a template role.

All the available evidence indicates that the second possibility is correct: DNA acts as a template. The initial evidence came from studies that showed that the (A+T)/(G+C) ratio of in vivo synthesized DNA was the same as the DNA used to start the reaction. However, an objection to this is that the results could be a coincidence since base ratios do not indicate the arrangement of bases along the DNA. A more exacting comparison of newly synthesized and template DNA can be made by nearest neighbour analysis, which shows the relative frequency with which each of the four bases is located adjacent to each of the four bases in the DNA chain (i.e., A next to A, G, C, and T; C next to A,G,C and T; G next to A,G,C, and T; and T next to A,G,C, and T). Nearest neighbour analysis does nto give some information about the sequence of bases along the DNA, but it does give some information about the general arrangement of the bases in the DNA.

In nearest neighbour analysis DNA is isolated from the organism of interest and then is used as the template in Kornberg's in vitro synthesizing system. In this case, only one of the four deoxyriboncleoside 5'-triphosphates (dATP in the figure) is labeled with ^{32}P, a radioactive isotope of phosphorus. The result DNA is treated with a mixture of micrococcal deoxyribonuclease and spleen diesterase. These enzymes break the backbone of DNA between the 5'-carbon and the phosphoryl group, producing deoxyribonucleoside 3'-monophosphates, which can be analyzed by paper electrophoresis to determine the amount of radioactivity each contains.

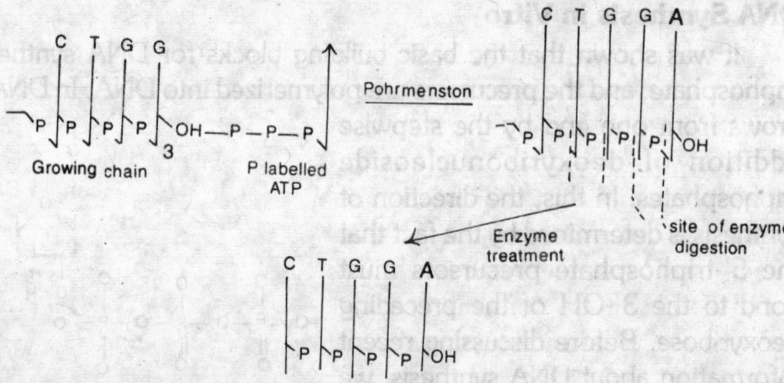

Fig. 29.11. The procedure of nearest neighbour analysis.

In the example shown, the ^{32}P is now found in the form of DG 3' MP (deoxyguanosine 3'-monophosphate), which was the nearest neighbour to the 5' side of the adenine nucleotide in the DNA. Extrapolating this discussion to the entire DNA, the method provides data concerning the relative amount of radioactive label transferred from ^{32}-labeled dATP to the four possible neighbours. This gives information about the relative frequency with which A is next to A,G,C and T in the DNA. These four nearest neighbours are usually written as 5'-ApA-3', GpA, CpA and TpA. The experiment is then repeated with each of the other three DNA precursors such that in the end a matrix of 16 nearest neighbour values is produced. The 16 frequencies found are generally unique to the DNA is question. An example of the type of data that is obtained from nearest neighbour analysis is shown in Table.

Table 29.1. Nearest Neighbour Frequencies of Mycobacterium phlei DNA

	Deoxyribonucleoside 3'-monophosphate isolated			
Labelled triphosphate	*Tp*	*Ap*	*Cp*	*Gp*
dATP	TpA 0.012	ApA 0.024	CpA 0.063	GpA 0.065
dTTP	TpT 0.026	ApT 0.031	CpT 0.045	GpT 0.060
dGTP	TpG 0.063	ApG 0.045	CpG 0.139	GpG 0.090
dCTP	TpC 0.061	ApC 0.064	CpC 0.090	GpC 0.122
Sum	0.162	0.164	0.337	0.337

The data presented illustrate several points:

1. The sixteen possible nearest-neighbour frequencies occur with a large number of frequencies.
2. The sums of the vertical columns show that the amount of A is equal to the amount of T and that G equals C, thus indicating the DNA was probably replicated correctly.
3. The two DNA strands are of opposite polarity. This is borne out by the frequency equivalence of the pairs of sequences: CpT and ApG, GpT and ApC, GpA and TpC, and CpA and TpG, as predicted by antiparallel DNA strands. Were the two strands of the same polarity, different matching sequences would have been predicted, for example, TpA and ApT, GpA and CpT, CpA and GpT etc.

To show that newly synthesized DNA is made using a DNA template requires two rounds of nearest neighbour analysis. In the first round the originally isolated DNA is used in the reaction and a set of 16 nearest neighbour frequencies are obtained as described. In the second round the enzymatically synthesized DNA is used in the reaction and a second set of frequencies is produced. The results show good agreement between the two sets of frequencies, thus indicating that DNA plays a template role in DNA replications in vitro.

Since the early, comparatively crude in vitro DNA synthesis experiments, a large number of refinements have been made. For example, the reaction mixtures now duplicate the conditions that are present in growing cells such that it is possible to synthesize DNA in vitro that is identical in all respects with DNA produced in vivo. The reaction mixtures required for efficient DNA synthesis vary, depending on the DNA used as the template. In general, there is a basic set of enzymes and proteins required for the synthesis of all DNA sources. Beyond that, each DNA requires a number of specific proteins for new synthesis to occur, and the actual set of proteins needed depends on the DNA in question.

EVIDENCES IN SUPPORT OF SEMICONSERVATIVE REPLICATION OF DNA

1. Meselson and Stahl's Experiment

(i) *Meselson and Sthal* (1958) cultured a species of bacteria (*Escherichia coli*) in a culture medium containing N^{15} isotopes of nitrogen. After these had replicated for a few generations in that medium both the strands of their DNA contained N^{15} as constituent of purines and pyrimidines. When these bacteria with N^{15} were transferred in cultural medium containing N^{14}, it was found that DNA separated from fresh generation of bacteria possesses one strand heavier than the other. The heavier strand represents the parental strand and lighter one is the new one synthesized from the culture indicating semi conservative method of DNA replication.

(ii) Density determination experiments were also conducted. The DNA of parental generation (labelled with N^{15}) was found to be heaviest. In first generation it was lighter than the first one. In second generation two bands were observed indicating the presence of DNA with two different densities, as is expected according to semiconservative method of DNA replication. These two bands were of equal intensity in the second generation. In the subsequent generations the intensity of light density bands gradually increased and the intensity of hybrid density band gradually decreased.

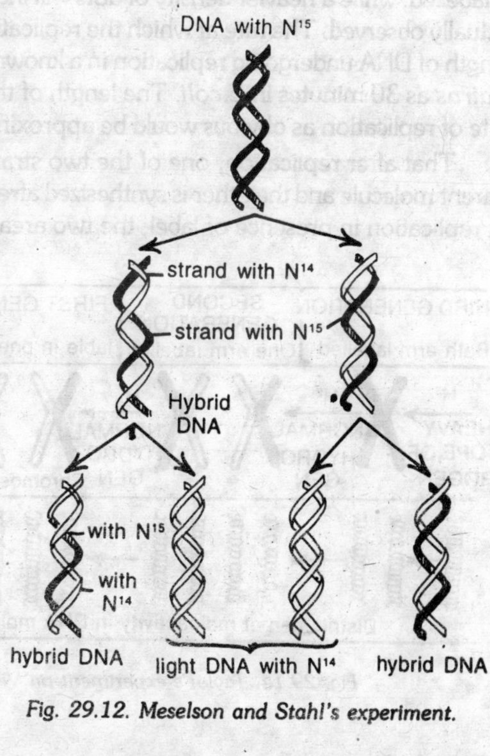

Fig. 29.12. Meselson and Stahl's experiment.

2. Cairn's Autoradiography Experiment

J. Cairns demonstrated the semiconservative type of replication by using a technique of autoradiography as given in the following lines:

In autoradiography technique, the material is first supplied with a suitable radioactive material like tritiated thymidien (H^3-TdR; H^3 is heavy isotope of hydrogen and it replaces normal hydrogen in thymidine to give rise to tritiated thymidine). Tritiated thymidine is used since this will selectively label only DNA and will not label RNA, since thymine base is absent in RNA. The tritiated thymidine gets incorporated into DNA and replaces ordinary thymidine. The material is then sectioned or else the cells may be broken down to release the intact bacterial chromosome on slides. These slides are then covered by photographic emulsion and stored in the dark. During this storage the particles emitted by tritiated thymidine will expose the film, which can be developed. This photograph will then show the regions of the presence of tritium and thus indirectly the presence of labelled DNA.

Using this above technique, the replication of DNA could be easily followed by *J. Cairns* and the results were reported in 1963. During incorporation of tritiated thymidine, autoradiographs could be prepared at regular known intervals. Autoradiographs from this replicating material prepared at regular known intervals demonstrated semi-conservative mode of replication. It would be expected that in regions, where one cycle of replication of DNA has been completed, the density of dots will be higher than in the region where the replication has not taken place. The lighter density of dots is considered to indicate that only one of the two strands is labelled, while a heavier density of dots will indicate that both strands are labelled. Such a situation was actually observed. The rate at which the replication proceeds could also be worked out by measuring the length of DNA undergoing replication in a known interval of time. The generation time was worked out by *Cairns* as 30 minutes in *E.coli*. The length of the chromosome was worked out to be about 1 mm. The rate of replication as obvious would be approximately 30μ to 40μ per minute (1 mm = 1000μ).

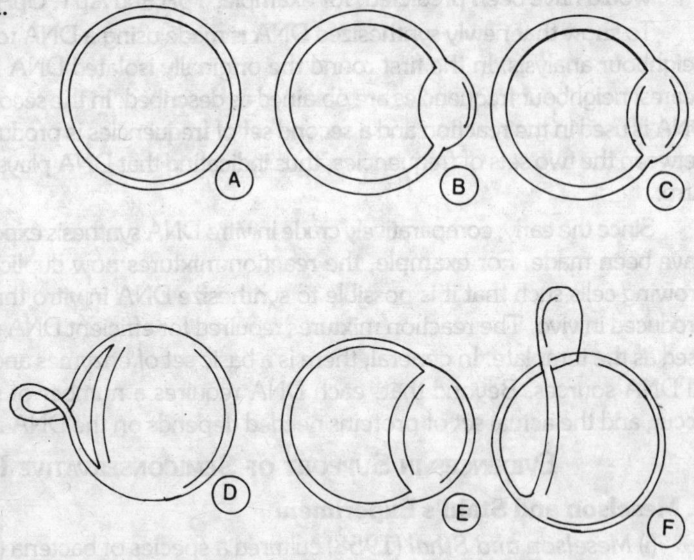

Fig. 29.13. Carn's model for circular DNA replication.

That after replication, one of the two strands in the daughter DNA molecules is derived from the parent molecule and the other is synthesized afresh. In θ shaped figure, which is obtained the second cycle of replication in presence of label, the two areas in the split region would never be equally labelled. For instance, one arc would be twice as heavily labelled as the other arc. This is what actually observed by *Cairns*. Thus the observation support the semiconservative type of replication of DNA.

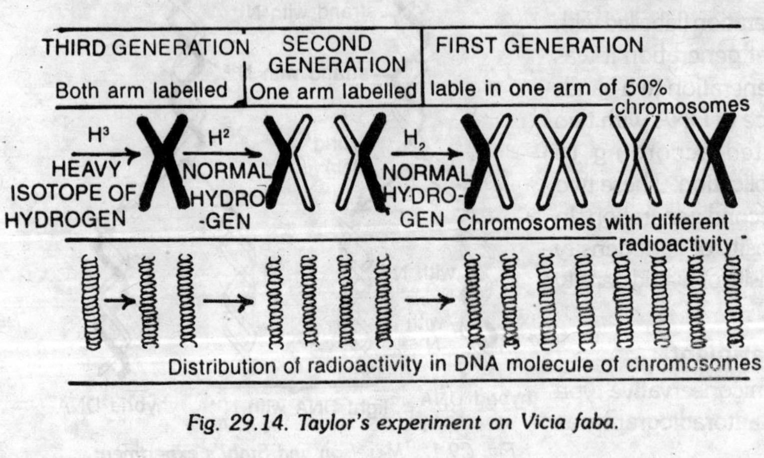

Fig. 29.14. Taylor's experiment on Vicia faba.

3. Taylor's Experiment on Vicia faba Root Tips

J.H. Taylor (1957) and his coworkers also demonstrated the semiconservtive method of DNA duplication in the root tip cells of *Vicia faba* by autoradiography. The roots were grown in a medium containing radioactive thymidine, so that the radioactivity is incorporated in the DNA of these cells. The outline of this labelled chromosome on a photographic film appears in the form of scattered black dots of silver grains. When these root tips with labelled chromosomes were transferred to the unlabelled medium containing colchicine and studied for radioactivity, the following observations were made:

1. In the chromosomes of first generation the radioactivity was found to be uniformly distributed in both the chromatids, because in them the original strand of DNA double helix was labelled with radioactivity and the new one was non-labelled.
2. In the chromosomes of second generation only one of the two chromatids in each chromosome was radioactive.

This experiment demonstrates the semiconservative method of chromosome replication. But as it is known that each chromatid is composed of only one double helical molecule of DNA, it is the replication of DNA.

Enzymes and Proteins Involved in DNA Replication

The complicated process of DNA replication in *E.coli* requires many different proteins and enzymes, some of which also function in other cellular process such as the repair of DNA damage and genetic recombination. The key enzyme in the synthesis of new DNA is DNA polymerase.

In *E. coli* there are three DNA polymerases. Each catalyzes the DNA template-directed condensation of deoxyribonucleoside 5'-triphosphates. There are significant differences in their activity in DNA synthesis and in the nuclease (DNA or RNA break-down) activities associated with them.

All three polymerases catalyze new DNA synthesis in the 5'-3' direction. The rate of DNA synthesis catalyzed by the three enzymes varies circular prokaryotic DNA molecules replicate bidirectionally (in both directions away from the origin of replication). In these cases there are two replication forks that are mirror images of one another picture two Ys joined head-to-head by the two pairs of arms. The area of a double-stranded DNA molecule that has denatured for replication is called a *replication bubble*.

In *E. coli* the unwinding of the DNA is catalyzed by an enzyme called helicase, the product of a gene called *rep*. For each 10 base pairs of DNA unwound, one turn of the helix becomes untwisted. As a result, as the helix unwinds a torsional strain is imposed on the DNA. That is, since the DNA molecule is circular, the pulling apart of the helix at one location will cause increased tightening of the molecule elsewhere, much like what happens when the strands of a piece of rope that is fixed at each end are pulled apart. This torsional strain is relieved by the action of enzymes called topoisomerases, which cause single-stranded breaks in DNA away from the replication forks and then allow one strand to rotate relative to the other strand.

The unwinding of the DNA produces single-stranded regions, which are stablized by DNA *single-stranded binding* (SSB) *proteins*. Each SSB protein is a tetramer of a 74,000 dalton subunit, and this tetramer binds to eight bases of single stranded DNA. Over 250 tetrameres bind to each replication fork. Once the strands have started to unwind, the internal bases become available for the formation of bonds with bases in new chain.

Initiation of DNA replication next takes place. In *E. coli* primases bind to the single-stranded DNA of both arms of the replication fork and synthesize short primers. The primers are lengthened by the action of DNA polymerase III, which synthesizes the complementary DNA chains to the template strands. Note that, because of the opposite polarities of the two DNA template strands and the requirement for a 5'- to -3' direction of new DNA synthesis, the primers are located at different relative positions on the two template strands.

The DNA helix continues to unwind. On the template on the left the new strand continues to be synthesized continuously (in the same direction as the direction of undwinding). However, because DNA polymerases can only synthesize DNA in the 5'-to-3' direction, the new synthesis on the template on the right has gone as far as it can. Thus, a new initiation of DNA synthesis occurs on the right-hand template on the newly exposed single-stranded template close to the site of helicase activity. As before, a primer is made, which is lengthened by DNA polymerase III action. The overall result is that the new DNA strand being made on the left-hand tempate is synthesized continuously while the new DNA being made on the right-hand template is synthesized discontinuously so that Okazaki fragments are produced.

Eventually the Okazaki fragments are joined together into a continuous DNA strand. This process requires the activities of the enzymes DNA polymerase I and DNA ligase. If we consider two adjacent Okazaki fragments, the 3' end of the newer DNA fragment is adjacent to but not joined to the primer 5' end of the previously synthesized fragment. The DNA polymerase III that has just completed the synthesis of the newer fragment now dissociates from the DNA and DNA polymerase I takes its place. This enzyme continues the 5'- to -3' synthesis of the newer DNA fragment, at the same time removing the primer of the older fragment through the action of the 5'-to-3' exonuclease function of DNA polymerase I. When DNA polymerase I is finished, there is a gap between the two new DNA fragments, and this gap is sealed in a sealed in a reaction catalyzed by DNA ligase, thereby producing a longer DNA strand. The whole process is completed until all the DNA is replicated.

30

Ribonucleic Acid (RNA)

Ribonucleic acid usually occurs in nature as long unbranched polymeric molecule formed of single chain. RNA is generally present in the cytoplasm as well as in nucleolus. *Casperson* 1950 and *Brachet* in 1955 suggested that there is an association between RNA and protein synthesis. A large number of studies have been undertaken to characterise physically the nature and structure of RNA and their biochemical properties and by now we have acquired a lot of knowledge about the various types of RNA and their molecular structures but still there are various gaps to be filled to make a complete picture.

Some plant viruses (e.g., TMV, turnip yellow mosaic viruses, wound tumour viruses, etc.), animal viruses (e.g., influenza viruses, foot and mouth viruses; rous sarcoma viruses, poliomyelitis viruses reoviruses, etc.) and bacteriophages (e.g., MS2, etc.) contain ribonucleoic acid (RNA) as their genetic material. That is why in these cases it is called the *Genetic RNA*.

Table 30.1. Genetic RNA in Different Viruses

Types of Viruses	*Name of Viruses*	*Type of RNA*
1. Plant viruses	1. TMV	1. Single stranded
	2. Wounded tumour	2. Double stranded
2. Animal viruses	3. Influenza virus	3. Single stranded
	4. Poliomyelitis	4. Single stranded
	5. Reo virus	5. Double stranded
3. Bacteriophages	6. MS_2	6. Single stranded
	7. F_2	7. Single stranded
	8. r_{17}	8. Single stranded

Molecular Weight of RNA

The molecular weight of RNA may be determined by light scattering measurements, by sedimation velocity measurements, by viscosity measurements, by equilibrium ultracentrifugation or by polyacrylamide gel electrophoresis in formawide.

Bacteria, *Actinomycetes,* blue green algae and higher plant chloroplasts all have rRNA's of molecular weight 1.1 X 10^6 (23S) and 0.55 X 10^6 (16S) whereas the corresponding values for higher plants, ferns, algae, fungi and some protozoa are 1.3 X 10^6 (26S) and 0.7 X 10^6 (18S). The 18S components is common to all animals, but the large 28S component has evolved from 1.4 X 10^6 daltons in sea urchins to 1.75 X 10^6 daltons is mammals. In mitochondria (mammalian), the rRNA's are somewhat smaller's the 21S species is 0.56 X 10^6 daltons and the 12S species 0.36 X 10^6 daltons. Whlst the ribosomal RNAs form most organisms possess broadly similar base compositions with G plus C content between 50 and 60 percent, there are some exceptions to this, e.g. *Drosophila* (40%) and *Tetrahymena* (43%).

Primary Structure of RNA

RNA usually occur in nature as long unbranched polymeric molecule made up of single chain or helix of polynucleotide strand: Ribnucleic acid is composed of pentose sugars, phosphoric acid and nitrogenous bases. The pentose sugar of RNA is ribose sugar. The sugar forms the back bone along with the phosphoric acid just in a manner studied in DNA.

Phosphoric acid is also one of the constituent of the RNA which along with the pentose sugar forms the back bone, on which the nitrogenous bases are linearly arranged, as shown in the Fig. On the bases of phosphate or phosphoric acid, the chemical constituents of RNA can be written in two ways:

Nitrogenous Base + Ribose Sugar + Phosphate = Nucleotides

Nitrogenous Base + Ribose Sugar = Nucleoside

The nitrogenous bases of ribonucleic acid are of two types:

(i) Purine

(ii) Pyrimidine.

Purine bases comprise mainly adenine (A) and guanine (G) while Pyrimidine bases comprise cytosine (C) and Uracil (U).

These bases are joined to the pentoses by N-C *glycosidic bonds.* For the purine bases the glyclosidic bond is between the C_1 position of the pentose sugar and the N_9 position of the base. For the pyrimidines, the linkage joins the C_1 and N_3 positions in figure given below:

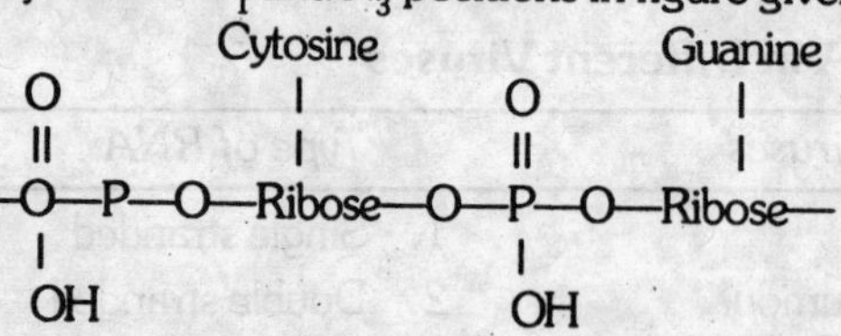

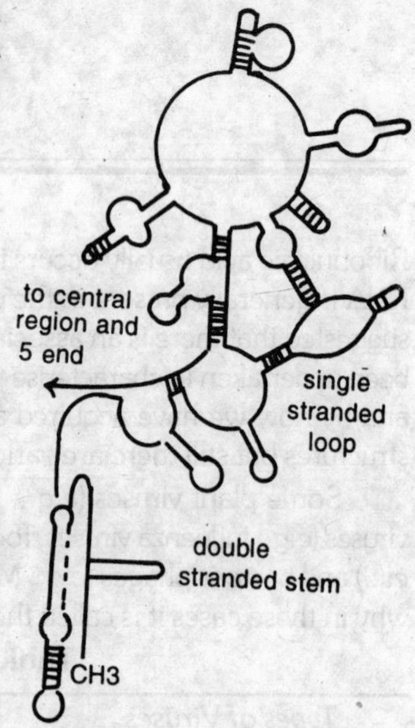

Fig. 30.1. Showing the secondary structure of RNA.

The bases are arranged on the back bone of the chain which is composed of alternating sugar and phosphate. The bases are attached to the ribose sugar in the same way, the bases in DNA are attached. The arrangement of the bases on the back bone is shown in the following diagram:

Secondary Structure of Ribonucleic Acid

Generalization as to the secondary structure of RNA is probably less justified than in the case of DNA, as the available information is much less extensive. The available information does permit the following conclusions:

1. There is no evidence for the existence of more than one polyribonucleotide strand in the natural RNAs.
2. Hence the helical regions of natural RNA arise from the hair pin like bending of the single strand back upon itself.
3. The helical regions are stabilized by A—U and C—G interactions similar to those which stabilize the confirmation of DNA. However, the irregular base composition of RNA does not allow the incorporation of the entire molecule into double perfect helix of this type, and the helical zone co-exist with the random segments.
4. Two alternative models for the structure of RNA may be postulated:

(a) The non-helical fraction of RNA may take the form of extended amorphous zones separating segments of nearly perfect helical organization. This model pre-suppose that extensive complementary sequences exist within the single strand. These tend to seek each other out to form perfect or nearly perfect, helical regions.

(b) Alternatively, nucleotides which do not participate in the helical organization may occur largely as numerous small loops external to the helically paired structure.

TYPES OF RIBONUCLEIC ACID

Four types of ribonucleic acids have been studied and they are as follows:

1. mRNA
2. tRNA
3. rRNA
4. vRNA

1. Messenger RNA (mRNA)

Studies by *Huxley* and later by *Volkin* and *Astrachan* with T_2 bacteriophage infected (bacterial virus, *E.coli*) in 1956 led to the first existence for the unique kind of the RNA, later called messenger RNA. *Jacob* and *Monod* (1961) who first proposed this name, believed messenger to be necessary to convey the message of the genetic code from the DNA of the nucleus to the cytoplasm. This messenger is known as mRNA and it acts as a template for the translation of the DNA code into a specific protein. Thus it helps in determining the structure of protein.

mRNA molecules exist in a large variety of lengths, depending on the length of the polypeptide chain for which they code. Its molecular weight and sedimention value vary considerably. Generally the life span of mRNA is relatively short, although in some eukaryotic cells there is evidence of some degree of stability.

Further mRNA is always complementary to a portion of strand of DNA, with the exception that uracil replaces thymine. It should therefore hybridize with the DNA from which it was transcribed.

It is now an established fact that when the code has been transcribed from DNA on to mRNA, the later leaves the nucleus passes through the nucleus membrane into the cytoplasm. Here it moves to the ribosomes, the site of protein synthesis. mRNA molecules attaches reversibly to the surface of ribosome always binding to the smaller sub-units. The magnesium cation, Mg^{++} is involved in the formation of a complex between mRNA and ribosomes.

Types of mRNA

On the basis of size and the coding two types of mRNA has been described:

(a) *Monocistronic mRNA.* Mostly the mRNA carries the codons of single cistron (i.e., codes for one complete protein molecule) of the DNA. Such mRNA molecule is called monocistronic mRNA. For example, for the synthesis of a polypeptide chain of 300 to 500 amino acid residues, a monocistronic mRNA of *E.coli* contains 900 to 1500 nucleotides in its molecule.

(b) *Polygenic or polycistronic mRNA.* Sometimes a mRNA molecule carries the codes from several adjacent DNA cistrons and become much longer in size. This type of mRNA is called polygenic or polycistronic mRNA. For example, for the metabolism of the histidine protein the cell synthesizes about 10 specific enzymatic proteins and a mRNA in this case may carry codons for all the 10 enzymes.

Informosomes

In eukaryons mRNA is associated with protein forming ribonucleo-protein complexes. Some of these complexes remain free in the cytoplasm without being attached to the polyribosomes. These are named as informosomes by *Spirin.* These are very stable and can remain in the cell cytoplasm for several days. The stability is attributed by the presence of protein sheath around mRNA molecule. In informosomes proteins and mRNA are found in the ratio of 4:1. These occur in the form of particles of 80S or less.

Since informosomes are very stable, these are used by the cell is protein synthesis only when the is delay in translation. In embryonic cells, genetic apaparatus starts functioning only late during organogenesis.

Therefore, in early stages of cleavage when there is no rRNA, informosomes govern the process of protein synthesis.

Stability

The cell does not contain quantities of mRNA. This is because mRNA, unlike other RNAs is constantly undergoing breakdown. It is broken down to its constituent ribonucleotides by ribonucleases. In bacteria, mRNA may be so short-lived that while one end is translating protein the other end may be undergoing breakdown. In *E.coli* the average half-life of some mRNA is about two minutes.

Eukaryotic cells, on the other hand, contain metabolically stable mRNA. Mammalian reticulocytes are immature RBC which have lost their nuclei but have retained ribosomes. These cells can synthesize haemoglobin for many hours and even days by utilizing the mRNA transcribed by the nuclei when they were present.

2. Transfer Ribonucleic Acid

In the literature, t-RNA molecule have been variously termed tRNA, sRNA for soluble RNA, or adaptor RNA. It forms about 10-15% of the total RNA in most of the existence. Each tRNA contains approximate 80 nucleotides. The soluble or t-RNA forms large varieties of different cells, is homogenous in its molecular size having a molecular weight of about 25000.

The nucleotide sequence of tRNA was first worked out by *Holley et.al.* (1965) for yeast alanine tRNA. Since then the sequence of about 75 different tRNAs have been established. According to the Holley's leaf model the tRNA is a polynucleotide chain which is folded upon itself to form 5 arms. As a result of the folding the 3' and 5' ends of the chain come near each other. An arm consists of a stem and a loop. Bases are arranged in pairs in arms. There is no base pairing in the loops.

The 3'end of the polynucleotide chain ends in CCA base sequence. This represents site for the attachment of activated amino acid. The end of the chain terminates with guanine base. The bent in the chain of each tRNA molecule contains a definite sequence of three nitrogenous bases which constitute the anticodon. It recognizes the codon on mRNA.

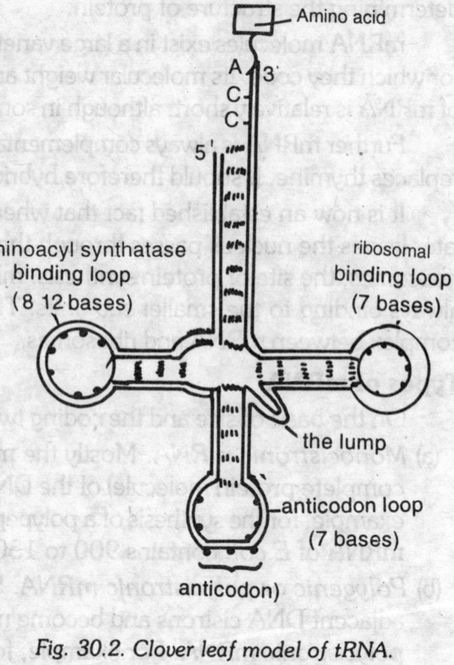

Fig. 30.2. Clover leaf model of tRNA.

Four different regions or special sites can be recognized in the molecule of tRNA. These are:

(i) *Amino acid attachment site.* It occurs at the 3'end of the tRNA chain and has -OH group of ribose molecule free to pair. This -OH group combines with the specific amino acid in the presence of ATP forming amino-acyl tRNA. It is common to all the tRNA molecules.

(ii) *Recognition site.* It contains a specific base sequence which dictates the attachment of correct amino acid to the tRNA molecule. It matches with the amino acid activating enzyme through which attachment of amino acid to tRNA takes place.

(iii) *Anticodon or codon recognition site.* This site has three unpaired based (triplet of base) whose sequence is complementary with a codon (triplet) in mRNA. Therefore, it determines the pairing of tRNA with the specific codon (triplet) of mRNA. It is therefore, the most specific region of tRNA molecule.

(iv) *Ribosome recognition site.* This helps in the attachment of tRNA to the ribosome. This site is common to all the molecules of tRNA.

Three dimensional structure of tRNA. In order to understand the structure-function relationship of tRNA, its three dimensional structure (TDS) should be known. With the availability of tRNA in crystalline form in 1968, TDS of tRNA could be studied through X-ray crystallography as was done for DNA double helix. About a dozen models for TDS were proposed, but the latest and most acceptable model was proposed in 1973 by *S.H. Kim*. He suggested that TDS of tRNA takes the shape of letter L with a thickness 20Å. This can be easily derived from two dimensional clover leaf model. The extra arm which varies in different tRNA molecules can be extended without distoring TDS conformation. The CCA stem projects out and can take different orientations.

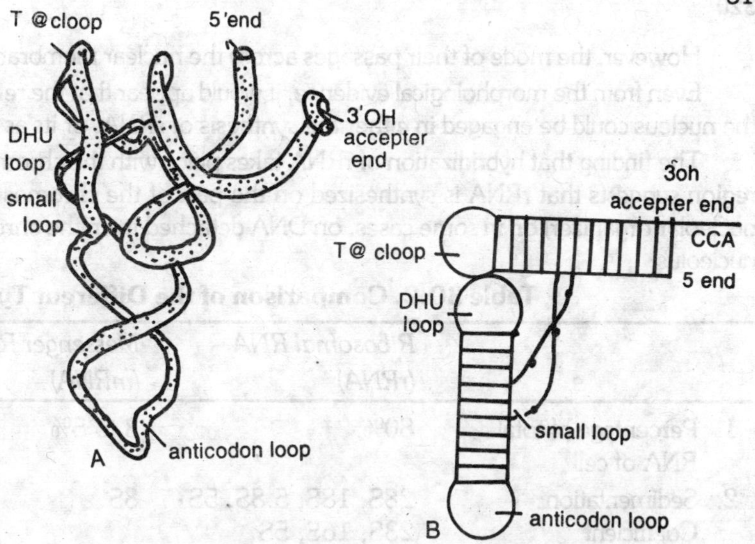

Fig. 30.3. Three dimensional structure of yeast phenylalanine tRNA.

Unusual bases

In addition to the usual basses, A, U, G and C, tRNA contain a number of unusual bases, and in this respect differs from mRNA and rRNA. The unusual bases of tRNA account for 5-20% of the total RNA of cell. Most of the unusual bases are formed by methylation (addition of $-CH_3$ or methyl group to the usual bases) e.g., cytosine and guanine on methylation yield methylcytosine and methylguanine, respectively. Precursor tRNA molecules transcribed on the DNA template contain the usual bases. These are then modified to unusual bases. The unusual bases are important because they protect the tRNA molecule against degradation by RNAse. This protection is necessary because RNA is found floating freely in the cell.

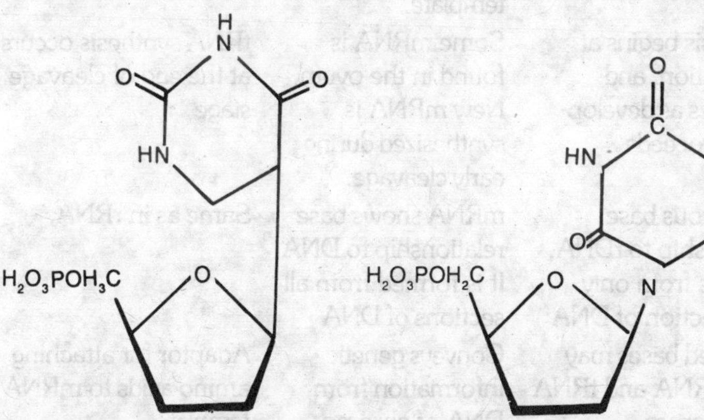

pseudo uridylic acid ribo thymidylic acid

Fig. 30.4. Showing the structural formula of Pseudo-Uridylic acid and Ribo-thymidylic acid.

Some of the unusual bases of tRNA are methylguanine (GMe), dimethylguanine (GMe_2), methylcytosine (Me), ribothymine (T) pseudouridine (ψ), dihydrouridine (DHU_2, H_2U, UH_2), inosine (I) and methylinosine (IMe, MeI).

3. Ribosomal Ribonucleic Acid rRNA

Ribosomes are small particles of about 200Å to 250Å in diameter, and are generally present in the cytoplasm either lying free in the cytoplasm or attached to the outer membrane of endoplasmic reticulum or nuclear membrane. These particles are ribosomes containing about 60% ribonucleic acid or RNA which is ribosomal RNA ro rRNA and 40% protein. It is generally agreed that cytoplasmic ribosomes or their subunits are made in the nucleus and are subsequently transferred to the cytoplasm.

However, the mode of their passages across the nuclear membrane is still a matter of controversy.

Even from the morphological evidence, it would appear that the reticular or microvesicular spheres of the nucleus could be engaged in either the synthesis of rRNA, or its assembly, or both.

The finding that hybridization of rRNA takes place with the chromatin associated with the nucleolar region suggests that rRNA is synthesized on the part of the chromosome traversing the nucleolus (the nucleolar organizer) or, in some cases, on DNA detached from the chromosome and associated with the nucleolus.

Table 30.2. Comparison of the Different Types of RNA

	Ribosomal RNA (rRNA)	*Messenger RNA (mRNA)*	*Transfer RNA (tRNA)*
1. Percentage of total RNA of cell	80%	3 to 5%	10 to 20%
2. Sedimentation Coefficient	28S, 18S, 5.8S, 5S 23S, 16S, 5S	8S	3.8S
3. Number of nucleotides	5S RNA: 120 nucleotides 16- 18S RNA: 1,600 to 2,500 nucleotides 23-28S RNA: 3,200 to 5,500 nucleotides	*E.coli*: 900 to 1,500 nucleotides	73 to 93 nucleotides.
4. Molecular weight	23S RNA: 1.1×10^6 30S RNA: 0.55×10^6	500,000	25,000-30,000
5. Unusual bases	Small amount of methylated bases.	Small amount of unusual bases. 1 per 30-40	High content unusual bases. (*E.coli*: 1 per 100-150 nucleotides).
6. Site of synthesis	Derived from nucleolar DNA	Synthesized in nucleus on DNA template.	Synthesized in nucleus on DNA template.
7. Beginning of synthesis	Synthesis begins at gastrulation, and increases as development proceeds	Some mRNA is found in the ovum. New mRNA is synthesized during early cleavage.	tRNA synthesis occurs at the end of cleavage stage
8. Base relationship of DNA	No obvious base relationship to rDNA. rRNA is from only small section of DNA	mRNA shows base relationship to DNA. It is formed from all sections of DNA	Same as in rRNA.
9. Function	Unpaired bases may bind mRNA and tRNA to ribosomes.	Conveys genetic information from DNA of chromosomes where it takes part in protein synthesis.	Adaptor for attaching amino acids to mRNA template.

Ribosomal RNA in size and base contents from tRNA and from other RNA classes of most cells. rRNA from most organism is relatively rich in guanine and cytosine, and it lacks many of the rare base found in tRNA, although some methylated bases are present.

The function of rRNA is protein synthesis is not yet completely clear. The unpaired bases in the molecule may bind mRNA and tRNA to ribosomes possibly by Mg^{++} linkages between phosphate groups on the two molecules.

Ribosomal RNA also gives some evidence for developing the double helical structure.

4. Virual RNA (vRNA)

Viruses may contain either DNA or RNA. However, each particular virus generally contains nucleic acid of only one type. Some examples of RNA viruses are influenza, turnip yellow mosaic, tomato bushy stunt, tobacco necrosis, and the particularly well studied tobacco mosaic virus (TMV).

Table 30.3. Comparison between DNA and RNA

Deoxyribonucleic acid	*Ribonucleic acid*
1. DNA is the usual genetic material	1. RNA is the genetic material of some viruses.
2. DNA is usually double-stranded. (In certain viruses DNA is single stranded $\phi\chi$ 174)	2. Most cellular RNA is single stranded. (Some viruses e.g., reovirus, have double stranded RNA).
3. The pentose sugar is deoxyribose	3. The pentose sugar is ribose.
4. The common organic bases are adenine, guanine, cytosine and thymine.	4. The common organic bases are adenine, guanine, cytosine and uracil.
5. Base pairing: adenine pairs with thymine and guanine with cytosine.	5. Adenine pairs with uracil and guanine with cytosine.
6. Pairing of bases is throughout the length of the molecule.	6. Pairing of bases is only in the helical region.
7. There are fewer uncommon bases.	7. There are more uncommon bases.
8. DNA is only of one type.	8. There are three types of RNA, messenger RNA, ribosomal RNA and transfer RNA.
9. Most of the DNA is found in the chromosomes. Some DNA is also found in the cytoplasm e.g. in mitochondria and chloroplasts.	9. Messenger RNA is formed on the chromosomes, and is found in the nucleolus and cytoplasm. rRNA and tRNA are also formed on the chromosomes, and are found in cytoplasm.
10. Denaturation (melting) is partially reversible only under certain conditions of slow cooling (renaturation).	10. Complete and practically instantaneous reversibility of the process of melting.
11. Sharp, narrow temperature interval of transition in melting.	11. Board temperature interval of transition in melting.
12. DNA on replication forms DNA, and on transcription forms RNA.	12. Usually RNA does not replicate or transcribe. (In certain viruses RNA can synthesize on RNA chain).
13. Genetic messages are usually encoded in DNA.	13. The usual function of RNA is translating messages encoded in DNA into proteins.
14. DNA consists of a large number of nucleotides, upto 4.3 million.	14. RNA consists of fewer nucleotides, upto, 12,000.

31

THE BIOSYNTHESIS OF RNA : TRANSCRIPTION

In the process of DNA transcription the positioning of nucleotide units in the RNA molecules that are being made is under the control of the DNA which acts as template.

Each RNA chain is initiated at a specific site on the DNA template and subject to termination at another unique type of site on the template. In other words there are defined units of transcription. Specific signals in the DNA template are recognized by the transcription apparatus. Initiation is governed by *promoter* regions in the DNA, and a region governing termination is designated a *terminator.*

Transcription mediated by DNA-dependent RNA polymerases. The properties of the enzyme from *E.coli* have however been the most widely explored.

The products of DNA-dependent RNA polymerases have ribonucleotide sequences complementary to *one* of the strands of the DNA which was used as 'template'. A guanine residue in the DNA template strand dictates the insertion of a cytosine nucleotide in the RNA strand under construction, whilst a cytosine in the DNA causes a guanine nucleotide to appear in the new RNA strand. Similarly, a thymine in the DNA results in an adenine nucleotide in the RNA, and an adenine in the DNA results in a uracil nucleotide in the RNA.

The DNA-dependent RNA polymerase has now been extensively purified form *coli.* The holoenzyme is a complex zinc-containing protein of molecular weight 480 000-500 000 which can be dissociated, for instance in 6M-urea, into a number of polypeptide chain subunits as follows:

Two α-chains each of mol. wt. 39 000. One β-chain of mol. wt. 155 000. One β'-chain of mol. wt. 165 000. One molecule of σ(sigma)-factor, mol. wt. 95 000.

The holoenzyme may therefore be represented as $\alpha_2\beta\beta'\sigma$. The hole enzyme without α-factor is termed 'core enzyme', or $\alpha_2\beta\beta'$. If the holoenzyme is separated into the core enzyme and the α-factor, the core enzyme will catalyse the synthesis of RNA chains from random sites on a DNA template *in vitro.*

Addition of σ-factor however restores the synthetic activity of the core enzyme by activating selective initiation of RNA synthesis. It complexes with the core enzyme to yield holoenzyme which interacts specifically with promotor regions on the DNA template.

Binding and Initiation

The sequence of molecular events for accurate initiation of RNA synthesis can be envisaged as follows. After a series of unproductive random interactions with the DNA template, the holoenzyme 'recognizes' a specific structure, or sequence, within a promoter region. This enables the holoenzyme to bind to the DNA in this region at least an order of magnitude more tightly than in the non-specific interaction mentioned above.

The next step in the Initiation process is a change in the conformational state of the DNA in this promoter. Once this transition has occurred, the holoenzyme is poised to commence the manufacture of an RNA chain, but actually using only *one* of the strands as template.

Although it is the β'-subunit that is implicated in the binding of the polymerase of the DNA and perhaps has the required specificity, σ appears to bind to the β-subunit. It may be best to consider the σ-factor as an allosteric effector of a multi-submit enzyme. The β-subunit itself is probably most concerned with at least the initial catalytic reactions. Once the initial internucleotide link between either GTP or ATP and the next nucleotide specified by the template has been made, the σ-factor is released, thus possibly reducing the affinity of the polymerase from the promoter site and allowing the remaining core to move along the DNA template, spinning off a strand of RNA as it proceeds. The process can be visualized in the electron microscope.

Elongation of RNA Chains and Direction of Transcription

Basically the new chain grows by the subsequent addition of ribonucleotides to the free 3'- hydroxyl group. For instance, the RNA has a triphosphate group at position 3' on the other, end. Alkaline hydrolysis of this particular RNA will yield a molecule of the nucleoside cytosine from the 3'-hydroxyl end of the molecule, uridine and adenosine monophosphates and a molecule of guanosine tetraphosphate from the 5'-end. Synthesis proceeds from the 5'-end to the 3'-end of the RNA molecule.

Still unclear are the precise characteristics of DNA structure at the promoter where the DNA is actually used as template rather than just for the tight binding of the RNA polymerase. In any even the nucleotide residue at the 5'-end of the nascent chain always contains a purine base; thus any DNA sequence involved in the actual commencement of template-directed RNA synthesis must contain a thymine or cytosine nucleotide.

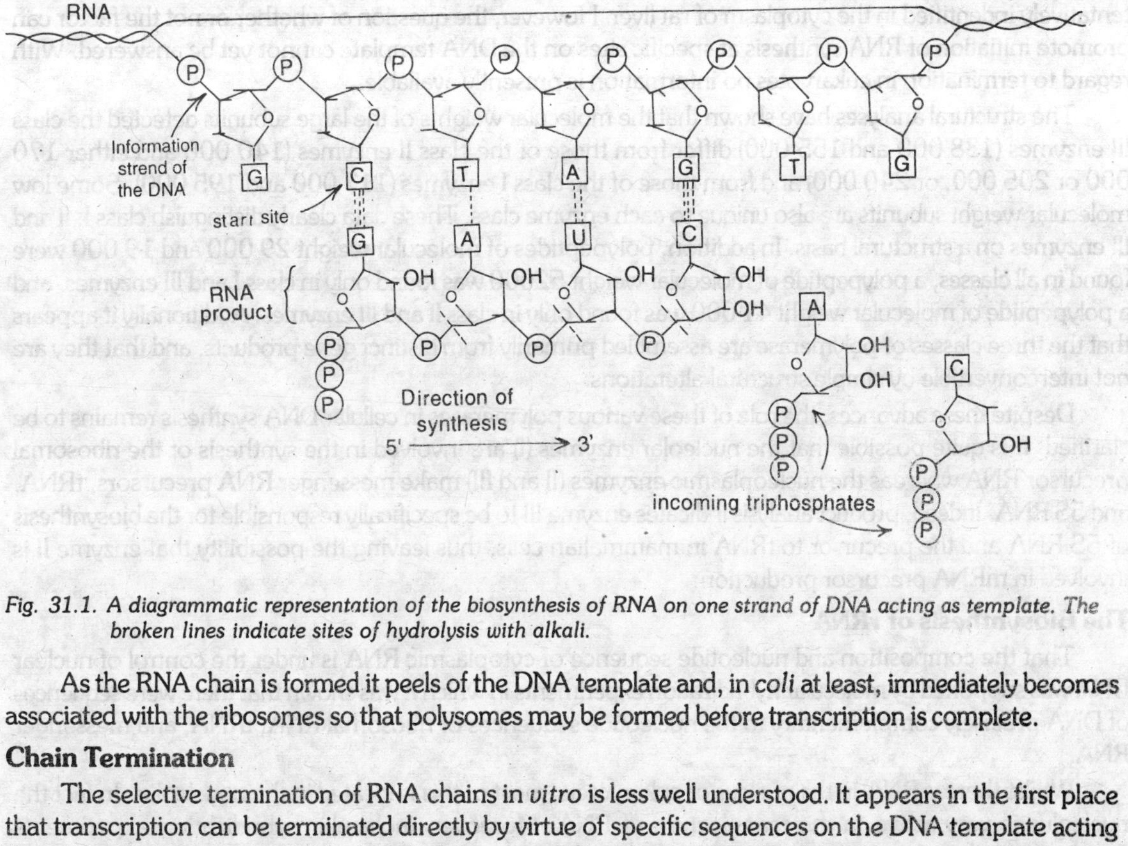

Fig. 31.1. A diagrammatic representation of the biosynthesis of RNA on one strand of DNA acting as template. The broken lines indicate sites of hydrolysis with alkali.

As the RNA chain is formed it peels of the DNA template and, in *coli* at least, immediately becomes associated with the ribosomes so that polysomes may be formed before transcription is complete.

Chain Termination

The selective termination of RNA chains in *vitro* is less well understood. It appears in the first place that transcription can be terminated directly by virtue of specific sequences on the DNA template acting directly on the core enzyme. In principle the process of termination involves (a) the cessation of RNA chain elongation, (b) the release of the newly formed RNA, and (c) the release of the RNA polymerase from the DNA. All three of these reactions can occur at 'direct terminator' sites *in vitro*. Little is known regarding the structure of such sites. A second type of termination can be induced *in vitro* by rho (ρ) factor. This factor from *coli* is an oligomeric protein (monomer molecular weight 50 000) which depresses the amount of RNA formed in the *in vitro* reactions by causing RNA chain termination. The resulting RNA chains can be released from the DNA but the RNA polymerase remains bound to the DNA.

As already mentioned, relatively little is known about the structure of terminators; however, by *in vitro* chemical and enzymic methods, some information has been gained about the nucleotide sequences of a possible *terminator* region of the *coli* tyrosine suppressor tRNA gene.

The Eukaryotic RNA Polymerase

In eukaryotic cells the three classes of RNA polyerases have been isolated after solubization from chromatin, chromatography of the enzyme activity from variety of animal tissues on DEAE-Cellulose. These are RNA polymerase I, II and III. RNA polymerase III has been further resolved into IIIA and IIIB.

Class I RNA polymerases has been established to be nucleolar origin, whereas classes II and III are of nucleoplasmic origin. Additionally these enzymes operate optimally under somewhat different conditions. The nucleolar enzymes work best at low ionic strength and utilize Mn^{++} and Mg^{++} equally well. On the other hand, higher Mn^{++} concentrations and ionic strengths are required for maximum activity of the nucleoplasmic enzymes.

Since factors such σ and ρ appear to play an important role in the control of transcription, a search for such factors was made in mammalian systems. A protein factor (mol. wt. 70 000) which can specifically bind to, and stimulate, the activity of nucleoplasmic class II enzymes from rat liver and calf thymus has been tentatively indentified in the cytoplasm of rat liver. However, the question of whether or not the factor can promote initiation of RNA synthesis at specific sites on the DNA template cannot yet be answered. With regard to termination in eukaryotes no information is presently available.

The structural analyses have shown that the molecular weights of the large subunits detected the class III enzymes (138 000 and 155 000) differ from those of the class II enzymes (140 000 and either 170 000 or 205 000, or 240 000) and from those of the class I enzymes (117 000 and 195 000). Some low molecular weight subunits are also unique to each enzyme class. These data clearly distinguish class I, II and III enzymes on a structural basis. In addition, polypeptides of molecular weight 29 000 and 19 000 were found in all classes, a polypeptide of molecular weight 52000 was found only in class I and III enzymes, and a polypeptide of molecular weight 41 000 was found only in class II and III enzymes. Additionally it appears that the three classes of polymerase are assembled primarily from distinct gene products, and that they are not interconvertible by simple structural alterations.

Despite these advances, the role of these various polymerases in cellular DNA synthesis remains to be clarified. It is quite possible that the nucleolar enzymes (I) are involved in the synthesis of the ribosomal precursor RNA whereas the nucleoplasmic enzymes (II and III) make messenger RNA precursors, tRNA, and 5S RNA. Indeed, product analysis indicates enzyme III to be specifically responsible for the biosynthesis of 5S RNA and the precursor to tRNA in mammalian cells, thus leaving the possibility that enzyme II is involved in mRNA precursor production.

The Biosynthesis of rRNA

That the composition and nucleotide sequence of cytoplasmic RNA is under the control of nuclear DNA was supported by molecular hybridization experiments in which it was shown that there were sequences of DNA precisely complementary to the nucleotide sequences of ribosomal RNA, tRNA, and messenger RNA.

Whilst there is RNA in the nucleus of eukaryoteis, most of this is located in the nuucleolus. In fact the nucleolus is very active in the manufacture of RNA. Moreover, molecular hybridization experiments demonstrate that the parts of the DNA genome which contains the deoxyribonucleotide sequences complementary to the sequences that make up ribosomal RNA are located in the nucleolus, or nucleolar organizer region of the genome, of certain eukaryotes.

Siherrer and *Darnell* (1962) have demonstrated that these exists a short-lived species of RNA in the cultures of human cells which were exposed for a short time to ^{3}H-uridine. This RNA was distinguishable from tRNA and rRNA. It has a sedimentation constant of 45S and having a molecular weight of about 41 000 000 daltons. The site of its synthesis found to be nucleolus. Shortly after its synthesis, or transcription from the nucleolar DNA, this long single-stranded molecule undergoes a series of molecular 'tailoring' events in the nucleolus whereby its molecular dimensions are progressively reduced to give rise eventually to the 28S and 18S ribosomal RNA species found in cytoplasmic ribosomes.

Whilst the 45S species is found exclusively in the nucleolus, after longer labelling times radioactivity begins to appear in other species of RNA, firstly in 41S and 32S RNA (which are also confined to the nucleolus) and 20S RNA, and then in 18S RNA which passes rapidly to the cytoplasm'. Finally, radioactivity appears in 28S RNA initially in the nucleolus, then in the nucleoplasm, and shortly afterwards in the cytoplasm. A precursor-produced relationship between these RNA's has been proposed based on a variety of data including recent electron microscopic examination of the various RNA species.

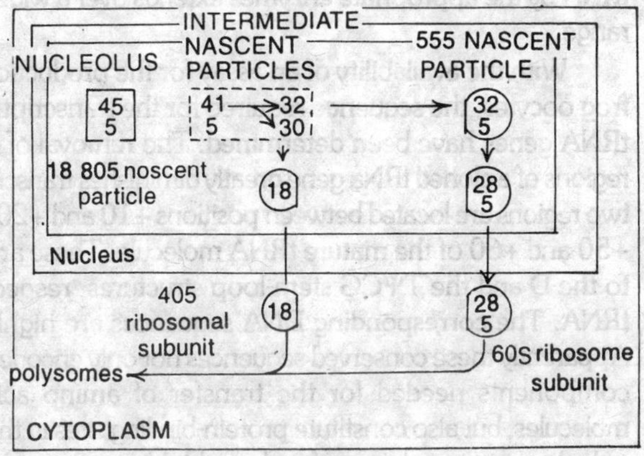

Fig. 31.2. An outline of eukaryotic ribosome formation illustrating the processing of RNA taking place within 'nascent ribosomal particles'. The number within the particles refer to the sedimentation coefficients of their constituent RNA's. Note the occurrence of 5S RNA even within '80S nascent particles'.

The '5.8S RNA' fragment appears to be generated simultaneously with the 32S→28S RNA transition. This has been interpreted by some as indicating the possibility that, in the conversion, a loop of 'spacer' RNA located on the precursor chain between the 28S and 5.8S species is degraded.

With regard to the sequence modifications to rRNA (i.e. the methylation and pseudouridine formation), the majority of these occur rapidly in the nucleolus on the 45S precursor. Interestingly, the methylation occur only in the portions of the precursor destined to become either the 28S RNA or 18S RNA components, and are mostly in the form of 2'-O-methylribose.

5S rRNA originates independently in the non-nucleolar part of the nucleus and later becomes permanently associated with the large ribosomal subunits.

Although the ribosomal proteins appear to be synthesized in the cytoplasm, it is reasonably certain that they become associated with the ribosomal precursor RNA's in the nucleolus where the actual assembly of the ribosome takes place. It is believed that the cleavage of the precursors takes place within 'nascent ribosomal particles'.

Biosynthesis of tRNA

Next to rRNA, tRNA is most plantiful RNA by weight; it makes up about 15% of the total RNA of the cell, compared to 80% for rRNA. The total tRNAs average only 75-80% nucleotides in length, however, so there are actually many more tRNA than rRNAs. Many tRNA genes exist in multiple copies in chromosomal DNA. Although they are not so highly repeated as 5S rRNA genes are. All of the primary transcription products of tRNA genes studied so far are, like pre-rRNA, larger than the final products. Thus, precursor tRNA molecules require shortening as they mature. They also require many other chemical modifications, such as the addition of methyl and isopentenyl groups and the conversion of uridines into pseudouridines. In fact, about 1 base in every 10 of a tRNA molecule is modified in some way.

Some tRNA genes in eukaryotic cells to not exist as a consecutive stretch of DNA. For example, to form the yeast phenylalanine and tyrosine tRNAs, a segment must be removed from the middle of the primary transcripts, and the two ends must be spliced together.

The transcription of tDNA (the genes that encode tRNA) is calalyzed by RNA polymerase III. As has been done with 5S rDNA, the tDNA sequence from a number of different eukaryotic organisms have been cloned as recombinant DNA molecules in *E. coli*. The processing steps include the addition of CCA to the 3' ends of the yeast tRNAs, and the addition of various other chemical structures such as methyl and isopentenyl groups. Thus the recognition of the tRNA by the appropriate enzymes extends over a wide evolutionary range.

With the availability of an assay for the production of tRNA in frog oocytes, the sequence required for the transcription of several tRNA genes have been determined. The removal of either of two regions of a cloned tRNa gene greatly diminishes transcription. These two regions are located between positions +10 and +20 and between +50 and +60 of the mature tRNA molecule. These are correspond to the D and the TΨCG stem-loop structures, respectively, in the tRNA. The corresponding DNA sequences are highly conserved. Apparently these conserved sequences not only encode the structural components needed for the transfer of amino acids by tRNA molecules, but also constitute protein-binding sites in the DNA. Thus a dual role is now known for these highly conserved regions: one role in structuring functional tRNA and one role in promoting the transcription of the tRNA genes.

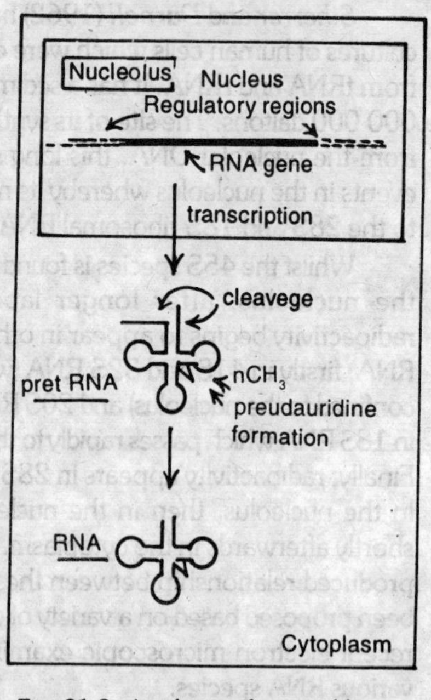

Fig. 31.3. A scheme the possible steps involved in the production of mammalian cell tRNA.

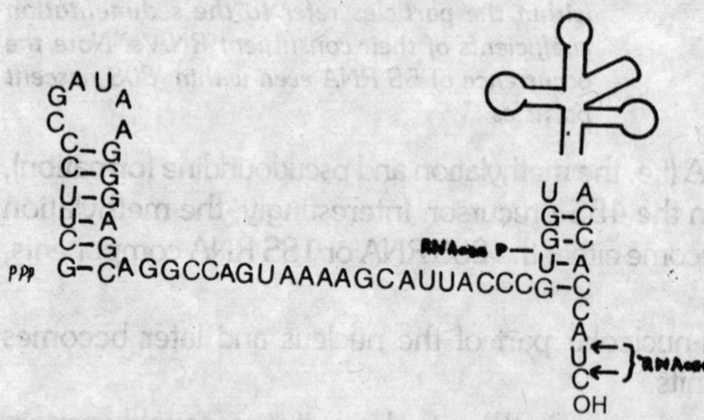

Fig. 31.4. The structure of Esch. Coli tyrosine suppressor tRNA, showing sits of cleavage (->).

In *E.coli*, where there is also evidence for precursors, a sepcific study of *E.coli* tyrosine suppresor tRNA and its formation has been made by *Smith et. al.* A specific precursor to this special tRNA found in the number suppressor su_3^+ has been isolated and characterized. Joined to the actual tRNA sequence at its 5'-end are 41 extra nucleotides beginning with pppG, and at the 3'-end are two nucleotides. This precursor is cleaved in *Esch. coli* to give the mature tRNA sequence.

It appears that, for processing to be efficient, the tRNA part of the tyrosine tRNA precursor must be folded up in the correct conformation, and that base-substitutions that prevents this, or favour an alternative structure, make processing inefficient. Nucleotide modifications apparently take place after the cleavages have occurred.

Biosynthesis of mRNA

Despite the great interest in the studies on the synthesis of rRNA and tRNA, it was considered of even greater importance to learn how mRNA in a cell determines which proteins the cell makes. The discoveries in the 1960s that tRNA and most rRNA were formed by the cleavage of precursor RNAs suggested that nuclear precursors to mRNA might also exist. Newly labeled nuclear RNA had been divided into pre-rRNA molecules (located in the nucleoplasm) and the hnRNA molecules (located in the nucleoplasm) whose lengths ranged from 2 to 30 kb. Also the hnRNA fraction had an average base composition like that of genomic DNA, suggesting that hnRNA was made up of many different molecules, as also might be expected of the total mRNA of a cell. Were these hnRNA molecules precursors to mRNAs? Early in the 1960s, cell free translation of eukaryotic mRNA had not been achieved, so direct tests of mRNA function could not be attempted. In their place, however, experiments designed to detect mRNA as a labeled RNA species were begun, so that the sizes and average base compositions of mRNA and of hnRNA could be compared.

In eukaryotes, heterogeneous nuclear RNA (HnRNA) also known as high molecular weight RNA (HMWRNA) or DNA like RNA (dRNA) is first synthesized in the nucleus. HnRNA molecules are much bigger in size than mRNA molecules which are utilized in cytoplasm. In the early 1960s, an enzyme termed poly A polymerase had been isolated; this enzyme did not copy a template, but simply added adenylate residues to the 3' end of RNA chains. The length of a poly A segment in the nucleus is quite consistent, it is between about 200 and 250 nucleotides. In the cells of lower plants and animals, the initial length of poly A may be shorter than in mammalian cells, but the size of the segment appears to be fairly uniform in each species. The specificity of poly A length is only one example of the nuclear RNA processing events that have been clearly demonstrated but that are not yet fully understood. After the poly A containing mRNA reaches the cytoplasm, the poly A segment is shortened so that its size is not longer uniform, but varies between 30-250 nucleotides.

Poly A at the 3' end of eukaryotic mRNA is called the tail and the other 5' terminus *cap*. The cap consists of a terminal nucleotide, 7-methylguanylate (m^7G) in a 5'→5' linkage with the initial nucleotide of mRNA chain. Thus there is no free 5' end of mRNA molecule; in fact it has 3' hydroxyl groups on the ribose sugar rug of the nucleotides at both ends. The cap probably aid in translation after a mRNA molecule is completely synthesized, but the structure is made in the nucleus by the series of enzymatic steps. It could be shown that removal of the cap leads to loss of translational activity due to loss of the formation of mRNA—ribosome complex. It suggests that cap can helps in recognition of ribosome.

32

PROTEIN SYNTHESIS

The proteins are highly organized and complex macromolecules. They have significant role in the structural and functional organisation of the cells. The structural proteins constitute various cellular components (such as plasma membrane, nuclear membrane, organelles etc.) as well as some extracellular parts such as cuticle, fibres etc. The functional proteins are enzymes and some hormones which control almost all the metabolic, regulatory, sensory and reproductive activities of the cells. They also control the expression of phenotypic traits.

Chemically, the proteins are the polymers of *amino acids* and at a primary level these can be considered as polypeptide chains. The polymer in a protein of amino acids linked with the help of peptide bonds (–Co –NH). The peptide bond consists of a convalent bond between carboxy group (COOH) of one amino acid and the amino group (NH_2) of another amino acid.

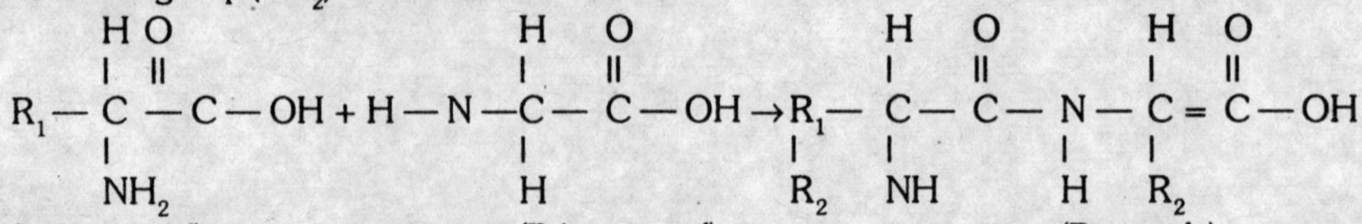

(I Aminoacid) (II Aminoacid) (Dipeptide)

Only 20 amino acids are biologically important and they participate in the protein synthesis forming polypeptide chains.

Deoxy ribonucleic acid (DNA) initiates the mechanism of protein synthesis since it carries the genetic message for protein synthesis. The information is transcribed or copied over messanger ribonucleic acid (mRNA) from one DNA strand. The mRNA translates the coded information by bringing amino acids in a particular sequence during protein synthesis through the agency of transfer ribonucleic acid (tRNA). It occurs over the ribosomes. The first step is called *transcription* which involves copying of genetic information from DNA into mRNA. The second step is called *translation* which involves the translation of information encoded on mRNA with newly synthesized protein molecules.

Flow of Genetic Informations

The protein synthesis is under direct control of DNA in most of the cases. Information of a polypeptide chain is stored in mRNA chain. The relationship between DNA, RNA and protein is called *central dogma*. From time to time dogma has been variously proposed.

1. *One way flow of information. Crick* in 1958 suggested that DNA transmits the genetic informations by transcribing mRNA. This mRNA translates the nucleotide sequence into the sequence of amino acid.

$$\text{DNA} \xrightarrow{\text{transcription}} \text{mRNA} \xrightarrow{\text{translation}} \text{RNA}$$

2. *Circular flow of information.* It was suggested by *Barry Commoner* in 1968. The DNA transcribes mRNA and mRNA translates information into protein. Further proteins directs the synthesis of mRNA and this mRNA dictates DNA synthesis.

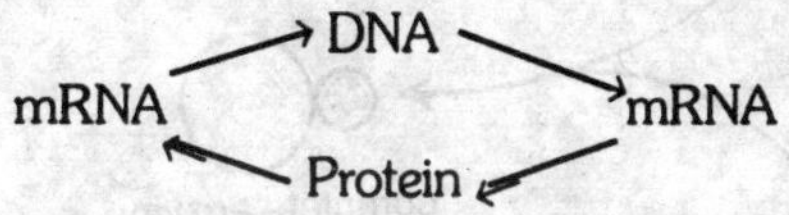

3. *Inverse flow of information. Temin* (1970) has described the presence of RNA dependent DNA polymerase enzyme, which controls the synthesis of DNA from single stranded RNA. The synthesis of DNA on RNA template in tumor virus was demostrated simultaneously by *Temin* and *Baltimore.* This discovery was considered significant for understanding of cancer. Therefore, *Temin, Baltimore* and *Dulbecco* were awarded Nobel Prize in medicine for 1975.

$$\text{DNA} \xrightarrow{\text{transcription}} \text{mRNA} \xrightarrow{\text{translation}} \text{Proteins}$$

Mechanism of Protein Synthesis in Prokaryotes

Transcription of mRNA from DNA

The copy of a complementary mRNA strand on DNA strand is called transcription. RNA polymerase is the main eyzyme involved in the transcription. All types of RNA (mRNA, t-RNA and rRNA) are synthesized with the help of this enzyme. In eukaryotic system three types of RNA polymerases are known to be present in the nucleus. In bacterial system only one type of RNA polymerase is known which is responsible for synthesis of all types of RNA. Besides RNA polymerase, transcription requires a template activated precursors and a divalent metal ion (Mg^{++} or Mn^{++}).

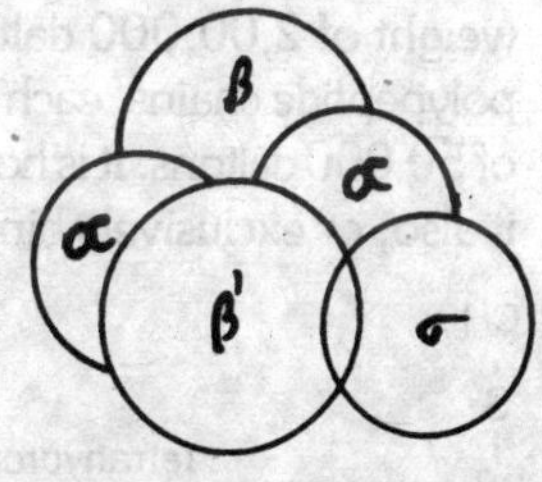

Fig. 32.1. A model of the structure of prokaryotic RNA polymerase showing association of five polypeptides (α_2 ββ σ).

As earlier state that in bacterial system a single RNA polymerase is responsible for synthesis of all kinds of RNAs. It has been purified and its structure and function is now known. It consists of four polypeptide chains. There are two alpha (α) chains, one beta (β), one beta desh (β) and one sigma (σ) chains. The first-four form core enzyme and sigma factor units to form holoenzyme. The molecular weight of α-chain is 40,000 daltons, that of β chain 1,55,000 daltons, β' chain 1,65,000 daltons and σ chain is 95,000 daltons. The attachment of σ factor is not very firm sot hat core enzyme ($\alpha_2\beta\beta'$) can be easily isolated. Function of different polypetide chains is known β and β' help RNA polymerase in unwinding the DNA molecule for the process of transcription. The σ factor helps in recognition of start signals on DNA molecule and directs the enzyme in selecting the initiation sites. Once the mRNA synthesis is initiated, the σ factor dissociates and core enzyme brings about elongation of mRNA.

For transcription entire chromosome or DNA molecule does not unwind but a local denaturation of the DNA helix takes place. Thus only those bases are exposed that constitute the gene to be transcribed. After unwinding only one of the two strands is copied. For synthesis of mRNA on DNA template, there are transcription units, each having a starting point and a termination signal. The start signals within DNA molecule are called *promotor sites.* These promotor sites are not identical to each other, though all attract the same RNA polymerase.

RNA polymerase initially binds to a promotor site for RNA synthesis. It then diffuses to a binding site consisting of a sequence of six bases, which never differs by more than two bases from the following sequence 5' TATAAT. It is called *Pribnow box* and is available in almost all promotors.

The first base of RNA is always purine, either G or A. This means that the first base of the gene will be a pyrimidine either C or T in DNA. After synthesis the terminal triphosphate group may be cleaved by endonucleases. This results in RNA chain without terminal triphosphate groups starting with pyrimide nucleosides. The RNA chains can be formed *de novo* and do not require primers, as necessary for DNA replication.

Growth or elongation of RNA chain that proceeds in the 5'→3' direction, just as the growth of DNA chain does. Chain elongation takes place by addition of activated ribonucleoside triphosphates (ATP, GTP, CTP and UTP) to one strand of DNA template. From each activated ribonucleoside triphosphate, added to the growing mRNA chain, pyrophosphate (PP) is given out. During mRNA synthesis energy is released in the RNA synthesis is initiated, the σ factor dissociates and core enzyme brings about elongation of mRNA.

Termination of mRNA chain is brought about by certain *termination* or *stop signals* on DNA. Stop signals, like the promotor site, is a sequence of bases. Some of the stop signals may be recognized by the RNA polymerase, other appear to require the help of additional proteins. In *E. coli* such protein is known as *rho factor* (P). The P factor is discovered by *Roberts* (1962). It has a molecular weight of 2,00,000 daltons. It is made up of four polypeptide chains, each having a molecular weight of 50,000 daltons. It is however, not certain whether it also, or exclusively binds to DNA.

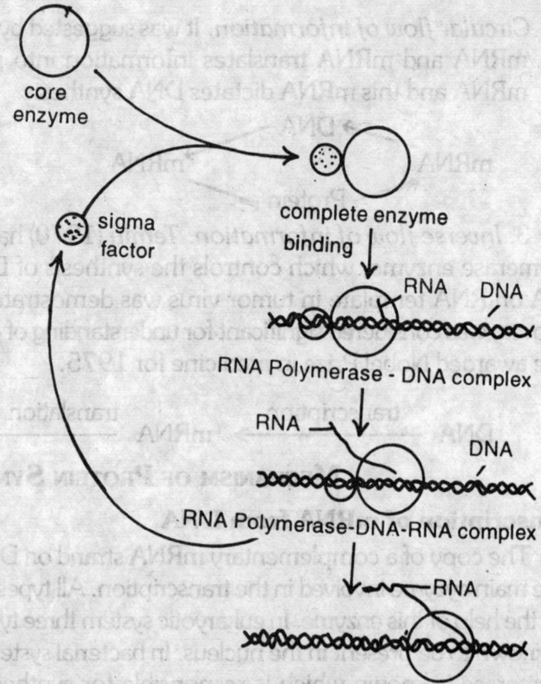

Fig. 32.2. Role of sigma factor and core enzyme of RNA polymerase during transcription.

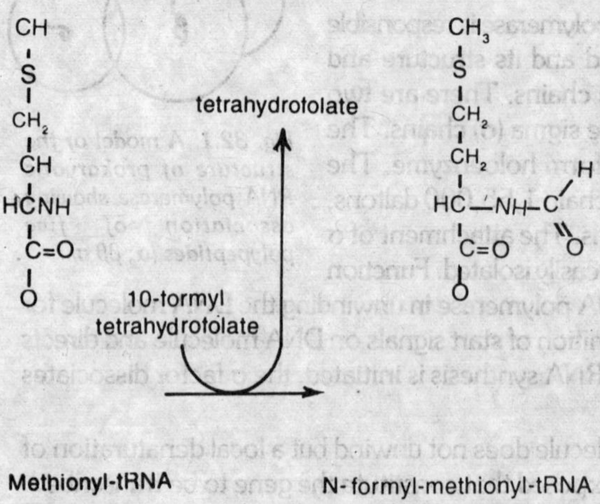

Fig. 32.3. The initiator N-formyl-methionyl-tRNA (1Met-tRNA$_1$) is generated by formylation of methionyl-tRNA$_1$ using formyl-tetrahydrofolate as cofactor.

Translation in Prokaryotes

During translation the genetic information present in mRNA directs the order of specific amino acids to form protein molecule or polypeptide chain. This step is much more complex than transcription and involves the following steps:

1. Activation of amino acids

The 20 amino acids used in protein synthesis are activated by adenosine triphosphate (ATP). The activation is mediated by specific activating enzyme known as *amino acyl RNA synthetases.* As a result of this reaction between amin acid and ATP, a complex is formed. This complex is known as amino acyl-AMP-enzyme or amino acyl AMP complex is formed. Pyrophosphates (PP) are released. It should be realized that amino acyl RNA synthetases are specific with respect to amino acids. Therefore, 20 such enzymes are required for 20 amino acids.

$$\underset{\text{Amino acid}}{AA_1} + ATP + \underset{\text{Enzyme}}{E_1} \longrightarrow \underset{\text{Amino acyl AMP enzyme complex}}{(AA_1 - AMP)\, E_1} + PP$$

2. Attachment of activated amino acid to tRNA

The complex formed in first step above, reacts with a particular tRNA and transfer the amino acid to tRNA with the release of AMP and enzyme amino acyl RNA synthetase.

$$(AA_1\text{-AMP})\,E_1 + tRNA_1 \longrightarrow AA_1\,tRNA_1 + AMP + E_1$$

For each amino acid, there is a specific tRNA. As there are 20 different amino acids so 20 different tRNAs are there. There are cases where more than one kind of tRNA may be available for the same amino acid, but no case is known where same tRNA can recognize two or more different amino acids.

3. Initiation of polypeptide chain

Initiation requires the ribosome subunits, mRNA, guanosine triphosphate (GTP) as energy source, amino acid t-RNA complex and initiation factors (IF).

Initiation of polypeptide chain is always brought about by the amino acid, *methionine.* It is coded by the codon AUG. In *E.coli,* there are two different tRNA for methionine. The first one is $tRNA_m^{met}$ or nonformylatable tRNA. The second one is $tRNA_1^{met}$ or formylatable tRNA. The methionine amino acid carried by $tRNA_m^{met}$ cannot be formylated while methionine that is carried by $tRNA_t^{met}$ can be formylated. $tRNA_t^{met}$ deposits formylated amino acid as the first amino acid of the polypeptide chain. $tRNA_m^{met}$ deposits amino acid methionine at intercalary position in the polypeptide chain.

In *E. coli* the initiation involves the following steps:

(i) *Formation of formyl-methionine-$tRNA_t^{met}$.* In the presence of transformylase enzyme and formyl-tetrahydrofolic acid that takes place.

$$\text{met-}tRNA_t^{met} + \text{formate} — \text{f-met-}tRNA_t^{met}$$

(ii) *Binding of 30 'S' subunit of ribosose with mRNA.* mRNA becomes attached to the 30 'S' subunit to form mRNA-30 'S' complex. This process requires the initian factor 3 (IF-3). Only one molecule of IF-3 binds to each 30 'S' subunit. (There are three 'F' in prokaryotes namely IF-1, IF-2 and IF-3. These factors are found in 30 'S' subunit of ribosomes. Their molecular weights are 9,200, 80,000 and 23,000 daltons respectively. These factors perform different functions). This 30 'S' subunit attaches at the 5' end of mRNA carrying AUG codon.

$$\text{IF-3-30 'S' + mRNA} \longrightarrow \text{IF-3-30 'S' -mRNA}$$

Besides binding 30 'S' subunit with mRNA, the IF-3 is also necessary for (a) affecting 30 'S' subunit conformation and (b) prevents reassociation of 30 'S' and 50 'S' subunits of ribosome.

(iii) *Binding of f-met-$tRNA_f^{met}$ with IF-3-30'S' -mRNA complex.* IF-2 and IF-1 are necessary for this step. Besides GTP is also necessary. IF-2 may either attached to f-met-$tRNA_t^{met}$ to form a binary complex or may directly bind to IF-3-30 'S' -mRNA complex. In the first case IF-2-f met-$tRNA_t^{met}$ may locate or recognize AUG codon on the IF-3-30 'S' mRNA complex and associate with it. In the second case IF-2 recognizes fmet-$tRNA_t^{met}$ and later associates to form the initiation complex.

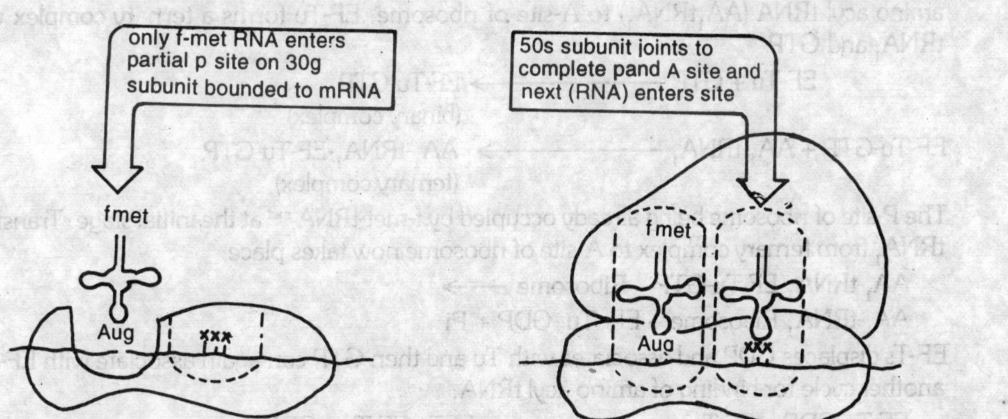

Fig. 32.4. Only tMet-tRNA can be used for initiation by 30S subunits; only other aminoacyl-tRNAs can be used for elongation by 70S ribosomes.

IF-3-30 'S' mRNA + fmet-$tRNA_t^{met}$ + IF-2—
IF-3-30 'S' -mRNA-IF-2fmet-$tRNAf^{met}$,

The mRNA has many AUG codons. Only one of these functions as an initiation codon. What factor determines that which AUG will function as the initiation codon? It is probably a specific sequence of nucleotides near AUG which is required for it to function as an initiation codon.

(iv) ***Attachment of 50 'S' subunit of ribosome with above complex.*** The complex formed in the above step associate with 50 'S' subunit of ribosome, to form a complete ribosome. This action depends on IF-2 activity. Energy is supplied by GTP. The IF-2 and IF-3 are now released.

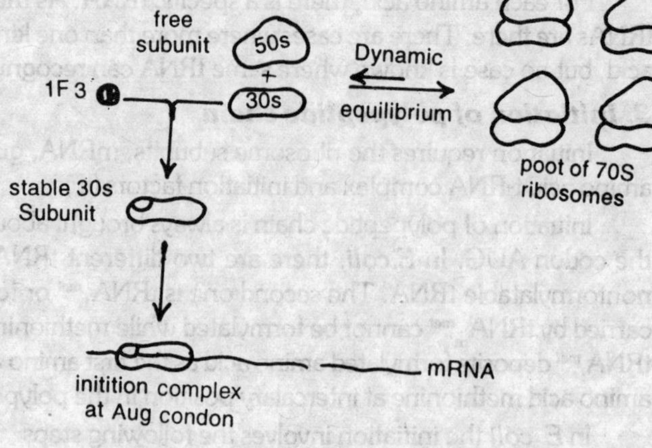

Fig. 32.5. IF3 is involved in upsetting the equilibrium with 70S ribosomes to generate stable 30S subunits with the ability to bind at Intlation sites on mRNA.

IF3-30 'S' -mRNA - IF-2-fmet-$tRNA_t^{met}$ + GDP—
70 'S'-mRNA-fmet-$tRNA_t^{met}$ + GDP + P_i + IF-2 + IF-3

The bacterial ribosome has two binding sites for tRNA. The first one is *A-site* or *amino acyl site* or *acceptor site* and the othr site is *P-site* or *peptidyl site* or *polymerisation site.* Thus each ribosome accomodates two codons or six nucleotides at a time. Both the sites are located in 50 'S' subunit of ribosome.

4. Elongation of polypeptide chain

After the formation of 70 'S' -mRNA-fmet-$tRNA_t^{met}$ complex, elongation of polypeptide chain takes place. Soluble protein *elongation* factors (EF) are required. Two EF are isolated called EF-T and EFG. Later EF-T was separated into two components, EF-Tu (unstable when heated) and EF-Ts (stable when heated). FE-Tu and FE-Ts are required for binding the amino acyl tRNA to the ribosomes. EF-G is involved in the translocation of mRNA.

The elongation process takes place in following steps:

(i) ***Binding of AA_1tRNA, to the A-site of ribosome.*** EF-Tu and EF-Ts are requried for binding the amino acyl tRNA (AA_1tRNA_1) to A-site of ribosome. EF-Tu forms a ternary complex with AA_1 -$tRNA_1$ and GTP.

EF-Tu + GTP ⟶ EF-Tu-GTP (binary complex)

EF-Tu-GTP + AA_1-$tRNA_1$ ⟶ AA_1-$tRNA_1$-EF-Tu-GTP. (ternary complex)

The P-site of ribosome being already occupied by f-met-$tRNA_t^{met}$ at the initial stage. Transfer of AA_1-$tRNA_1$ from ternary complex to A-site of ribosome now takes place.

AA_1-$tRNA_1$-EF-Tu-GTP + Ribosome ⟶
AA_1-$tRNA_1$ Ribosome + EF -Tu -GDP + Pi

EF-Ts displaces GDP and associates with Tu and then GTP can again associate with EF-Tu to start another cycle for binding of amino acyl tRNA.

EF-Tu-GDP + EF-Ts ⟶ EFTu-EF-Ts + GDP

EF-Ts - EF-Tu + GTP ⟶ EF-Tu - GTP (Binary complex)

(ii) ***Peptide bond formation.*** Bond formation occurs by the transfer of the first amino acid i.e. f-met-$tRNA_t^{met}$ from its tRNA at the P-site to the amino acyl tRNA (AA_1–$tRNA_1$) just brought to the A-site. The peptide bond is formed between the carboxyl group of first amino acid (f-met) and the amino group of second one (AA_1). The reaction is catalysed by *peptidyl transferase.* This enzyme is the intergral part of the 50 'S' subunit of ribosome. After the formation of a peptide bond, the tRNA at P-site is deacylated and the tRNA at A-site now carries the polypeptide.

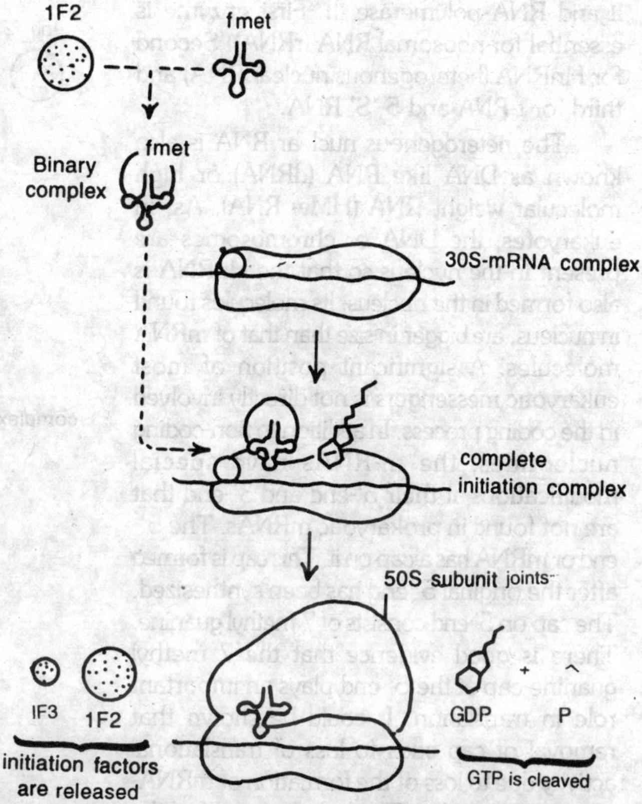

Fig. 32.6. Bacterial initiation factors are needed for 30S subunits to bind mRNA and for fMet-$tRNA_i$ to bind to the 30S-mRNA complex; they are released with cleavage of GTP when the 50S subunit arrives.

(iii) ***Translocation.*** The movement of ribosome relative to mRNA is called translocation. During translocation the ribosome moves along mRNA, a codon at a time, from 5' to 3' direction (5'→3'). The peptidyl tRNA present A-site now translocated to P-site. During this process, the deacylated tRNA is liberated out from P-site. The translocation of peptidyl tRNA from A-site to P-site requires EF-G and GTP. Now the A-site of ribosome is occupied by another AA-tRNA-EF-Tu-GTP complex and the cycle continues till the ribosome comes across the termination codon i.e. UGA or UAA or UAG.

There exists a great ambiguity about relative movement of ribosome and mRNA. According to a view during translocation, the energy is provided by GTP which drive a pulsating ribosome contraction after the formation of a peptide bond between two amino acids. This contraction would move the ribosome along mRNA and change the status of two bound tRNAs is one concerted movement. According to *Dyson* (1978) the mRNA is moved past the ribosomes three nucleotides at a time so that a tRNA and a mRNA codon occupy each of two ribosomal positions in turn.

(iv) *Termination.* As mentioned above UGA, UAA and UAG are trinucleotide codons serve in the process whereby polypeptide chains are terminated. There are no tRNAs with anticodons complementary to these triplets. When the ribosome reaches one of these codons, the signel is received to stop further elongation and terminate the polypeptide. How the termination is accomplished is poorly understood, it is known to involve *release factors* and GTP. The release factors are soluble proteins. There are three release factors RF-1, RF-2 and RF-3. Factors RF-1 and RF-2 are involved in the recognition of the termination codons. RF-3 is somehow involved in the release of the completed polypeptide from its attachment of its C-terminal to the last tRNA that had entered.

Protein Synthesis in Eukaryotes

Transcription of mRNA from DNA

RNA polymerases are the main enzymes involved in transcription of DNA in the forms of RNAs. Unlike prokaryotes, in eukaryotes there are three RNA polymerases—RNA polymerase I, RNA polymerase

II and RNA polymerase III. First enzyme is essential for ribosomal RNA (rRNA). Second for HnRNA (heterogenous nuclear RNA) and third for t-RNA and 5 'S' RNA.

The heterogenous nuclear RNA is also known as DNA like RNA (dRNA) or high molecular weight RNA (HMw RNA). As, in eukaryotes, the DNA or chromosomes are present in the nucleus so that the HnRNA is also formed in the nucleus. Its molecules found in nucleus, are bigger in size than that of mRNA molecules. A significant position of most eukaryotic messengers is not directly involved in the coding process. In addition to non-coding nucleotides, the m-RNAs have special modifications at their 5'-end and 3'-end that are not found in prokaryotic mRNAs. The 5'-end or mRNA has a cap on it. This cap is formed after the original 5' end has been synthesized. The cap on 5' end consists of 7 methyl guanine. There is good evidence that the 7 methyl guanine cap at the 5' end plays an important role in translation. It could be shown that removal of cap bads to loss of translational activity due to loss of the formation of mRNA-ribosome complex. The base sequence of the 3' end of most mRNA is very unusual. It is composed exclusively of adenine. These adenosine residues, which make up the last 50-200 nucleotides, are referred to as *poly A*. The poly A is also added to mRNA molecule after it has been synthesized. The poly A at 3'-end is usually called *tail*. Conclusive evidence of function of tail has been very difficult to obtain. Some evidence point to a role in stabilizing the mRNA in cytoplasm, protecting it from premature degradation. Other evidence points out its role in processing the primary transcript within the nucleus.

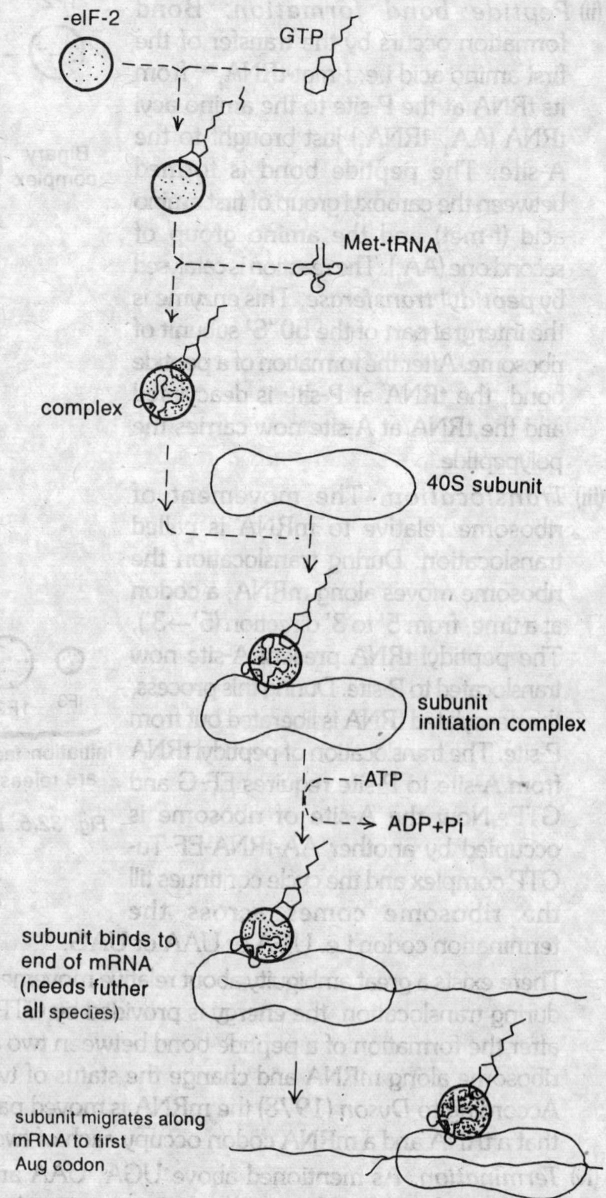

Fig. 32.7. In eukaryotic initiation, eIF2 forms a ternary complex with $tRNA_1$ that binds to free 40S subunits; then the initiation plex binds to mRNA at its 5' end, from where it must migrates to the AUG initiation codon.

mRNA is derived from HnRNA

The ends of some HnRNA molecules carry the same modifications that are seen in mRNA. The 5'-end gains a methylated cap and the 3'end is polyadenylated. The similarities of these end modifications suggest that HnRNA and mRNA may have a *precursor-product relationship*. This relationship can be demonstrated by *pulse-chase* experiment.

A radioactive precursor is added for a very short time. This is the pulse label. Then it is removed and replaced by an unlabelled precursor. This is the chase. The amount of label in HnRNA and in mRNA is measured at intervals of time.

Because the end-modifications are stable, they can be chased from HnRNA to mRNA. This demonstrates that ends of mRNA are derived from HnRNA. Probably about 20% of HnRNA is capped following its transcription. At least a large proportional of the cap can be chased from HnRNA into mRNA. At the 3' end of the molecule the length of the nuclear poly A is slightly greater than that found in mRNA. A few bases are removed at the time of passage from nucleus to the cytoplasm.

In eukaryotes the termination of RNA polymerase II activity during transcription may take place through the termination sites. Obviously the 3'end of mRNA will be generated due to post-transcriptional cleavage of poly A. This cleavage is acheived by *snurp* (small nuclear RNA-protein complex). How the snurp helps in post-transcriptional cleavage is unknown.

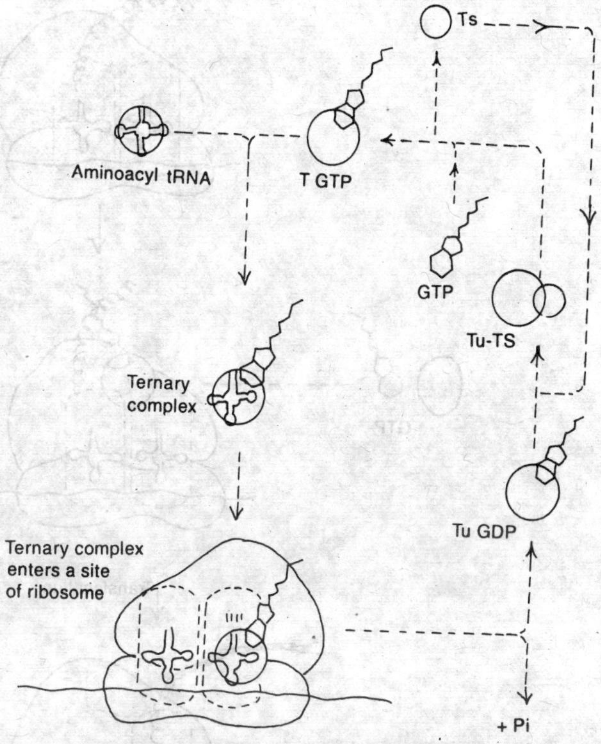

Fig. 32.8. EF-Tu. GTP places aminoacyl-tRNA on the ribosome, is released as EF-Tu. GDP, and then is regenerated as EF-Tu. GTP by the action of EF-Ts.

Translation

This step involves translation of the language of nucleic acid into the language of proteins translation occurs in following steps:

(i) *Activation of amino acids.* This step in eukaryotes is similar to that of prokaryotic activation.

(ii) *Transfer of amino acid to tRNA.* This step in eukaryotes is also similar to that of prokaryotic transfer.

(iii) *Initiation of polypeptide chain.* This step in eukaryotes is similar to that in prokaryotes. In eukaryotes there are more initiation factors as compared to prokaryotes where three factors are present. They are named by putting prefix e to signify eukaryotic origin. These factors are eIF-1, eIF-2, eIF-3, eIF-4A, eIF-4B, eIF-4C, eIF-4D, eIF-5 and eIF-6.

In eukaryotes the methionine formylation does not occur. Following steps are involved in the initiation.

(a) *Formation of met-tRNA-elongation factor 2-GTP ternary complex.* This complex is formed in two steps—in the first GTP binds to eIF-2 forming binary complex

GTP + EIF-2 ⟶ GTP eIF-2

In the second step above complex associates with met-tRNA

GTP-eIF-2 + met-tRNA ⟶ met-tRNA - eIF-2-GTP.

In mammals eIf-2 has three subunits eIF-2 α-which binds to GTP eIF-2, γ-which binds to met-tRNA and eIF-2β which may be a recycling factor.

(b) *Binding of above complex with 40 'S' ribosomal subunit.* The above complex associates with 40 'S' subunits of ribosome to form 40 'S' initiation complex.

met-tRNA-eIF-2-GTP + 40 'S' ⟶ 40 'S'-met-tRNA-eIF-2-GTP

(c) *Association of mRNA with 40 'S' initiation complex.* For this reaction eIF-3 is necessary. The binding of mRNA is assisted by other factors like eIF-1, eIF-4A, eIF-4B and ATP.

40 'S' -met-tRNA-eIF-2-GTP+mRNA ⟶ mRNA-40 'S'-met-tRNA-eIF-2-GTP.

(d) *Association of above complex with 60 'S' ribosomal subunit.* For this reaction eIF-5 is necessary. It releases eIF-2. The 40 'S' and 60 'S' subunits joining reaction depends on eIF-4C and GTP is hydrolysed to GDP. The above complex moves on mRNA strand towards 3'-end in the search of initiation codon i.e. AUG.

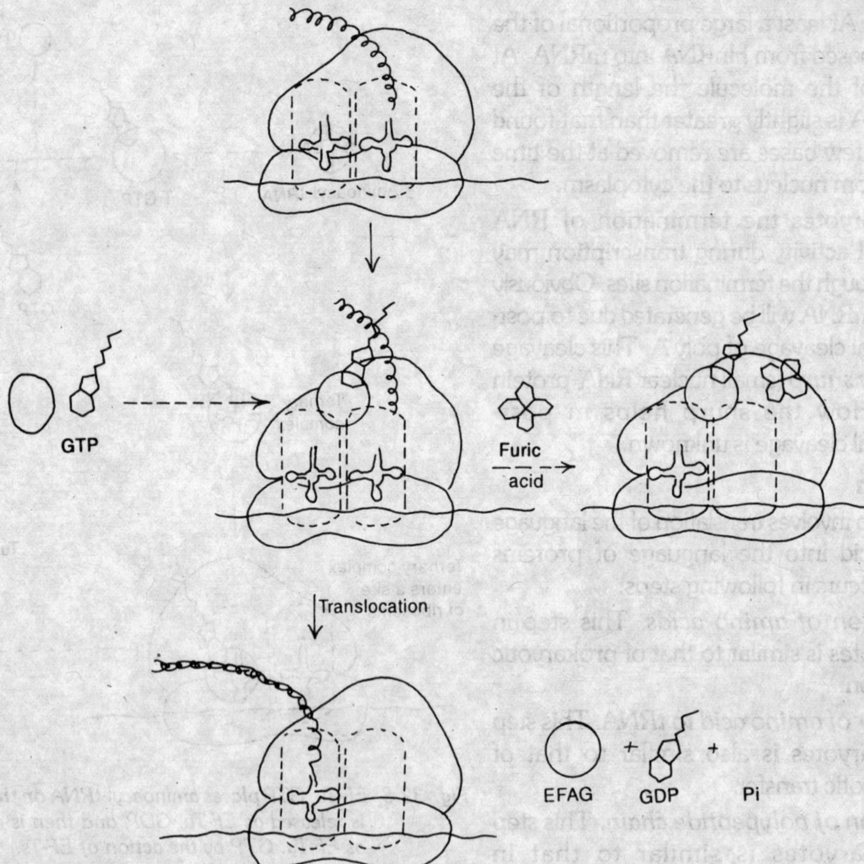

Fig. 32.9. After peptide bond formation, EF-G and GTP are needed for translocation, but then dissociate from the ribosome (unless fusidic acid is present to stabilize the post-translocation state).

mRNA-40 'S'-met tRNA-eIF-2-GTP + 60 'S' $\longrightarrow$ mRNA-80 'S'-met-tRNA + eIF-2 + eIF-3 + GDP + Pi.

The eIF-3 remains present in 40 'S' ribosomal subunit, eIF-2 and eIF-3 inhibit the association of 40 'S' and 60 'S' ribosomal subunits.

(iv) *Elongation of polypeptide chain.* This step in eukaryotes is similar to that of prokaryotes. In eukaryotes the elongation factor is called eEF-2. This factor is needed for translocation.

(v) *Termination of polypeptide chain.* Only one release factor is known called eRF-1. GTP is necessary for binding of eRF-1 to the ribosome.

33

GENETIC CODE

Genetic code is a sort of relationship between the nucleotide sequence in mRNA and amino acid sequence in protein molecule. The word "code" has several meanings. Out of them one is "a system of signals communicating." A pertinent example is the Morse code. In this chapter too we are dealing with the code, for messenger RNA having a means for communicating message that directs the combination of amino acids in specific sequence. *George Gamow* first proposed the coding scheme in 1954.

Several theories were porposed to explain the mechanism by which a particular sequence of nitrogenous bases in DNA by transcribing complementary bases in mRNA determines the position of specific amino acid in the protein molecule. The theory which is widely accepted till now was proposed by *F.H.C. Crick*. The theory holds the existence of a genetic code and its smallest unit which codes for one amino acid is known as *codon*. Thus, we see that the code in DNA is copied in the form of mRNA. But since little was known about the base sequences in DNA, the genetic code seemed a remote goal never to be attained. But in few years the genetic code become known as a result of ingenious experimentation mainly in the laboratories of *M.W. Nirenberg, S. Ochoa* and *H.G. Khorana*.

The Genetic Code is a Triplet Code

DNA contains four kinds of nucleotides (of A, T, G and C) and proteins are synthesized from 20 different types of amino acids. A basic problem regarding the genetic code was; how many bases of DNA specify one amino acid? In a *singlet code* each base or latter would specify one amino acid. Only 4 of the 20 types of amino acids would be coded unambiguously by a singlet code. In a two-letter or *doublet code* two bases would specify one amino acid. Here 16 (4 x 4) of the 20 amino acids can be specified, but there would be ambiguous determination of a number of amino acids. *A triplet* or three-letters code was first suggested by the physicist *Gamow* in 1954. According to the triplet code three letters or bases specify one amino acid. Thus 64 (4 x 4) distinct triplets of purine and/or pyrimidine bases determine the 20 amino acids. These triplets has been called *codons*. Since there are 64 codons and only 20 amino acids it is obvious that there are many more codons than there are amino acids, i.e. the code degenerate. Experimental evidence shows that the code is a triplet one and that 61 of the 64 codons code for individual amino acids during protein synthesis.

A
G
C
T

Single code

AA	AG	AC	AT
GA	GG	GC	GT
CA	CG	CC	CT
TA	TG	TC	TT

Double code

AAA	CAA	CAA	TAA
AAG	GAG	CAG	TAG
AAC	GAC	CAC	TAG
AAT	GAT	CAT	TAT
AGA	GQA	CGA	TGA
AGG	GGG	CGG	TGG
AGC	GGC	CGC	TGC
AGT	GGT	CGT	TGT
ACA	GCA	CCA	TCA
ACG	GCG	CCG	TCG
ACC	GCC	CCA	TCC
ACT	GCT	CCT	TCT
ATA	GTA	CTA	TTA
ATG	GTG	CTG	TTG
ATC	GTC	CTC	TTC
ATT	GTT	CTT	TTT

Fig. 33.1. Singlet code, a double code and a triplet code.

Properties of Genetic Code

Following properties of the genetic code have been studied and proved experimentally:

(i) Genetic code is non-overlapping

When *Gamow* first proposed the genetic code, he has suggested it to be an overlapping code. No doubt *Crick* had suggested a non overlapping code but the direct experimental evidence for a non-overlapping triplet code was given by *Khorana* and his co-workers in 1965.

On the surface, it appears wasteful for 64 triplets to stand for 20 amino acid, so attempts were made to hypothesize a triplet code in which the code was overlapping and so to reduce the number of codons to about 20. But a lot of shortcoming are there in this proposal. Let us examine the case of lysine, whose codons appear to be AGG and AAA, and that of phenylalanine, whose codons appear to be UUU and UUC. Since the code letters of these two amino acids are mutually exclusive, lysine could not follow phenylalanine, or *vice versa*, in any protein; this would consitute a forbidden combinations of amino acids in proteins. Any two or three or so on amino acid combinations are possible. Therefore it can be concluded that the genetic code is non-overlapping.

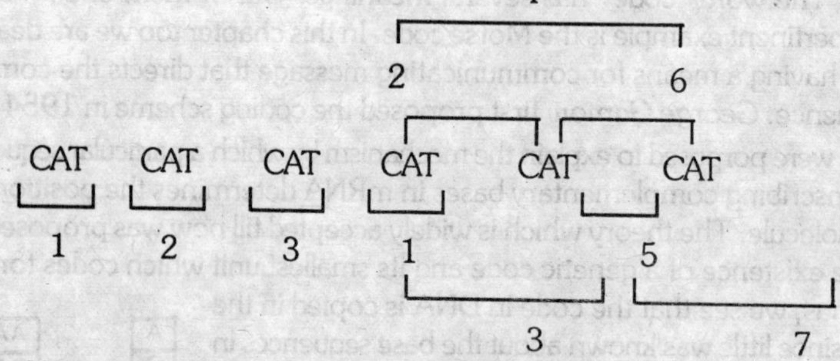

Non-overlapping Overlapping

(ii) Genetic code is degenerate

The code contains many synonyms, in that almost all amino acids are represented by more than one codon. For example, the three amino acids-arginine, serine and leucine-each has six synonymous codons. However, for many of the synonym codons specifying the same amino acid the first two bases of the triplet are constant, whereas the third can vary; for example, all codons starting with CC specify proline (CCU, CCC, CCA and CCG) and all codons starting with AC specify threonine. This flexibility in the third nucleotide of a codon may well help to minimize the consequences of errors.

(iii) Genetic code is commaless

A commaless code means that no punctuations are needed between any two words. In other words, we can say that after one amino acid is coded, the second amino acid will be automatically coded by the next three letters and that no letters are wasted for telling that on amino acid has been coded and that now second should be coded.

(iv) Universality of the Code

The code is bsed on work on the bacterium *E. coli* but it is valid for other organisms: This important characteristic of the genetic code is called its *universality.* Consider any codon that codes for the same amino acid from the smallest organism to the largest, plant or animal. Thus, UUU codes for phenylalanine and GUC for valine in all living organisms, from *Amoeba* to ape, bacteria to the banyan tree and from cabbage to kings. The genetic code which was first developed in the bacteria about 3 billion (300 crore) years ago has not undergone any change and has been preserved in its almost original form in the course of evolution. In other words, the code is conservative one.

The best evidence for the universality of the code comes from studies in which amino acid-changed tRNA from *E. coli* was placed with mRNA and ribosomes derived from rabbit reticulocytes. The result was the production of rabbit haemoglobin. Obviously, the bacterial tRNA was able to recognize perfectly the mammalian mRNA for haemoglobin and could work perfectly with mammalian ribosomes. Studies on amino acid sequence of homologous proteins from widely divergent and unrelated species (cytochrome—C from man, chicken, several bacteria and yeast etc.) show great uniformity although genetic divergence took place hundreds of millions of years ago. Similarly, studies with synthetic messenger RNAs and binding studies with tRNAs also point towards a universal code, though slight variations occasionally occur. Organisms with high G/C content in the DNA appear geared for more common reading of codons ending in C or G. Their proteins also are relatively rich in amino acids coded by G- and C-rich codons, for example Ala, Arg and Gly and relatively poor in Ile, Lys, Phe and Tyr. Certain mutant microorganisms show exceptions to the absolute universality of the code. These are mutants that have acquired ability to read nonsense triplets. However, this accessory ability is a only partial and nonsense triplets are nonsense much of the time. Some natural differences may exist among organisms. Initiation of polypeptide chains is encoded differently in bacteria and their bacteriophages (viruses) as compared to higher organisms. Thus, tRNA for cystine binds well with mRNA triplet UGA (nonsense in bacteria). Biological singificance of these observations remains to be seen as there is more to biology than the 64 triplets.

(v) The code has polarity

If a gene is to specify the same protein repeatedly it is essential that the code must be read between fixed start and end points. These points are the initiation and the termination codons, respectively. It is also essential that the code must be read in a fixed direction. In other words, the code must have polarity. It is obvious that if the code is read in opposite directions it would specify two different proteins, since the codons would have reversed base sequences.

For example:

	Leu	Ile	Val	Ser →
Codon-	UUG	AUG	GUG	UCG
	← Val	Leu	Leu	Ala.

(vi) Initiation codons

The starting amino acid in the synthesis of most protein chains is *methionine* (eukaryotes) or *N-formyl methionine* (prokaryotes). Methionyl or N-formyl methionyl-tRNA specifically binds to initiation sites containing the AUG codon. This codon is therefore called the *initiation codon*. Less often, GUG also serves as the initiation codon in bacterial protein synthesis.

(vii) Non-sense codons

The UAA (also called *ochre*), UAG (also called *amber* because in investigator who studied the properties of this codon belonged to the Bernstein family, and Bernstein means "amber" in German) and UDA codons do not specify any amino acid so, are called *non-sense codons*. They are also called *termination codons*, because, these codons are used by the cell to signal the natural end of translation of a particular polypeptide. However, their inclusion in any mRNA results in the abrupt termination of the message at the point of their location even though the polypeptide chain has not been completed.

Deciphering or Cracking of Genetic Code

We may briefly describe the techniques which ultimately resulted in determining the code words for each of the 20 proteinogenic amino acids. Most of our current knowledge of the general nature of the genetic code and of the nucleotide composition and sequences of the various codons has been obtained from the following five main types of experimental approaches:

1. *The first method* is analysis of the primary structures of the wild type and mutant proteins. In each of the mutant proteins so far analysed a mutant amino acid replaces a particular wild type amino acid. But the data so far accumulated on such amino acid replacements is very meagre and sets restrictions on the genetic code.

2. *The second method* for establishing the genetic code consists of determining the sequence of nucleotides in the mRNA and also of aminoacids in the protein molecule synthesized under its influence.

The sequence analysis for proteins is now a standard biochemical technique but unfortunately the determination of the base sequence of large nucleic acid molecules (like mRNA) is yet to be achieved. Biochemists, therefore, adapted indirect methods to crack the genetic code. A cell-free amino acid incorporating system and the polymerising enzymes were known which could polymerise single mononucleoside triphosphates or their mixtures to produce long nucleotide chains of random base sequence. Initial break-through in this direction was achieved by *Nirenberg* and *Mathei* (1961) who adopted the following technique. Using the enzyme *polynucleotide phosphorylase* they prepared a synthetic polyribonucleotide containing only one kind of base and added it to a cell-free amino acid-incorporating system from *E. coli*. In this way they prepared, for example, a poly uracil (UUU...UUU) to which they added different amino acids labelled with C^{14}. They found that this poly-U system specifically stimulated the formation of a polypeptide which contained only the amino acid phenylalanine. The conclusion was drawn that the code word for the amino acid phenylalanine was a sequence of three uracil nucleotides (UUU) in mRNA. This was the first demonstration that synthetic polyribonucleotides could act as a messenger for polypeptide synthesis. Later on, *Nirenberg* and *Ochoa* became deeply involved in this problem. They prepared poly-G, poly-C and poly-A polynucleotides from the riboside diphosphate GDP, CDP and ADP, and employing the same technique. They observed that these homoribonucleotides stimulated the incorporation of amino acids glycine, proline and lysine to the corresponding homopeptides respectively. *Crick* has already shown that the code was triplet in nature. It was, therefore, concluded that the base triplet GGG, CCC and AAA coded respectively for the amino acids glycine, proline and lysine.

Later on, mixtures of two more ribonucleotide diphosphates were used by *Nirenberg* and with the help of the enzyme mentioned above polyribonucleotides were prepared. Thus, using UDP and CDP in the ratio of 3:1 he obtained a polynucleotide which contained the following triplets in the descending order of frequency: UUU, UUC, UCU and CUU. The triplets containing 2Cs and 1U were the least frequent. With such a poly-UC system *Nirenberg* obtained a poly-peptide containing the amino acids phenyalanine and serine in the ratio 3:1. He concluded that the code word for serine contains 2Us and 1C. This method, however, did not give the exact sequence of the three bases. *Ochoa* also used this technique and determined the code words for most of the amino acids. The exact sequence of the three bases for any amino acid was not known because all these polynucleotides had random base sequence. The assigning of the exact triplet for every amino acid was not possible at this stage.

3. *The Third method.* The determination of the exact base sequence of mRNA triplets which coded for each amino acid was made possible by a later discovery of *Nirenberg* that each trinucleotide (triplet) caused a specific aminoacyl-tRNA to bind to the ribosome and to the specific base sequence of synthetic mRNA. With the help of this elegant technique and using trinucleotides of known sequences. *Nirenberg* and *Lader* (1965) observed that amino acyl-tRNA could be bound to ribosomes in the presence of specific trinucleotide phosphates (triplet). In the absence of a specific triplet the amino acyl-tRNA could not bind to the ribosome and therefore, passed through the nitrocellulose filter while the tertiary complex containing amino acyl-tRNA-ribosome-mRNA was retained on the filter. Using this technique, *Nirenberg* and his associates tested 64 triplets for binding of the 20 amino acyl-tRNA's. They succeeded in assigning 50 out of the 64 codons with their exact base sequences with which each of the amino acyl-tRNAs formed a union. This method was based on the assumption that binding required the same code words as did protein synthesis. This method, however, did not prove to be very satisfactory as binding stimulations observed in a number of cases were very poor and artifactual. Moreover, several authentic codons did not bind the amino acyl-tRNA.

4. *The fourth method-Khorana's direct method,* of confirming the genetic code is the most direct method and was devised by the Nobel Prize winner *Dr. H.G. Khorana*. This method consists *in vitro* chemical synthesis of short segments of DNA of known base sequence with the help of DNA *polymerase*. From this synthetic DNA a polyribonucleotide (RNA) of strictly defined base sequence is transcribed under the catalytic influence of *RNA polymerase*. A polypeptide is then synthesized under the direction of RNA as represented in the following schematic diagram:

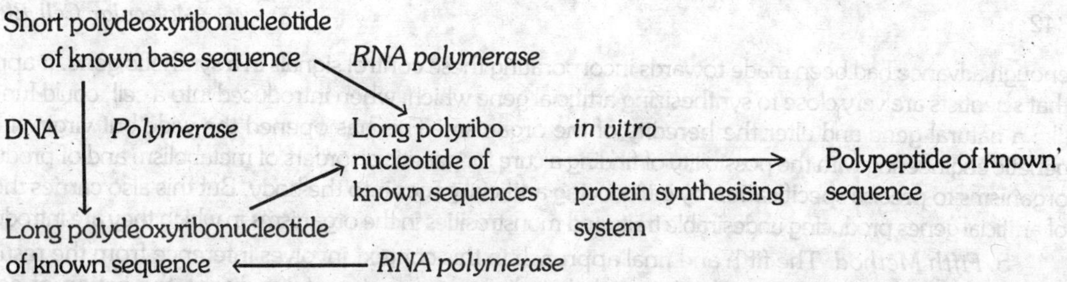

The amino acid sequence of the polypeptide so formed is then determined and correlated with the base sequence of DNA and RNA. These studies of *Khorana* and his co-workers with chemically defined messengers proved very conclusively that (1) the base sequence in DNA specifies the sequence of amino acids in proteins; (2) the information contained in DNA is conveyed through RNA and (3) the genetic code is triplet and non-overlapping in nature. The results obtained from these polypeptide synthesis in combination with the results obtained from the binding technique of *Nirenberg* and co-workers led to the deciphering and establishment of the genetic code.

First Letter	Second Letter: U	C	A	G	Third Letter
U	UUU(Phe)	UCU (Ser)	UAU (Tyr)	UGU (Cys)	U
	UUC (Phe)	UCC (Ser)	UAC (Tyr)	UGC (CYS)	C
	UUA (Leu)	UCA (Ser)	UAA (Ochre)	UGA ?	A
	UUG (Leu)	UCG (Ser)	UAG (Amber)	UGG (Tryp)	G
C	CUU (Leu)	CCU (Pro)	CAU (His)	CGU (Arq)	
	CUC (Leu)	CCC (Pro)	CAS (His)	CGC (Arq)	C
	CUA (Leu)	CCA (Pro)	CAA (Glun)	CGA (Arq)	A
	CUG (Leu)	CCG (Pro)	CAG (Glun)	CGG (Arq)	G
A	AUU (Heu)	ACU (Thr)	AAU (ASpN)	AGU (Ser)	U
	AUC (Heu)	ACC (Thr)	AAC (ASPN)	AGC (Ser)	C
	AUA (Aeu)	ACA (Thr)	AAA (LYS)	AGA (Arg)	A
	AUG (Met)	ACG (Thr)	AAG (Lys)	AGG (Arg)	G
	GUU (Aal)	GCU (Ala)	GAU (Asp)	GGU (Glu)	U
	GUC (Val)	GCC (Ala)	GAC (Asp)	GGC (Gly)	C
	GUA (VaL)	GCA (Ala)	GAA (Glu)	GGA (Gly)	A
	GUG (Val)	GCG (Ala)	GAG (Glu)	GGG (Gly)	G

Fig. 32.2 The triplet genetic code.

By this technique, *Khorana* not only worked out and confirmed the exact sequence of code words for all the amino acids but also clearified the roles of the 5' and 3' terminals of mRNA molecule. He showed that the translation of mRNA proceeded from its 5' hydroxyl end towards its 3'-hydroxyl end. His investigations also led to the artificial production of small segments of DNA molecule, thereby paving the way for the synthesis of artificial genes which could function in a living cell. In 1970, a Japanese student of *Khorana, Dr. Miss Otsuka*, succeeded in chemically linking six nucleotides by attaching an extra phosphoric acid artifically to a nucleotide in addition to the one it already had. By this process *Otsuka* and her co-workers at the Osaka University synthesized alanine tRNA. This synthetic RNA had exactly the same sequence of nucleotides as the one that occurs in natural cells. The same year Dr. *K.L. Agarwal,* another student of *Khorana,* succeeded in making an artifical gene. This gene, however, failed to function in a living cell because it did not possesed the necessary "start" and "stop" signals for its functioning. He reported that enough advance had been made towards incorporating these control signals in a synthetic gene. It appears that scientists are very close to synthesizing artificial gene which, when introduced into a cell, could function like a natural gene and alter the heredity of the organism. This has opened the wide but virgin field of genetic engineering with the possibility of finding a cure for genetic disorders of metabolism and of producing organisms to precise specification by introducing artificial genes into the body. But this also carries the risk of artificial genes producing undesirable traits and monstrosities in the organisms in which they are introduced.

5. *Fifth Method.* The fifth and final approach in this context involves inference from the results of purely genetical experiments. These included studies on mutant proteins due to the action of certain mutagens in reversing particular kinds of mutation; others involve incorporating two or more mutations in the same DNA strand and observing the effect on the phenotype. Such observations give us confidence in the genetic code as deduced from *in vitro* experiments. Moreover, extraction of tRNAs from cells and studies of the nucleotide sequence found in certain regions of these isolated tRNAs all coverage to give single, consistent picture of the code.

Wobble Hypothesis

It is sometimes argued that the third base of a codon is not very imporant and that specificity of a codon is particularly determined by the first two bases. It has been shown, for instance that the same tRNA can recognize more than one codon differing only at the third position. This pairing is not very stable and is allowed due to wobbling in base pairing at the third position. *Crick,* in 1965 proposed a hypothesis called *Wobble hypothesis* to explain this phenomenon. He discovered that if U is present at first position of anticodon, it can pair with either A or G at the third position of codon. Similar is the case with U and I (Inosone, modified base in tRNA) found in anticodon. The pairing relationships between first base of anticodon and third base of codon are given in Table 33.1.

Table 33.1. Wobble Pairing

Anticodon (first base)	Codon (third base)
U	A, G
C	G
A	U
G	U, C
I (inosine, resembles G)	U, C, A

This kind of wobbling allows economy of the number of tRNA molecules, since several codons meant for same amino acid are recognized by same tRNA. For instance, anticodon IGC can recognize codons GCU, GCC and GCA (coding for alanine); IGA can recognize UCU, UUC and UCA (coding for serine) and IAC can recognize GUU, GUC and GUA (coding for valine).

Contribution of *Khorona*

India born biochemist *Dr. Khorana* has devised an ingenous technique for artificially synthesizing mRNA with repeated sequences of known nucleotides. For this valuable contribution he was awarded Nobel Prize in 1970.

By using synthetic DNA, *Khorana* and his co-workers prepared chains of polyribonucleotides with known repeating sequences of two or three nucleotides as follows:

(i) Poly CUC UCU CUC UCU ...

(ii) Poly CUA CUA CUA CUA ...

In first case CUA and UCU are two codons arranged alternately in the polynucleotide chain. This dicates the formation of a polypeptide chain having two amino acids (leucine and serine) arranged alternatively. The second case is an example of *copolymers chains.* The polynucleotide chain is formed of repeated linkage of codon CUA. This dictates the formation of a polypeptide chain consisting of only one amino acid-leucine.

34

Regulation of Gene Action or Protein Synthesis

Genes are arranged in a linear fashion on the chromosomes. The number of genes present on a single chromosome varies from 1000-10,000 depending upon the size of the chromosome. In human being there are 90,000 functional genes. But all genes do not become functional in every cell or the body. Some genes function in one type of cell while some others in other type of cells. This regulation of genes was best described in certain microorganisms such as *E. coli* and certain higher organisms. *Jacob* and *Monod* proposed *Operon* to explain this sort of gene regualtion. Genes show their activities by synthetizing their related proteins. Thus the regulation of gene action or protein synthesis is one and the same thing. The entire mechanism can be studied at three stages.

(i) transcription stage,

(ii) translation stage,

(iii) at the release of polypeptide chain.

Certain enzymes are produced continuously by an organism regardless of the presence or absence of its substrate in the environment. Such enzymes are called *constitutive enzymes* and they require no regulation. The constitutive enzymes are generally those that are needed at all times. Thus the enzymes involved with glucose metabolism, enzymes of glycolysis and the hexose monophosphate pathway—are constitutive, and one can imagine that the metabolism of glucose is sufficiently central to the physiology of *E. coli* that these pathways should be continuously functional.

There are certain enzymes which are synthesized in certain specific conditions and require one or other types of gene regulation. For example, the metabolism of less common sugars such as lactose, arabinose, and glactose, is effected by *regulated enzymes,* and it is clearly economical to elaborate such enzymes only on those occasions when their substrates are present. Regulated genes also specify enzymes of most biosynthetic pathways-pathways for the synthesis of amino acids, purines, pyrimidines, and so on—and these enzymes are synthesized according to the demand for their biosynthetic products.

Induction and Repression

The substrate whose addition induces the synthesis of an enzymes is called *inducer.* The phenomenon involved in the production of new enzyme in the presence of inducer is called *induction*. For example in *E. coli* β galactosidase enzyme hydrolysis the lactose into glucose and galactose. A bacteria culture of *E. coli* growing in a medium with glucose produces only minute quantities of the enzyme β galactosidase. On addition inducer (lactose), the production of β galactosidase is, however, increased as much as 10,000 times and can easily be detected. The formation of β galactosidase can be slowed down quickly if the lactose is removed. Such as enzyme is called an *inducible* or *adaptive enzyme,* because its synthesis is induced by the inducer which is lactose in this case. Reverse activities have also been observed in *E. coli*. The amino acid *histidine* is synthesized from carbohydrate precursors in a culture of bacteria. All enzymes necessary for the synthesis of histidine are produced. If, however, histidine is added to the medium, synthesis of histidine stops. Such enzymes whose synthesis can be checked up by adding repressor are known as *repressible enzymes*. This phenomenon is called the *feedback mechanism*.

Operon Hypothesis

Jacob and *Monod* in 1961 put forwarded a hypothesis called *Operon hypothesis* to explain the regulating mechanism of expressions of regulated genes at transcriptional level. They analysed the genetic basis for induction and repressions. The hypothesis has the following components:

Structural Genes

The structural genes direct the synthesis of cellular proteins through messenger RNA (mRNA) and determine the sequence of amino acids in the proteins synthesized. There are as many structural genes in an operon as there are proteins or polypeptides under common control. Each structural gene may be controlled independently and transcribes a sequence mRNA molecule, or all the structural genes of an operon may form one long polycistronic or polygenic mRNA molecule, or all thestructural genes of an operon may form one long polycistronic or polygenic mRNA molecule. Polygenic RNAs are common in bacteria and bacteriophages. All known eukaryotes, however, contain only a single functional site for initiating protein synthesis.

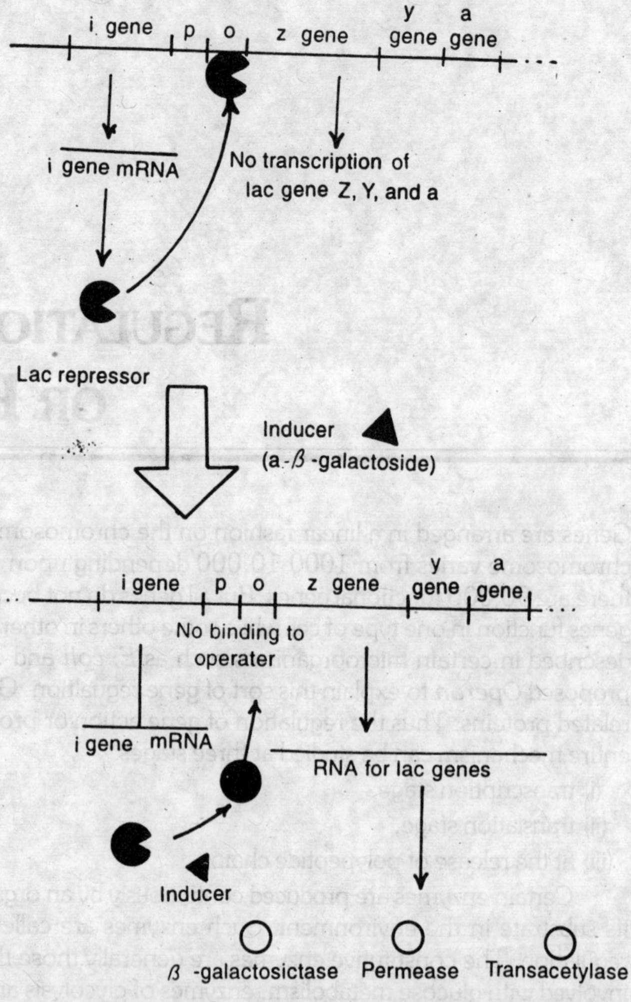

Fig. 34.1. The Jacob-Monod model for end product repression.

The lactose (lac) operon of the bacterium *E. coli* has been particularly well-studied. It is a small segment of a circular DNA molecule which constitutes most of the genetic material or the bacterium. It has three structural genes, z., y and a which transcribe one long polycistronic RNA molecule. This controls the synthesis of three different proteins (enzymes), β *galactosidase, galactoside permease* and *galactoside transacetylase*. These consists of 4, 1 and 2 units, respectively. β *galactosidase* is the enzyme that splits lactose into glucose and galactose. *Galactoside permease* facilitates the entry of lactose into the cell. *Galactoside transacetylase* appears in small quantities upon lactose induction. The initiation codon of structural gene z is TAC (corresponding to AUG of mRNA). It is located 10 base pairs away from the end of the operator gene.

Operator Gene

An operator gene present adjacent to the structural gene and represents a sequence of bases that interacts with a protein. The protein is also called a regulator molecule, and operator-regulator interaction will either prevent or promote transcription of the regulated gene, depending on which operator is being considered. The major distinction between an operator and a promotor is that the operator is not normally the site of attachment of RNA polymerase. Instead, it ordinarily lies between the promotor and the beginning of a regulated gene of an operon and thus it comes to influence whether a polymerase is able to progress from its attachment site to the gene itself.

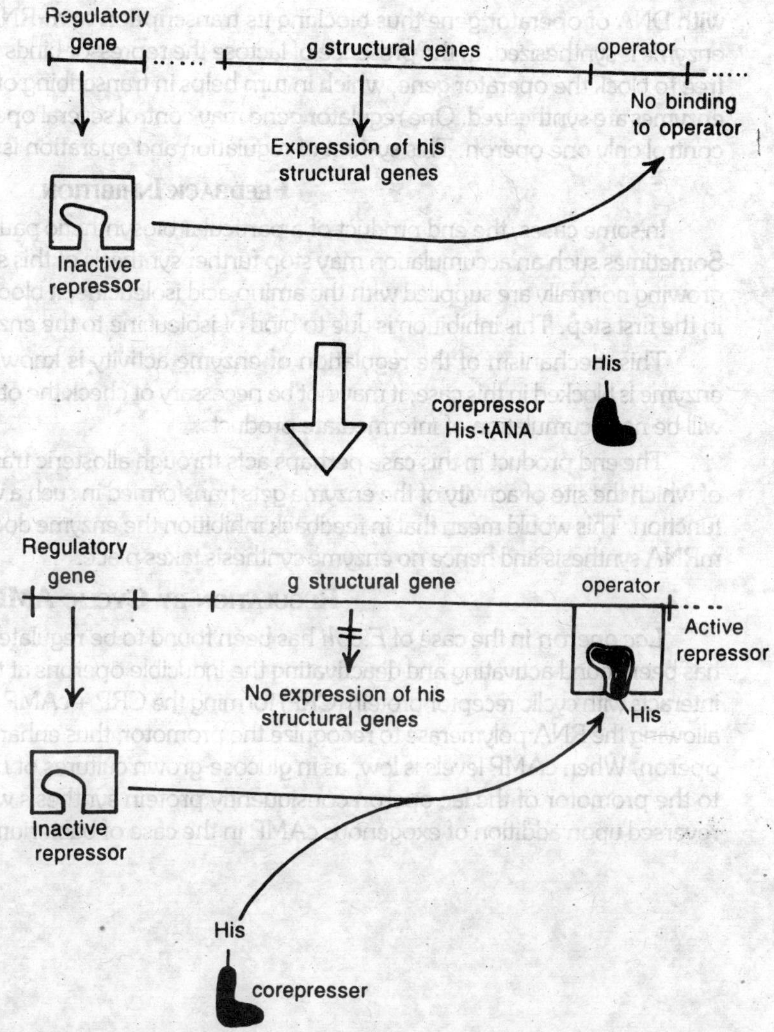

Fig. 34.2. Repression of gene expression in the histidine operon of salmonella typhimurium.

Promotor Gene

While operator gene is the site of action of repressor, the actual site of transcription initiation is different and is known as *promotor gene,* which lies to the left of operator gene. It is also believed, therefore, that perhaps RNA polymerase, binds to and moves from the promotor site. The repressor, when attachment to operator prevents the movement of RNA polymerase. The presence of the promotor gene was inferred from the mutations at this locus, which were found to affect the maximum rate of transcription.

In the recent years, promotor region has been studied in some detail and three sites are proposed (a) catabolite gene activator site (Cga), (b) initial binding site (ibs) and (c) operator gene (op). It is obvious that operator gene makes a part of promotor region.

Two proteins are known to be involved in the regulation of lactose operon, (i) lac repressor, (ii) catabolite gene activator protein (Cga protein).

While repressor binds to operator site, Cga protein binds to Ca site. Moreover, repressor exercises positive control by checking RNA polymerase to travel through operator. Cga protein exercises positive control. Only when cyclic AMP molecule activates Cga protein, the latter will allow RNA polymerase to bind.

Regulator Gene

The activity and working of operator gene depends on the activity of another gene called regulator gene. The regulator gene produces a substance called *repressor substance.* The repressor is an allosteric protein which has affinity of binding with the DNA. This substance controls the switching on and switching off the operator gene. The activity of this substance in turn depends on the presence or absence of lactose substrate on which β galactosidase enzyme acts. In the absence of lactose the repressor substance binds with DNA of operator gene thus blocking its transcription to m-RNA. Since no m-RNA is produced, no enzyme is synthesized. In the presence of lactose the repressor binds to the lactose, thus it does not remain free to block the operator gene, which in turn helps in transcribing of m-RNA by structural genes and thus enzymes are synthesized. One regulator gene may control several operons and several regulator genes may control only one operon. This system of regulation and operation is called a *Regular.*

FEEDBACK INHIBITION

In some cases, the end product of a particular biosynthetic pathway gets accumulated if not utilized. Sometimes such an accumulation may stop further synthesis of this substance. For instance, if *E. coli* cells growing normally are supplied with the amino acid isoleucine, it blocks the activity of the enzyme involved in the first step. This inhibition is due to bind of isoleucine to the enzyme threonine deaminase.

This mechanism of the regulation of enzyme activity is known as *feedback inhibition.* If the first enzyme is blocked in this case, it may not be necessary ot check the other intermediate enzymes, since there will be no accumulation of intermediate products.

The end product in this case perhaps acts through allosteric transformation of the enzyme as a result of which the site of activity of the enzyme gets transformed in such a way that it cannot perform its catalytic function. This would mean that in feedback inhibition the enzyme does not function, while in repression no mRNA synthesis and hence no enzyme synthesis takes place.

REGULATION BY CYCLIC AMP

Lac operon in the case of *E.coli* has been found to be regulated by cyclic AMP. Cyclic AMP (cAMP) has been found activating and deactivating the inducible operons at the level of the promotor. Cyclic AMP interacts with cyclic receptor protein (CRP) forming the CRP + cAMP complex which binds to the promotor, allowing the RNA polymerase to recognize the promotor, thus enhancing the transcription of the inducible operon. When cAMP levels is low, as in glucose-grown cultures of *E. coli,* the cAMP + CRP will not bind to the promotor of the lac operon consiquently protein synthesis will not be "switched on". This can be reversed upon addition of exogenous cAMP in the case of bacterium.

35

THE PROKARYOTES : BACTERIA

The main function of each individual bacterial cell, as for any organism, is growth an eventual replication. There is no tissue differentiation, hence no division of function. The ratio of surface area to volume is highest in the smallest organisms, hence bacterial chemical and metabolic activities are particularly efficient. Many of these activities comprise fundamental links in the geochemical processes that allow the presence and continuation of other forms of life and have considerable geochemical and evolutionary significance. Among these important bacteria processes are the fixation of atmospheric nitrogen, and the decomposition of organic matter to release fixed carbon as a requisite step in the carbon cycle.

Some bacteria are photosynthetic, their pigments localized in simple rounded particles or lamellae that lack the surrounding membrane of the complex plastids of other algae and green plants. Bacterial photosynthesis is less efficient than that of green plants, or even of the blue-green algae, and does not produce oxygen.

The Cell

The simplest and smallest bacteria are the true bacteria or Eubacteriales (*Schizosporales herein*). Morphologically they are the least complex of any living organism visible at the limits of the compound light microscope. Other bacteria variously resemble protozoa, algae, fungi, or *slime* molds.

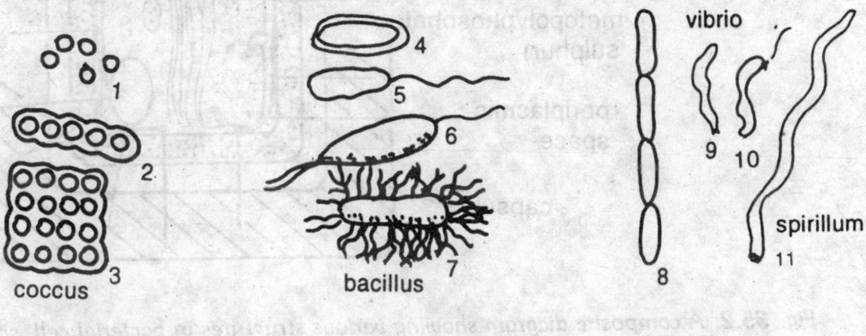

Fig. 35.1. Bacteria cell form.

There cell forms are characteristic of the true bacteria (1) the *coccus,* an ellipsoidal or spherical cell, averaging 0.5 to 1.0 µm in diameter; (2) the *bacillus* or rod, a cylinder or elongate ellipsoid about 0.5 µm in length; but rarely up to 10 times that length; and (3) the *vibrio*, a slightly larger elongate curved rod of a half turn, which is termed a spire or *spirllum* when it consists of several helices. The latter may be 1.5 µm in diameter and up to 13 to 14 µm in length.

Cell increase is normally by transverse fission, but the fission may not result in complete separation, and various types of cell aggregates are formed.

The protoplasm is chemically complex. The *mucleoid* (a small particle somewhat resembling a nucleus), cytoplasm, and cell wall contain a wide variety of organic compounds, largely protein, RNA (ribonucleic acid), carbohydrates, and lipids. The nuclear material is concentrated in the central part of the cell, although there is no well defined membrane-bounded nucleus. The remainder of the cell consists of uniform granular cytoplasm, with some *small* particles of protein and ribonucleic acid, the *ribosomes* that are the site of protein synthesis. Granules of fat, protein, or starch or globules of minerals such as sulphur or calcium carbonate also may be present.

Bacterial Size

The small size of bacterial cells is certainly obvious from microscopic examination. The dimensions of the smallest bacterial species border on the limits of the resolution of the bright-field microscope. The majority of pathogens ranges in size from approximately 0.2 to 1.2 μm in diameter, and 0.4 to 14 μm in length. The dimensions of particular causative agents of disease are given in the chapters in which they are discussed.

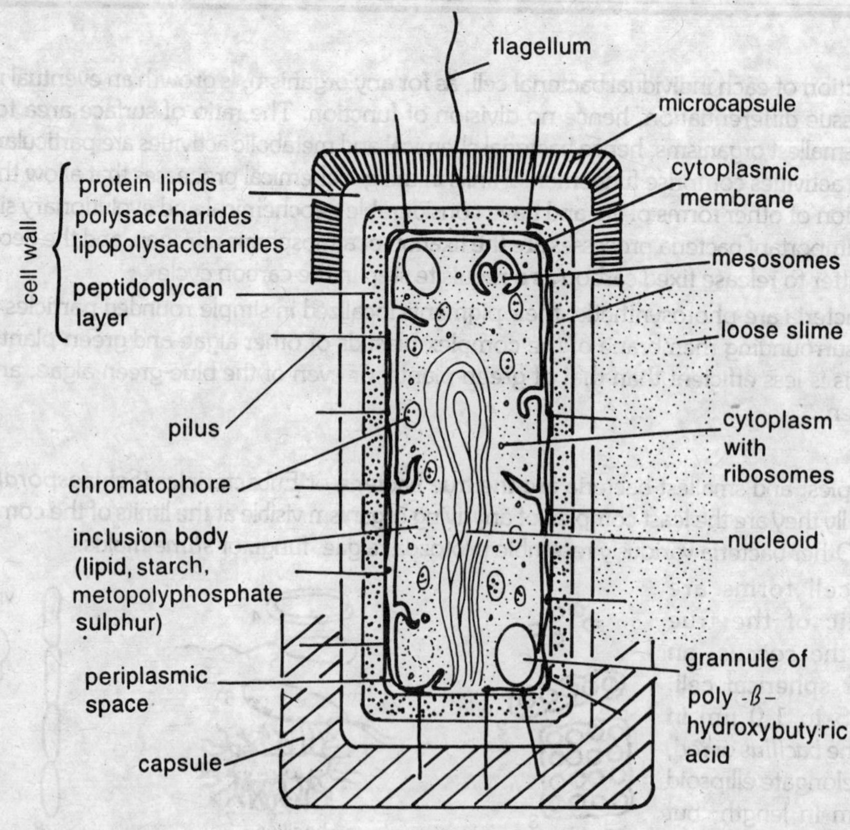

Fig. 35.2. A composite diagram showing various structures in bacterial cell, some of which are present only in certain groups of bacteria.

The size relationships of bacteria to the cells of higher forms of life. A comparison of certain cellular components also is made here. It is readily seen that bacterial cells approximate the size of mitochondria.

Structure Interior of the Cell Wall

Bacterial Plasma Membrane

This structure lies just beneath the organism's cell wall. It is a discrete and differentiated cellular component. The plasma membrane, in addition to serving as an osmotic barrier that passively regulates the passage of materials into and out of the cell, participates in the active transport of various substances into the bacterial cell. Active transport provides bacteria as well as other microorganisms with certain advantages, including the ability to maintain a fairly constant intercellular ionic state in the presence of varying external environmental ionic concentrations and the means with which to capture nutrient materials present in low concentrations in media. Membranes contain several types of transport systems for such substances as amino acids, mineral, ions, sugars and related compounds.

The bacterial plasma membrane also participates in the outward transport of molecular waste products, certain "building block" substances necessary in the formation of surface structures, and exoenzymes. The membrane - associated structures, e.g., mesosomes, are important to the energy - producing reactions of the cell.

Chemically, the plasma membrane of bacteria consists of both proteins and lipids. Sterol, characteristically is not found in the bacterial or blue - green algal membranes, but is a common constituent of higher forms. The exact structural arrangement of bacterial membrane remains undetermined at this time. However, they range from 5 to 8 mμ in thickness and comprise 10 to 20 percent of a bacterium's dry weight.

Mesosomes

Bacterial membranous structures were observed in *Mycobacterium avium* (the cause of tuberculosis in birds) by *Shinohara, Fukushi,* and *Suzuki* in 1957. Since the, many other investigators have found organelles of this type in various bacterial species. Several terms have been used to designate such membranous structures; these include chondroid, intracytoplasmic membrane, plasmalemmosome, and mesosomes. The last named terms thas been widely accepted the consequently will be used in the text book.

Mesosomes have been reported to occur primarily in species of various Gram-positive bacterial genera. Included among this group of organisms are members of *Bacillus, Lactobacillus Staphylococcus* and *Streptococcus*. Membranous structures similar to the mesosomes of Gram-positive are seldom seen in Gram-negative bacteria. The appearance of their organelles differs substantially. Mesosomes in Gram-positive bacteria generally appear as pocket-like structures and contain tubules, vesicles of lamellae (folded membrane arrangement). Gram-negative organisms primarily exhibit the lamellar form.

It is important to distinguish mesosomes from the extensive intercellular membranous system associated with the specific functions performed by nitrogen-fixing, nitrifying, and photosynthetic bacteria. The discussion which follows will concentrate on mesosomal structures of Gram-positive organisms.

Fine Structure and Function

The variations in the morphology of mesosomes observed in electron micrographs appear to be caused by the fixative and fixation procedure used in specimen preparation. In general, using *Lactobacillus plantarum* as an example, a mesosome pocket results from the in-folding of a membrane (invagination). Frozenetched preparations of bacilli have clearly shown that mesosomes are morphologically different from the cell's outer membrane. Chemical and functional differences have also been noted. A single coiled tube is believed to occupy the membrane invagination.

Functionally, mesosomes appear to be utilized in several bacterial processes, essentially, to increase the cell's membrane surface, which in turn increases enzymatic content. A relationship seems to exist between the "enzymatic needs" of cells and the formation of new mesosomes.

The activities with which mesosomes have been associated include: (1) cell wall synthesis; (2) division of nuclear material; (3) cellular respiration; and (4) sporulation. With respect to nuclear division, which involves the segregation of DNA, and cell wall synthesis, mesosomes do not seem to be indispensable. These procedures have been observed to occur in the absence of such membranous organelles.

Bacterial Nucleus (Chromatin Body, Nucleoid)

Prokaryotic cells lack of distinct structure known as the nucleus in eukaryotic cells. In addition, bacteria and blue-green algae, i.e., protokaryons, do not possess other features and structures characteristically associated with higher forms, such as a mitotic apparatus, nuclear membrane, and nucleus. The bulk of the DNA in prokaryons appears to be concentrated in one or several regions within them. No membrane specifically surrounds this concentration of nucleic acid. Because of this difference in the internal organization of the nuclear region, the DNA-containing structure are not considered to be true nuclei. The terms used to designated them vary, and include *bacterial nucleus, nucleoid, chromatin* and *nuclear body.*

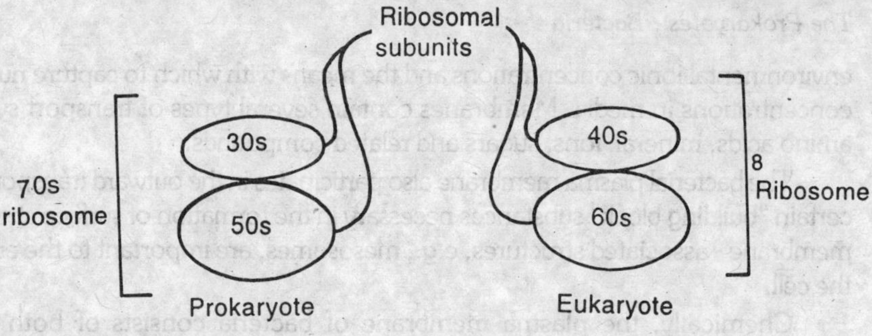

Fig. 35.3. Diagram of the structure of a ribosome.

Nucleoids can be observed by several means. With light microscopy such structures can be observed using special procedures such as the use of DNA-specific stains (Feulgen, or the fluorescent dye acridine orange) and the use of hydorchloric acid (HCl) or ribonuclease (RNAse) for the hydrolysis of the ribonucleic acid (RNA) in fixed bacterial cells. The latter technique tends to remove the RNA which is densely packed around the DNA of the cell's nuclear region. Thus, when a basic dye is applied after this treatment, the nucleoids can be readily recognized.

Ribosomes

In bacteria and in related microorganisms, e.g., blue-green algae, a large portion of the ribosomes appear to be free in the cytoplasm. In ultrathin sections these submicroscopic structures appear as fine granules within the cell's interior. Ribosomes can be obtained from bacteria by first disrupting them with acid then centrifuging the resulting homogenate at high speeds (100,000 x g) for 1 to 3 hours. The recovered bacterial spherical particles generally have a sedimentation coefficient of 70 S. Each ribosome is composed of one 30 S and one 50 S particle. By weight these ribosomes are somewhat smaller than those found in eukaryotic cells and other microbial types. They are estimated to account for approximately 40 per cent of a bacterium's dry weight. The ribosomes of green algae, fungi, and protozoa have values of 80 S. Such structures are made up of one 40 S and one 60 S particle.

Chemically, ribosomes are composed of approximately 40 percent protein and 60 percent RNA; 90 percent of bacterial cell's total RNA is associated with these structures.

As with other forms of life, ribosomes are important components of the protein synthesizing process. Groups of these units, held together by strands of messenger RNA, i.e., polyribosomes or polysomes are the sites where such activity takes palce.

Chromatophores, the Photosynthetic Apparatus of Protokaryons

Blue green algae and certain algae-like species of bacteria carry out the essential steps of photosynthesis in pigment-bearing granules. These structures differ from the complex cellular component, the chloroplast which is found in eukaryotic plants: photokaryons lack true chloroplasts. The photosynthetic membranes of bacteria, called *chromatophores,* and those of blue-green algae, *thylakoids* are not membrane-enclosed nor are they self-duplicating structures. Furthermore, they are heavier and larger than ribosomes. The organelles of a similar pigment-containing nature is brown algae, diatoms, and in dinoflagellates and called *phaeoplasts.* Red algae have *rihodoplasts.* The photosynthetic membranes of blue-green algae and several photosynthetic bacteria extend through these organisms in a parallel arrangement. Spherical vesicles (i.e. spindle-shaped bodies) connected by narrow constrictions also have been observed in some photosynthetic protokaryons. The photosynthetic apparatus of these micro-organisms in apparently oxidative and synthetic enzymes.

Photosynthetic bacteria do not contain pigments which are characteristically found in the blue-green algae or in higher plants, but possess *bacteriochlorophylls.* The chlorophylls found in these organisms are bacteriochlorophylls a—the main chlorophyll found in photosynthetic bacteria— and *b* and *chlorobium* chlorophylls 650 and 660. The latter two pigments are found in the green sulphur bacteria.

SELECTED CYTOPLASMIC INCLUSIONS

During respective growth cycles certain bacteria can accumulate several kinds of reserve materials, water-insoluble and water-soluble. Such accumulations, which are "non-living bodies," are called *inclusions,* and include metachromatic granules, lipids, polysaccharides, and certain inorganic substances. Some of these cellular inclusions are common to a wide variety of microorganisms, while others appear to be limited to a few, and can be extremely useful in the identification of certain bacterial species. For example, metachromatic granules are quite conspicuous as characteristically found in *Corynebacterium diptheriae,* the etiologic agent of diptheria. Obviously, such cytoplasmic inclusions are of diagnostic importance. The demonstration of these various accumulations can be achieved by several methods. The choice of procedures is generally determined by the chemical nature of the inclusion type in question.

Metachromatic Granules

These inclusions are found in several types of microorganisms, including algae, bacteria, fungi and protozoa. The actual chemical composition of metachromatic granules (also called Babes-Ernst granules or volutin) has been disputed for sometime. Certain investigators hold that they are polymerized inorganic phosphate. Others contend that the granules are composed of nucleic acid, lipid and protein. Metachromatic granules are considered to be used for the temporary storage of reserve food material.

Lipid Inclusions

Lipid droplets have been reported to occur in numerous bacteria, including species of the general *Azotobacter, Bacillus, Corynebacterium, Microbacterium, Mycobacterium* and *Spirillum*. Generally speaking, fat granules appear most regularly in Gram-positive species. These inclusions becomes more prominent as the bacterial cell ages. Lipids may take the form of either neutral fats or granules of polymerized β-hydroxybutric acid. The latter is an important "sole storage lipid" in certain organisms.

Polysaccharide Granules

Examples of these granules include glycogen and starch. Iodine solutions generally have been used for their demonstration. Those areas producing a bluish colouration are considered to be starch-containing, while those with as reddish-brown colour generally are designated as glycogen.

DORMANT BACTERIAL STRUCTURES (ENDOSPORES, CYSTS AND CONIDIA)

The production of dormant structures by certain bacteria is comparable to the process of feed formation in plants. According to several authorities, this type of activity may represent cellular differentiation at a primitive level. In bacteria, dormant structures of three kinds can be produced: *endospores, cysts* and *conidia.* The first two of these resting bodies are formed asexually, i.e., without the union of nuclear material from different types of cells. Usually only one such structure is produced per cell. Conidia present a different situation, as several are formed and are utilized for purposes of reproduction.

Several species of four bacterial genera are capable of forming endospores. However, this property is more common among the members of *Bacillus* and *Clostridium.* The rare occurrence of endospore formation in other genera suggests that the process was not a characteristic of very many organisms early in their evolutionary development. This point is borne out by the results of genetic studies which have demonstrated *Sporosarcina* and *Sporovibrio* to be more closely related to active spore-forming species than to the members of the genera of *Sarcina* and *Vibrio.*

A comparison of cross sections of endopores, cysts and conidia is present in Figure 34.4. The gross morphology of these structures appears to be quite similar, with their central cytoplasmic regions surrounded by thick outer layers.

Bacterial Endospores

An *endospore* is a dormant or resting structures formed within individual bacterial cells. This term refers to the specific structure present in a parent cell or sporangium during the spore formation period. Upon disintegration of these cells, the endospores are released or freed. The terms *exo-spore* and *free spore* are used to describe such structures.

The dehydrated state of a bacterial spore is such that its refractility, i.e., ability to bend light rays, and its physical density approach those characteristic of dried protein. Additional properties of spores include an unusual resistance to the effects of heat, drying, chemical disinfection and radiation, and an impermeability to common strains. Several of these characteristics are used not only to detect the presence of spores, but to determine when these structures give rise to bacterial cells.

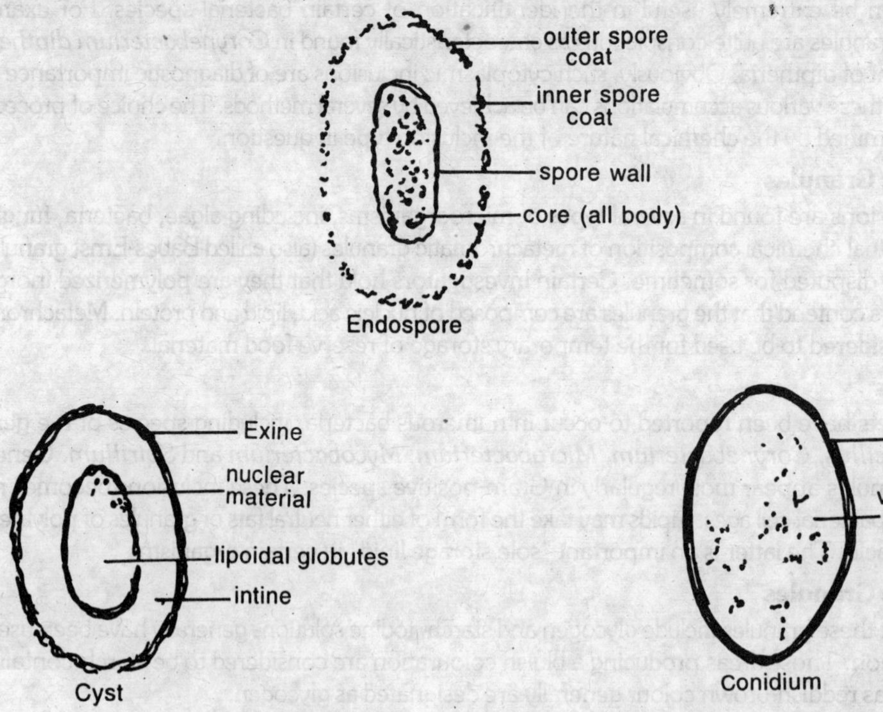

Fig. 35.4. Representative cross-sections of bacterial dormant structures.

The process by which bacteria form spores is called *sporulation,* whereas the process by which spores develop into bacteria is called *germination*.

In certain species of the genus *Bacillus,* the process of sporulation is accompanied by the deposition of an inclusion known as a *parasporal body* or *Restkorper.* One such inclusion body occurs in a cell and usually appears as a single large crystal. The chemical composition of such structures is protein. The function of parasporal bodies to the bacterial cells in which they form and their relationship to the process of sporulation is unknown. However, these structures have been used in the control of certain insects.

Bacterial Cysts

The cysts produced by *Azotobacter* sp. and various myxobacteria in general are somewhat spherical and possess a contracted cytoplasm and thick structured wall. The cytoplasm of *Azotobacter* sp. usually contains evident nuclear material, lipoidal globules and electron-dense bodies. The thick wall of this organism consists of an inner layer, or *intine,* and an outer layer, or *exine*. Cysts of *A. agilis* are not heat-resistant, while those of another species, *A. chroococcum*, exhibit resistance comparable to bacterial spores. As in the case of bacterial spores, cysts are resistant to drying and are formed singly within vegetative cells. Unlike bacterial spores, they are not highly refractile, and do not possess an obvious cortex or dipicolinic acid (DPA).

Conidia

The dormant structure of the actinomycetes is the *conidium,* an asexual spore which is formed at the end of aerial mycelia by a process of fragmentation. These resting bodies are comparable to the conidia produced by various genera of fungi (e.g., *Aspergillus* and *Penicillium*). A mild thermal resistance appears to be the only common characteristic of conidia and of bacterial spores. Among the differences know to exist between these two types of dormant bodies are the absence of a cortex, DPA, a lack of refractility, and the external formation of multiple conidia.

The Bacterial Surface

Within recent years, the surface of bacterial cells have attracted the attention of various microbial specialists, including biochemists, geneticists, immunologists, molecular biologists, and physiologists. The results of numerous diverse and intriguing studies have provided knowledge of the transport of substance across surface barriers, the formation of surface structures, the effects of antibiotics and the respective roles of bacterial surface components in the causation, diagnosis, and prevention of infectious diseases. This section will describe the currently recognized surfaces, which *Salton* place into the following categories: (1) surface appendages (flagella and pili); (2) surface adherents (capsules and slimes); (3) the cell wall; and (4) the protoplast membrane.

FLAGELLA

These filamentous appendage of bacteria are responsible for their motility. It was the movements of bacteria which aroused the interest of *Leeuwenhoek* and drew his attention to the very existence of these microorganisms. In addition to being observed microscopically, motility today can be demonstrated in various semisolid media i.e., those preparation which contain approximately 0.3 percent agar. Motile organisms are recognized by the visible spreading of their growth throughout the medium.

In studying or identifying a bacterial species, determining whether the organisms are motile or not can be significant. Care must be taken to distinguish true movement from the quivering ro to-and-fro motion known as *Brownian movement.* The later is caused by a bombardment of the bacteria by various molecules of the fluid in which they are suspended.

Table 35.1. Selected Characteristics of Bacterial Dormant Structures

	Structures		
Property	*Endospore*	*Cyst*	*Conidia*
Heat resistance	Characterstically Present	May be present	To a limited degree
Cortex	Present	Absent	Absent
Dipicolinic acid (DPA)	Present	Absent	Absent
Number formed per cell	1	1	Formation in chains

Table 35.2. The Occurrence of Dormant Structures in Bacteria Genera

	Structures	
Endospores	*Cysts*	*Conidia*
Bacillus	Azotobacter	Norcardia
Clostridium	Myxococcus	Actinomyces
Sporosarcina	Sporocytophaga	Streptomyces
Sporovibrio		Micromonospora

The formation or synthesis of flagella and their associated activity do not necessarily correlate with other bacterial physiological properties. However, it appears that flagellation does bear a direct relationship to growth rate. Both the logarithmic and maximum stationary phases of a growth cycle favour flagellar synthesis. Furthermore, *Leifson* indicated that several factors may affect flagellation, among them the chemical composition of media, pH, and the liquid or solid state of the medium. More flagellation occurs in liquid preparations. This finding also holds true for the formation of another type of filamentous appendage, pili (fimbriae).

The existence of locomotor organelles, i.e., flagella, has been inferred, according to *Van Iterson,* since the mid- 1800s. Yet their visualization was not possible without the aid of electron microscopy and the Liefson flagella stain, used in light microscopy.

The flagella stain of *Einar Leifson* generally has been satisfactory for the demonstration of the majority of bacteria. In this procedure tannic acid, basic fuchsin, methylene blue, and sodium chloride are applied to bacterial smears. The methylene blue is used as a counterstain. In preparations utilizing this dye, bacteria stain blue and the flagella are red. However, in general all portions of cells appear red.

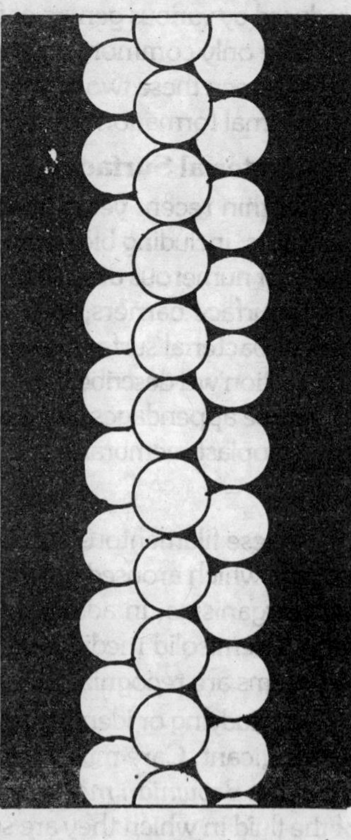

Fig. 35.5. A model showing three helically wound strands in a bacterial flagellum.

Flagellar Shape and Arrangement

These locomotor organelles are extremely delicate structures, as indicated by the relative case with which they can be detached from their bacterial cells. Flagella in a stained preparation generally exhibit an undulating appearance, and are long and slender. With most bacterial species the shape of the organelles is fairly uniform and constant.

The thickness of flagella has been observed to vary from species to species. In addition, differences as to the number and arrangement of these locomotor organelles exist between bacterial species. Some organisms do not passes flagella, a condition referred to as *atrichous* while others may have one (*monotrichous*) or several (*Multitrichous*).

The terminology of classification has varied. Table 35.3 presents two sets of terms which are currently in use. It will be observed that classification scheme I is general, while scheme II is somewhat more specific. Note that certain designations are common to both systems.

Flagellar Ultrastructure

A typical flagellum possesses a relatively uniform diameter along its length. However, near the region where this locomotor organelle attaches to a *basal granule* or *body,* just beneath the cytoplasmic membrane, a thickened, hook-shaped portion, the *basal hook,* can be observed. Flagella are believed to originate in the basal body. In certain negatively stained preparations of various bacterial species, basal flagellar ends were found to be connected with a broadened body, designated by *Hoeniger* and *Van Iterson* as a *collar,* which in turn was connected by means of a constricted region, or neck, to a disc- or cup-shaped part. The latter structure, which may have a paired disc appearance, is considered by certain investigators to be a detached segment of the basal body. The collar is believed to be a cell wall fragment. It should be noted here that a flagellum originate in the cytoplasmic regions and pierces the cell wall as it emerges from the bacterial cell. In short, the locomotor organelles of bacteria are not appendages of the cell wall.

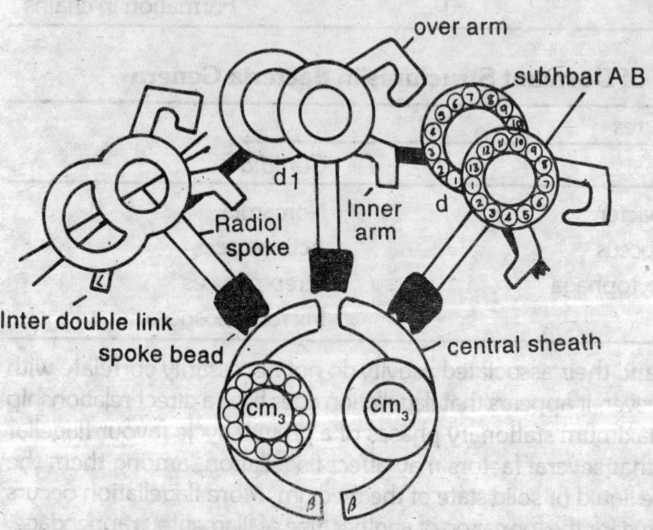

Fig. 35.6. Diagram representation of a cross section of a cilium or flagellum.

Bacterial flagella are much thinner than the cilia of vertebrates or the flagella of protozoa. The organelles of bacteria generally range in thickness from 12 to 15 nm (120 to 150Å). Furthermore, bacterial flagella do not possess the 9 (peripheral) + 2 (central) subfibrils characteristically found in the cilia and flagella of higher forms. Electron micrograph has demonstrated that the flagella of bacteria consist of an interwining of three parallel protein fibers in a triple helical structure.

Chemically, the aforementioned fibers are composed of a single kind of protein, called *flagellum*. The molecular weight of flagellin is relatively low, approximately 20,000 to 40,000. A unique amino acid, ε (epsilon) -N-methyl-lysine has been associated with this protein compound. Although a similarity in the amino acid composition and molecular weight exists among the flagellins obtained from different bacterial species, these compounds are by no means identical. This fact is demonstrated by the immunological specificity, i.e., the production of different antibodies, associated with flagella preparation from different bacterial species, sub-species, and strains. From a medical diagnostic standpoint, such differences are important in the identification of certain pathogens.

Table 35.3. A Comparison of Classification Schemes Based on the Number and Arrangement of Flagella.

Scheme Terminology	*Brief Explanation*
I Atrichous	Absence of flagella
Monotrichous	One flagellum at one end of a bacterial cell
Amphitrichous	One flagellum located at each end of the organism
Lophotrichous	Two or more flagella at one or both ends of a bacterium
Peritrichous	Flagella surrounding the bacterial cell
II Polar	Two categories exist: (a) *Monotrichous,* the presence of a single flagellum at one or both ends of the bacterium. (b)*multitrichous,* two or several flagella at one or both ends of the cell In both categories, the bases of individual flagella are parallel to the long axis of the bacterium to which they belong.
Subpolar (rarely found)	Two categories also are recognized in this division. However, the distinguishing point is that the bases of individual flagella are situated at right angles to the long axis of the bacterial cell (a) *monotrichous,* a single flagellum located near the pole or end of the cell (b) *multitrichous,* several flagella near the pole of the bacterium
Lateral	Flagella associated with this category appear to arise predominantly from the middle portion of the bacterial cell (a) *monotrichous,* one flagellum (b) *multitrichous,* several flagella in the form of a tuft originate from the mid-portion of the bacterium
Peritrichous	A random, haphazard arrangement of flagella scattered around the bacterial cell
Mixed	The presence of two or more flagella exhibiting distinctly different physical properties in different regions of the bacterial cell.

The motility observed with flagellated bacteria is believed to be associated with certain mechanical changes in the basal body. The mechanism or mechanisms responsible for movements of the nonflagellar species, such as spirochetes, certain blue-green algae, and some slime bacteria (the myxobacteria), and not well understood.

Chemotaxis

Bacterial motility is controlled by a process called *chemotaxis*, whereby cells tend to swim toward higher concentrations of certain nutrients (chemoattractants) or away from gradient of repellents. This movement is controlled by the direction of flagellar rotation. Bacteria swim in a straight line when the flagella are rotating counterclockwise (viewed toward the cell) and tumble randomly when the flagella rotate in a clockwise direction. The ability to change the direction of flagellar rotation permits the organism to respond to attractants and repellants. A swimming bacterium can be seen to travel for a short time in a straight line (counterclockwise rotation) followed by a period of tumbling (clockwise rotation), which is then followed by the cell swimming off in a new direction. The period of clockwise rotation probably does not cause the flagellar bundle to fly apart, but rather it causes a different sort of helical wave to pass down the bundle, thereby flipping the cell in a new direction.

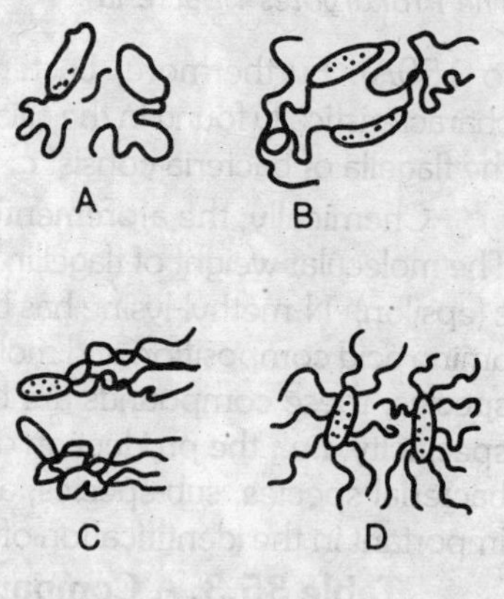

Fig. 35.7. Diagram representation of a cross section of a cilium or flagellum.

When a cell is moving toward a source of a chemoattractant (or away from a repellent), tumbling occurs less frequently and straightline motion predominates (i.e., clockwise rotations are suppressed). Conversely, a cell that finds itself moving into a lowering concentration of attractant will tumble more frequently until a more favourable direction is found. Bacteria sense and respond only to changes in concentration. If the attractant is present at the same concentration is the medium, the bacteria will swim as randomly as if the attractant were absent.

The mechanism whereby bacteria sense and adapt to their environment is a fascinating example of simple sensory control, involving transmembrane signaling and protein methylations. A few of the sensory system for chemotaxis have been well defined. These system include a protein that spans the cytoplasmic membrane and has binding sites for specific chemoattractants and repellents on its exterior face. Somehow, the binding of chemoeffectors to their binding sites transmits a signal to the internal portions of the protein. This change in proteins conformation is the signal that provides an assay for the presence and concentration of the chemoeffectors. The mechanism whereby this signal is transmitted to the flagellar basal body to control the frequency of the switching of direction of flagellar rotation is not yet clear. The signal appears to be transmitted through the cytoplasm, rather than being a change in membrane potential.

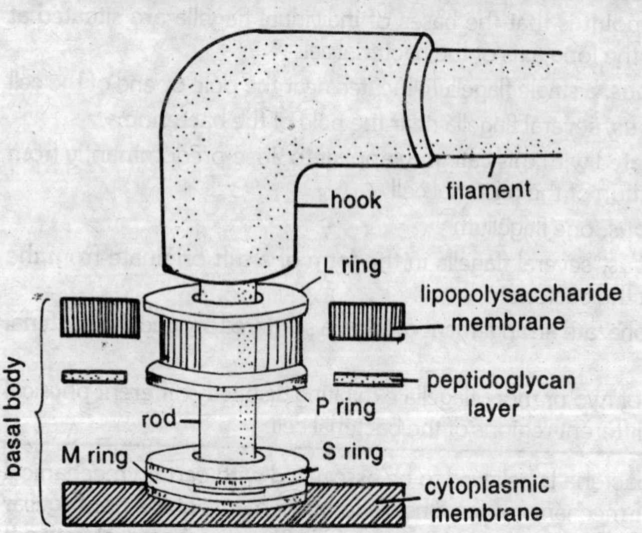

Fig. 35.8. Interpretive sketch to show the attachment of rings to cell wall and cell membrane of a Gram-negative bacterial cell.

Chemotaxis senses are gradient of chemicals; in other words, it adapts to the presence of a constant concentration of any chemoeffector. The mechanism of adaptation involves the methylation of the signal receptor proteins described above. These proteins are termed MCPs (for methylated chemotaxis protein) because the extent of their methylation increased when attractants are present and decreases when they are removed. Thu, the ability of these MCPs to serve as a substrate for a methylation system, using S-adenosyl-methionine as donor, is determined by the conformation of the protein, which was itself determined by the level of chemoeffector present at that time. Thus, the degree of methylation of the MCPs provides a measure of the external concentration of the chemoeffectors specific for that protein at an earlier time. If the concentration has changed because the cell swam into a region of different concentration, then the conformation of the protein will be different from that which would be appropriate for its degree of methylation. The degree of methylation is continuously updated to reflect the concentration of its chemoeffectors. In this way, the MCPs serve both as signal for the presence of a specific chemoeffector and as the site for comparison, through the degree of methylation, with the concentration of the effector at a previous time.

Phase Variation

Some bacteria are able to switch from the production of one antigenic type of flagella to another. This reversible genetic change, called *phase variation*, was first observed in *Salmonella*. This periodic alteration of a surface structure helps the bacterium to avoid elimination by the host's immune system.

Fimbriae (Pili)

Fimbriae are straight filaments, thinner and shorter than flagella, extending out form the surface of the cell. They are present almost exclusively on gram-negative cells where they provide a means for adherence to other cells, either bacterial or animal. Fimbriae are an example of a class of surface structures termed *adhesins* that allow attachment of the bacterial cell to many cell surfaces. As will become clear in the descriptions of virulent bacteria, they are extremely important for the bacterium's survival in an animal host. They originate in the cytoplasmic membrane and are composed of self-aggregating protein monomers (pilin).

Although their basic functions are the same, there are several types of fimbriae. Fimbriae or common pili are distributed over the entire cell surface, often densely (100-200 per cell). They are lectins, which means they possess the ability to bind to specific sugar residues present on glycoproteins and to other surface components on other cells. These functions can be detected by measurement of the adherence of the bacteria to epithelial cells or by their ability to promote hemaggulatination. In some cases, one can experimentally prevent the adhesion of the bacteria to red blood cells, cultured cells, or tissue surfaces by the addition of large amounts of the specific sugar recognized by the fimbriae. Monnose is one of the targets for the more common fimbriae, and the adherence of some bacteria to cell surfaces can be competitively blocked by the presence of mannose or a mannoside. In these cases, the bacterium is said to possess a mannose-sensitive hemagglutinin or adherence factor. Other bacteria exhibit mannose-resistant adherence.

Table 34. 4. Monosaccharide Components from *Streptococcus Pneumoniae* Capsules

Types	*Constituent Monosaccharides*
1.	Galacturonic acid and amino sugars
2.	L-Rhamnose, D-glucose, D-glucuronic acid
3.	Glucose, glucuronic acid
4.	Glucose, N-acetylamino sugar
5.	Galactose, glucose, rhamnose
6.	Galactose, glucose, rhamnose, amino sugar
7.	Glucose, glucuronic acid, galactose
8.	Glucose, amino sugar, uronic acid
9.	Galactose, glucose, N-acetylglucosamine
10.	Rhamnose, glucose.

As discussed earlier, the presence of fimbriae is intimately associated with the ability to cause disease, and the loss of fimbrial production may result in a loss of virulence. Examples include *Neisseria gonorrhoeae,* in which freshly isolated virulent strains produce fimbriae that bind to specific ganglioside, receptors on the surface of buccal mucosal or cervical epithelial cells and thereby prevent the bacteria from being washed out by bulk flow. A similar situation occurs in the case of enterotoxigenic strains of *Escherichia coli* in which the ability to cause diarrhea requires the production of two factors, an enterotoxin and a colonization factor representing a type of fimbriae whose binding is not inhibited by mannose.

Fimbriae also allow adherence of bacteria to one another. This is evident in the clumping of cells on the surface of a liquid medium, allowing the cells better access to air. Another important process in which bacteria adhere to each other is the genetic exchange mechanism called conjugation. In this case, cell adherence is mediated by a pilus, which is a plasmid-encoded surface structure called a sex pilus. These pili are present in much smaller numbers per cell and are usually longer than the adhesion fimbriae described above.

The Capsule or Sheath

The cell is enclosed by three layers, the inner or cytoplasmic membrane, the cell wall (which may have more than one layer), and an outermost covering of slime. The innermost, flexible, easily deformed and differentially permeable cytoplasmic membrane is composed of lipids and lipoproteins. A rigid cell wall of distinctive composition surrounds the membrane. The outermost layer of slime varies in thickness and composition with the environmental conditions or from species to species. A thick, gummy, and mucilaginous slime layer around a single cell is termed a *capsule;* that surrounding a filament is termed the *sheath,* Eubacterial capsules contain different chemical substances, ranging from a complex of polysaccharides to a simple polypeptide of a single amino acid. One species has a slime layer of cellulose, forming a loose fibrillar network. Separate strains of a single species may have capsules of several different strain-specific polysaccharides.

In habitats with a paucity of dissolved nutrients, the slime layer may form a *stalk*. The simplest stalks are for the attachment of a single cells; others may be much ramified or form a rosette with a cell at the end of each of the radially arranged stalks. Stalks may be very elongate and slender, or may be twisted. They may be wholly organic, organic with an iron-encrusted surface, or composed entirely of ferric hydroxide. An example of the latter is *Gallionella*, which grows only in iron-bearing waters, either fresh or marine. Other species deposit magnanese compounds on the slime surface.

Cell Wall

A cell wall is characteristic of the plant kingdom. In higher algae and plants it is composed of cellulose, a polysaccharide or polymer of glucose. The cell wall of fungi consists of chitin, a polymer of an amino sugar. In contrast, the unique bacterial cell wall consists of a mucocomplex substance of two amino sugars (one of these, muramic acid, was known only in the bacteria until recently identified in blue-green algae), and three or four amino acids, and may also contain lipids and proteins. Two or three layers in the wall may have different compositions. Gram-negative species (see succeeding section on the Gram reaction) have very thin walls only about 10 nm in thickness. The thin wall is seen in electron microscopy to consist of two electron-opaque layers separated by a less dense layer. The wall consists of 5 to 10 percent mureins (mucopeptide) and 20 to 30 percent lipids, with varied amounts of protein, lipoprotein, and lipolysaccharide, the latter component having important endotoxic and antigen properties

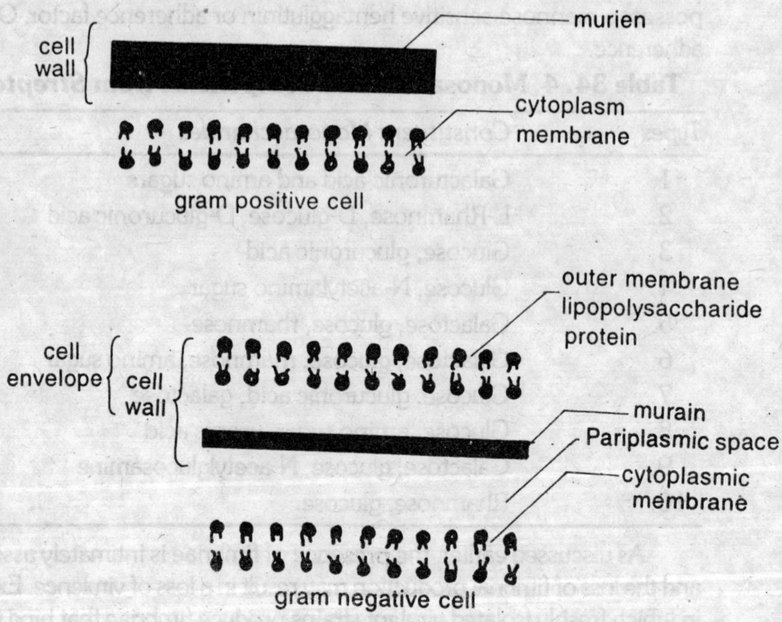

Fig. 35.9. The Gram positive cell wall has a relatively thick murein layer. The Gram negative cell wall has a thin murein layer but also has an outer membrane and additional lipopolysaccharides and proteins not present in Gram positive cell walls.

(*Glauert* and *Thornley*, 1969). Gram-positive species have thicker walls (20 to 35 nm) that appear three-layered and resemble those of blue-green algae. They contain no lipids, have a much smaller number of amino acids, but have two amino sugars; mucopeptides range from 50 to as much as 90 percent of the dry weight.

Amino acids are important in the formation of proteins, and all amino acids known to occur in proteins are found in the bacterial wall. In higher organisms, the amino acids occur in the cytoplasm and nucleus, but not as major contituents of the wall.

The rigid cell wall is not deformed by an osmotic pressure of up to 300 pounds per square inch, and its function appears to be wholly mechanical, giving the cell its shape. If the cell wall is removed by laboratory procedures, the naked protoplast will become spherical; it will continue to carry out all metabolic activities, although the semipermeable cytoplasmic membrane then will be very sensitive to osmotic changes.

In electron microscopy the different layers of the wall may have different and distinctive surface patterns. Excluding the plasma membrane at the inside and the slime layers at the outside, the outside, the other wall layers have patterns ranging from the wavy, undulating surface of the Gram-negative *Escherichia coli* (Migula) Castellani and Chalmers, to a linear pattern in *Caulobacter* and hexagonal patterns in *Spirillum* and *Azotobacter.* Some *Bacillus* have a hexagonal pattern and others have tetragonal one; *Lampropedia* has an outer echinate layer held by fine fibers to an inner punctuate membrane, both layers showing different hexagonal surface patterns. The extent of such surface features in the bacteria is not yet known as delicate and readily lost in the process of preparation for microscopy.

Bacterial respiratory enzymes are found in the plasma membrane, whereas in other organisms these enzymes occur in special organelles, the *mitochondria,* within the cell. Intrusive tubules or lamellae may arise from the plasma membrane of some species; the peripherally arrangement chromatophores of the purple nonsulphur bacteria arise in this way. Similar invagination of portions of the wall and plasma membrane, containing the amino acids, enzymes, and so forth, may have given rise to such cell organelles of the higher protists as the mitochondria and plastids, although an alternative theory suggests that they arose by endosymbiosis of entire prokaryote cells.

Peptidoglycan

Peptidoglycan, a structure unique to bacteria, is responsible for the shape and structural rigidity of the cell. Its basic structure is shared by all procaryotic cells except the mycoplasmas and some archaebacteria. Peptidoglycan is an enormous polymer composed of long polysaccharide chains cross-linked by peptide bridges. Each polysaccharide chain consists of repeating disaccharides of N-acetylglucosamine and N-acetyl-muramic acid. The latter amino sugar is composed of N-acetyl-glucosamine carrying a lactyl group attached to the C_3 hydroxyl group. To the carboxyl group of the N-acetymuramic acid is linked a peptide, which, although originally synthesized and attached as a pentapeptide, is normally found in the completed peptidoglycan as a tetrapeptide. The tetrapeptides are often cross-linked to one another, thereby joining adjacent glycan chains into a strong, rigid structure.

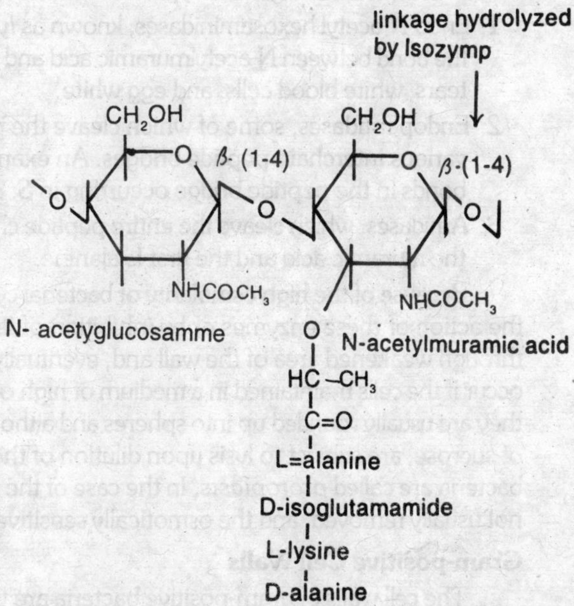

Fig. 35.10. ***Structure of the repeating unit of cell wall peptidoglycan as it exists in Staphylococus aureus.***

The compositions and linkages of the peptide chain varies in different organisms, but several common features are apparent. The peptide consists of alternating L and D amino acids. Its amino end (linked to the N-acetyl muramic acid) comprises L-alanine linked to D-glutamic acid, which is joined through its γ-carboxyl in an iso-peptide bond to the third amino acid. The third amino acid contains two amino groups; in gram-positive bacteria, it is usually lysine or, less frequently, ornithine, In gram-negative bacteria, this third diamino acid is most commonly diaminopimelic acid. The fourth amino acid is always D-alanine.

The peptide is initially synthesized as a pentapeptide containing an additional D-alanine residue, which is removed when one peptide chain becomes cross-linked to another. The bridge between peptide joins the free carboxyl group of the remaining terminal D-alanine on one glycan chain to the free amino group of the third diamino acid on the adjacent chain. In all gram-negative bacteria and some gram-positive ones (mycobacteria, the nocardia, *Bacillus* and *Corynebacterium*), the peptide bond is formed directly between these two amino acids. In other gram-positive bacteria, however, these amino acids are linked together by an additional peptide bridge, which allows more distant glycan chains to be cross-linked together. The composition of these cross-bridges is quite variable; some eamples are shown in Table 35.5.

Table 35.5. Representative Compositions of Peptidoglycan Peptide Bridges

Organism	*Peptide Bridge*
Staphylococcus epidermis	$Gly_{2\text{-}4} \rightarrow$ L-Ser $\rightarrow$ Gly
Streptococcus sp.	L - Ala $\rightarrow$ Gly $\rightarrow$ L-Ala
Streptococcus salivarius	Gly $\rightarrow$ L-Thr
Lactobacillus viridescens	L-Ser $\rightarrow$ L-Ala
Lactobacillus cremoris	L-Ala $\rightarrow$ L-Ser
Arthrobacter cirreus	L-$Ala_2 \rightarrow$ L-Thr
Micrococcus roseus	L-$Ala_3 \rightarrow$ L-Thr

Lytic Enzymes

There are a number of enzymes that cleave the peptidoglycan, resulting in bacterial lysis. These enzymes fall into three general types:

1. Endo-N-acetyl hexosaminidases, known as *lysozymes,* which cleave the glycan chains by hydrolyzing the bond between N-acetylmuramic acid and N-acetylglucosamine. Such enzymes are found in serum, tears, white blood cells, and egg white.
2. Endopeptidases, some of which cleave the peptide cross-links at D-alanine, while others hydrolyze various interchain peptide bridges. An example of the later is *lysostaphin*, which splits the glycine bonds in the peptide bridge occurring in *S. aureus.*
3. Amidases, which cleave the entire peptide chain from the glycan chain by splitting the bond between the muramic acid and the first L-alanine.

Because of the high osmolarity of bacterial cytoplasm, disruption of the peptidoglycan layer, either by the action of these enzymes or by inhibition of its synthesis, results in the bulging of the cell membrane through weakened area of the wall and, eventually, lysis of the cell. Under these conditions, lysis does not occur if the cells maintained in a medium of high osmolarity. Thus, cells lacking their walls can be obtained; they are usually rounded up into spheres and although stable in the presence of high concentrations of salts of sucrose, are subject to lysis upon dilution of the osmoprotectant. The wall-less forms of gram-positive bacteria are called *protoplasts.* In the case of the gram-negative organisms, all envelope components are not usually removed, and the osmotically sensitive forms are called *spheroplasts.*

Gram-positive Cell Walls

The cell walls of gram-positive bacteria are thick (20 nm - 80 nm) and relatively homogenous. They consist largely of peptidoglycan (50%-90%) that is extensively cross-linked in all directions. The remainder of the wall comprises primarily cell wall polysaccharides. These include teichoic acid and related polymers as well as non-teichoic acid and acidic and neutral polysaccharides. Proteins are occasionally present and may have considerable biological importance.

Other Components

As described, the basic structure of the gram-positive cell wall is a thick mesh of peptidoglycan interlaced with teichoic acids. Other components present in certain organisms may have important antigenic or medical properties. Example are the branched, group-specific rhamnose polymers present in many streptococci and the M protein of the group A streptococci.

Gram-negative Cell Walls

The cell envolpe of gram-negative bacteria is considerably more complex than the wall of gram-positive bacteria. It is a multi-layered structure composed of an outer membrane lying external to a thin peptidoglycan layer, which is separated from the cytoplasmic membrane by the periplasmic space. The whole envelope contains four major constituents: peptidoglycan and, in the outer membrane, phospholipids, proteins, and a unique glycoplipid called lipopolysaccharide (LPS).

ORGANISATION OF THE OUTER MEMBRANE

Various components are organized into a membrane of unusual, highly ordered structure and a typical functional properties. Like all membranes, the outer membrane consists of two lipid leaflets with opposing hydrophobic regions. LPS is located exclusively in the outer leaf and is essentially the only lipid in that layer, whereas phospholipid is present almost entirely in the inner leaf. LPS is much less fluid than are phospholipids, partly because the diglucosamine backbone is relatively large and the fatty acids tens to be more highly saturated than in the phospholipids, and partly because the LPS chains are cross-linked to one another through Ca^{++} or Mg^{++} salt bridges (or polyamines) existing between the phosphate groups in the core sugar region. Moreover, the major outer membrane proteins are complexed to each other, to the LPS and to the murein lipoprotein, resulting in a very stable structure. All of these properties make the outer membrane surprisingly rigid for a membrane, and the movement of proteins of LPS molecules in the plane of the membrane is unusually slow.

Properties of the Outer Membrane

The structure of the outer membrane imparts two unusual properties to most gram-negative bacteria that are not observed in other cells. Since the membrane is so rigid and because regions of phospholipid-phospholipid bilayer are rare, the membrane is not readily penetrated by hydrophobic compounds and is therefore resistant to dissolution by detergents. Thus, many gram-negative becteria are able to grow in the presence of detergents that are lethal for gram-positive cells. Treatment of gram-negative cells with metal chelators disrupts the salt bridges between outer membrane components and results in a decrease in the barrier function against antibiotics and detergents, concomitant with loss of about half of the cell's LPS. Similarly, mutations deleting certain major outer membrane proteins or those that affect synthesis of the inner core sugars or phosphate groups on LPS also decrease this barrier function. Presumably, in these cases, the outer surface is not as tightly cross-linked, or the absence of large amounts of protein is compensated by the appearance of phospholipid in the outer layer, providing a route for entry of hydrophobic molecules.

In contrast to the poor passage of hydrophobic compounds, the outer membrane is unusually permeable to hydrophilic compound. Nutrients diffuse through pores or channels formed by the porin proteins. The porin proteins associate in the outer membrane as trimers, and the diameter of the resulting channel imposes a size limitation on the nutrients that a cell can employ. In the enteric bacteria, this cut-off point is at a molecular weight around 600, allowing entry of molecules only up to this size of trisaccharides or tetrapeptides. These pores exhibit little steric specificity, and most appear to be negatively charged, since anions enter more slowly than cations. Some of the minor proteins in the outer membrane provide specific transport systems for nutrients whose passage through the pore might be too slow. Examples include nucleosides, dexitrins, iron chelates, cabalamin and phosphate.

Movement of Substances Across the Cytoplasmic Membrane

The primary function of the cytoplasmic membrane is to regulate the flow of material into and out of the cell. There are several different ways by which molecules can enter or leave a cell, including diffusion and active transport. It is the properties of the membrane that regulate these different transport mechanisms, establishing which materials can be translocated across the membrane. Organisms use different transport mechanisms for moving various biochemicals into and out of the cell and have different capabilities for transporting molecules across the cytoplasmic membrane.

When there is a difference in the concentrations of a solute on either side of a membrane, water will move across the membrane from the region of lower to the region of higher concentration of solute until the concentrations of the solute are equalized on both sides of the membrane, or until a pressure force prevents further flow of the water. The process by which the water crosses the membrane in response to the concentration gradient of the solute is known as osmosis. Osmosis may be viewed as a form of passive diffusion in which a small water molecule can cross a membrane, but the larger solute molecules cannot freely move across this barrier. This process exerts a pressure known as the osmotic pressure on the membrane, which represents the force that must be exerted to maintain the concentration differences between solution on opposite sides of the membrane.

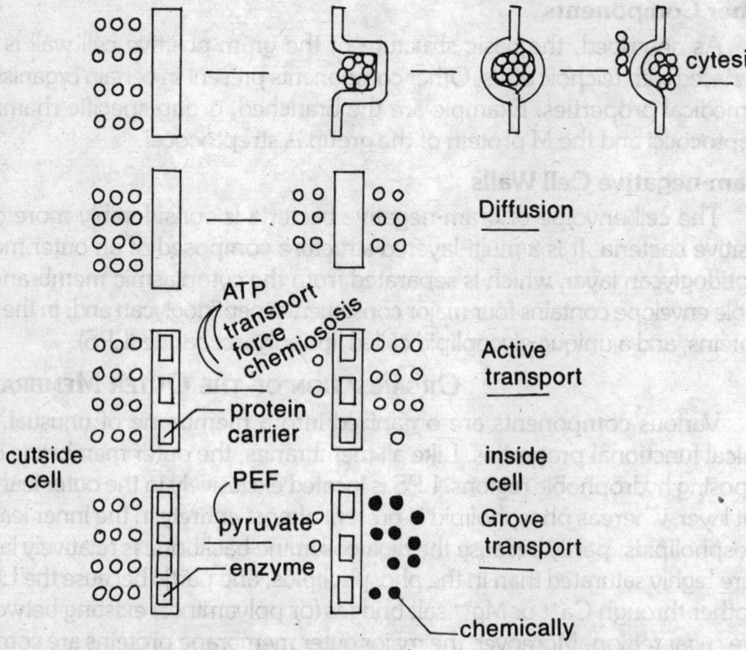

Fig. 35.11. This figure illustrates several ways for substances to cross a cell membrane and enter a cell.

Now let us examine the consequences of osmosis for a microbial cell. In a medium where the solute concentration inside the cell is equal to the solute concentration outside the cell, water will flow equally in both directions across the membrane. However, in a medium where the solute concentration is higher (hypertonic) outside the cell than inside the cell, water will tend to flow out of the cell and the cell will tend to shrink. The reverse is true if the cell is in a medium where the solute concentration is lower (hypotonic) outside of the membrane than inside of the cell, in which case water will tend to flow into the cell, causing the cell to expand and—if unrestricted—the cytoplasmic membrane to burst. Microoganisms usually find themselves in this later situation because the various concentration of the cell normally are in considerably higher concentrations than the solutes in the dilute aqueous medium in which most microorganisms exist.

In addition to water, various small solute molecules can diffuse passively across the membrane. Passive diffusion occurs when a solute molecule moves from an area of higher concentration on side of the membrane to an area of lower concentration on the other side of the membrane. The rates of passive diffusion are determined by the concentration gradient; the greater the gradient, the more rapid the movement. The rates of simple diffusion across the cytoplasmic membrane are not rapid enough for many of the exchanges between biochemicals and surroundings that must be accomplished by a cell, and microorganisms have developed additional membrane transport mechanisms.

Some molecules can move across a membrane at higher rates than would occur by simple diffusion in a process referred to as *facilitated diffusion*. Facilitated diffusion involved some of the proteins associated with the cytoplasmic membrane. Some of the membrane proteins— called carrier proteins or permeases— selectivity increase the permeability of the membrane for specific substances. In both simple and facilitated diffusion movement is form a region of higher concentration to a region of lower concentration.

Substances may also move across the membrane against a concentration gradient by active transport. Active transport can be likened to a pump that requires energy to move water uphill. The cell must actively work to move substances up a free energy gradient and therefore, unlike diffusion, active transport requires that the cell expand energy to move biochemicals across the membrane against a concentration gradient. As with most energy-requiring activities of living systems, the energy for active transport may come from the conversion of ATP, (adenosine triphosphate) to ADP (adenosine diphosphate). ATP, a universal carrier of energy in living systems, is one of several energy sources used by different organisms to drive active transport. The active transport capabilities of a cell determine in large part what substrates can move more into a cell and thus help define the metabolic potential of the cell.

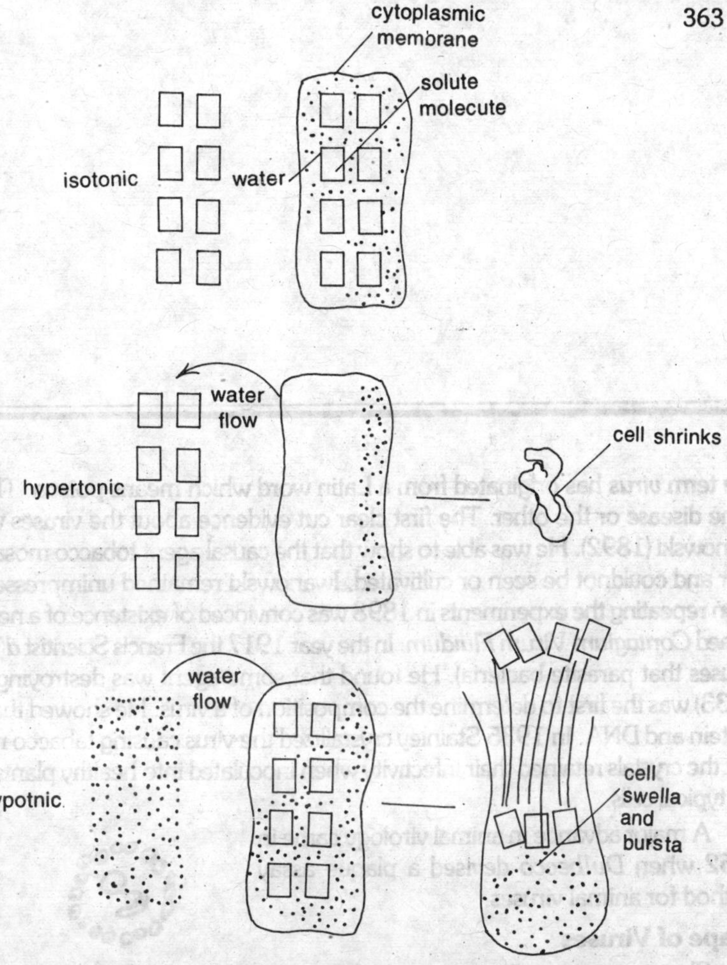

Fig. 35.12. Water flows into and out of cells because of osmosis.

Additional Functions of the Bacterial Cytoplasmic Membrane

Because the prokaryotic cell lacks internal membrane-bound organelles, several functions associated with such organelles in eukaryotic organisms are associated with the cytoplasmic membrane in prokaryotes. For example, the cytoplasmic membrane of bacteria plays a critical role in the generation of ATP during respiration. Specific proteins associated with the cytoplasmic membrane of the prokaryotic cell establishe channel through which hydrogen ions are pumped to establish a concentration gradient. In accordance with fundamental chemical principles, hydrogen ions will move by diffusion toward a region of lower concentration, which in this case would be to move back across the membrane into the cell. The return flow of ions into the cell is channelled through protein pores associated with the enzyme ATPase. This enzyme catalyzes the formation of the energy-rich compound ATP.

The process that drives the formation of ATP is known as *chemiosmosis*. It was first hypothesized by *Peter Mitchell* in generation of ATP by both cellular respiration and photosynthesis. In eukaryotic cells, the chemiosmotic generation of ATP during oxidative phosphorlyation—a process in cellular respiration—occurs within a specialized organelle—the mitochondrion - during respiration and the chemiosmotic generation of ATP during oxidative photophosphorylation—another process for generating ATP—occurs within the chloroplast during photosynthesis. The involvement of membranes in these ATP-generating processes is essential, as it provides the necessary semipermeable boundary across which an electrochemical gradient can be established. By performing work to pump hydrogen ions across the membrame, the return flow can be efficiently coupled with the synthesis of ATP. The process works like an electricity generating station, where water is pumped up across a dam and the return gravitational flow of water routed through specific channels so that it turns a turning. The work of turning the turbine generates electricity that can be used elsewhere. In the case of a cell, the ATP formed at the membrane by chemiosmosis can then be an energy source within the cell.

36

THE VIRUSES

The term *virus* has originated from a Latin word which means poison. This word is always associated with some disease or the other. The first clear cut evidence about the viruses was given by the Russian botanist Iwanowski (1892). He was able to show that the causal agent tobacco mosaic passes through a bacteria proof filter and couldnot be seen or cultivated. Iwanowski remained unimpressed by his discovery but *Beijerinck* upon repeating the experiments in 1898 was convinced of existence of a new form of infective agent which he named *Contagium Virum Fluidum*. In the year 1917 the Frencis Scientist *d' Herelle* discovered bacteriophages (viruses that parasite bacteria). He found that some agent was destroying his culture of bacilli. Schelsinger (1933) was the first to determine the composition of a virus. He showed that a bacteriophage consists of only protein and DNA. In 1935 Stainley crystallized the virus causing tabacco mosaic disease, and demonstrated that the crystals retained their infectivity when inoculated into healthy plants. He thus shared that viruses were like typical cells.

A major advance in animal virology came in 1952 when *Dulbecco* devised a plaque assay method for animal viruses.

Shape of Viruses

Shapewise, most viruses would fall in any one of the following categories: (a) *spheroidal*, also known as *cuboidal* (b) *elongated* or *rodlike* and (c) *mixed* or combination particles with shapes which are an admixture of the first two types. These groups in their turn may be with or without envelops. Apart from these possible groups there are a few viruses with shapes which do not fall into any of the above categories, like for instance, the 'brick shaped' pox viruses. Their shapes may be considered to be exclusive to a certain extent.

Shape of viruses species falling into a particular category may show variation amongst themselves. For instance rod shaped particles could be simple elongate (tobacco rattle virus), coir shaped (tobacco mosaic virus), flexuous (Coliphage fd, potato virus X) or much coiled (vesicular stomatitis virus). Similar variation within spheroidal types are also quite common.

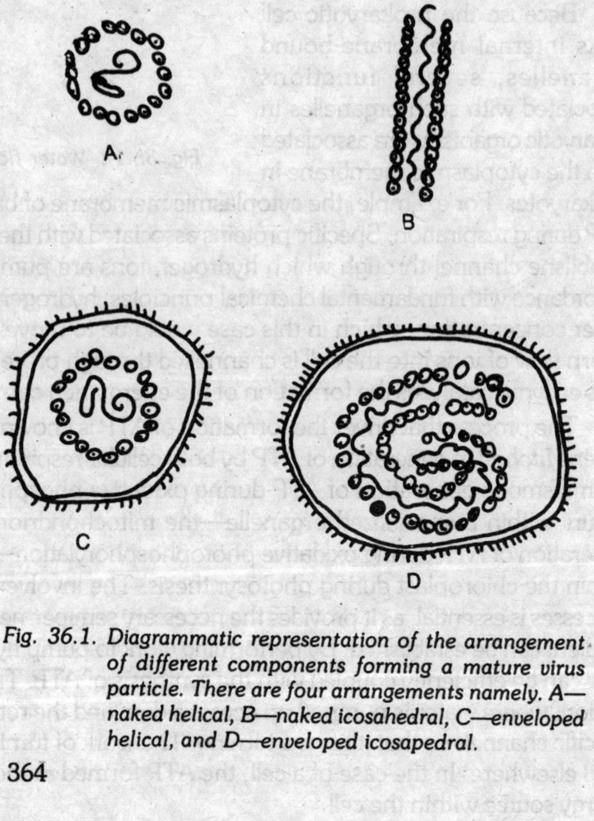

Fig. 36.1. Diagrammatic representation of the arrangement of different components forming a mature virus particle. There are four arrangements namely. A—naked helical, B—naked icosahedral, C—enveloped helical, and D—enveloped icosapedral.

The most widely used and effective method for the determination of shape of virus particles is electron microscopy. Utilising the various techniques involved in this method, particularly the technique of negative staining, it had been possible to gain an insight not only into the broad morphology but also into the internal arrangements of virus particles.

Physical information based on X-ray crystallography *Brenner et. al.* and sedimentation profile analyses chemical data have been of significant importance in the elucidation of ultra-structure of viruses.

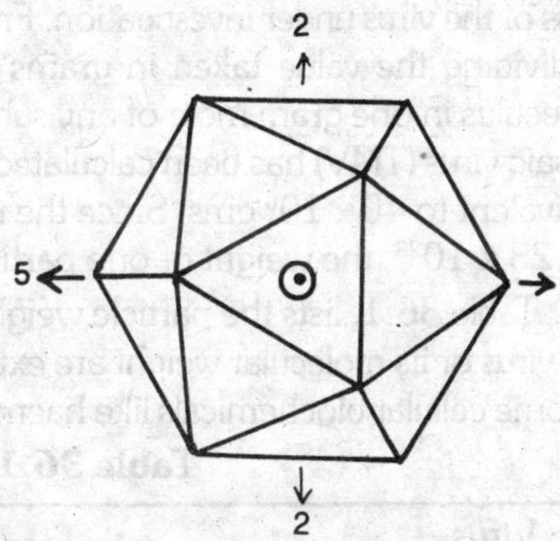

Fig. 36.2. Diagram of an icosahedron showing axes of rotation the reveal 2-, 3- and 5- face symmetry.

Size and Visibility

Except for certain "large" viruses (e.g., smallpox, vaccinia), a major distinguishing property of viruses is their invisibility under even the best optical microscopes (magnifications up to 2000 x). However, electron microscope, which can give clear resolutions at magnifications of several hundred thousand times, make even the smallest viruses and their innermost structures readily visible.

An idea of the degree of magnification implied by 200,000 x may be gained by imagining the letter O on this page to be so magnified. Approximately 2.5 mm. in diameter, the letter would appear as an oval 200,000 x 2.5 = 500,000 mm. = 500 meters or nearly one half mile in diameter: a large race track! Magnified 2000 x by an optical microscope the O would appear only about 16 feet in diameter.

Size and Filtrability

Iwanowski, *Walter Reed* and, later, many others demonstrated what was though, until about 1930, to be a uniquely distinctive property of viruses: the ability when suspended in fluids to pass through porcelains filters of such fine porosity as a retain all known bacteria. However, it is now known that under some experimental conditions bacteria can readily be made to pass through the same porcelain filters by special means, since size is not the only factor that determines filtrability through silicious filters. Surface electrical charge, which can be altered experimentally, is also very important. The "filtrable" viruses, however, cannot pass through the same procelain filters by special means, since size is not the only factor that determines filtrability through silicious filters. Surface electrical charge, which can be altered experimentally, is also very important. The "filtrable" viruses, however, cannot pass through filters made from collodion or cellulose acetate (called *membrane filters*) if the filters are made with sufficiently fine pores. Thus the term filtrable, once widely used to distinguish viruses, is not longer applicable without appropriate qualifications.

Certain details of structure as well as sizes of viruses are often inferred from the results of indirect methods. Graded ultrafiltration is sometimes used. It consists of passage of an aqueous suspension of virus particles through successive collodion or plastic membranes of known graded, decreasing pore sizes, the effect being like that of a gravel sorter. An important source of error is electrostatic adsorption of the virions to opposite charges on the membrane surfaces. Measurement by direct observation of virus particles with electron microscopes is valuable but subject to error resulting from dehydration of the specimen and distortion by the electron beam. Sedimentation rates and buoyant densities as determined in ultracentrifuges under carefully controlled conditions of specific gravity (e.g., density gradients in cesium chloride solutions), viscosity of temperature furnish useful data if expertly determined.

Virion size determinations are usefully accurate and are important in primary differentiations of virus types and classification of viruses.

There is a considerable range in size of viruses. Some, such as those of yellow fever, foot-and-mouth disease and poliomyelitis, are "small," with diameters about 25 mμ (mμ = millimicron, or one millionth of a millimeter). Others, like the virus of smallpox (vaccinia), are "large," about 250 mμ (0.25μ) in size. These are just within the range of resolution by the best optical microscopes. There are many intermediate sizes.

Particle Weight of Viruses

Another physical parameter of interest has been the particle weight of a virus. There values have mostly been determined by analytical ultra-centrifugation which gives the molecular weight or molecular mass of the virus under investigation. From this value the absolute weight of one particle could be calculated by dividing the value taken in grams by Avagadro's number (6.023×10^{23}, which is the number of molecules in one gram mole of any substance). Thus the molecular mass or molecular weight of tobacco mosaic virus (TMV) has been calculated to be 40×10^6. Therefore, one gram molecular weight of TMV is equivalent to 40×10^6 gms. Since the number of particles (molecules) inone gram mole of any substance is 6.23×10^{23}, the weight of one particle can easily be calculated to be 6.68×10^{-17}g.

Table 36.1. lists the particle weights of some viruses. It is interesting to note that the particle weight of a virus or its molecular weight are extremely small. The ranges are comparable to the molecular weights of some cellular biochemicals like haemocyanin or enzymes like glutamate dehydrogenase.

Table 36.1. Particle Weight of some Viruses.

Virus	*Mole weight of virus*	*Weight of one particle*
Adenovirus	6.7×10^7	11.06×10^{-17} g
Coliphage fd	1.1×10^7	1.84×10^{-17} g
Caliphage fr	4.1×10^6	0.665×10^{-17} g
Coliphage $\phi \times 174$	7.8×10^6	1.29×10^{-17} g
Coliphage T_2, T_1	24×10^7	39.84×10^{-17} g
Foot and Mouth Disease	5.0×10^6	0.82×10^{-17} g
Polio virus	5.5×10^6	0.82×10^{-17} g
Polyoma virus	2.3×10^7	3.7×10^{-17} g
Reo virus	5.8×10^7	9.75×10^{-17} g
Satellite tobacco necrosis	4.0×10^6	0.664×10^{-17} g
Tobacco mosaic	40×10^6	6.64×10^{-17} g
Cowpea Chlorilic mosaic	$4.6 \div 10^6$	7.4×10^{-17} g
Cauliflower mosaic	30×10^7	5.0×10^{-17} g
Cyanophage, LPPI	51×10^6	8.21×10^{-17} g
Vaccinia	1.33×10^9	221.0×10^{-17} g

Another obvious observation is the very wide variation, sometimes three hundred and twenty folds as between vaccinia and coliphage fd particles, found in viruses. The exact significance of such variation is not clear. Perhaps there is not any. However, it could be that size variation indicate their independent origin from different sources.

Subunit Structure

A fundamental feature concerning the morphology of viruses had gradually but surely been emerging during the last two decades. It is that a virus particles is in most instances composed of numerous identical protein subunits and one or rarely a few molecules of nucleic acid. It is also obvious that the shape of a virus particle is determined by the protein component. This is because, first, protein is the external viral component and secondly, it comprises the bulk of the particles. Therefore, it is clear that the configuration of and interactions between protein subunits ultimately decade the shape or form of the protein structure and consequently that of the virus particle.

The fact that such nucleoprotein particles as tobacco mosaic virus (TMV) could be crystalized and apparently had very high molecular weight led to the suspicion as to whether they were single unitary structures or made up of still smaller units. They were found to be too large to be considered as simple macromolecules. *Dorothy Hodgkin* (1949) was one of the first to point this out experimentally. On the basis of X-ray crystallographic analyses of various proteins, she suggested that in small spherical plant viruses, the protein was a composite macromolecule formed of smaller and identical subunits. *Crane* (1950) discussing the principles of biological growth predicted that any natural rod shaped object, like the TMV would be found to be built of a number of similar or even identical subunits arranged in a helix.

In the subsequent years this conception regarding the structure of viruses got confirmed on both theoretical and experimental grounds. In an illuminating summary *Francis Crick* and *James Watson* pointed out that the very low nucleic acid content of some of the smaller viruses could not possibly account for the coding of the large peptide chains required for the formation of the protein component, assuming the latter to be a single large entity. This is because the nucleic acid component in them is too small. Also because there would be no space in one virus particle for the amount of nucleic acid necessary for coding such large proteins molecules (assuming that three nucleotides are necessary for coding one amino acid).

Crick and *Watson* further pointed out that it identical protein subunits are used to constitute the pretein coat, then it would be possible for the small nucleic acid to code for the small subunit a number of times till the protein structure is built entirely.

Chemical investigations by *Knight* and his associates on TMV coat protein confirmed the above contentions. Using the enzyme carboxypeptidase (which can selectively cleave out only the C-terminal amino acid residues) they showed that one mole of TMV particles yielded 2320 threonine residues. This indicated that there are 2320 polypeptide chains, each with a threonine moeity at the C-terminal end, building up the protein coat of TMV.

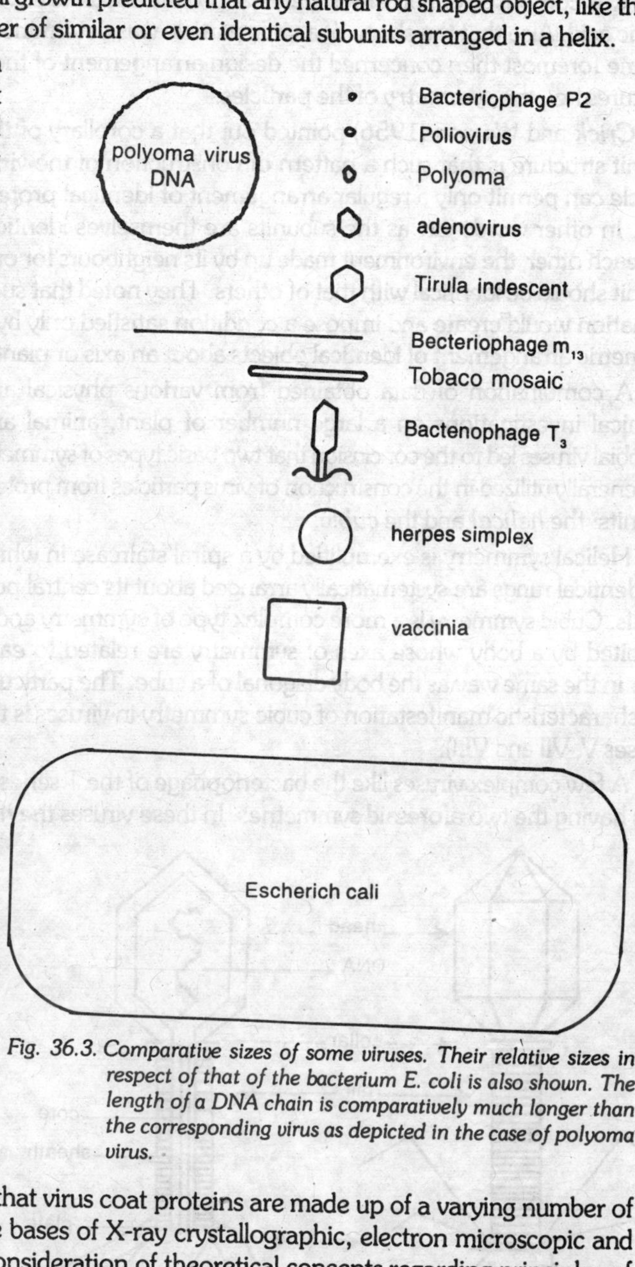

Fig. 36.3. Comparative sizes of some viruses. Their relative sizes in respect of that of the bacterium E. coli is also shown. The length of a DNA chain is comparatively much longer than the corresponding virus as depicted in the case of polyoma virus.

In the subsequent years, the basic idea that virus coat proteins are made up of a varying number of identical subunits was fully confirmed on the bases of X-ray crystallographic, electron microscopic and chemical experimentation and also on the consideration of theoretical concepts regarding principles of symmetry and medelling based on principles of solid geometry.

As interesting point raised by *Waterson* (1968) is whether the subunit revealed by various methods is the same. In other words, is the subunit determined by chemical analysis (the chemical subunit) the same as the one determined by electron microscopy (morphological subunit) or the one which is the basic unit as revealed by X-ray crystallography (Crystallographic subunit)? Although in some cases, for example the TMV, they appear to be the same, in some others they are not so. It is quite conceivable, therefore, that the subunits as revealed by electron microscopy and X-ray analysis are in themselves composite structures.

Symmetry

It was thus established that the basic feature of virus morphology was that a virus particle in most cases was composed of numerous identical protein subunits enveloping one or rarely more molecules of nucleic acid. Also that the shape of a virus particle was determined by the virus protein. The question that became foremost then concerned the design arrangement of these subunits into the composite protein structures i.e., the *symmetry* of the particles.

Crick and *Watson* (1956) pointed out that a corollary of the subunit structure is that such a pattern of construction of the virus particle can permit only a regular arrangement of identical protein units. In other words just as the subunits are themselves identical with each other, the environment made up by its neighbours for one subunit should be identical with that of others. They noted that such a situation would create and impose a condition satisfied only by a symmetric arrangement of identical objects about an axis or plane.

A combination of data obtained from various physical and chemical investigations on a large number of plant, animal and microbial viruses led to the conclusion that two basic types of symmetry are generally utilized in the construction of virus particles from protein subunits: the *helical* and the *cubic*.

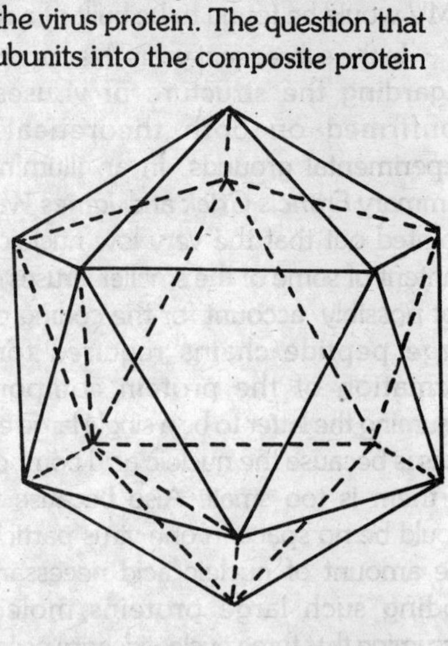

Fig. 36.4. Icosahedron. A twenty faceted structure. Cuboidal virus crystals have such twenty faceted design.

Helical symmetry is exemplified by a spiral staircase in which the identical rungs are systematically arranged about its central pole or axis. Cubic symmetry is a more complex type of symmetry and is exhibited by a body whose axes of symmetry are related to each other in the same way as the body diagonal of a cube. The particular and characteristic manifestation of cubic symmetry in viruses is the *Icosahedpon* (a twenty faceted cube) (Phases V, VII and VIII).

A few complex viruses like the bacteriophage of the T series exhibit a form which is a combination of units having the two aforesaid symmetries. In these viruses the *head* is icosahedral but the *tail* helical.

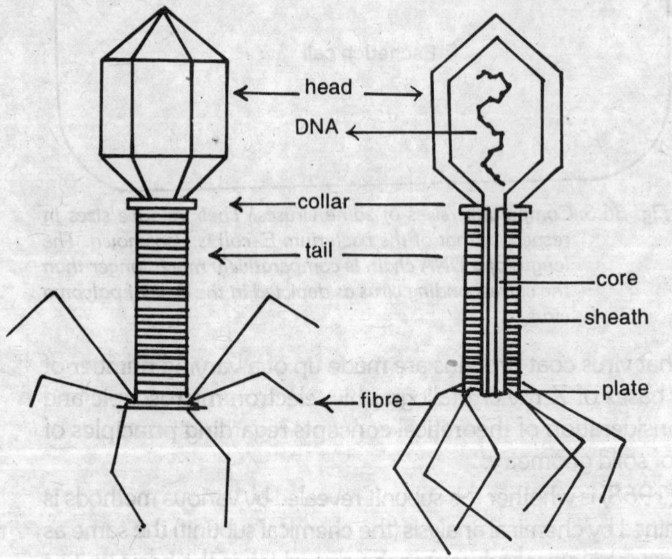

Fig. 36.5. Structure of Bacteriophage. Note the alignment of the head and the tail. The tail fibers are flexible and are attached at the bottom of the tail.

According to the symmetry they possess, the virus particles are categorized as either *anisometric* or *isometric*. These show helical and cubical symmetry respectively. A third group may be added to this comprising of particles showing dual or binal symmetry.

Architecture of Anisometric Particles

Anisometric particles with helical symmetry generally consist of elongate forms in which the central nucleic acid strand is coiled like a helical spring. The protein coat is closely coiled about its axis, arranged in a succession of rings. In other words, it may be said that in helical viruses the nucleic acid is located in a helical groove on inside of the cylindrical protein coat.

This orderly relation between the nucleic acid and the protein sheath has been beautiful elaborated in the well-known plant virus, the tabacco mosaic virus (TMV). These particles are rod shaped structures with the subunits of protein arranged in a helical pattern. It has been calculated that there are about 134 turns to one TMV protein coat. Utilising X-ray diffraction analyses it has been shown that there are 2130 subunits forming the protein coat of TMV. Detailed analysis has shown that there are 49 subunits distributed over three turns. In an average, these three turns from one structural unit which is repeated every 6.9 nm in the axial direction. In other words each succeeding ring is at a distance of 2.3 nm in the same direction (i.e. pitch). There is a central core, about 4 nm in diameter.

RNA, which forms about six percent of the TMV particles, is a single strand surrounded by the protein subunits. It has been shown that the RNA strand is arranged according to the turns of the protein sheath subunits. However, the RNA molecule is not located in the central hollow core. Rather it is deeply embedded amongst the successive protein subunit rings. In fact, the RNA molecule sort of fits in a helical groove formed by depressions on the successive protein subunits. This can be beautifully demonstrated in these particles by removing a limited portion of the protein coat thus exposing the central RNA core. Anisometric particles of some important animal viruses like the myxoviruses (influenza, parainfluenza etc.) are different in being enclosed within an envelop. However, the basic architecture of these particles is helical.

The tail of the railed bacteriophages can also be regarded as anisometric because of its helical configuration. *Brenner* and his co-workers demonstrated that the tail of bacteriophage of T even series has a multiple architecture in which a central hollow tube is surrounded by a protein sheath. With the aid of negative staining techniques it was revealed by them that the sheath showed 25 striations formed by parallel row of protein subunits. It has been estimated that about 200 *cog* like subunits make up these twenty-five striations.

Fig. 36.6. Tobacco Mosaic Virus. This much enlarged diagram of the helical virus is rod shaped having a central core surrounded by a helical protein sheath.

Architecture of Isometric Particles

The cubic symmetry exhibited by many viruses requires that there be specific number of axes of symmetry about which the particles can be rotated to give a series of identical appearances. One type of cubic symmetry is termed 5:3:2 because there are 5- fold, 3-fold and 2-fold axes of symmetry in it. Two polyhedral structures, the dodecahedron (12 sides) and icosahedron (20 sides) exhibit this type of symmetry. Isometric virus particles are universally found to be icosahedrons.

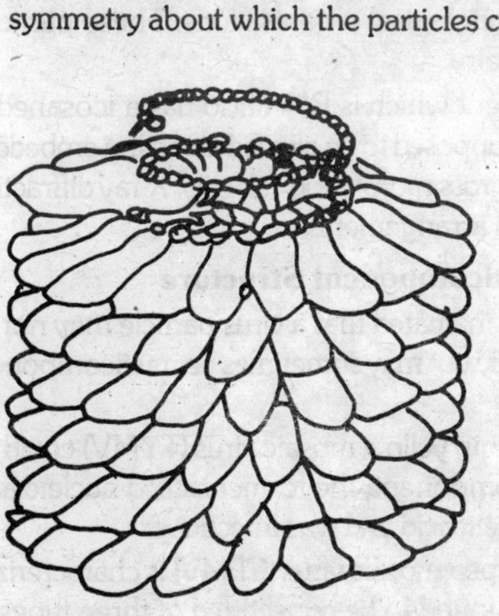

Fig. 36.7. Arrangement of nucleic acid and capsid in a helical virus (tobacco mosaic virus).

Detailed investigations into the icosahedral capsids reveal regularly spaced knob or ring like subunits or capsomeres separated by interstices. These subunit structures are also referred to as the major morphological units. The number of subunits forming the protein coats as icosahedrons in cuboidal viruses vary from 12 to 812 or even more. Each subunit in its turn, has been shown to be multimeric composed of a specific number of yet smaller basic units. Theoretical consideration and experimental data suggest the presence of two types of major morphological units; pentagonal *pentamers,* made up to five basic units of monomers and hexagonal *hexamers,* made up of six monomers. In either types the monomers from polygonal rings with central holes of various dimensions, up to 4 nm. The capsid is constructed by the assembly of capsomers according to specific geometrical pattern, wherein the pentamers from the corners of the icosahedron and the hexamers occupy the spaces in between.

Many plant viruses, like trunip yellow mosaic, tomato bushy stunt, tobacco necrosis and tobacco ringspot, which appear circular in outline under the electron microscope, have been shown to have a hexagonal contour. This indicated a icosahedral symmetry of these virus particles. X-ray crystallographic studies coupled with electron microscopy have also confirmed the icosahedral nature of such viruses as turnip yellow mosaic, tomato bushy-stunt and southern mosaic virus.

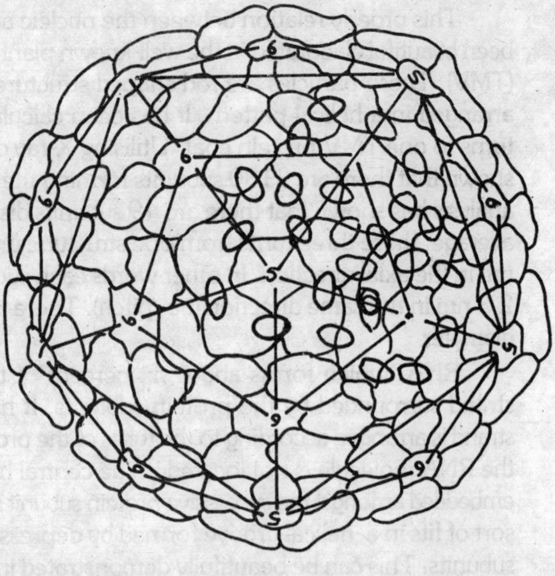

Fig. 36.8. *Architecture of an icosanhedral capsid. Note specifically the arrangement of the capsomers. Each capsomere is made up of specific number of monomer as indicated in the centre of each of them. Usually, these are either Pentamers (made up of five monomers) or Hexamers (made up of six monomers). Both may be present in the same capsid.*

Utilizing the technique of metal shadowing on freeze dried virus particles of *Tipula* iriddescent virus, *Williams* and *Smith* (1958) demonstrated them to be icosahedrons when viewed in the electron microscope.

Animal viruses like that of the encephalomyocarditis group (EMC) have been shown to be icosahedral with probably two distinct protein groups. Reoviruses have also been shown to be with incosahedral symmetry. Several other animal viruses like Polyoma, simian virus 40, herpes and adeno have been shown to be isometric with icosahedral particles.

Fig. 36.9. *A drawing of the outer surface of the turnip yellow mosaic virus particles is revealed by negative staining (approximately x 2 million).*

The most well-studied of the isometric viruses is the turnip yellow mosaic virus (TYMV). Studies have shown that the protein sheath of TYMV has a surface structures made up of 180 structural units. These were earlier supposed to be arranged uniformly but recent studies reveal that these are clustered into 20 rings of six and 12 rings of five. In other words, there are 32 large morphological organisations. These particles show a three fold axis of symmetry.

The nucleic acid component which is RNA also has a icosahedral distribution. The nucleic acid is supposed to be aligned along the embedded within the protein shell or its 32 gross morphological units. X-ray diffraction studies confirm such embedded arrangement.

Isometric Viruses with Multicomponent Structure

Some recent reports have indicated that a virus particle may not be entirely homogeneous structurally but may sometimes be multicomponent with a divided genome.

As early as in 1949 *Markham* and *Smith* reported that turnip yellow mosaic virus (TYMV) could be separated out by ultra-centrifugation into a *top* and a *bottom* component; the former lacked nucleic acid and was not infectious whereas the bottom component had nucleic acid and was infectious.

In a recent study Geelen *et. al.* (1972) have shown that cowpea mosaic virus (CPMV) is characterized by two types of heterogeneity. On the structural plane they are found to be constituted of three types of particles all icosahedral and of about the same size. However, they have different RNA contents and because of this have different sedimentation constants of 58S, 95S and 118S respectively. These are

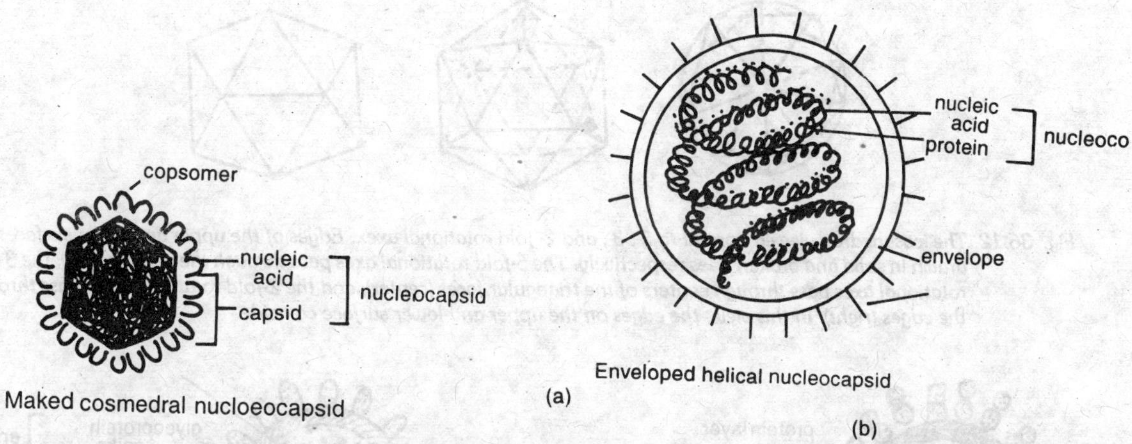

Fig. 36.10. The two basic patterns of animal virus structure, (a) the condensed genome is enclosed by a shell of capsomers arranged so as to display 5:3:2 rotational symmetry, (b) the extended genome is enclosed by protein molecules spaced so as to display helical symmetry. The resultant structure, the nucleocapsid, is enclosed in an envelope which is generally, but not always, studied with spikes on its outer surface.

referred to as the top (T), the middle (M) and the bottom (B) components. The middle and the bottom components are nucleoproteins consisting 25 and 34 percent of RNA respectively. Both these components are required for successful infection. Thus, there is some heterogeneity on functional plane as well.

Most viruses of CCPM group behave in the same fashion. Other viruses like bromegrass mosaic and respberry ringspot mosaic viruses have been shown to be multicomponent.

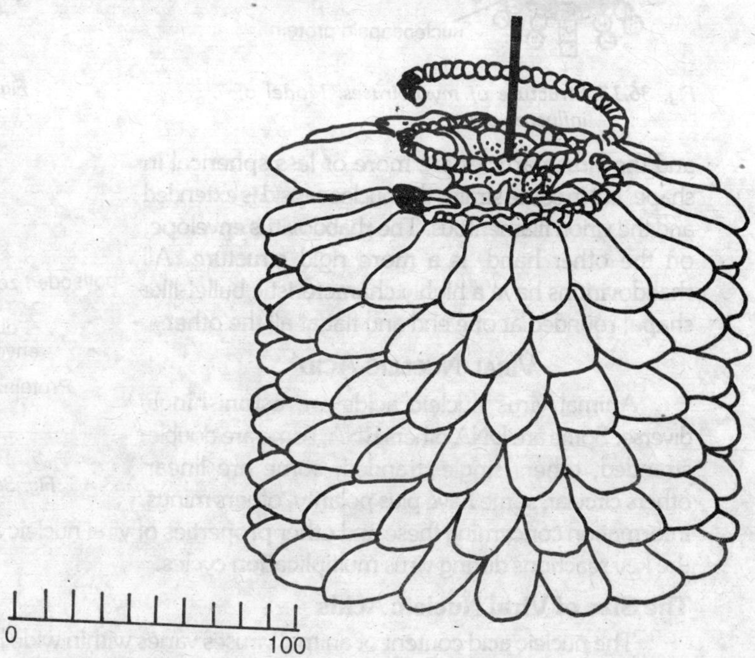

Fig. 36.11. Schematic representation of tobacco mosaic virus. As can be seen in the cutaway section, the ribonucleic acid helix is associated with protein molecules in the ratio of three nucleotides per protein molecule.

Envelopes

Only five groups of animal viruses exist as naked nucleocapsids : picornaviruses, reoviruses, adenoviruses, papovaviruses and parvoviruses. In all the remainder the nucleocapsids are enclosed by envelopes which are acquired as they bud through special patches of membranes on their way to the exterior of the cell. Viral envelopes therefore resemble cell membranes in their overall chemical composition.

Most envelopes consist of a layer containing both protein and lipid, to which spikes consisting of glycoproteins are attached in a regularly repeating manner. The envelopes of myxoviruses are rather flexible, causing them to appear pleomorphic, that is, heterogeneous with respect to both shape and size. Occasionally more than onę nucleocapsid becomes enclosed within one envelope; rare virus particles contain as many as 20 nucleocapsides. Usually myxovirus nucleocapsids are rolled up inside the envelope

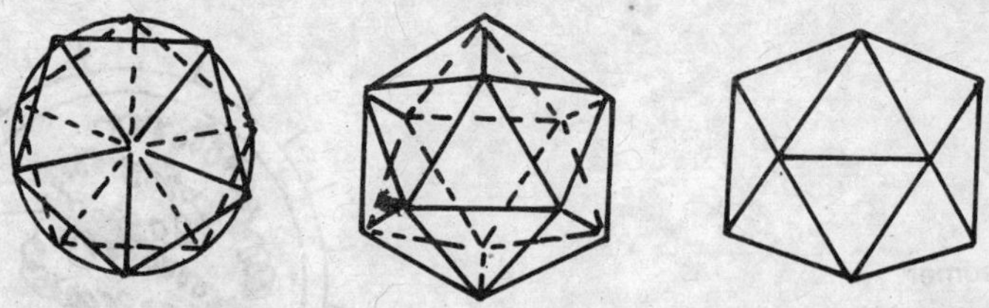

Fig. 36.12. The icosahedron viewed normal to 5-, 3-, and 2- fold rotational axex. Edges of the upper and lower surfaces are drawn in solid and broken lines respectively. The 5-fold rotational axes pass through the vertices (left); the 3 fold rotational axes pass through centers of the triangular faces (center); and the 2-fold rotational axes pass through the edges (right). In this view, the edges on the upper and lower surface coincide.

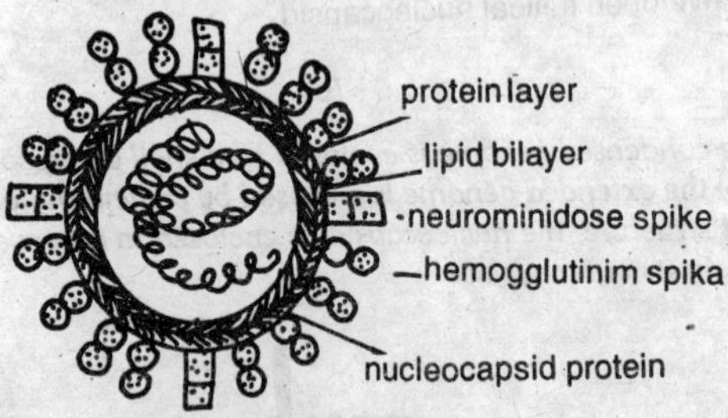

Fig. 36.13. Structure of myxoviruses. Model of influenza virus.

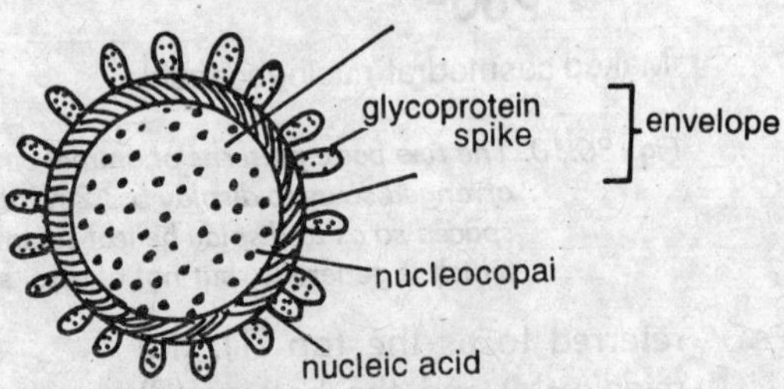

Fig. 36.14. The structure of arboviruses.

and the virus particles are more or less spherical in shape, but in some strains the nucleocapsid is extended and the virion filamentous. The rhabdovirus envelope, on the other hand, is a more rigid structure. All rhabdovirions have a highly characteristic bullet-like shape, rounded at one end and flat at all the other.

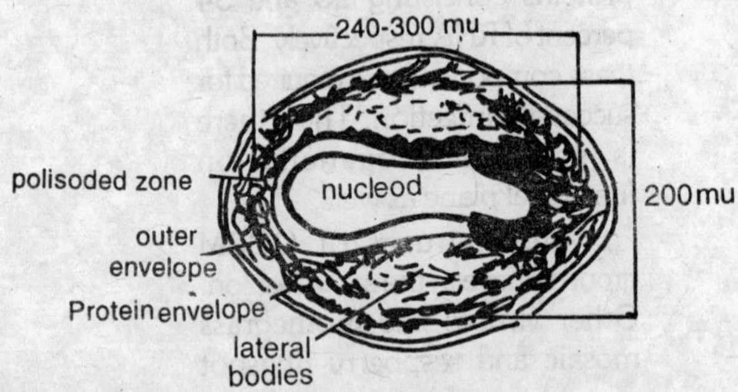

Fig. 36.15. The structure of vaccinia virus.

Viral Nucleic Acid

Animal virus nucleic acids are astonishingly diverse. Some are DNA others RNA; some are double-stranded, others single-stranded; some are linear, others circular, some have plus polarity, others minus. Information concerning these and other properties of viral nucleic acids is essential for an understanding of the key reactions during virus multiplication cycles.

The Size of Viral Nucleic Acids

The nucleic acid content of animal viruses varies within wide limits. At the lower end of the scale, only 1 or 2 percent of the myxovirion is RNA, and only about 5 percent of the poxiviron is DNA; at the upper end, about 25 percent of the picornavirion is RNA. However, the proportion of nucleic acid in virions is not as significant as its size, which determines the amount of genetic information it contains. The smallest animal virus genomes are those of the picornaviruses, arboviruses, inflenza viruses, paroviruses, and papovaviruses, the molecular weights of all of which are in the range from 1 to 3 x 10^6. Since the coding ratio, that is, the ratio of the molecular weight of single-stranded nucleic acid to the molecular weight of the protein for which it an code is about 9, this means that these viral genomes can code for protein with a total molecular weight of from 100,000 to 350,000, which is equivalent to 3 to 12 average size proteins. The largest animal virus genomes, those of the poxviruses, are about one hundred times larger, their molecular weight is 160 to 200 x 10^6. The molecular weights of viral genomes are listed in Table 35.3.

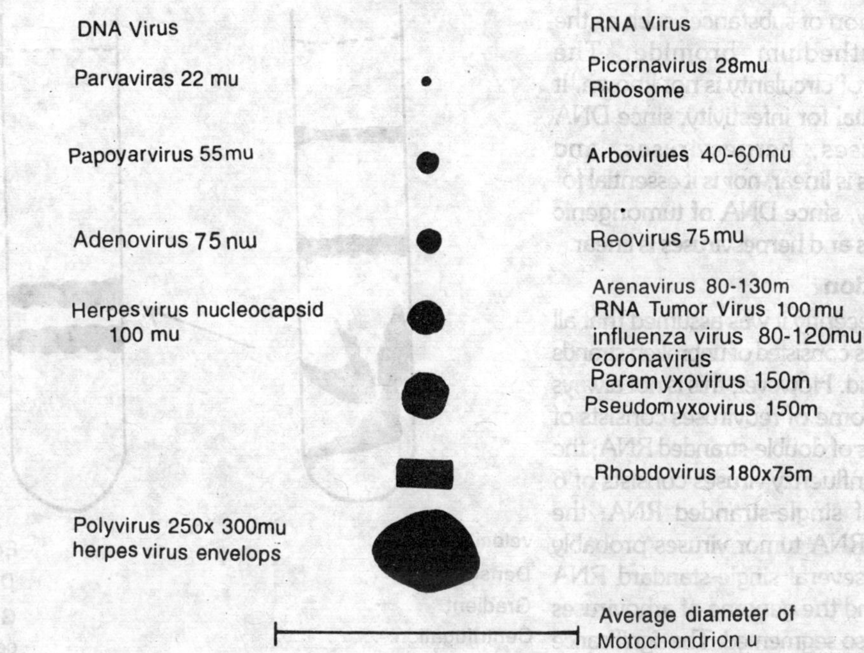

Fig. 36.16. The relative sizes of the principal groups of animal viruses. Unless otherwise indicated, the scals is the same for all.

Table 36.2. The Morphology of Animal Viruses

DNA Viruses	*Morphology*
Poxviruses	Complex
Herpesviruses	Enveloped icosahedral nucleocapsids
Adenoviruses	Naked icosahedral nucleocapsids
Papovaviruses	Naked icosahedral nucleocapsids
Parvoviruses	Naked icosahedral nucleocapsids
RNA Viruses	
Picornaviruses	Naked icosahedral nucleocapsids
Arboviruses	Enveloped nucleocapsids
Reoviruses	Naked double-shelled icosahedral nucleocapsids
Myxoviruses	Enveloped helical nucleocapsids
Rhabdoviruses	Enveloped helical nucleocapsids
RNA Tumor Viruses	Enveloped nucleocapsids
Arenaviruses	Enveloped nucleocapsids
Coronaviruses	Enveloped nucleocapsid

The Structure of Viral Nucleic Acids

Strandedness. Both double-stranded and single-stranded DNA as well as RNA can act as the genome of animal viruses.

Circularity. Most viral nucleic acids are linear, but papovavirus DNA exist in the form of double-stranded supercoiled circles. The reason for the supercoiling is a deficiency of 15 to 20 turns in their double helix; the reason for the deficiency is not clear. Supercoiling is relieved either by the introduction of a single break or "nick" in one of the two strands, thus permitting unwinding into closed nicked circles, or by intercalation of substances such as the acridine ethedium bromide. The significance of circularity is not known. It is not essential for infectivity, since DNA in poxviruses, herpesviruses, and adenoviruses is linear, nor is it essential for oncogenicity, since DNA of tumorigenic adenoviruses and herpesviruses is linear.

Segmentation

Until recently it was assumed that all viral genomes consisted of unbroken strands of nucleic acid. However, this is not always so. The genome of reoviruses consists of 10 segments of double-stranded RNA; the genome of influenzy viruses consists of 6 seganded of single-stranded RNA; the genome of RNA tumor viruses probably consists of several single-standard RNA segments and the genome of arboviruses is perhaps also segmented. The significance of segmentation is not yet clear. One consequence is highly efficient formation of genetic recombinants by random assortment of segments in multiply infected cells.

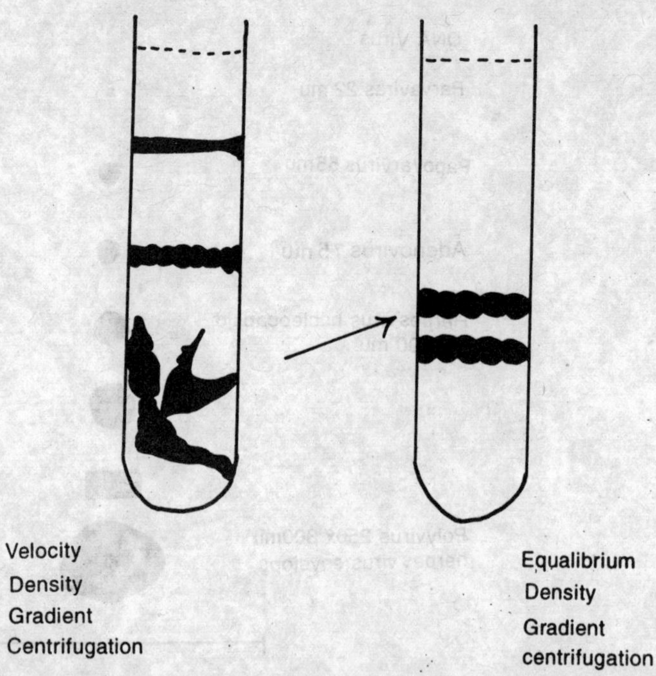

Fig. 36.17. Application of the technique of density gradient contrifugation to virus purification. A partially purified virus preparation is layered onto the density gradient on the left, which is then contrifuged so that particles with the same sedimentation coefficient form distinct bands. Three such bands are shown, one of which contains virus (black spheres). This band is then centrifuged to equilibrium in the gradient on the right, in which particles are separated according to their buoyant density.

The Polarity of Viral Nucleic Acids

The single-stranded DNA molecules present in picornavirus and arbovirus particles can combine with cell ribosomes and serve as messenger RNA; all the genetic information necessary for the formation of progeny virus can be directly translated from them. This is not the case for the RNA molecules present in myxovirus and rhabdovirus particles. These RNAs must first be transcribed into RNA strands of opposite polarity, and these transcripts are then translated by the cell ribosomes. The polarity of messenger RNA is generally designated as plus. Accordingly the polarity of picornavirus and arbovirus RNA is plus, and that of the RNA is myxovirus and rhabdovirus particles is minus.

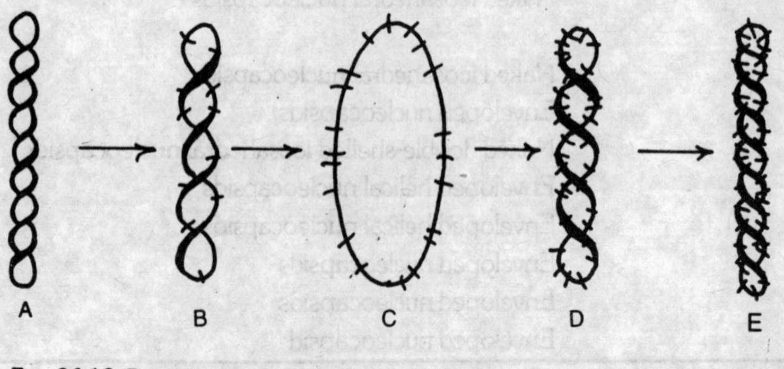

Fig. 36.18. Diagrammatic representation of the removal and reversal of supercoiling turns.

The parvoviruses, which contain single-stranded DNA, provide a unique situation. These are two kinds of parvovirus particles: one contains plus strands, the other minus strands. These two kinds of particles are always produced in equal amounts. When the DNA is extracted from them, it tends to hybridize rapidly, thus giving the impression that they contain double-stranded DNA. It is only when the DNA is extracted under conditions when the plus and minus strands are prevented from hybridizing with each other that the true situation is revealed.

The polarity of the RNA in RNA tumor virus, arenavirus, and coronavirus particles is not known.

Table 36.3. Characteristics of Viral Nucleic Acids.

Virus	*Nature of Nucleic Acid*	*MW x 10^6*	*Strandedness*	*Structure*	*Number of Segments*	*Polarity*	*Infectivity Naked Nucleic Acid*
Poxvirus	DNA	160-200	Double	Linear	1		–
Herpesvirus	DNA	100	Double	Linear	1		–
Adenovirus	DNA	23	Double	Linear	1		+
Papovavirus	DNA	3.5, 3	Double	Supercoiled Circular	1		+
Parvovirus	DNA	1.7	Single	Linear	1	+ and –	?
Picornavirus	RNA	2.5	Single	Linear	1	+	+
Arbovirus	RNA	3.2	Single	Linear	1 (or 2)	+	+
Reovirus	RNA	15	Double	Linear	10		–
Myxovirus							
Influenza virus	RNA	4	Single	Linear	6	–	–
Other Myxoviruses	RNA	6.5	Single	Linear	1	–	–
Rhabdovirus	RNA	3.5	Single	Linear	1	–	–
RNA Tumor Virus	RNA	10	Single	Linear	several (3.4)	+	–
Arenavirus	RNA	?	Single	?	?	?	
Coronavirus	RNA	?	Single	?	?	?	?

The Base Composition of Viral Nucleic Acids

The molar base composition of viral genomes varies within wide limits. It is frequently of taxonomic significance : thus all poxvirus DNAs so far examined contain about 36 percent guanine + cytosine (GC); all papovanvirus DNAs contain about 48 percent (GC): there are three classes of adenoviruses, highly, weakly, and nononcogenic, the DNA of which contains 48, 52 and 58 percent GC; and among the herpesviruses, some oncogenic ones contain 42 percent GC, while some which are not oncocogenic contain as much as 68 percent GC. In general, the GC content of the DNA of oncogenic viruses is closer to that of the DNA of mammalian cells (about 45 percent GC) than that of nononcogenic viruses, but there are notable exceptions where this is not the case.

The Genetic Relatedness of Viral Nucleic Acids

The genetic relatedness of animal virus nucleic acids is of interest for its taxonomic and evolutionary significance. For example, there are many viruses which are poxviruses, according to a variety of criteria. The question naturally arises as to how closely related they are.

The best measure of genetic relatedness is determination of similarity of nucleic acid-base sequence. This is accomplished readily for double-stranded nucleic acids by measuring how extensively they can hybridize with each other. The two nucleic acids the relatedness of which is to be determined are denatured tothe single-stranded state, then mixed and allowed to reanneal. Single strands derived from the two genomes are thus presented with the opportunity of pairing with each other and conditions are readily arranged so that such pairing can be quantitated. If the two genomes are very closely related, pairing will occur extensively; if there is no relatedness, no pairing will occur. To take an example, the genomes of the highly oncogenic adenoviruses referred to above hybridize with each other to the extent of 80 percent of more; in other words, they share over 80 percent of base sequences with the genomes of the nononcogenic adenoviruses, which, however, share over 80 percent of base sequences among themselves.

It is technically much more difficult to measure similarity of base sequence content among viral genomes which consist of single-stranded nucleic acids.

Another measure of genetic relatedness employs comparison of proteins rather that of nucleic acids. Similarily of the base sequences of nucleic acids signifies similarity of the amino acid sequences of the proteins coded by them, which in turn implies antigenic similarity, that is, ability of proteins to react with each others' antibodies. Ability of nucleic acids to hybridize and of the proteins coded by them to cross-react antigenically are thus measures of the same parameter.

The Infectivity of Animal Virus Nucleic Acids

Viral nucleic acids contain all the information necessary for the formation of virus particles. This was first shown by *Hershey* and *Chase* in 1952, when they found that infection by bacteriophage is initiated by injection of viral DNA (along with a small amount of protein which has no genetic function) into the host cell. Later, in 1956, Gierer and Schramm showed that the same was true for plant viruses when they found that the RNA extracted from tobacco mosaic virions was infectious.

The nucleic acids of four groups of animal viruses are also infections. These are picornavirus RNA, arbovirus RNA, papovavirus DNA, and adenovirus DNA. In all cases, the naked nucleic acids are less infectious by factors varying from 10^3 to 10^6 than the virions from which they were extracted. There are two principal reasons for this. First, naked viral nucleic acids are readily degraded by nucleases which are generally present in extracellular fluids, as well as on outer cell membranes; and second, naked nucleic acids are taken up very poorly by cells. Uptake can be increased significantly either by treating cells briefly with concentrated salt solutions, which no doubt damges cell membranes transiently thereby facilitating nucleic acid entry, or by complexing the naked nucleic acids with polycations like protamine, or DEAE-dextran. The inefficiency with which naked viral nucleic acids penetrate into cells emphasizes the role of the viral capsid in this vital function.

The host range of naked viral nucleic acid is very much broader than that of the respective virions. This stems from the fact that the latter's host range is restricted by the specificity of the interaction of the capsid with cell surface receptors. This is not the case for naked viral nucleic acids, which can apparently multiply in any cell into which they can penetrate without being degraded. For example, whereas poliovirions can only infect and therefore multiply in cells of human or primate origin. Poliovirus RNA can also multiply in chicken cells an mouse cells.

Infectious viral nucleic acid can be extracted not only from infectious virus particles, but from any virus particles which contain undamaged nucleic acid. Among these are virus particles inactivated by heat, proteolytic enzymes, or detergents.

The Presence of Host Nucleic Acids in Virions

As a general rule, viral capsids enclose viral nucleic acids. However, several instances are known where pieces of host nucleic acids become encapsidated in viral capsids. One example of this type is provided by the papovaviruses. Particles exist which contain, within a papovavirus capsid, a linear piece of host DNA roughly the same size as the papovavirus genome. As far as one can tell, each particle, known as a pseudovirion, contains a different piece of DNA. Generally, was pseudovirions make up only a small fraction of the yield; but in one cell line the majority of the particles formed as a result of infection with polyoma virus are pseudovirions. They have attracted attention because of their potential ability to transduce host genes from one cell to another, which could conceivably be exploited in correcting inborn errors of metabolism.

RNA tumor viruses provide another example. Some of these appear to contain not only the nucleic acid which coded for the proteins required for virus multiplication, but also nucleic acid which functions in some as yet unknown manner in carcinogenesis.

Morphological Categories in Viruses

We have seen in the previous paragraphs that the viruses exhibit but a few specific morphologic characteristics. Most of them can be grouped in one category or the other in accordance with these characteristics. These categories, however, cannot be considered to be taxonomic groupings since often different viruses may possess similar or even identical morphological attributes.

One of the convenient ways of categorising viruses would be by taking into consideration their symmetry and the presence or absence of envelop around them. Some viruses, notably the pox viruses and a few of the arboviruses would, however, be necessarily left out of such a scheme. This would be so because of the unusual symmetry possessed by virus particles belonging to these groups.

Accordingly the viruses could be categorised into the following principal morphological groups.

1. Naked elongated (helical) viruses.
2. Naked spheroidal (icosahedral) viruses.
3. Enveloped elongated (helical) viruses.
4. Enveloped spheroidal (icosahedral) viruses.
5. Tailed (binal) viruses.
6. Uncertain types.

Naked Elongated Viruses

Elongated, non-enveloped particles are observed mostly in plant and bacterial viruses. Common examples are, tobacco mosaic virus (TMV), tobacco rattle viruses (TRV), Potato virus.

Most studied of the naked elongated viruses in the tobacco mosaic virus. It is now established that it is a nucleo-protein particle of 40 x 10^6 daltons with one central single standed RNA. molecule surrounded by a protein sheath. The protein is a single species but is made up of about 23000 identical units each with a molecular weight for 1800. The nucleic acid has a molecular weight of about 2 x 10^6 and is about 3300 nm long (about 10 time the length of the particle) and is situated centrally.

Enveloped Elongated (Helical) Viruses

A large number of animal and plant viruses would fall under this category. All of them share the feature of maturing at cell membranes through which they bud. Thus they acquire an envelope around them. The envelope is composed of both viral and host cell components; the protein is usually virus specific while the lipids and carbohydrates are the characteristic of the host membrane. Quite often the membranes have spikes on them. An interesting features about these viruses is that while the nucleocapsid particles exhibit a helical configuration, the enveloped virus may be spheroidal e.g., some myxoviruses like sendai virus.

In these viruses and helical nucleo-capsid exists in a folded or coiled state within the envelope. Sometimes, however, they may show morphological plasticity leading to a sort of plemorphism. There may even be fragmented nucleocapsid as in sendai virus and in oncorna viruses.

Some plant viruses like the vesicular stomatitis virus and potato yellow dwarf virus are other examples of helical enveloped viruses.

Naked Spheroidal (Icosahedral) Viruses

Many plants, animals and microbial viruses have a naked spheroidal shape. As for example, turnip yellow mosaic virus (TYMV), brome mosaic virus (BMV), poliovirus, adenovirus-5, ϕ x 174 coliphage qβ.

Adenoviruses are one of the larger spheriodal viruses which have been extensively investigated into. These are about 80 nm in diameter. The other protein coat has 252 morphological units each consisting of either five or six subunits. The cluster formed by five subunits are termed *pentons* or *pentamers* and those with six subunits are known as *hexons* or *hexamers.* There are 32 hexons which all are roughly polygonal discs of 8 nm diameter with a central hole which 2.5 nm across. Each hexon is surrounded by six more hexons thus building up most of the icosahedral particle. There are 12 situated at the 12 vertices of the icosahedral and each is bounded by 5 hexamers. Detailed analyses revealed that each penton serves as base structure to whcih fibers called penton fibers are attached. Each penton fiber is about 2 x 20 nm in diameter and terminates in a spherical knob about 4 nm in diameter. These fibers have important serological properties.

Another well studied naked spheroidal virus is the turnip yellow mosaic virus (TMYV) which has been described earlier in this chapter. A special feature revealed by X-ray crystallographic studies on these concerns the alignment of nucleic acid with the protein. It has been suggested that there is a regular interlacing of the nucleic acid with the protein subunits deep enough below the surface.

Another interesting example of naked spheroidal viruses are the reoviruses. These viruses have been shown to be consisting of a segmented double stranded RNA surrounded by two successive layers of protein shells, an innr and an outer. It is not clear whether the outer shell is part of the nucleocapsid particle or is an independent covering.

Not much is known regarding the alignment of nucleic acid strands with the protein coat in these viruses excepting the turnip yellow mosaic virus. However, it is reasonable to assume a non-covalent association between the two.

Enveloped Spheroidal (Icosahedral) Viruses

This group also includes several plant and animal viruses and they share with the helical enveloped virus their mode of maturation and budding.

Herpes viruses are the most well-known example of enveloped icosahedral viruses. Some myxo virus like the influenza virus and oncorna viruses like Rous sarconna virus are the other examples. Some arbo viruses (Group A) are also spheroidal with surrounding envelopes.

The exact nature of the protein sheath in these viruses and their relationships with the nucleic acid component are not clear as yet.

The bacteriophage *Pseudomonas* 2(*PM*2) is icosahedral with an envelope covering it. This is known to have a double stranded circular DNA molecule.

Tailed (Binal) Viruses

While some bacterial viruses are spheroidal (ϕ x 174) and some elongated (coliphage fd) a large number of them are combinations in which head and tail structures are clearly discernible. The head is usually sphrodal (icosahedral) and the tail elongated (helical) in symmetry.

The most well-known example of tailed viruses are the bacteriophage of the *T* series. The head in these particles houses the nucleic acid and varies considerably in shape and size. The tail serves as an attachment organ in the initial steps of infection. The tail has tubular structure surrounded by a sheath like convering. The tube penetrates into the host cell during infection and being hollow allows the viral DNA to pass into the host cell. The tail sheath may (*T* even) or may not (*T* odd) be retractible. These particles are extremely complex and have an array of such accessory structures as collars, base plates, spikes, tail-fibers and so on.

Tailed viruses are also exemplified by infectious of blue-green algae (cyanophaes). Their structure appears to be similar to bacteriophages and exhibit a similar range of variation.

Uncertain Types

A few viruses possess different and more complex morphology than encountered in the members of any of the group of viruses mentioned earlier.

Poxviruses are the largest of most complex of such viruses. Their virions are usually described as brick or loaf shaped having a size of 250 nm by 300 nm. These viruses, irrespective of their respective sources, have a tubular, highly convoluted lipoprotein outer membrane. This surrounds an internal protein-nucleoprotein core also called the nucleoid. Between the central nucleoid and the outer membrane are two lateral proteinaceous bodies on either side. Two lateral bodies cover the nucleoid partially in the central region alone. The nucleoid itself appears to be surrounded by a membrane.

The viruses are reported to possess several enzymes including DNA destroying nucleases. These enzymes present with the nucleoid but are supposed to remain inactivated by the presence of the lateral bodies.

Some vaccinia viruses e.g., 'orf' (contagious pustular dermatitis) viruses have a crossed helical pattern on their surface.

Encapsulated or Occluded Viruses

A few insect viruses appear in their natural form in characteristic inclusion bodies. The nuclear polyhedroses and the cystoplasmic polyhedroses viruses are two well-known examples of such viruses.

These inclusion bodies are one generally crystalline protein packages which contain one or two or more virus particles. Some of these packages are called polyhedral bodies and are found characteristically in either the nucleus or the cytoplasm of host cells. Therefore, the names of the diseases caused by them i.e., nuclear polyhedrosis and cytoplasmic polyhedresis.

The occluded virions of these viruses may be either spherodal (cytoplasmic polyhedrosis) or elongated (nuclear polyhedroses). The number of particles in the crystalline matrices are many in these viruses. However, in another similar virus, the granuloses virus, the number of virus particle per inclusion body or capsule may be only one.

This section of polyhedral viruses and capsules reveal that two concentric membranes surround each particle. The exact nature and significance of these structures are not very clear.

37

AGEING

Some men live for more than 100 years; no mouse live more than 4 years. Why is it that two mammalian species, so similar in many ways, should have such very different life-spans? A man of 75 usually has a more wrinkled skin, more grey hair, less muscular strength, poorer vision and hearing than he had when he was 25. He has become less fit and vigorous. What is the cause of the deteriorative changes which have occurred during this 50-year period? Can the changes be reversed? Could they have been prevented? Why is it that the chances of a 25-year-old man dying in the next year are less than one in a thousand, whereas the chances that a 75-year-old will die in the next year almost one in ten? Is it possible to prolong the human life-span?

The aim of gerontology, the study of ageing and senescence, is to find the answers to questions like these. Senescence has been a subject of interest and speculation for centuries, and many causes and cures have been suggested for the deleterious changes whch accompany increasing age. Speculations about the causes of senescence and differences in longevity have frquently been based on inadequate or incorrect factual information, and have led to hypotheses which cannot be tested. The suggested cures for senescence simply have not worked. However, during the past twenty years, gerontology has attracted the interest of an increasing number of biologists, and knowledge about many aspects of the subject is beginning to accumulate rapidly. Although we still know very little about the nature and causes of ageing changes and senescence, it certainly can no longer be said, as Medawar did in 1946, that *the problems of old age and natural death are hardly yet acknowledged to be within the province of genuine scientific enquiry.* In this chapter we shall consider the different approaches which are being used by gerontologists to investigate the nature and causes of ageing processes and senescence.

The Definition of Ageing and Senescence

The terms *ageing* and *senescence* are familiar, but unfortunately they are rather imprecise. There is no general agreement among gerontologists about how they should be used. Usually *senescent* and *senescence* are used when talking about the changes which occur during the period of obvious functional decline in the later years of an animal's life-span. Some people use the term *ageing* for the same processes and period. Others use it in a much more general way, with *ageing* meaning simply growing older, and *ageing changes* being any changes related to age, regardless of when in the life-span they occur. Thus the onset of puberty might be described as an ageing change, but not as a senescence change. Although there may be advantages in using *ageing* in this general way, it is rather difficult to do so, because in common language *ageing* implies something more than simply getting older. For example, it would be unusual to talk of an ageing child. We would normally refer to a developing child, because in everyday English the word *ageing* carries with it the idea of decline and deterioration. Most biologists have tended to accept these connotations, and think of ageing as occuring only after maturity has been reached. In fact, as can be seen from some of the definitions given below, the terms *ageing* and *senescence* are frequently used interchangeably.

There have been many formal definitions of ageing processes and senescence. Medawar (1952) suggested that senescence could be defined as, "*that change of the bodily faculties and sensibilities and energies which accompanies ageing, and which renders the individual progressively more likely to die from accidental causes of random incidence. Strictly speaking, the word "accidental" is redundant, for all deaths are in some degree accidental. No death is wholly "natural"; no one dies merely of the burden of the years.*"

Strehler (1962) defined senescence as, "*the changes which occur generally in the post-reproductive period and which result in a decreased survival capacity on the part of the individual organism.*"

Ageing processes have been defined by Maynard Smith (1962) as, "*those which render individuals more suceptible as they grow older to the various factors, intrinsic or extrinsic, which may cause death.*"

Comfort (1960) said that ageing is, "*an increased liability ot die, or an increasing loss of vigour, with increasing chronological age, or with the passage of the life cycle.*"

It is worth while emphasizing and amplifying some of the points which are made or implied in these definitions. Firstly, the changes which occur during ageing are deleterious; they increase the chances that an animal will die. Ageing therefore involves a decrease in the ability of an animal to cope with its environment. Secondly, the deleterious age-related changes are cumulative. Death, the ultimate result of ageing, is sudden, but the process of ageing involves a progressive increase in the probability of dying. A third characteristics of ageing and senescence, which is implicit in most of the definitions which have been given, is that the processes involved are common to all members of a species and are inescapable consequences of getting older. That is to say, ageing and senescence are fundamental instrinsic properties of living organisms.

Ageing Changes in Man

Greying of Hair

One of the most obvious signs of human ageing is the greying of hair. As fig. shows, with increasing age more and more people have some grey hair. This figure is based on data obtained from a survey of Australian men and women, in which hair-greying was assessed in three categories: *complete greying, any greying,* and *no greying.*

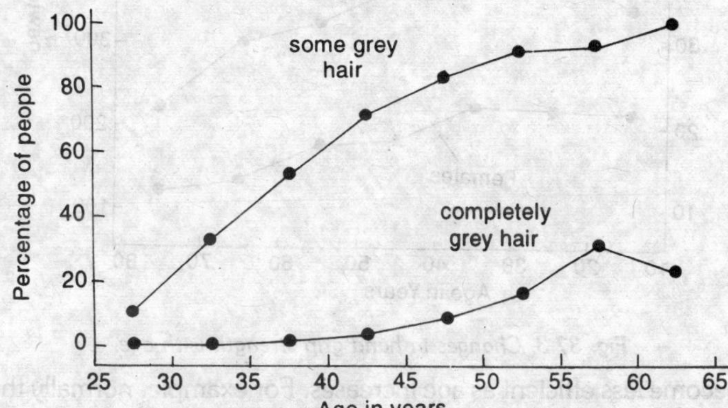

Fig. 37.1. Rate of hair greying for 3872 Australian men and women with medium colour hair.

Skin Elasticity

Figure shows the way in which skin elasticity changes with age. This measurable change in skin properties is probably associated with one of the obvious visible signs of ageing, namely changes in skin texture and appearance. The figure shows that in both sexes there is a gradual rise in the modulus of elasticity, but that this "stiffening up" is more marked in women than in men, in spite of the fact that women have the longer life-span.

Loss of Strength

Apart from visible changes in appearance, one of the most familiar signs of ageing is a general loss of strength. Several diferent methods have been used to measure this change. Figure shows the results from measuring hand-grip strength. The values plotted are the averages for each decade of life, and they show that after an initial rise there is a decline in grip strength for both men and women. These data come from a preliminary study of a number of physiological functions in over 400 people in Hiroshima. The aim of the study is to measure a number of variables which are known to change with age, and from these to attempt to determine "biological age" and ultimately to relate this to mortality in the subjects surveyed. Figure shows the results from two other series of tests made in this survey; both blood pressure and the vital capacity of the lungs show marked changes after the age of 40. Hearing, visual acuity, vibration perception, manual dexterity and mental reaction time were also tested, and all showed deteriorations, particularly beyond the age of forty.

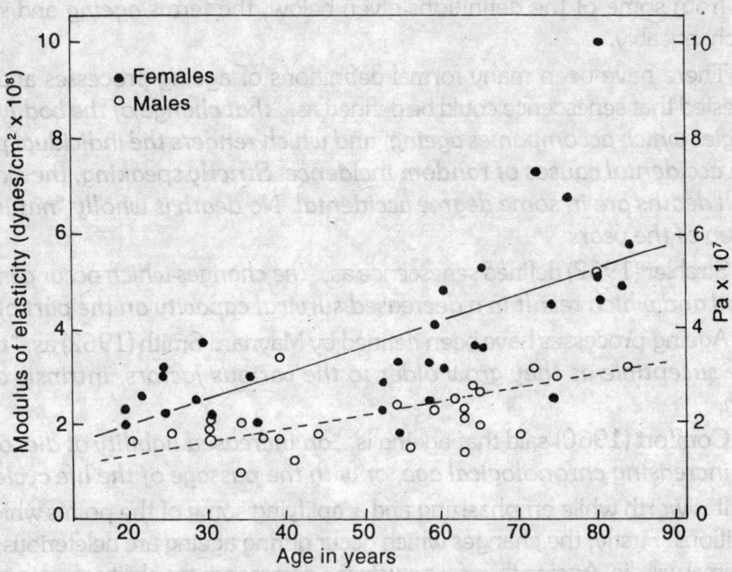

Fig. 37.2. Age-related changes in the in vivo elasticity fore-arm skin.

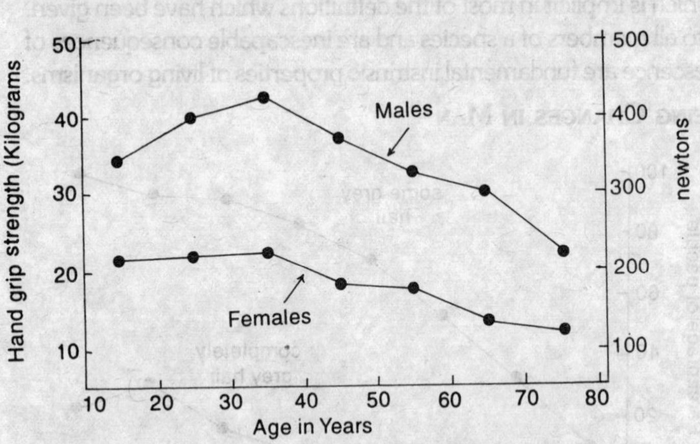

Fig. 37.3. Changes in hand grip strength with age.

pH

Some of the most striking age-related changes are revealed by studying the responses of the body to a physiological challenge. The pH, ion concentration, temperature, levels of metabolites and of food substances in the internal environment of mammals are normally maintained within very narrow limits. Numerous feedback mechanisms tend to restore the composition of the internal environment to normal if a natural or artificially applied stress causes it to change. It is clear, however, that these homeostatic control mechanisms become less efficient as age increases. For example, normally the glucose level in blood shows very little change with age, but if glucose is injected into the blood, the rate of restoration to the normal level shows a marked dependence on age: it is far slower in old people than in young. This decline in the efficiency and effectiveness of the homeostatic regulatory mechanisms could be due to deficiencies in one or all of a number of interdependent tissues and cells. In the case of glucose, the muscles and liver are important for its storage, the pancreas is involved in the production of insulin, which regulates storage and mobilization, and the kidneys are involved in getting rid of excess amounts.

Ageing, Injury and Disease

One of the signs of ageing which is all too familiar to many of us is the increasing incidence of many different types of disease. As we get older, we frequently begin to show symptoms of arthritis, osteoporosis, atheroscelerosis, or malignant tumours, for example. When old poeple die, it is often very difficult to pinpoint the exact cause of death, because pathological changes can be seen in so many tissues and organs.

In developed countries, the major causes of death are at present cardiovascular diseases and cancer, but it has been calculated that if it was possible to eliminate both of these, average life-span would be increased by only 6 or 7 years. If we did not die of cancer or cardiovascular disease, we would soon die of something else.

Most of the diseases which we associate with old age may well be initiated long before they become clinically obvious. For exmaple, the early signs of changes in the arteries which are thought to precede artherosclerosis—a condition which may ultimately cause death by a stroke or heart attack—can often be detected *post mortem* before the age of 20. With increasing age, the number and size of the lesions in the arteries increase, and their structure changes. the problem is, are diseases like this the causes or the consequences of ageing processes? Are they innate and inevitable, or are they caused by environ-mental factors?

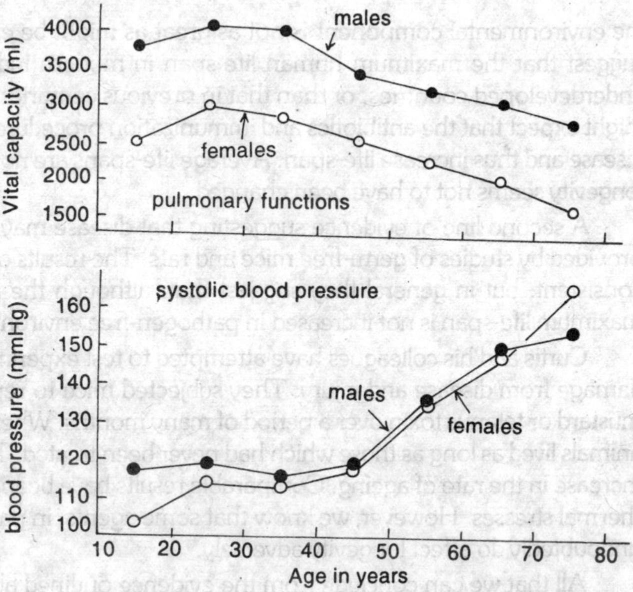

Fig. 37.4. Changes in pulmonary function and blood pressure with age.

From time to time it has been suggested that age-dependent loss of vigour is simply the result of pathological changes initiated by earlier disease or by other forms of wear and tear and injury to the body. In the early part of this century, Metchnikoff, the discoverer of phagocytosis, proposed that many of the adverse effects of increasing age would be prevented if harmful microbes could be eliminated from the body. In particular, he believed that the bacteria in the large intestine were harmful because they produced toxins which passed into the lymph and blood, and led to degenration of other parts of the body. He suggested that elimination of the harmful gut bacteria, elimination of syphilis, and suppression of alcoholism would prevent many of the pathological symptoms of old age and probably increase the length of life.

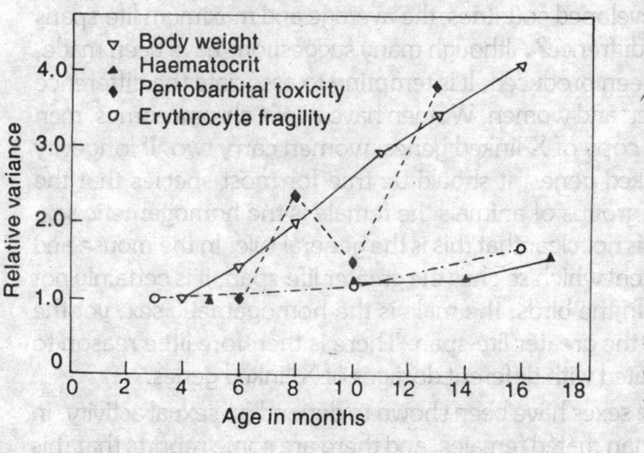

Fig. 37.5. Age-related changes in the variance of body weight, erythrocyte fragility, pentobarbital toxicity, and haematocrit for male mice.

Although in this extreme form the idea that ageing is due to disease now receives little support, it cannot be denied that previous disease and injury must make some contribution to the age-related increase in the probability of dying. Animals are constantly exposed to infections which are likely to cause minor damage to cells and tissues. In mammals, teeth are damaged and lost, and cannot be replaced. Injuries such as limb fractures occur; usually they heal, but the healed limb may not be as efficient as it was before it has lived longer—accumulate more and more of the type of damage associated with minor diseases and injuries.

At present, few people seem prepared to accept that this accumulation of damage due to earlier infection and injury of environmental origin is the sole cause of senescence. Rather, senescence is seen as an innate process which would occur even in the absence of injury and disease. Separation of the environmental and innate components of age-related deterioration is difficult, but several pieces of evidence suggest that the environmental component is not as great as might be expected. First of all, there is no evidence to suggest that the maximum human life-span in modern industrial societies is any greater than that in underdeveloped countries, or than that in previous centuries. If previous disease history is important, we might expect that the antibiotics and immunization procedures now used would reduce the damage due to disease and thus increase life-span. Average life-spans are now greater in western societies, but maximum longevity seems not to have been changed.

A second line of evidence suggesting that disease may not be an important cause of ageing is that provided by studies of germ-free mice and rats. The results obtained by different workers are not entirely consistent, but in general they suggest that, although the average age at death may be different, the maximum life-span is not increased in pathogen-free environment.

Curtis and his colleagues have attempted to test experimentally the idea that ageing is due to residual damage from disease and injury. They subjected mice to repeated doses of substances such as nitrogen mustard or tetanus toxin over a period of many months. When the stress was removed, they found that the animals lived as long as those which had never been treated. Thus these stresses did not appear to cause an increase in the rate of ageing. Comparable results have been obtained with *D. melanogaster* subjected to thermal stresses. However, we know that some agents, in particular sub-lethal doses of ionizing radiation, undoubtedly do affect longevity adversely.

All that we can conclude from the evidence outlined above is that there is no compelling reason to believe that ageing changes are caused solely by the cumulative residual effects of disease, injury, and wear and tear. Environmental factors such as disease and injury must play some part, but it is not clear how big a part it is.

Sex and Longevity

A number of people have stated that in most species the female sex outlives the male. This is certainly true in man where, in both developed and underdeveloped countries, the average and maximum life-spans are greater for women. What is the reason for this diffrence? Although many suggestions have been made, no really satisfactory explanation seems to have been produced. It is tempting to associate the difference with the different chromosomal constitutions of men and women. Women have two X chromosomes, men a single X and a Y; men therefore carry only one copy of X-linked genes, women carry two. If longevity differences are due to different dosages of X-linked genes, it should be true for most species that the females live longer than the males, since in most groups of animals the female is the homogametic sex. However, in spite of statements to the contrary, it is not clear that this is the general rule. In the mouse and in *Drosophila,* the strain determines to a large extent which sex has the greater life-span; it is certainly not true that females always live longer than males. In the birds, the male is the homogametic sex, yet the available data suggest that the female usually has the greater life-span. There is therefore little reason to believe that sex differences in longevity are correlated with different dosages of X-linked genes.

In some animals the relative longevities of the sexes have been shown to depend on sexual activity. In *Drosophila,* virgin females generally live longer than mated females, and there are some reports that this is also true for males. For females, the shorter life-span of mated flies is known to be associated with their high rate of egg-production. A mated female which is semi-sterilized by a heat shock, a chemosterilant, or a low dose of ionizing radiation, will live as long as or longer than a virgin female kept in the same conditions. The size of the difference between the life-spans of virgin and mated females depends on the environmental conditions in which they are kept.

In mice, too, virgin female usually live longer than mated females, but it is not clear whether this difference between virgin and mated animals is widespread in the animal kingdom. Marital status certainly affects longevity in human populations, but not in the way which would be expected from the data for experimental animals. The unmarried usually have shorter life-spans than the married. However, since widowed and divorced people have average life-spans which are shorter than those of unmarried poeple, it seems likely that these differences are associated with social and environmental factors rather than with strictly biological differences.

Molecular Mechanisms of Ageing

The Free-radical Hypothesis

Free radicals are chemical species which contain an unpaired electron in an outer orbital. This unpaired electron makes them very reactive. They are produced as transient intermediates in the course of normal metabolism, as for example in the oxidative processes found in mitochondria; but it has been suggested that some of these free radicals produced in metabolic reactions, as well as those of random spontaneous origin, may contribute to ageing processes. Since free radicals may attack important molecules such as DNA, proteins and lipids, and since they also tend to be self-propagating, they are capable of generating considerable damage.

The hypothesis that ageing changes are the result of accumulated free-radical damage was based in part on the observation that some types of radiation injury resemble the changes seen in natural ageing. Much of the damage caused by ionizing radiation is thought to be initiated by the ionization of water, and the formation and subsequent reactions of OH•, OOH• and H• radicals. These radicals can react with organic molecules, and lead to polymerization and peroxidation. Although both the damage cuased by radiation and that accumulating during natural ageing may be initiated by free-radical attack, there is no reason to expect the pathologies to be exactly the same, because the distribution of free radicals produced by irradiation will be different from that of free radicals produced in non-irradiated cells.

Free radicals can attack proteins and DNA, and may lead to cross-linkages within or between these molecules. However, it seems to be generally agreed that intracellular membranes are likely to be particularly susceptible to free-radical damage because they are so rich in unsaturated fatty acids. Free-radical action induces peroxidation of unsaturated fatty acids, and this could affect the functioning of the membranes. For example, reactions producing cross-linking or lipid-chain splitting in mitochondrial membranes could render mitochondira non-functional. The single membranes of lysosomes might, if damaged, break down or become leaky, so that hydrolytic enzymes are released; these enzymes might in turn cause damage to other cytoplasmic and nuclear components, and eventually to the extracellular material. Peroxidation reactions in membrane lipids might affect transport processes across cell and organelle membranes, and thus affect cell functioning.

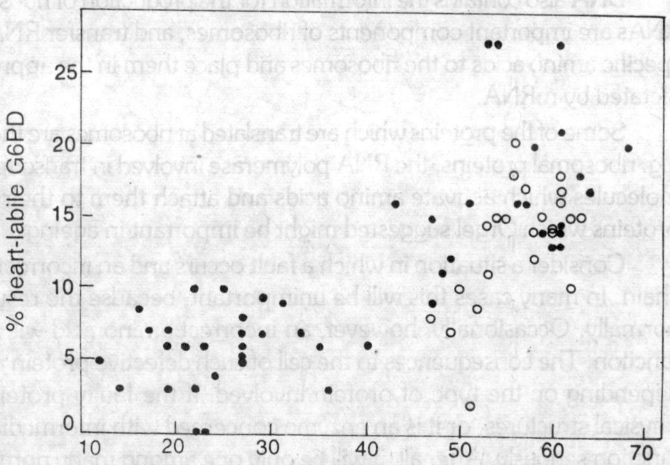

Fig. 37.6. Changes in the amount of heat-labile enzyme in human fibroblast cultures. Closed circles are measurements with assorted cell lines of different ages; open circles are successive measurements on a single cell line.

According to Gordon (1974), many of the changes that lipid peroxidation injury would be expected to produce are in fact found in old animals. For example, in blowflies the flight-muscle mitochondria show ultrastructural changes with age, and they also have a decreased ability to respire normally. The accumulation of lipofuscin pigment in cells is an excellent indicator of age, and some people believe that this substance is the end-product formed when membranes are damaged and cannot be completely metabolized by the cells. It is worth noting that one of the richest deposits of lipofuscin is in the rodent testis, which also has a very high unsaturated lipid content. Furthermore, vitamin E deficiency and hyperoxia both increase the rate of lipofuscin accumulation, and they are also known to increase free-radical action on unsaturated lipids. Conversely, animals fed on high vitamin E diets accumulate lipofuscin more slowly.

If ageing changes are caused by free-radical damage to membranes and other cellular components, substances whch scavenge free radicals or stabilize cell membranes would be expected to prolong the life-span of cells and animals. There is some evidence which suggests that they do. Vitamin E has been added to the diet of *Drosophila* and mice, and in some cases it has been found to produce an increase in life-span. Some artificial antioxidants have also been found to have longevity enhancing effects, although the results of this type of experiment have not been entirely consistent, and they can often be interpreted in the other ways, e.g. the antioxidants may affect longevity because they affect food intake by the animals.

The Error Catastrophe Hypothesis

In 1963 Orgel suggested that a progressive decrease in the accuracy of protein synthesis might be one of the factor contributing to the age-related deterioration of cells. He pointed out that metabolic errors are inevidtable and, although most of them are unlikely to be important, some might lead to further errors and eventually result in the cell containing so many faulty molecules that it is no longer able to function normally.

Information about proteins is coded in DNA in the nucleus. Protein synthesis involves the transcription and translation of this information. For any particular protein, the base sequences in DNA which code for the polypeptide chains are first transcribed into messenger RNA. At the ribosomes in the cytoplasm, mRNA is translated into the sequences of amino acids which make up the polypeptide chains.

DNA also contains the information for the production of ribosomal RNAs and transfer RNAs. Ribosomal RNAs are important components of ribosomes, and transfer RNAs are adaptor molecules which transport specific amino acids to the ribosomes and place them in the appropriate positions in the polypeptide chain dictated by mRNA.

Some of the proteins which are translated at ribosomes are themselves part of the translation machinery, e.g. ribosomal proteins, the RNA polymerase involved in transcription, and the aminoacyl-tRNA synthetase molecules which activate amino acids and attach them to their appropriate tRNA molecules. It is these proteins which Orgel suggested might be important in ageing.

Consider a situation in which a fault occurs and an incorrect amino acid is inserted into a polypeptide chain. In many cases this will be unimportant, because the resulting protein will still be able to function normally. Occasionally, however, an incorrect amino acid will result in the protein having an impaired function. The consequences to the cell of such defective protein molecules may be negligible or disastrous, depending on the type of protein involved. If the faulty protein is one which is involved in short-lived physical structures, or it is an enzyme concerned with intermediary metabolism, it is unlikely to affect cell function seriously. After all, it will be only one among many normal protein molecules, and it will probably be degraded soon after it is formed. Its existence will have no long-term consequences. If, however, the faulty protein molecule is one which is involved in the transcription and translation machinery of the cell, its presence could have very serious secondary effects. For example, a defective RNA polymerase molecule could cause mistranscription of the DNA message and hence induce further errors; a mistake in a tRNA synthetase molecule could lead to incorrect amino acids being inserted into growing polypeptide chains. Since the enzymes involved in transcription and translation will play a part in the formation of many new protein molecules,and since some of these new protein molecules may themselves be part of the transcription and translation machinery, the original defective protein might lead to what Orgel called "an error catastrophe" Such an error catastrophe might be responsible for the deterioration and death of a cell.

If Orgel's hypothesis is correct, then it leads to at least three testable predictions. The first is that the degenerative changes accompanying ageing should be primarily to cytoplasmic rather than nuclear origin. The second is that abnormal proteins should be detectable in old cells and animals. The third prediction is that any natural or artificially induced increase in the frequency of errors in protein synthesis should lead to premature senescence. We shall consider each of these predictions in turn.

The second prediction of Orgel's hypothesis is that with increasing age there should be an increase in the amount of defective protein. Direct evidence for old cells containing higher levels of proteins with incorrect amino acids is not available, because detecting these errors is technically very difficult. Nevertheless, it has been shown that with advancing age tissues and cells contain defective enzymes which could have been produced by mistranslation.

It is known that random amino-acid substitutions frequently make enzymes more susceptible to heat denaturation. Holliday and Tarrant (1972) therefore measured the size of the heat-labile fraction for the enzymes glucose-6-phosphate dehydrogenase (G6PD) and 6-phosphogluconate dehydrogenase (6PGD) extracted from fibroblast cells from cultures of different ages. Some of their results are shown in figure. Young cells had only about 5% of heat-labile enzyme but, after passage 40, the amount increased rapidly and reached 27% in some late-passage cells. Both of the enzymes studied behaved in a similar way.

There is also some evidence that the longevity difference within a species may be associated with different rates of accumulation of defective enzyme. Fibroblasts from the skin of people with Werner's syndrome have a reduced doubling potential in culture. They also show an increased amount of heat-labile enzyme at an early passage level. Therefore this type of genetically determined premature senescence is accompanied by a relatively early increase in the level of defective protein molecules. It has been suggested that the mutation involved here may be in a gene coding for part of the machinery of protein synthesis.

The third prediction of Orgel's hypothesis, namely that increasing the protein error frequency should increase the rate of ageing ánd shorten the life-span, was in fact the first to be studied experimentally. Orgel suggested that this prediction could be tested by feeding animals with analogues of amino acids or RNA bases. Analogues are substances which are sufficiently similar to normal amino acids or RNA bases to be incorporated into newly synthesized protein or RNA molecules, but are different enough to impair the function of some of the molecules which contain them. Since they cause defects in function, incorporation of low levels of these analogues should, if Orgel's hypothesis is correct, initiate an error catastrophe which results in premature senescence.

It was found that feeding *Drosophila* larvae with amino-acid analogues for short periods did indeed shorten the life-span of the imagoes. However, later work showed that, contrary to prediction, feeding adult flies with amino-acid analogues had no effect on their life-span, even though there was evidence that the analogues had been incorporated into proteins. It was suggested, therefore, that larval feeding may have been effective in shortening adult life-span, not because an error catastrophe was initiated, but rather because some of the long-lived proteins which are formed in the pre-adult stages are not replaced in adult flies contained analogue and were therefore less efficient.

Elimination of Error-Containing Cells

If errors occur spontaneously and can lead to further errors, why is it that *all* cell lineages do not die out? How is it HeLa cell cultures can continue to multiply indefinitely? More importantly, how do germ cells escape from the apparent inevitability of senescence?

It seems likely that many cells which contain defects are selected against both *in vivo* and *in vitro* because the grow more slowly and divide less frequently than non-defective cells. A cell line dies out only if, on the average, each cell produces less than one viable descendent. Considerable selection between defective and non-defective cells can therefore occur in cultures or *in vivo*. However, it is probably possible for cells to contain defective molecules which do not affect their ability to divide and compete with normal cells. in simple symmetrical division during logarithmic growth cells containing such faulty molecules, or the breakdown products of faulty structures, will pass them to both of their daughter cells. Gradually, therefore, a whole clone of cells should come to contain more and more defective molecules and waste products, and die out.

Holliday and his colleagues have made some interesting calculations about the behaviour of cells in culture which show how (i) culture size, (ii) the probability that a critical error committing a clone to mortality will occur, and (iii) the number of cell divisions between initiation of the critical event and the ultimate death of the clone, all affect the likelihood of a culture continuing. One of the rather paradoxical conclusions that came from these calculations is that cultures of transformed cells may be immortal because errors can build up more rapidly in these cells than in normal diploid cells. The reason for this is that if the number of divisions between the initiation of the critical event which irreversibly commits a cell's descendents to death and the actual death of the clone is high, the culture will come to consist of more and more committed cells. Selection between committed and uncommitted cells cannot occur until death or a reduction in proliferation rate takes place. By the time that this reduction in division rate occurs, it may be too late for the culture to survive, because all the uncommitted cells will have been diluted out by the subculturing technique. On the other hand, if after the initiating event the build-up of errors is so rapid that the cells die shortly afterwards, selection between committed and uncommitted cells will occur. The uncommitted cells will continue to grow and divide and will predominate in the culture. Thus, in cultures of limited size, for the same rate of initiation of critical errors, rapid build-up of further errors will be associated with immortality, whereas slow build-up of errors will be associated with a limited culture life-span. Transformed cells might build up errors more rapidly if, for example, they are less efficient at repairing or degrading defective molecules. Thus "inferior" cells could have the longest clonal life-span.

Repair and Elimination of Error-Containing Molecules

Although in renewing or expanding tissues selection between cells may enable those with defects to be eliminated and replaced by new cells, cell replacement obviously cannot occur in non-renewing tissues. For this reason alone it would be expected that, since it seems likely that metabolic errors are inevitable, selection would favour mechanisms which repair damaged molecules or structures, or break them down into metabolically inactive forms.

There is a growing body of evidence suggesting that cells are indeed capable of eliminating or repairing damaged molecules. We have already considered the accumulation of lipofuscin in cells of old animals. This pigment is possibly the end-product left after the breakdown of cell organelles containing membranes which have suffered peroxidation damage. Although the accumulation of this pigment may ultimately affect cellular function, it seems likely that it is less detrimental than damaged mitochondria or endoplasmic reticulum would be.

There is evidence that *Escherichia coli* has a mechanism for the selective degradation of abnormal proteins. In this bacterium incomplete polypeptide chains, proteins which contain translation errors induced by mutations affecting ribosomes or tRNA specificity, and proteins containing amino-acid analogues, are all degraded more rapidly than normal cell proteins. It is becoming clear that the cells of eukaryotes also have a comparable system for scavenging faulty proteins (Goldberg and Dice, 1974).

It has been known for a long time that the cells of eukaryotes are able to repair damaged DNA. Largely as a result of studies of the repair of damage produced by exposure to ionizing and ultraviolet radiation, several different mechanisms for DNA repair have been identified. The details of these mechanisms are not important for our present discussion, but it is important to realize that different enzyme systems are involved in the various types of repair. Thus a cell may be capable of repairing ultraviolet-induced damage, but not damage produced by X-radiation, and *vice versa*.

Recently a great deal of interest has been shown in the possibility that defective or deficient repair of DNA damage may be one of the causes of ageing. It has been found that old tissues do show evidence of an accumulation of DNA damage. For example, Price *et.al.* (1971) found that nuclear DNA in sections from a number of different tissues from old mice had a higher template activity for calf thymus DNA polymerase than that in young tissues. DNA polymerase catalyses the synthesis of a complementary strand to single-stranded regions of DNA. Therefore, if tissues are incubated with DNA polymerase and radioactivity-labelled DNA precursors, the amount of radioactive label incorporated into the nuclei is a measure of the amount of single-stranded DNA present. Table shows that the amount of label incorporated was greater for nuclei in old tissues than it was for those in young tissues. This indicates that senescent-cell DNA contains more single-strand breaks than young-cell DNA.

Confirmation of this observation has come from work with beagles. By using a completely different technique, Wheeler and Lett (1974) found that the molecular size of the DNA extracted from tissues of old dogs was smaller than that from young dogs. This too suggests that old DNA has more breaks than young DNA. However, when additional breaks were induced by radiation, it was found that old and young dogs were equally capable of restoring these breaks. Therefore, although there is evidence that DNA in old animal contains more strand breaks than that in young animals, this may not be because old cells are less capable of repairing DNA damage. It might be that the DNA in old cells suffers more damage due to increased lysosomal enzyme activity or free radical attack. Whatever the cause of the increase in strand breaks, these breaks are probably detrimental to cell function, since they may affect the ability of RNA polymerase to transcribe some messages from DNA.

Table 37.1. DNA polymerase catalysed incorporation of radioactive DNA precursors into cells of old and young mice.

Cell type	*Mean grain count per nucleus*			
	Young		*Old*	
	Undenatured	*Acid denatured*	*Undenatured*	*Acid denatured*
Neurones and astrocytes	8.92	30	22.13	very high
Liver Kupffer cells	2	4.5	6.28	13.9
Cardiac muscle	0.67	1.58	3.5	8.63

One case where an association between ageing and defective DNA repair has been found is in fibroblast cells from people suffering from Hutchinson-Gilford progeria (Epstein *et. al.*, 1973). These cells have been shown to be defective in their ability to repair X-ray induced DNA damage, and the cultures also have very short life-spans. However, it may be incorrect to deduce from this that deficent DNA repair is the cause of the reduced life-span of the patients or of their cells. Xeroderma pigmentosum is another genetic disease which is associated with defective DNA repair. Patients usually die from multiple skin cancers, since their skin is abnormally sensitive to sunlight. It has been shown that their cells cannot repair the DNA damage induced by ultraviolet radiation. Nevertheless, sufferers from this disease do not show signs of progeria, and the life-span of cultures of their fibroblasts is not abnormally short. Therefore deficient DNA repair is certainly not always associated with premature senescence.

In normal fibroblast cultures, the ability to repair DNA damage does not decline until relatively late in culture life-span. Since it occurs so late in the life-span, this decline is probably a consequence of some of the other changes which we have discussed in this chapter, rather than the cuase of the culture death.

At present, therefore, although we know that DNA damage increases with age, there is little reason for believing that there is an age-related deterioration in repair capacity. It must be remembered, however, that although old cells may still be able to repair their DNA, the quality of the repair may not be the same as in young cells.

38

CANCER

Cancer is an informal term for a diverse class of diseases marked by abnormal cell proliferation. Control of normal cell development is lost, and cell proliferate at an inappropriate rate to form growths known as *tumors* (*neoplasms*). Benign tumors grow only in one place and do not invade other tissues. The cells of malignant turmors not only continue to proliferate, but also invade nearby tissues or, by a process called *metastasis,* spread to distant parts of the body through blood or lymph vessels and start new centers of uncontrolled cell growth wherever they go.

Cancers are generally divided into four groups, dependent on the type of cells originally involved. Two types of cancer cause overproduction of white blood cells. *Leukemias* are diseases of the bone marrow that cause excessive production of leukocytes, which originate in the bone marrow. *Lymphomas* are diseases of the lymph nodes and spleen that cause excessive production of lymphocytes, which originate in the lymph nodes and spleen. *Sarcomas* are tumors of tissue that arise from the embryological mesoderm such as muscle, bone, and cartilage. About 85% of cancers are *carcinomas,* tumors arising from epithelial tissue such as glands, breast, skin, and linings of the urogenital and respiratory systems.

Most evidence indicates that cancers are *clonal*—that is, they arise from a single aberrant cell that then proliferates. Analyzing the cause, or causes, of cancer comes down to trying to understand how one cell is changed, or *transformed,* from a normal cell to cancerous one.

Currently, there are two theories regarding the causes of cancer: the mutation theory and the virus theory. The mutation theory states that most cancers are caused by genetic mutation. The viral theory states that most cancers are viral in origin. These theories are becoming one unified theory. In essence, cancers are genetic diseases; they result from the inappropriate activity of certain genes whether those genes changed from mutation or were imported or activated by viruses.

TYPES OF CANCER

About two hundred different types of cancers have been studied. These are broadly classified ten groups:

1. Carcinoma

Carcinoma includes tumours of brain, breast, skin and cervical region. These are derived from epithelial tissue, originating either from ectoderm or endoderm. These occur as solid tumours, located in the nervous tissue on body surfaces or in the associated glands.

Carcinoma are the commonest cancer. About 85 percent of cancers are carcinomas. Carcinomas may be:

(i) *squamous carcinomas*—The cancers of epithelial tissue and

(ii) *adeno carcinoma*—The tumours of glands.

2. Sarcoma

Sarcoma are tumours of connective tissue, cartilage, bone and muscles which are of mesodermal origin. These constitute about 2 percent of human cancers. Sarcomas are named according to the tissue in which these are formed:

(i) *fibrosarcomas*—Sarcomas of fibrous connective tissue (fibroblast cells).

(ii) *liposarcoma*—Cancer of adipose tissue (more often seen in thigh and buttocks).

(iii) *chondrosarcoma*—Cancer of cartilage cells (chondroblast cells).

(iv) *osteosarcoma*—Cancer of bone.

(v) *leiomyosarcoma*—Sarcoma of smooth muscles.

(vi) *rhabdomyosarcoma*—Sarcoma of striped muscles.

3. Lymphomas

In lymphomas, the lymph nodes, bone-marrow, liver and spleen produce excessive lymphocytes. These constitute about 5 percent of human cancers. *Hodgkin's disease* and *mycosis fungoides* are examples of lymphoma.

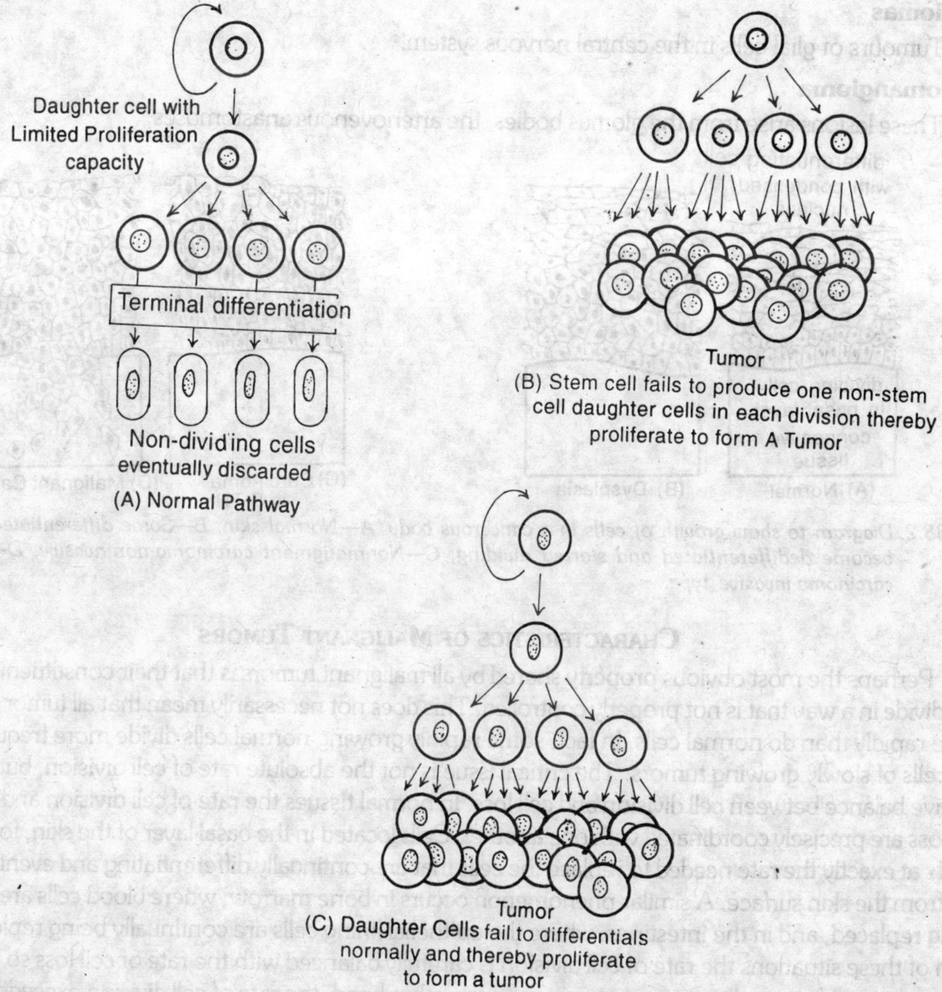

Fig. 38.1. Diagram showing difference in normal cells and neoplastic cells of tumour. A—Normal pathway where nondifferentiated stem cells divide and produce dividing stem cells and nondividing cells which differentiate according to the tissue. B—All the daughter cells are nondifferentiated dividing neoplastic cells which divide unchecked and form tumour or neoplastic growth. Nonstem cells are not formed. C—Daughter cells fail to differentiate and undergo rapid division to form a mass of neoplastic cells.

4. Leukaemia

These include the neoplastic proliferation of leukocyte precursors in the bone marrow and are described as *myeloproliferative disorders.* These lead to increase in the number of leukocytes that infiltrate into various other tissues and organs, producing general enlargement of organs or formation of tumours. Leukaemia constitutes about four percent of human cancers and can be:

(i) Monocytic leukaemia
(ii) Megakaryocytic leukaemia
(iii) Erythraemic myelosis
(iv) Promyelocytic leukaemia
(v) Lymphoblastic leukaemia
(vi) Myeloblastic leukaemia.

5. Melanomas

There are tumours of pigment cells.

6. Gliomas

Tumours of glial cells in the central nervous system.

7. Glomangioma

These lesions arise from the glomus bodies, the arteriovenous anastomoses.

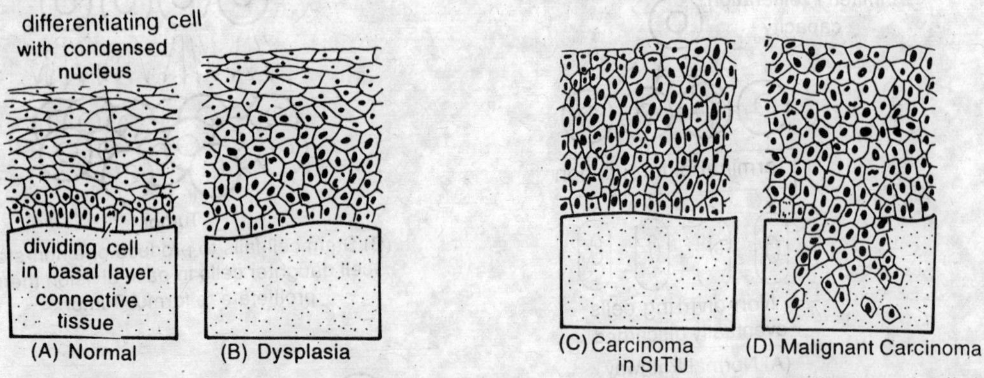

Fig. 38.2. Diagram to show growth of cells in a cancerous body. A—Normal skin, B—Some differentiated cells have become dedifferentiated and started dividing, C—Nonmaligmant carcinoma-noninvasive; D—Malignant carcinoma-invasive type.

Characteristics of Malignant Tumors

Perhaps the most obvious property shared by all malignant tumors is that their constituent cells grow and divide in a way that is not properly controlled. This does not necessarily mean that all tumor cells divide more rapidly than do normal cells. In fact, some rapidly growing normal cells divide more frequently than the cells of slowly growing tumors. The critical issue is not the absolute rate of cell division, but rather the relative balance between cell division and cell loss. In normal tissues the rate of cell division and the rate of cell loss are precisely coordinated with one another. Cells located in the basal layer of the skin, for example, divide at exactly the rate needed to replace the cells that are continually differentiating and eventually being lost from the skin surface. A similar phenomenon occurs in bone marrow, where blood cells are constantly being replaced, and in the intestines, where the epithelial lining cells are continually being replenished. In each of these situations the rate of cell division is carefully balanced with the rate of cell loss so that no net accumulation of new cells occurs. In tumors, on the other hand, the rate of cell division exceeds the rate of cell loss, causing the cell population to gradually become larger and larger. If the rate of cell division is rapid, the tumor will enlarge rapidly; if the rate of cell division is relatively slow, tumor growth will be slower. In either case the tumor continually increases in size. Because new cells are being produced in greater numbers than required, the normal organization of the tissue gradually becomes disrupted as more and more dividing cells accumulate.

Uncontrolled cell division is an attribute of malignant and benign tumors alike. The property that distinguishes these two types of tumors from each other is the ability of malignant tumors to spread throughout the body. Such spreading occurs in two ways: direct invasion of tumor cells into surrounding tissues, and penetration of tumor cells into vessels of the circulatory or lymphatic system. This latter process permits cancer cells to move throughout the body, where they establish secondary tumors referred to as *metastases.* Tumors often tend to establish metastases in characteristic locations. Breast cancer, for example, commonly metastasizes to bone and ovary, while lung tumors metastasize principally to the brain.

It is the ability of malignant tumors to invade and metastasize that makes them potentially life threatening, since tumor cells have often spread throughout the body by the time the cancer is diagnosed. Although malignant tumors as a class are therefore more serious than benign tumors, individual exceptions do occur. Basal cell carcinoma of the skin, for example, is a common type of cancer that is rarely lethal because it is very slow to metastasize and the tumors are readily removed surgically. Some benign brain tumors, on the other hand, are surgically inaccessible and are therefore lethal, in spite of the fact that they do not metastasize.

The question of why cancer cells do not behave like their normal counterparts has occupied cancer biologists since the turn of the century. Although some information has been obtained from observations made directly on human cancers, tissue culture and animal model systems have played important roles in helping us to decipher the cellular and molecular alterations underlying the malignant state. In the following section we shall summarize the major conclusions derived from these various kinds of analyses.

Profile of the Cancer Cell

There are several different ways in which the properties of cancer cells can be studied. The most straight forward approach is to analyze tissue taken directly from a naturally occurring tumor. Although such material is a reliable source of malignant cells, intact tumors usually contain a heterogeneous population of cell types, and more variability is encountered when one tries to compare tumors obtained from different individuals. The alternative to the direct examination of tumor tissue is to grow cancer cells in culture. Unfortunately, long periods in culture may cause the behaviour and makeup of cell populations to change, and the question therefore arises as to whether the properties exhibited by longterm cultures of cancer cells accurately reflect the behaviour of malignant cells in *vivo*. In spite of this potential drawback, the ability to study a defined population of cells under carefully controlled laboratory conditions has made tissue culture a popular approach among cancer biologists.

Cancer cell lines employed for tissue culture studies have been obtained by a variety of means. The most obvious approach is to remove cells from malignant tumors and culture them directly. Hundreds of human cancer cell lines, encompassing most of the major tissue types, have been established in this way. Alternatively, normal cells can be transformed into malignant ones while in culture. The first indication that malignant transformation can take place under such conditions came from studies carried out by George Gey and his associates in the early 1940s. These investigators made the rather remarkable discovery that rat fibroblasts maintained in culture for prolonged periods of time eventually acquire the ability to form sarcomas when injected back into healthy rats. This tendency of cultured normal cells to undergo spontaneous transformation into malignant cells has subsequently been documented for numerous cell types derived from mouse, hamster, and rat. Cells maintained in culture can also be induced to undergo malignant transformation by exposing them to cancer-causing agents such as chemicals, viruses, and radiation.

A vast amount of data has been acquired concerning the properties of malignant cells obtained in the various ways described above. We shall consider those generalizations that are most useful, keeping in mind that each has its own limitations and exceptions, and that none is completely adequate in itself as a criterion for the malignant state.

1. Tumorigenicity in Experimental Animals

The most obvious property distinguishing malignant cells from normal cells is the ability of malignant cells to produce tumors when injected into appropriate test organisms. For animal tumors this criterion is relatively easy to test, but with human tumor cells the situation is more complicated. Injection of cancer cells into human beings is clearly out of the question, and injecting human tumor cells into laboratory animals is not generally reliable because an animal's immune system may reject the human cells, independent of whether or not they are malignant. One way around this obstacle is to inject tumor cells into *nude mice*, an immunologically deficient strain of animals lacking a thymus gland. Because a thymus gland is required for production of the T-lymphocytes involved in cell-mediated immunity, nude mice are incapable of rejecting foreign cells. Hence human tumor cells injected into nude mice grow into tumors without immunological rejection. Unfortunately, failure to grow under those conditions does not absolutely rule out the possibility that the same cells might be capable of forming tumors under other conditions.

2. Indefinite Lifespan

Most normal cells maintained in culture have a limited life expectancy. Human fibroblasts, for example, multiply for about 50-60 generations and then begin to deteriorate and die. In contrast, malignant cells appear to have an indefinite lifespan. The human *HeLa* cell line, which was derived from a uterine carcinoma in 1953, has been dividing in culture ever since. Cells that have undergone transformation in culture behave in a similar fashion, a fact first revealed by studies carried out on mouse embryo fibroblasts by George Todaro and Howard Green. These investigators noted that within a few weeks after placement in culture the growth rate of these embryonic cells slowed down, suggesting that the cells would eventually stop dividing and die, as occurs with normal human fibroblasts. After two to three months, however, the growth rate of the cells was unexpectedly found to increase again. The newly emerging cell population, which had apparently undergone spontaneous transformation, formed permanently dividing cell lines that have now been growing in culture for over 20 years.

Although all malignant cells appear to have an unlimited lifespan when grown in culture, it does not necessarily follow that all nonmalignant cells have a limited lifespan in culture. Some of the "transformed" cell lines isolated by Todaro and Green are capable of indefinite growth in culture, yet do not form tumors when injected back into animals. This observation suggests that acquisition of an indefinite lifespan may be an early stage in the development of an malignancy, but that subsequent events must ensure before the malignant state is actually established.

3. Decreased Density-Dependent Inhibition of Growth

When normal cells are placed in culture, they move about and divide until the surface of the culture vessel is covered by a monolayer of cells. At this point, cell movement and cell division usually cease. In the early 1950s Michael Abercrombie and Joan Heaysman introduced the term *contact inhibition* to refer to the decrease in cell motility that occurs when cells contact one another in culture. This same term has also been used to refer to the inhibition of cell division that generally takes place when culture conditions become crowded. Because of the confusion that can result from this double meaning of the term contact inhibition, we shall use the phrase *density-dependent inhibition of growth* to refer to the inhibition of cell division that occurs in crowded cultures. This phrase simply describes the fact that cell division is inhibited as the population density increases, without implying anything about the underlying mechanism.

Cells that have undergone malignant transformation typically grow to higher population densities in culture than do their normal counterparts. Instead of growth ceasing when a complete monolayer of cells covers the surface of the culture vessel, malignant cells continue to divide and pile on top of one another, forming multilayered aggregates that contrast markedly with the uniform monolayer typically formed by normal cells. Tumor cell growth in culture is thus less density dependent than growth of normal cells.

The connection between the loss of density-independent growth *in vitro* and the ability to form malignant tumors *in vivo* has been difficult to establish. The first thorough investigation of this issue was carried out by Stuart Aaronson and George Todaro, who examined the properties of several fibroblast cell lines derived from the same batch of mouse embryonic tissue. By manipulating the culture conditions under which these cells were grown, it was found that cell lines differing in their susceptibility to density-dependent inhibition of growth could be established. One type of cell line was produced by growing cells under uncrowded conditions; every time the cell population density increased and crowding imminent, the cells were diluted and transferred to a new culture flask. The cell line produced under such conditions was found to be very sensitive to density-dependent growth inhibition. Another set of cell lines was established by growing cells under extensively crowded conditions, thereby selecting for cells able to grow under these conditions. The cell lines established by this second procedure were found to be less susceptible to density dependent inhibition of growth, reaching much higher population densities before growth ceased. When these various cell lines were tested for their ability to produce tumors when injected into mice, tumor-forming ability was found to be inversely correlated with the susceptibility to density-dependent inhibition of growth.

The question of why transformed cells are less susceptible to density-dependent inhibition of growth than nromal cells requires an understanding of the basic mechanism underlying this phenomenon in normal cells. Although the term *contact inhibition* was used for many years to refer to the tendency of cells to stop dividing in culture when conditions become crowded, there is no definitive evicence that cell-cell contact per se is responsible. The degree of crowding tolerated before cell division stops in normal cells is subject to considerable variation, depending on the culture conditions employed. For example, if the serum, concentration in the growth medium is too low, cells will stop dividing before cell-cell contacts have become established. On the other hand, cell cultures that have stopped dividing because of crowded conditions can be made to start proliferating again if additional serum is added to the medium. Such observations suggest that density-dependent growth inhibition is caused by depletion of nutrients or growth factors, rather than by cell-cell contact.

Transformed cells generally require less serum in the growth medium than do normal cells. An apparent explanation for this difference is suggested by experiments carried out in 1978 by Joseph DeLarco and George Todaro that revealed that transformed mouse fibroblasts manufacture and secrete their own growth factors. The existence of these factors was discovered by taking the culture fluid in which transformed fibroblasts had been growing, and adding it to a flask containing normal rat kidney cells whose growth had been arrested by serum deprivation. Under these conditions, the rat kidney cells began to synthesize DNA and divide. Eventually the culture was converted from a regular monolayer of normal cells to an unorganized mass of cell aggregates more typical of a transformed culture. Such observations suggest that malignant cells are less susceptible to density-dependent inhibition of growth because they produce their own growth factors. The growth factors involved, referred to as *transforming growth factors* (TGFs), have subsequently been isolated from the growth medium of transformed cell cultures. At least two distinct types of TGF have been identified in such media; TGF-α, a single polypeptide chain 50 amino acids long that acts by binding to the plasma membrane receptor for opidermal growth factor, and TGF-β, a larger protein containing two polypeptide subunits that is present in many normal tissues as well as transformed cells. Stimulation of cell division requires the synergistic action of TGF-β with TGF-α (or epidermal growth factor). The fact that tumor cells can manufacture their own TGF-β and TGF-α may explain why tumors growing within intact organisms manage to grow autonomously without being subject to the normal feedback systems that regulate normal cell growth and division.

4. Anchorage-Independent Growth

Most normal cells derived from solid tissues do not grow well in culture if they are suspended in a liquid medium or a semisolid material such as soft agar or methylcellulose. If provided with an appropriate surface to which they can adhere, however, normal cells will attach, spread out, and commence growth. This type of growth is therefore referred to as *anchorage dependent*. In contrast to the behaviour of normal cells, most transformed cells are capable of growing in liquid suspension or a semisolid medium, and are hence said to be *anchorage independent*.

An extensive set of experiments carried out in the mid 1970s in the laboratory of Robert Pollack suggest that the property of anchorage-independent growth is closely correlated with a cell's ability to form malignant tumors *in vivo*. In these studies, a series of transformed fibroblast cell lines were tested for their ability to induce tumors *in nude* mice. The parent cell line exhibited many of the typical characteristics of transformed cells: decreased density-dependent inhibition of growth, low serum requirement for growth, expression of new cells surface antigens, and anchorage-independent growth. By isolating single cells from the parent population and growing a group of clones, Pollack established individual cell lines that had lost one or more of the properties of the parent cell line. Of the four properties studied, only the retention of anchorage-independent growth was consistently associated with the ability to form tumors in animals.

This relationship between anchorage-independent growth and tumorigenicity is not, unfortunately, without exceptions. Some human cell lines exposed to carcinogenic chemicals exhibit anchorage-independent growth, yet are unable to induce tumors when injected into *nude* mice. Conversely mouse 3T3 fibroblasts, which are anchorage-dependent and normally unable to induce tumor formation, gain the capacity to produce tumors when they are attached to glass beads prior to implantation. Such observations suggest that in spite of its general correlation with tumorigenicity, anchorage-independent growth in culture is not an absolute prerequisite for tumour formation *in vitro*.

5. Loss of Restriction Point Control

When normal cells are exposed to suboptimal growth conditions (e.g., inadequate nutrients or growth factors), they stop at a specific point in the cell cycle situated near the end of the G_1-phase. The point in G_1 at which they are blocked is referred to as the *restriction point*. Cells held up at this stage of the cell cycle are said to be in the G_0-phase. Once cells emerge from G_0 or G_1, they are committed to continuing through S_1 G_2, and M. In contrast to normal cells, transformed cells do not exhibit restriction point control. They continue to cycle and divide under conditions of high cell density, low serum concentration, or suboptimal nutrient concentration that would cause normal cells to stop at the restriction point. Under conditions of severe nutritional deprivation, transformed cells stop at random points in the cell cycle and die, rather than entering into G_0.

Attempts have been made to design therapeutic schemes that might exploit this difference in the behaviour of normal and transformed cells. For example, Arthur Pardee and his associates have devised a simple scheme employing a combination of caffeine, which blocks normal cells at the restriction point, and *cytosine arabinoside*, which interferes with DNA synthesis and therefore kills cells situated in the S-phase. When normal cells are treated with a combination of caffeine and cytosine arabinoside, the caffeine blocks cells in G_1 and therefore prevents them from entering S-phase and from being killed by the S-phase specific agent, cytosine arabinoside. If the two drugs are later removed, the cells will begin dividing again. When transformed cells are treated with this drug combination, on the other hand, the caffeine does not stop the cells in G_1 because of the loss of restriction point control. The cells therefore progress into S-phase and are killed by the cytosine arabinoside. The fact that this approach selectively destroys transformed cells in culture suggests that a similar stratgegy might eventually be adapted to the treatment of human cancer.

6. Cell Surface Alterations

Changes in the behaviour and biochemical composition of the plasma membrane are almost universally observed in tumor cells. The significance of such changes is difficult to assess, however, because plasma membrane alterations also occur in normal cells stimulated to divide. For example, active transport systems for the uptake of sugars, amino acids, and nucleosides are frequently activated in tumor cells, but a similar increase in transport activity also occurs in normal cells stimulated to divide by the addition of appropriate nutrients or growth factors. It is therefore important to determine which (if any) membrane changes are restricted to the malignant state and which are characteristic of dividing cells in general. Most of the distinctive biochemical changes that occur in the plasma membranes of malignant cells involve alterations in glycoproteins and glycolipids. Among the functional properties of the cell surface that appear to be influenced by these changes are adhesiveness, agglutinability, cell-cell communication, and antigenicity.

Most tumor cell are less adhesive than normal because they are deficient in the cell surface glycoprotein, *fibronectin*. In addition to its role in promoting cell-cell adhesion, fibronectin is known to influence cell shape and motility. Hence the loss of cell surface fibronectin may underlie some of the behavioural properties of tumor cells, such as their tendency to invade neighbouring tissues and metastasize. Addition of fibronectin to malignant cells restores cell adhesion, flattened cell morphology, and contact inhibition of movement. It does not however, restore normal growth control (i.e., density-dependent inhibition of growth), indicating that depletion of fibronectin is not a primary factor in the loss of growth control.

A second cell surface property characteristic of malignant cells is their enhanced tendency to *agglutinate*, or clump together, when exposed to lectins. *Lectins* are carbohydrate-binding proteins whose multiple binding sites allow them to link adjacent cells together. A close correlation between loss of growth control and susceptibility to lectin-induced agglutination is suggested by the fact that transformed cells that have spontaneously regained density-dependent growth control simultaneously lose their enhanced agglutinability by lectins. The simplest explanation for the enhanced agglutinability exhibited by tumor cells would be an increased number of lectin receptors, but careful measurements have led to the surprising conclusion that the total number of lectin receptor sites is similar on normal and transformed cells. What does differ, however, is the mobility of the lectin receptors. In tumor cells these receptors diffuse more rapidly within the plane of the membrane, a phenomenon that appears to promote lectin-mediated cell agglutination. Because the plasma membranes of transformed cells do not generally appear to be more fluid than the plasma membranes of normal cells, it has been proposed that a decreased interaction between lectin receptors and the underlying cytoskeletal network is responsible for the enhanced mobility of lectin receptors in tumor cells.

A third cell surface property altered in malignant cells is the ability to establish communicating junctions. This conclusion is based on the observation that malignant cells generally exhibit a decreased number of gap junctions, as well as the discovery that fluorescent dyes injected into normal cells will diffuse rapidly into surrounding cells that are normal, but not into cells that are malignant. Experiments carried out in the laboratory of Werner Loewenstein suggest that this inability to communicate through gap junctions is intimately involved with the development of malignancy. In these experiments normal cells were fused with malignant cells that had lost the ability to form gap junctions. In most cases cell fusion resulted in restoration of both normal growth control and the ability to form gap junctions. Eventually, however, some of the fused cells reverted to uncontrolled growth. When this occurred, the ability to form gap junctions was also lost. From these experiments Loewenstein concluded that normal growth control requires the ability to communicate through gap junctions.

The final cell surface property altered in malignant cells is antigenic specificity. Although this area of research has had a controversial history, it is now well established that some cancers exhibit tumor-specific cell surface antigens. Part of the difficulty in reaching a consensus on this issue has derived from the fact that multiple classes of antigens are involved. In human malanomas, for example, at least three types of antigens have been identified. Antigens of the first type are specific both for melanomas and for the individual from whom a particular melanoma is obtained. Antigens of the second type are specific for melanomas, but not for the individual from which the tumor is obtained. Antigens of the above two kinds are not detectable on normal cells, and can therefore be classified as tumor specific. Antigens of the third type are present on normal cells as well as melanoma cells, but their concentration on melanoma cells is greater.

The existence of tumor-specific antigens (the first two classes described above) raises the question of why individuals afflicted with cancer do not reject their tumors. According to the *immune surveillance theory,* immune rejection of newly forming malignant cells does occur in healthy individuals, and cancer simply reflects the occasional failure of an adequate immune response to be mounted aberrant cells. If correct, this theory suggest that one possible approach to cancer therapy might involve attempts to stimulate an organism to immunologically reject its own tumor cells. Although this approach has yielded promising results in some animal tumor systems, it is yet to be developed into a practical approach for treating human cancer.

7. Other Molecular Changes

A great deal of effort has recently been expended in searching for specific molecular "markers" unique to malignant cells. Such markers would be useful aids for cancer diagnosis, and might also provide insights into the molecular origins of this disease. Though this search has led to the identification of dozens of molecular properties that appear to be altered in cancer cells, many of these modifications are also exhibited by rapidly dividing normal cells and therefore are not unique to cancer. In this section we shall restrict our discussion to a few molecular changes that have contributed the most to our understanding of malignancy.

It has been known for many years that malignant tumors secrete proteolytic enzymes in high concentration. Although similar enzymes are secreted by some normal tissues, enhanced production of proteases may be responsible for certain properties of cancer cells. For example, the decrease in adhesiveness and loss of anchorage dependence commonly exhibited by tumor cells may be caused by protease-mediated digestion of cell surface components. Secretion of proteases *in vivo* might also provide a partial explanation for the invasive properties of malignant tumors. The main protease secreted by cancer cells is *plasminogen activator,* an enzyme that catalyzes the proteolytic cleavage of an enzyme precursor known as *plasminogen.* This cleavage reaction converts plasminogen to *plasmin,* which is itself a protease. Plasminogen is a component of plasma and extracellular fluids, and its conversion to plasmin in the presence of the plasminogen activator may be responsible for many of the proteolytic effects associated with malignant tumors.

Some transformed cells also manufacture proteins typical of embryonic cells. One such molecule is *alphafetoprotein*, a protein synthesized by embryonic liver. The amount of alphafetoprotein present in the blood-stream of normal adults is quite low, but in patients with liver cancer the concentration increases dramatically. Another embryonic marker occasionally present in tumor cells is *carcinoembryonic antigen,* a glycoprotein normally made in the fetal digestive tract. Fetal hormones, such as chorionic gonadotropin and placental lactogen, are also produced by some human tumors. The production of embryonic components by malignant tumors may relate to the fact that certain properties of cancer cells, such as invasiveness and rapid proliferation, are also manifested by embryonic cells. Testing for the presence of embryonic markers such as alphafetoprotein and carcinoembryonic antigen in the bloodstream has been touted as a potential tool for the early diagosis of cancer, but the fact that these markers are far from universally present in cancer patients limits the usefulness of this approach.

Some tumors also secrete a substance, known as *tumor anginogenesis factor* or *angiogenin,* which stimulates the formation of new blood-vessels. Since the presence of an adequate blood supply is a critical factor in determining a tumor's ability to grow and invade surrounding tissue, production of such angiogenesis factors may significantly influence tumor behaviour. Certain nonvascular tissues, such as cartilage, have been found to produce inhibitors whose ability to block tumor-induced angiogenesis may eventually be exploited for therapeutic uses. In preliminary studies on animal tumors such inhibitors have been found to depress the growth rate of tumor cells by interfering with the development of their blood supply.

One of the most striking molecular alterations observed in cancer cells involves changes in the genetic information encoded in their DNA. This alteration has been most dramatically demonstrated by exposing normal cells to DNA extracted from cancer cells. For example, if normal mouse fibroblasts are exposed to DNA extracted from human bladder cancer cells, the fibroblasts undergo malignant transformation. If the fibroblasts are exposed to DNA obtained from normal cells, on the other hand, transformation does not occur. Taken together, such observations suggest that the DNA of the tumor cells is somehow altered.

In order to ascertain whether the same gene sequence is always altered in malignant transformation. Robert Weinberg and his associates have investigated the properties of the DNA isolated from four different lines of transformed fibroblasts. In terms of their susceptibility to inactivation by restriction endonuclease cleavage, the transformign DNAs obtained from all four cell lines were found to behave similarly. It was therefore concluded that the same gene is altered in each of the four transformed cell lines. However, when experiments of the same type were carried out on DNAs obtained from three kinds of human cancer cells (colon cancer, bladder cancer, and leukemia), three different genes were found to be altered. Thus a different gene is involved when each of these cell types undergoes malignant transformation. As we shall see later, studies on the genes involved in malignant transformation and the proteins they encode are beginning to provide profound insights into the molecular basis of cancer.

8. Morphological Changes

Although no single morphological criterion is adequate for distinguishing cancer cells from normal cells, alterations in cell structure do occur in malignant tumors. Most notable is the tendency of malignant cells to undergo *anaplasia*, a process involving a loss of differentiation and proper orientation of cells to one another. Cancer cells commonly exhibit large nuclei, prominent nucleoli, and a cell surface covered with microvilli and lamellopodia. The number of cells undergoing mitosis is generally elevated in malignant tumors, and abnormal mitosis and multinucleated giant cells are frequently encountered. Irregularities in chromosome number and structure are also common. For example the *Philadelphia* chromosome, an abnormally shaped chromosome caused by translocation of a piece of chromosome 22 to chromosome 9, occurs in nearly 90 percent of the individuals suffering from chronic granulocytic leukemia.

Although none of these morphological attributes is sufficient in itself for unequivocal identifying cancer cells, in combination they are quite useful for diagnosing the presence of malignancy. It is important to note, however, that many of the above-mentioned characteristics are also manifested by some normal cells, especially in tissues where rapid cell division is occurring. As a result, the final diagnosis of malignancy may depend on criteria other than cell morphology, such as the occurrence of invasion and metastasis, or the presence of embryonic or tumor antigens.

MUTATION THEORY OF CANCER

Mutation, both point and chromosomal, have been implicated in carcinogenesis. For example, the disease *xeroderma pigmentosum* in human beings is caused by mutations that inactivate the UV-mutation repair system so that exposure to the sun results in skin lesions that often become malignant. A related disease, *ataxia-telangiectasia*, is a defect in X-ray-induced repair mechanisms. Persons with this defect are at risk for acute and chronic leukemia, lymphomas, and, in women, ovarian cancers.

People with certain other diseases have a higher than normal risk of developing cancer. *Fanconi's anemia* is a syndrome of malformations of the heart, kidney, and extremities, pigmentary changes of the skin, and changes in the bone marrow. It is associated with acute leukemia and with cancers of the skin, liver, and esophagus. Relatives of people with Fanconi's anemia re also prone to malignancies; the risk of cancer mortality before age forty-five for heterozygotes is increased three- to sixfold. On the basis of risk and gene frequency, as many as 1% of persons dying before age forty-five from any malignancy may be heterozygous for Fanconi's anemia, and as may as 5% may be heterozygous for ataxia-telangiectasia.

Most cancers are associated with chromosomal defects; improved chromosomal banding techniques have demonstrated that a specific chromosomal defect is often associated with a specific cancer. The implication is that because of the new location of a gene (by translocation or deletion of intervening material) that gene may fall under the control of more powerful promoters or hybrid promoters outside the range of that gene's normal control. As we shall see, very often genes that are known to be able to tranform cells (*oncogenes*) are the ones that are relocated into regions of new control. These oncogenes then become more active and transformation follows.

Cancer-Family Syndromes

Except for malignancies that are known to be controlled by a single gene, such as xeroderma pigmentosum, clinicians disagree as to whether or not a predisposition ofr malignancies is inherited. One view holds that the human population is heterogeneous, with a low-risk group and a high-rish group. When four thousand clinic registrants were interviewed, almost half reported virtually no family history of cancer, whereas about 7% reported that many family members had cancer. This 7% was considered cancer prone on the basis of the fact that three or more close relatives of the interviewed person had cancer. The interpretation of the study is that some families are predisposed toward cancer but most are not, rather than that everyone in the population has a uniform and low probability of developing cancer. Lending support to this interpretation are the *cancer-family syndromes,* in which family members seem to inherit a nonspecific predisposition toward tumors of various types. At least two cancer-family syndromes are known. A pedigree of a cancer-family syndrome in which the predisposition for several different types of cancers, rather than a particular type of cancer, seems to be inherited. Women get breast, colon, and pancreatic cancers, whereas men get colon, prostate, and pancreatic cancers. There also seem to be genes that suppress cancer.

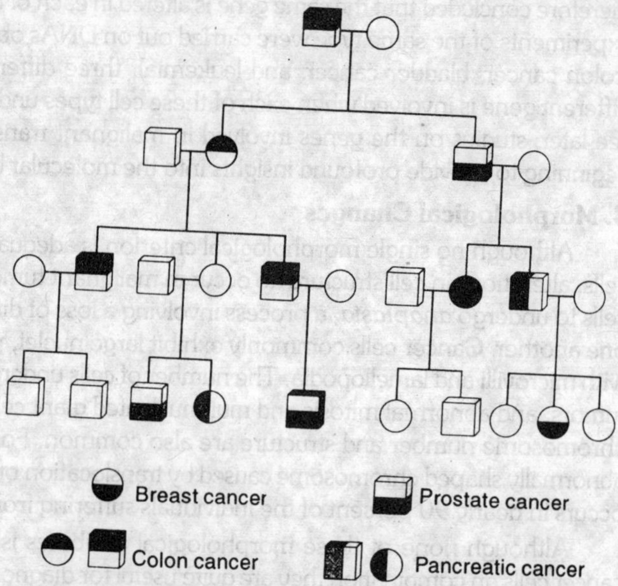

Fig. 38.3. Pedigree of a type I cancer-family syndrome. This is interpreted as the inheritance of the propensity toward cancer rather than the inheritance of any specific type of cancer.

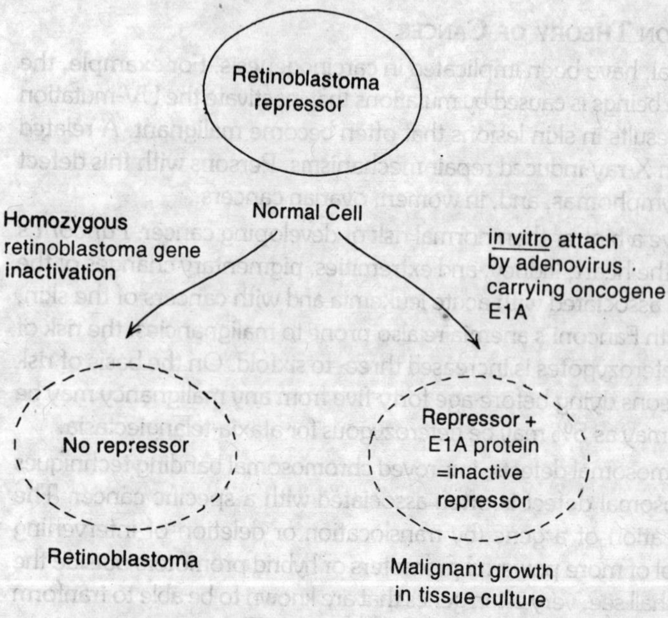

Fig. 38.4. Malignancy can result from lack of the retinoblastoma repressor protein either through homozygous deletion (or other mutagenesis) or through inactivation of the repressor by a viral oncogene protein. If the retinoblastoma gene itself is inactivated, retinoblastoma will result during early development of the individual. If the repressor is inactivated later in development, the inactivating agent (e.g., adenovirus) can cause other types of malignancies. For example, the E1A protein of adenovirus will inactivate the retinoblastoma repressor, leading to malignant growth in tissue culture.

Tumor-Supressor Genes

Recently, a new class of cancer-related genes has been discovered, the so-called *tumor-suppressor genes* (also called anti-cancer or *anti-oncogenes*). These genes seem to act by suppressing malignant growth; in the homozygous recessive state, however, cancer ensues. The first tumor-suppressor gene to be isolated was the gene for *retinoblastoma,* a tumor of retinoblast cells, which are precursors to cone cells in the retina of the eye. This is a disease of young children because after the retinoblast cells differentiate, they no longer divide and apparently can no longer form tumors. The disease occurs both in a hereditary and a sporadic form. Both forms are presumably due to the recessive homozygous (or, in some cases, hemizygous) state of the locus. In the hereditary form, individuals inherit one mutant allele; a second mutation generates the disease. In the sporadic form (with identical symptoms), apparently both alleles have mutated spontaneously in the somatic tissue of the retina. Alfred

Knudsen made this suggestion in 1971, that two mutations would be required for retinoblastoma. The retinoblastoma gene has also been implicated in other cancers including sarcomas and carcinomas of the lung, bladder, and breast.

How do we know that retinoblastoma results from the loss of suppression rather than simply the activity of an oncogene? This question was answered by J. Yunis, who examined cells from several retinoblastoma patients. He found that there was frequently a deletion of part of chromosome 13, specifically

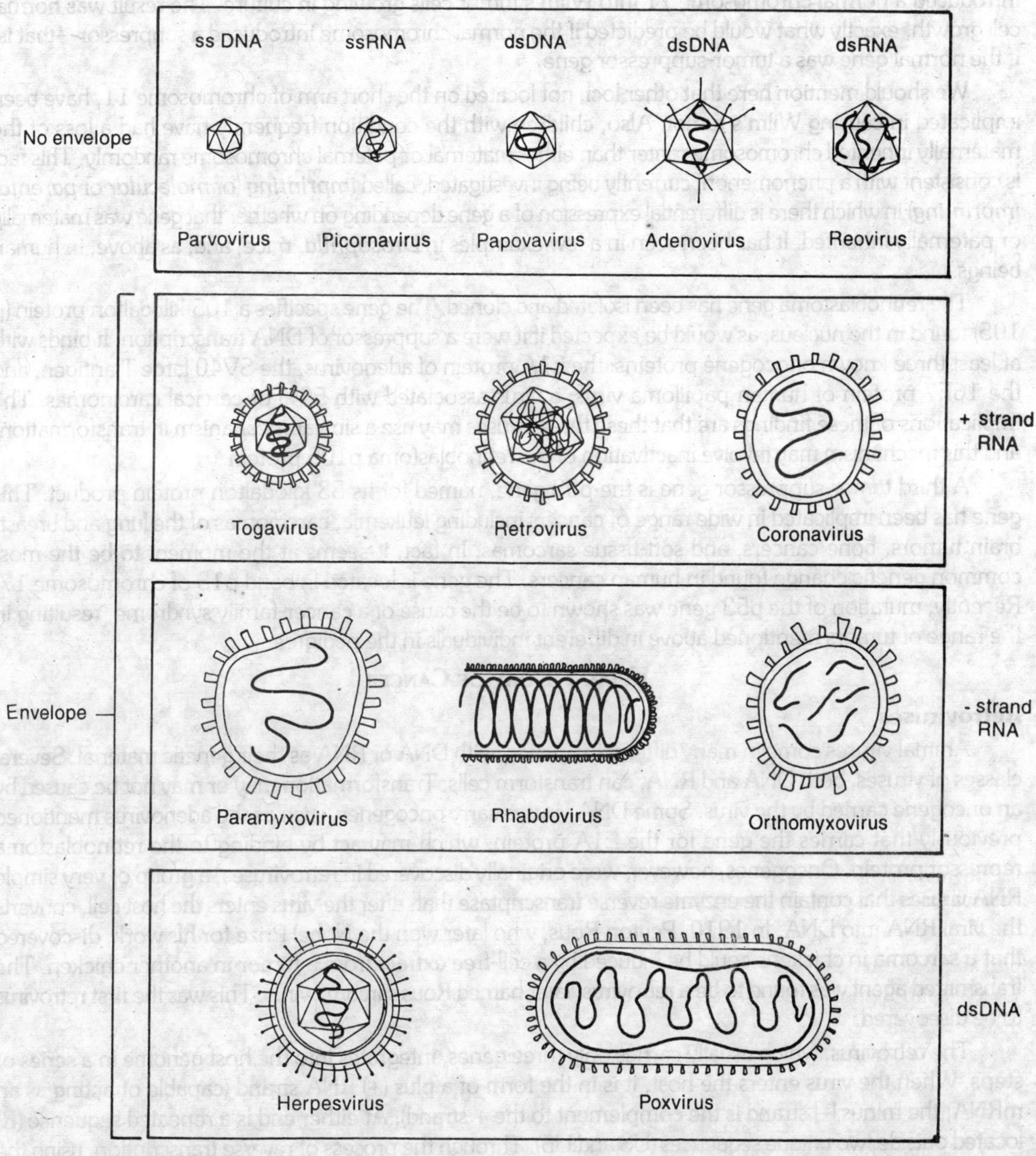

Fig. 38.5. Representatives of families of animal viruses. The abbreviations ss and ds refer to single-stranded and double-stranded, respectively. The viruses are drawn to size. The plus (+) and (–) signs define the strand of an RNA virus. Plus strands can serve as messengers; minus strands are complementary to the messenger strand. The retrovirus is about 100 nm in diameter.

band q14. Yunis noticed that the exact points of deletion varied from individual to individual, indicating that the phenomenon was due to loss of gene action rather than enhancement of gene activity due to the new placement of gene previously separated by the deleted material.

Further support of tumor-suppressor genes came from work by E. Stanbridge and his colleagues with another childhood cancer, *Wilm's tumor,* a kidney cancer that is also believed to be caused by the loss of activity of a tumor-suppressor gene. It is associated with the loss of band p13 on chromosome 11. Researchers introduced a normal chromosome 11 into Wilm's tumor cells growing in culture. The result was normal cell growth, exactly what would be predicted if the normal chromosome introduced a suppressor—that is, if the normal gene was a tumor-suppressor gene.

We should mention here that other loci, not located on the short arm of chromosome 11, have been implicated in causing Wilm's tumor. Also, children with the condition frequently have had a loss of the maternally inherited chromosome rahter than either maternal or paternal chromosome randomly. This fact is consistent with a phenomenon, currently being investigated, called *imprinting* (or *molecular* or *parental imprinting*) in which there is differential expression of a gene depending on whether that gene was maternally or paternally inherited. It has been seen in a few examples in *Drosophila,* mice, and, as above, in human beings.

The retinoblastoma gene has been isolated and cloned. The gene specifies a 105-kilodalton protein (p 105) found in the nucleus, as would be expected if it were a suppressor of DNA transcription. It binds with at least three known onocogene proteins: the E1A protein of adenovirus, the SV40 large T antigen, and the 16E7 protein of human papilloma virus, a virus associated with 50% of cervical carcinomas. The implications of these findings are that these three viruses may use a similar mechanism in transformation, and this mechanism may involve inactivation of the retinoblastoma p105 protein.

A third tumor-suppressor gene is the p53 gene, named for its 53 kilodalton protein product. This gene has been implicated in wide range of cancers including leukemia, carcinomas of the lung and breast, brain tumors, bone cancers, and soft-tissue sarcomas. In fact, it seems at the moment to be the most common genetic change found in human cancers. The gene is located in band p13 of chromosome 17. Recently, mutation of the p53 gene was shown to be the cause of a cancer-family syndrome, resulting in the range of tumors mentioned above in different individuals in the pedigree.

Viral Theory of Cancer

Retroviruses

Animal viruses come in many different varieties, with DNA or RNA as their genetic material. Several classes of viruses, both DNA and RNA, can transform cells. Transformation may or may not be caused by an oncogene carried by the virus. Some DNA viruses carry oncogenes, such as the adenovirus mentioned previously that carries the gene for the E1A protein, which may act by binding to the retinoblastoma repressor protein. Oncogenes, however, were originally discovered in retroviruses, a group of very simple RNA viruses that contain the enzyme reverse transcriptase that, after the virus enters the host cell, converts the viral RNA into DNA. In 1910, Peyton Rous, who later won the Nobel Prize for his work, discovered that a sarcoma in chickens could be induced by a cell-free extract from a tumor in another chicken. The transmitted agent was found to be a retrovirus, later named Rous sarcoma virus. This was the first retrovirus to be discovered.

The retrovirus, which usually carries only three genes, integrates into the host genome in a series of steps. When the virus enters the host, it is in the form of a plus (+) RNA strand (capable of acting as an mRNA; the minus [–] strand is the complement to the + strand). At either end is a repeated sequence (R) located outside two unique sequences (U3 and U5). Through the process of reverse transcription, using the reverse transcriptase brought in by the virus, the viral RNA is converted to a double-stranded DNA. During that process, the ends of the DNA take on the configuration of long terminal repeats (LTRs), repetitions of U3-R-U5. The linear DNA then circularizes and intergrates into the genome in the same manner that a transposon does, generating short direct repeats at either end. Since some transposons have been found to involve an intermidate RNA stage, like retroviruses, together these transposons and retroviruses are referred to as *retrotransposons* (*retroposons*).

We outline the relatively complex steps by which the viral RNA is converted to double-stranded DNA by reverse transcription. In the process, the long terminal repeat is created at each end. In a novel twist, a tRNA acts as the primer to DNA replication, being complementary to the primer-binding (PBS) of the viral genome. This tRNA is usually brought in with the viron from the previous host, already base paired at the PBS. In frame 1 of figure, we show the viral RNA in circular form, although the ends are not connected. The tRNA primer is base paired at the PBS, and the first region of DNA has been synthesized (minus strand). The R and U5 regions of the viral RNA are then degraded by RNase activity. The R region of the newly synthesized DNA then hybridizes to the R region of the 3' end of the viral RNA in what is called the first "jump". The DNA strand is then elongated all around the viral RNA, ending in the PBS, thereby displacing the tRNA primer. You can see that the DNA strand now has a U3-R-U5 sequence at one end, the sequence referred to as the long terminal repeat (LTR).

The original RNA is now further degraded from both ends. The remaining RNA acts as a primer for the beginning of synthesis of the plus DNA strand. The LTR is first replicated, followed by the PBS using the PBS complementary portion of the tRNA as a template. The tRNA and the rest of the original viral RNA is now degraded. The two DNA strands are now formed into a double helix at their PBSs (the second "jump"). Held together only at the PBS region, both strands of DNA undergo replication, each using the other as a template. In frame 10 we see the process completed, and in frame 11 the double-stranded DNA is again depicted linearly. Circularization followed by integration then converts the double-stranded form of the retrovirus to an integrated provirus. As a provirus, the retrovirus is replicated from generation to generation with the host chromosome. By transcription, the provirus produces RNA that either acts as an mRNA for viral function or is incorporated into new viral particles that are then released from the infected cell.

A retrovirus minimally contains only the *gag*, *pol*, and *env* genes. Depending on the particular virus, different mechanisms of gene expression are possible. The viral mRNA is translated starting with *gag*. There is a translation termination signal at the end of the *gag*. There is a translation termination signal at the end of the *gag* gene that is occasionally read through, resulting in a *gag-pol* protein. The *env* gene is translated only after the viral RNA is spliced to remove the *gag-pol* region. The protein products of all three genes are further modified by cleavage and other changes (phosphorylation and glycosylation), resulting in core virion proteins from *gag*, reverse transcriptase, protease, and integrase from *pol*, and envelope glycoproteins from *env*.

As we mentioned before, retroviruses can cause cellular transformation directly from their integration or from the oncogenes the carry. Transformation from integration comes about presumably because the provirus somehow either inactivates a tumor-suppressor gene or activates an oncogene in a process called *insertion mutagenesis*. The U3 region of the retrovirus contains both an enhancer and a promoter. Since there is an LTR at either end of the provirus, cellular genes can be turned on when the virus integrates. Currently, however, most research effort in this area is centered on the oncogenes themselves.

Oncogenes

By genetic analysis and recombinant DNA studies, Rous sarcoma virus was found to transform cells through the action of a single gene. This gene, called *src* for sarcoma, was the first viral oncogene discovered. Since then at least fifty have been discovered, and each has been given a three-letter designation. Unlike tumor-suppressors, which lead to cancer when in the homozygous mutant condition, oncogenes act in a dominant fashion. Only one copy of the activated gene need by present for transformation to occur.

With the viral oncogene in hand, researchers could create a probe for the gene and look with the DNA of the host organism. To the surprise of virtually everyone, these oncogenes were found in untransformed cells. Since cellular oncogenes have introns and viral oncogenes do not, and since transforming viruses can function quite well as viruses without their oncogenes, geneticists generally accept the theory that these oncogenes originated in the host and were picked up, presumably as mRNAs, by the retroviruses. In other words, genes that function nromally within a host cell into a cancerous cell.

To distinguish oncogenes within viruses and hosts, we prefix the name of a viral oncogene, such as *src*, with a v (v-*src*) and a cellular oncogene with a c (c-*src*). Cellular oncogenes within a nontransformed cell are also called *proto-oncogenes.* How are proto-oncogenes normally do in the cell?

Oncogene Induction

Proto-oncogenes can be induced, we believe, three different ways. First, a mutation can cause a proto-oncogene to transform its host cell. For example, a *ras* proto-oncogene was converted to an oncogene when one codon, GGC (glycine), was converted to GTC (valine). Second, a proto-oncogene can be activated if it is moved to a region with a strong promoter or enhancer. Burkitt's lymphoma, for example, is associated with a translocation involving the proto-oncogene c-*myc,* which is normally located on chromosome 8. When translocated to chromosome 14, c-*myc* is placed contiguous with the immunoglobulin IgM constant gene (*Ig*-Cμ). This gene is very active in lymphocytes. Hence c-*myc* is now transcribed at a much higher rate than normal, resulting in cellular transformation. The c-*myc* gene normally occur near a *fragile site,* a region of a chromosome that has a tendency to break. Many proto-oncogenes occur near fragile sites on chromosomes. The simple capture of a gene by a retrovirus might be enough for transformation since the gene is brought under the influence of viral transcriptional control. However, not all genes captured this way are oncogenes. Third, a proto-oncogene can be activated if it is amplified. Several cases are known in which amplified genes (e.g., c-*ras* and c-*abl*) or genes on trisomic chromosomes are related to transformation.

Viral oncogenes can cause transformation by the same mechanisms. Either a mutation of the oncogene itself or the placement of the gene next to an active viral promotoer can cause high levels of transcription of the oncogene and hence transformation of the cell. What are the gene products of proto-oncogenes?

Oncogene Function

We know that proto-oncogenes are important to the cell because they have been conserved evolutionarily. For instance, c-*src* is found in fruit flies (*Drosophila*) as well as in vertebrates; c-*ras* is found in yeasts and in human beings. We believe that they are all genes that can promote growth of cells. At the present moment, the known protein products of oncogenes can be classified into at least four categories: tyrosine kinases, growth factors, GTP-binding proteins, and DNA-binding proteins. As proto-oncogenes, they seem to function normally at high levels. Can these proteins explain cancerous growth?

Tyrosine kinases are enzymes that add a phosphate group to tyrosine residues in proteins. Most kinases phosphorylate serine and threonine. (These three amino acids have OH groups available for phosphorylation.) Proteins that are phosphorylated at their tyrosine residues appear to be involved in cytoskeleton shape (transformed cells are shaped directly than normal cells) and in glycolysis (cancer cells tend toward the anaerobic glycolytic pathway). Overactivation of the cellular oncogene could result in inappropriate kinase activity, thereby changing many of the cellular activities leading to the condition we call cancer.

The c-*sis* gene seems to encode platelet-derived growth factor, which stimulates cells to grow. Its potential in transformation is obvious. GTP-binding proteins, the product of v-*ras,* for example, can play a role in transmitting endocrine signals across membranes. Perhaps increased quantities of GTP-binding proteins send continuous or amplified signals to certain cells and thus enhance growth. The v-*myc* gene product is a protein that binds to DNA. Its role is not known, but if it were to signal continuous DNA replication, it too could induce transformation.

We seem to be at an exciting threshold in our understanding of cellular transformation (cancer). Although clinicians recognize over a hundred types of cancers, almost all may be the result of inappropriate activity of oncogenes or photo-oncogenes whose mechanism of action may fall into as few as four different categories. Chromosomal translocations, deletions, or simple mutations may be the major factor in activating these oncogenes or proto-oncogenes, thus tying together the mutation and viral theories of transformation.

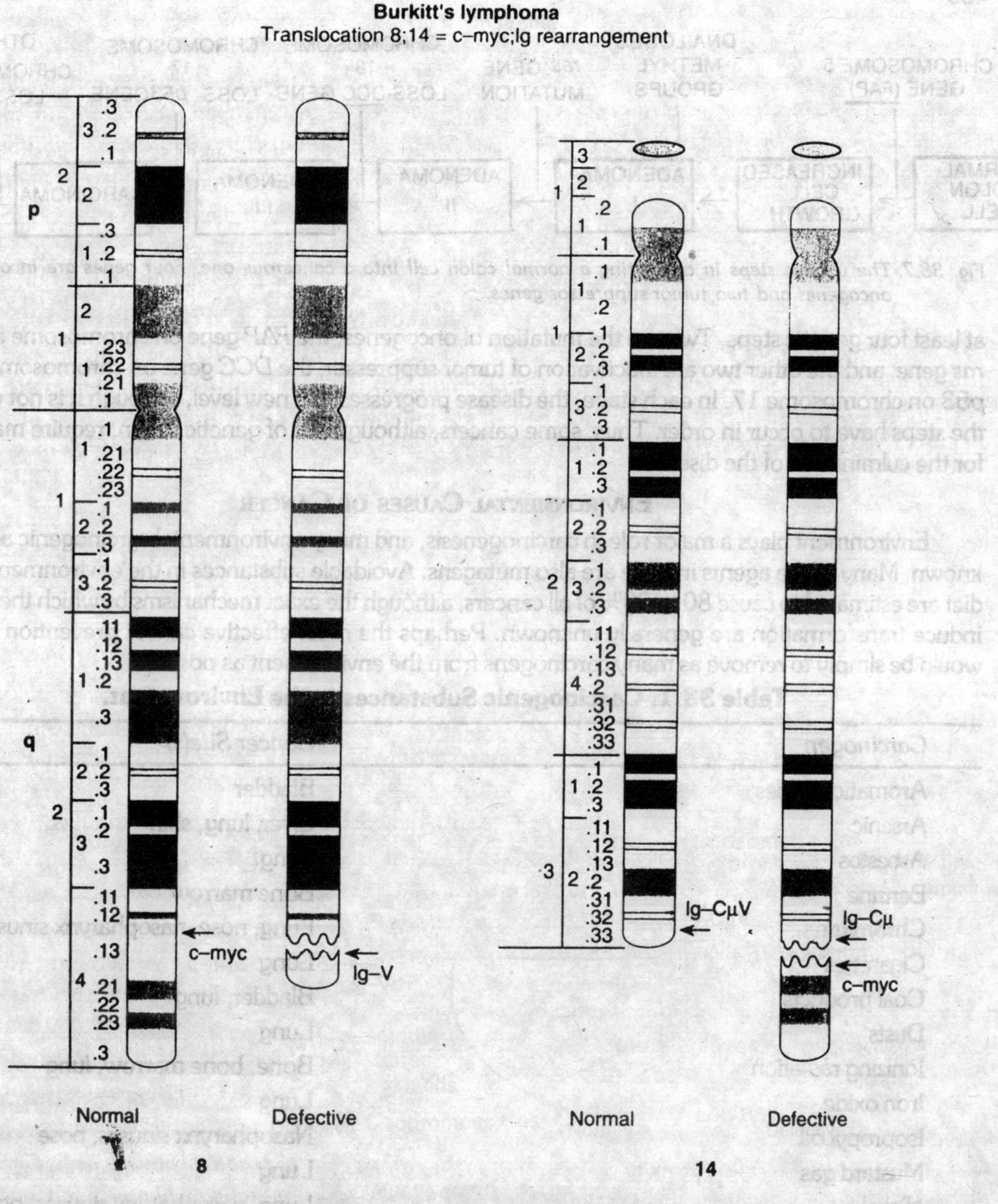

Fig. 38.6. Translocation associated with Burkitt's lymphoma. The proto-oncogene c-myc, on the end of chromosome 8q, is translocated to the end of chromosome 14q, adjacent to the immunoglobulin constant gene, Ig-C_μ. Presumably, location of c-myc next to the active immunoglobulin promoter (in B cells) results in transformation of the cell.

Cancer-family syndromes may be the result of the inheritance of the propensity for chromosome breakage, or they may be the result of the inheritance of one mutation in a two- or multistep process. However, knowing the action of transforming genes may not make for easy cures of cancer since they are "normal" genes that function properly in the untransformed cell.

On the other hand, we should not oversimplify the problem. Many cancers are the results of several different steps that lead eventually to full-blown cancerous growth. This is especially true in many of the cancers that produce solid tumors. The colon cancer named familial adenomatous polyposis (FAP) requires

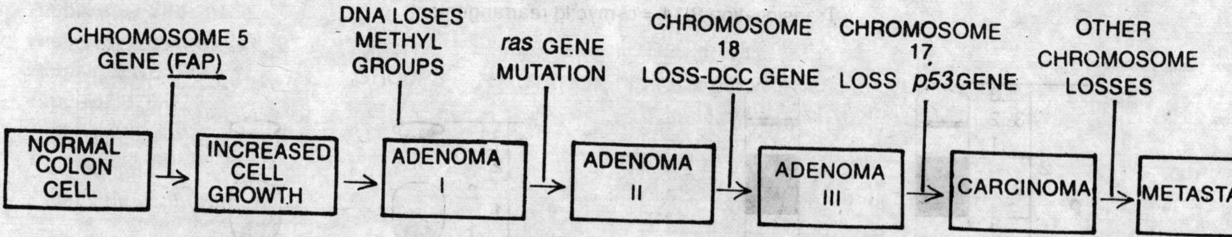

Fig. 38.7. The various steps in converting a normal colon cell into a cancerous one. Four genes are involved, two oncogenes and two tumor-suppressor genes.

at least four genetic steps. Two are the mutation of oncogenes, the *FAP* gene on chromosome 5 and the *ras* gene, and the other two are inactivation of tumor suppressor, the *DCC* gene on chromosome 18 and *p*53 on chromosome 17. In each stage, the disease progresses to a new level, although it is not clear that the steps have to occur in order. Thus, some cancers, although still of genetic origin, require many steps for the culmination of the disease.

Environmental Causes of Cancer

Environment plays a major role in carcinogenesis, and many environmental carcinogenic agents are known. Many of the agents in table are also mutagens. Avoidable substances in the environment and the diet are estimated to cause 80 to 90% of all cancers, although the exact mechanisms by which these agents induce transformation are generally unknown. Perhaps the most effective cancer prevention program would be simply to remove as many carcinogens from the environment as possible.

Table 38.1. Carcinogenic Substances in the Environment.

Carcinogen	*Cancer Site(s)*
Aromatic amines	Bladder
Arsenic	Liver, lung, skin
Asbestos	Lung
Benzine	Bone marrow
Chromium	Lung, nose, nasopharynx sinuses
Cigarettes	Lung
Coal products	Bladder, lung
Dusts	Lung
Ionizing radiation	Bone, bone marrow, lung
Iron oxide	Lung
Isopropyl oil	Nasopharynx sinuses, nose
Mustard gas	Lung
Nickel	Lung, nasopharynx sinuses, nose
Petroleum	Lung
Ultraviolet irradiation	Skin
Vinyl chloride	Liver
Wood and leather dust	Nasopharynx sinuses, nose

INDEX

A

B

D

N